INSTRUCTOR'S MANUAL

INTRODUCTION *to* MATERIALS SCIENCE *for* ENGINEERS

FOURTH EDITION

JAMES F.
SHACKELFORD

PRENTICE HALL
Upper Saddle River, NJ 07458

Production Editor: *James Buckley*
Production Supervisor: *Joan Eurell*
Acquisitions Editor: *Bill Stenquist*
Production Coordinator: *Donna Sullivan*

Simon & Schuster/A Viacom Company
Upper Saddle River, NJ 07458

Printed in the United States of America

10 9 8 7 6 5 4 3

ISBN 0-13-230111-3

Prentice-Hall International (UK) Limited, *London*
Prentice-Hall of Australia Pty. Limited, *Sydney*
Prentice-Hall Canada, Inc., *Toronto*
Prentice-Hall Hispanoamericana, S.A., *Mexico*
Prentice-Hall of India Private Limited, *New Delhi*
Prentice-Hall of Japan, Inc., *Tokyo*
Simon & Schuster Asia Pte. Ltd., *Singapore*
Editora Prentice-Hall do Brasil, Ltda., *Rio de Janeiro*

PREFACE

This **Instructor's Manual** is a companion volume to the Fourth Edition of the text, **Introduction to Materials Science for Engineers.** The three parts of the **Instructor's Manual** provide three distinct, supplementary components to the text. Part I is a complete solutions manual for all practice problems and chapter-ending problems.

Part II provides an introductory discussion of thermodynamics. Many instructors have indicated that they like to provide some background on thermodynamics to students prior to introducing phase diagrams in Chapter 5 of the text. As such, Part II serves as an additional component for the treatment of materials science fundamentals.

Part III is a general laboratory manual for the introductory materials science course. This specific document has been developed over several years in the laboratory component of the introductory materials science course at the University of California, Davis. Special acknowledgment is due to the work of Jerrold Franklin and Michael Meier in developing the laboratory manual in its original form. Subsequent use in quarterly offerings of the course has led to further refinements. The contributions of Professor Amiya Mukherjee and the other faculty in the Division of Materials Science and Engineering during this process is especially appreciated.

Finally, I wish to thank those whose contributions were both critical and much appreciated. Bill Stenquist and the editorial staff of Prentice-Hall were generous in their support for this supplementary volume. Scott Shackelford did an excellent job of page layout and final preparation for camera-ready copy.

J.F.S.

Davis, California

CONTENTS

Section 2.1 – Atomic Structure

PP 2.1 Calculate the number of atoms contained in a cylinder 1 μm in diameter by 1 μm deep of (**a**) magnesium and (**b**) lead. (See Sample Problem 2.1.)

PP 2.1

(a) $$N_{Mg\ atoms} = 6.64\times10^{10}\ \text{atoms Cu} \times \frac{(1.74\ g/cm^3)Mg}{(8.93\ g/cm^3)Cu} \times \frac{63.55\ g\ Cu/N_{AV}\ \text{atoms Cu}}{24.31\ g\ Mg/N_{AV}\ \text{atoms Mg}}$$

$$= \underline{\underline{3.38\times10^{10}\ \text{atoms Mg}}}$$

(b) $$N_{Pb\ atoms} = 6.64\times10^{10}\ \text{atoms Cu} \times \frac{(11.34\ g/cm^3)Pb}{(8.93\ g/cm^3)Cu} \times \frac{63.55\ g\ Cu/N_{AV}\ \text{atoms Cu}}{207.2\ g\ Cu/N_{AV}\ \text{atoms Pb}}$$

$$= \underline{\underline{2.59\times10^{10}\ \text{atoms Pb}}}$$

PP 2.2 Using the density of MgO calculated in Sample Problem 2.2, calculate the mass of an MgO refractory (temperature-resistant) brick with dimensions: 50 mm × 100 mm × 200 mm.

PP 2.2

$$m = \rho V = (3.60\ g/cm^3)(10^{-3}\ cm^3/mm^3) \times (50)(100)(200)\ mm^3$$

$$= 3.60\times10^3\ g = \underline{\underline{3.60\ kg}}$$

PP 2.3 Calculate the dimensions of (**a**) a cube containing 1 mol of copper and (**b**) a cube containing 1 mol of lead. (See Sample Problem 2.3.)

PP 2.3

(a) $$\text{edge} = \left(\frac{63.55\ g/mol}{8.93\ g/cm^3}\right)^{1/3} \times 10\ mm/cm = \underline{\underline{19.23\ mm}}$$

(b) $$\text{edge} = \left(\frac{207.2\ g/mol}{11.34\ g/cm^3}\right)^{1/3} \times 10\ mm/cm = \underline{\underline{26.34\ mm}}$$

2.1 A gold O-ring is used to form a gastight seal in a high-vacuum chamber. The ring is formed from an 80-mm length of 1.5-mm-diameter wire. Calculate the number of gold atoms in the O-ring.

2.1

$$N_{atoms} = \rho V\left(\frac{N_{AV}}{at.\ wt.}\right) = 19.28\times10^6\ g\ Au/m^3 \times 80\times10^{-3}\ m$$

$$\times\ \pi\left(\frac{1.5\times10^{-3}\ m}{2}\right)^2$$

$$\times\left(\frac{0.6023\times10^{24}\ atoms}{196.97\ g\ Au}\right)$$

$$= \underline{\underline{8.33\times10^{21}\ atoms}}$$

2.2 Common aluminum foil for household use is nearly pure aluminum. A box of this product at a local supermarket is advertised as giving 75 ft^2 of material (in a roll 304 mm wide by 22.8 m long). If the foil is 0.5 mil (12.7 μm) thick, calculate the number of atoms of aluminum in the roll.

2.2

$$N_{atoms} = \rho V\left(\frac{N_{AV}}{at.\ wt.}\right) = 2.70\times10^6\ g\ Al/m^3 \times 12.7\times10^{-6}\ m$$

$$\times\ 304\times10^{-3}\ m \times 22.8\ m \times\left(\frac{0.6023\times10^{24}\ atoms}{26.98\ g\ Al}\right)$$

$$= \underline{\underline{5.31\times10^{24}\ atoms}}$$

2.3 In a metal-oxide-semiconductor (MOS) device, a thin layer of SiO_2 (density = 2.20 Mg/m^3) is grown on a single crystal chip of silicon. How many Si atoms *and* how many O atoms are present per square millimeter of the oxide layer? Assume that the layer thickness is 100 nm.

2.3

$$V = 100\ nm \times 1\ mm^2 = 10^2\times10^{-9}\ m \times (10^{-3}\ m)^2 = 1\times10^{-13}\ m^3$$

$$m_{SiO_2} = 2.20\times10^6\ g/m^3 \times 1\times10^{-13}\ m^3 = 2.20\times10^{-7}\ g$$

1 mol SiO_2 has $(28.09 + 2[16.00])\ g = 60.09\ g$ for

N_{AV} atoms Si and $2N_{AV}$ atoms O

$$\therefore N_{Si\ atoms} = \frac{2.20\times10^{-7}\ g}{60.09\ g} \times 0.6023\times10^{24}\ atoms$$

$$= \underline{\underline{2.21\times10^{15}\ atoms}}$$

$$N_{O\ atoms} = 2\times N_{Si\ atoms} = \underline{\underline{4.41\times10^{15}\ atoms}}$$

2.4 A box of clear plastic wrap for household use is polyethylene, $-(C_2H_4)-_n$, with density = 0.910 Mg/m³. A box of this product contains 100 ft² of material (in a roll 304 mm wide by 30.5 m long). If the wrap is 0.5 mil (12.7 μm) thick, calculate the number of carbon atoms *and* the number of hydrogen atoms in this roll.

2.4

$V = 12.7\,\mu m \times 304\,mm \times 30.5\,m = 12.7\times10^{-6}\,m \times 0.304\,m \times 30.5\,m$

$= 1.18\times10^{-4}\,m^3$

$m_{C_2H_4} = 0.910\times10^6\,g/m^3 \times 1.18\times10^{-4}\,m^3 = 107\,g$

1 mol C_2H_4 has $(2[12.01]+4[1.008])\,g = 28.05\,g$ for $2N_{AV}$ atoms C and $4N_{AV}$ atoms H

$\therefore N_{C\,atoms} = \dfrac{107\,g}{28.05\,g}\times 2\times0.6023\times10^{24}\,atoms$

$= \underline{\underline{4.60\times10^{24}\,atoms}}$

$N_{H\,atoms} = 2\times N_{C\,atoms}$

$= \underline{\underline{9.20\times10^{24}\,atoms}}$

2.5 An Al_2O_3 *whisker* is a small single crystal used to reinforce metal-matrix composites. Given a cylindrical shape, calculate the number of Al atoms *and* the number of O atoms in a whisker with a diameter of 1 μm and a length of 30 μm. (The density of Al_2O_3 is 3.97 Mg/m³.)

2.5

$V = \pi\left(\dfrac{1\,\mu m}{2}\right)^2 \times 30\,\mu m = \pi\left(0.5\times10^{-6}\,m\right)^2 \times 30\times10^{-6}\,m$

$= 23.6\times10^{-18}\,m^3$

$m_{Al_2O_3} = 3.97\times10^6\,g/m^3 \times 23.6\times10^{-18}\,m^3 = 9.35\times10^{-11}\,g$

1 mol Al_2O_3 has $(2[26.98]+3[16.00])\,g = 101.96\,g$ for $2N_{AV}$ atoms Al and $3N_{AV}$ atoms O

$\therefore N_{Al\,atoms} = 9.35\times10^{-11}\,g \times \dfrac{2\,(0.6023\times10^{24}\,atoms)}{101.96\,g}$

$= \underline{\underline{1.11\times10^{12}\,atoms}}$

$N_{O\,atoms} = 3/2 \times N_{Al\,atoms}$

$= 3/2\,(1.11\times10^{12}\,atoms) = \underline{\underline{1.66\times10^{12}\,atoms}}$

2.6 An optical fiber for telecommunication is made of SiO_2 glass (density = 2.20 Mg/m^3). How many Si atoms *and* how many O atoms are present per millimeter of length of a fiber 10 μm in diameter?

2.6

For 1 mm section of fiber:

$$V = \pi \left(\frac{10\,\mu m}{2}\right)^2 \times 1\text{ mm} = \pi (5\times10^{-6}\text{ m})^2 \times 1\times10^{-3}\text{ m}$$

$$= 7.85\times10^{-14}\text{ m}^3$$

$$m_{SiO_2} = 2.20\times10^6\text{ g/m}^3 \times 7.85\times10^{-14}\text{ m}^3 = 1.73\times10^{-7}\text{ g}$$

1 mol SiO_2 has $(28.09 + 2[16.00])$ g = 60.09 g

for N_{AV} atoms Si and $2N_{AV}$ atoms O

$$\therefore N_{Si\text{ atoms}} = \frac{1.73\times10^{-7}\text{ g}}{60.09\text{ g}} \times 0.6023\times10^{24}\text{ atoms} = \underline{\underline{1.73\times10^{15}\text{ atoms}}}$$

$$N_{O\text{ atoms}} = 2\times N_{Si\text{ atoms}} = \underline{\underline{3.46\times10^{15}\text{ atoms}}}$$

2.7 Twenty-five grams of magnesium filings are to be oxidized in a laboratory demonstration. (a) How many O_2 molecules would be consumed in this demonstration? (b) How many moles of O_2 does this represent?

2.7

(a) $Mg + \frac{1}{2}O_2 \rightarrow MgO$

i.e., 1 gm-atom Mg is oxidized by 0.5 mol. O_2

or $\dfrac{25\text{ g Mg}}{24.31\text{ g/g·atom}} = 1.028$ g·atom Mg will be

oxidized by $\dfrac{1.028}{2}$ mol. O_2 = 0.514 mol. O_2

$\therefore$ no. molecules O_2 = 0.514 mol × 0.6023×10^{24} molec./mol.

$= 3.10\times10^{23}$ molec.

$= \underline{\underline{0.310\times10^{24}\text{ molec.}}}$

(b) no. moles = 0.310×10^{24} molec × 1 mol/0.6023×10^{24} molec.

$= \underline{\underline{0.514\text{ mol }O_2}}$

2.8 Naturally occurring copper has an atomic weight of 63.55. Its principal isotopes are Cu^{63} and Cu^{65}. What is the abundance (in atomic percent) of each isotope?

2.8

$$x\ Cu^{63} + y\ Cu^{65} = Cu^{63.55}$$

or

$$63x + 65y = 63.55$$

or

$$63x + 65(1-x) = 63.55$$

or

$$65 - 2x = 63.55$$

or

$$2x = 65.00 - 63.55$$

or

$$x = 0.725$$

and

$$y = 1 - x = 0.275$$

giving:

$$\underline{\underline{72.5\%\ Cu^{63} \text{ and } 27.5\%\ Cu^{65}}}$$

2.9 A copper penny has a mass of 2.60 g. Assuming pure copper, how much of this mass is contributed by (a) the neutrons in the copper nuclei? and (b) electrons?

2.9

(a) Compared to neutrons and protons, the mass of an electron is negligible. The mass of neutrons can be determined from the average number of neutrons in an isotope:

$$n_{neutrons} = \text{atomic weight} - \text{atomic number}$$

$$= 63.55 - 29.00 = 34.55$$

$$\therefore \text{mass}_{neutrons} = \frac{34.55}{63.55} \times 2.60\,g$$

$$= \underline{\underline{1.41\ g}}$$

(b) For an "average" copper atom,

$$mass_{electrons} = (\text{atomic number})(m_{e^-})$$
$$= 29 \times 0.911 \times 10^{-27} g$$
$$= 2.64 \times 10^{-26} g$$

$$mass_{atom} = (\text{atomic weight})(\text{amu})$$
$$= 63.55 \times 1.661 \times 10^{-24} g$$
$$= 1.056 \times 10^{-22} g$$

$$\therefore \text{wt. fraction electrons} = \frac{2.64 \times 10^{-26} g}{1.056 \times 10^{-22} g}$$
$$= 2.50 \times 10^{-4}$$

$$\therefore mass_{electrons} = 2.50 \times 10^{-4} \times 2.60 g$$
$$= \underline{\underline{6.50 \times 10^{-4} g}}$$

2.10 The orbital electrons of an atom can be ejected by exposure to a beam of electromagnetic radiation. Specifically, an electron can be ejected by a photon with energy greater than or equal to the electron's binding energy. Given that the photon energy (E) is equal to hc/λ, where h is Planck's constant, c the speed of light, and λ the wavelength, calculate the minimum wavelength of radiation necessary to eject a 1s electron from a ^{12}C atom. (See Figure 2-3.)

2.10 From Figure 2-3, $|E| = 283.9 eV$.

Then, $\lambda_{min} = \frac{hc}{E}$

$$= \frac{(0.6626 \times 10^{-33} J \cdot s)(0.2998 \times 10^{9} m/s)}{(283.9 eV)(1 J / 6.242 \times 10^{18} eV)}$$
$$= 4.37 \times 10^{-9} m \times 1 nm / 10^{-9} m$$
$$= \underline{\underline{4.37 nm}}$$

Note: We use the magnitude of the electron binding energy rather than the arbitrary negative sign convention to provide a physically meaningful positive wavelength.

2.11 Once the $1s$ electron is ejected from a ^{12}C atom, as described in Problem 2.10, there is a tendency for one of the $2(sp^3)$ electrons to drop into the $1s$ level. The result is the emission of a photon with an energy precisely equal to the energy change associated with the electron transition. Calculate the wavelength of the photon that would be emitted from a ^{12}C atom. (You will note various examples of this concept throughout the text in relation to the chemical analysis of engineering materials.)

2.11 From Figure 2–3 and again using the magnitude of the energies involved,

$$|\Delta E| = |-283.9 - (-6.5)| \text{ eV}$$

$$= 277.4 \text{ eV}$$

or

$$\lambda = \frac{hc}{\Delta E}$$

$$= \frac{(0.6626 \times 10^{-33} \text{ J}\cdot\text{s})(0.2998 \times 10^{9} \text{ m/s})}{(277.4 \text{ eV})(1 \text{ J}/6.242 \times 10^{18} \text{ eV})}$$

$$= 4.47 \times 10^{-9} \text{ m} \times 1 \text{ nm}/10^{-9} \text{ m}$$

$$= \underline{\underline{4.47 \text{ nm}}}$$

2.12 The mechanism for producing a photon of specific energy is outlined in Problem 2.11. The magnitude of photon energy increases with the atomic number of the atom from which emission occurs. (This is due to the stronger binding forces between the negative electrons and the positive nucleus as the numbers of protons and electrons increase with atomic number.) As noted in Problem 2.10, $E = hc/\lambda$, which means that a higher-energy photon will have a shorter wavelength. Verify that higher atomic number materials will emit higher-energy, shorter-wavelength photons by calculating E and λ for emission from iron (atomic number 26 compared to 6 for carbon), given that the energy levels for the first two electron orbitals in iron are at −7,112eV and −708eV.

2.12 $|\Delta E| = |-7{,}112 - (-708)| \text{ eV} = 6404 \text{ eV}$

or $\lambda = \dfrac{hc}{\Delta E} = \dfrac{(0.6626 \times 10^{-33} \text{ J}\cdot\text{s})(0.2998 \times 10^{9} \text{ m/s})}{(6404 \text{ eV})(1 \text{ J}/6.242 \times 10^{18} \text{ eV})}$

$= 1.94 \times 10^{-10} \text{ m} \times 1 \text{ nm}/10^{-9} \text{ m} = \underline{\underline{0.194 \text{ nm}}}$

Section 2.2 – The Ionic Bond

PP 2.4 **(a)** Make a sketch similar to Figure 2-4 illustrating Mg and O atoms and ions in MgO. **(b)** Compare the electronic configurations for the atoms and ions illustrated in part (a). **(c)** Show which noble gas atoms have electronic configurations equivalent to those illustrated in part (a). (See Sample Problem 2.4.)

PP 2.4 (a)

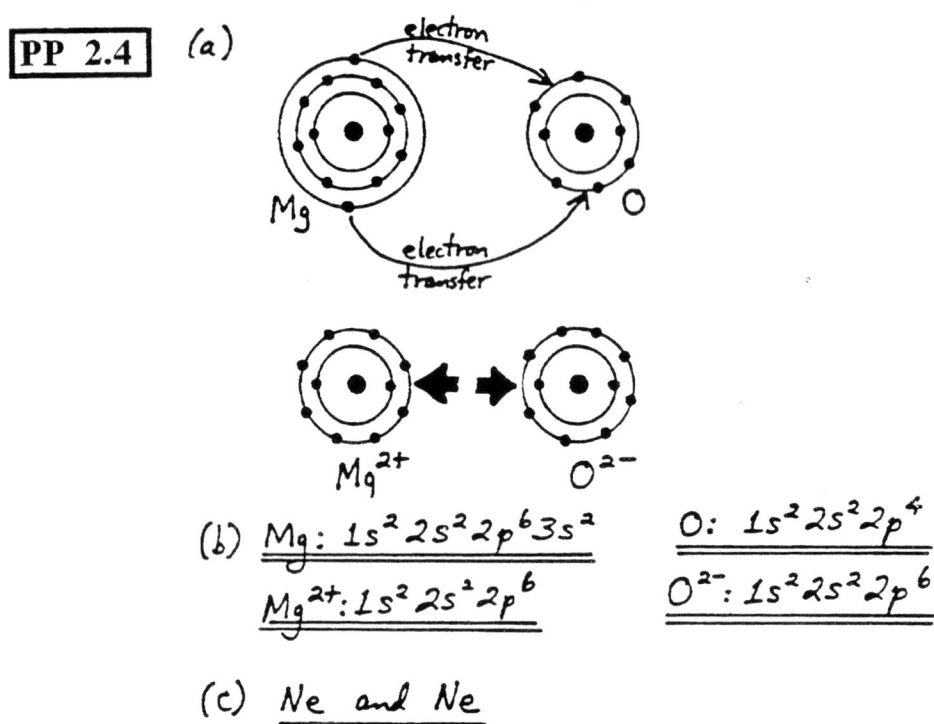

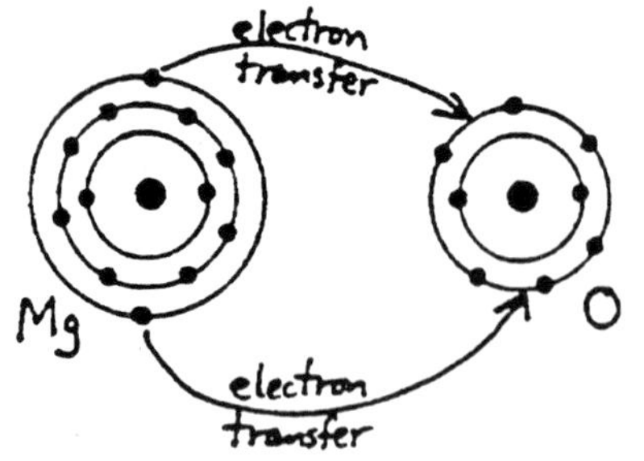

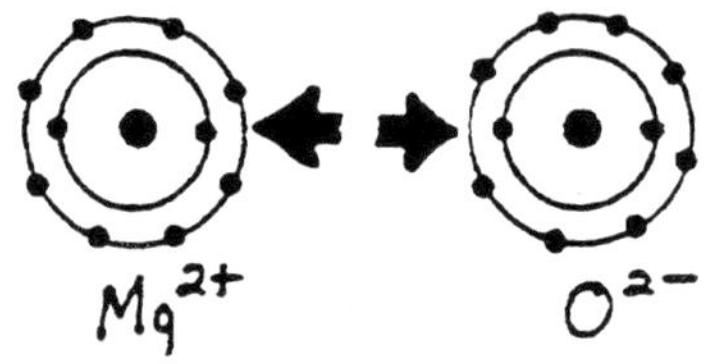

(b) Mg: $1s^2 2s^2 2p^6 3s^2$ O: $1s^2 2s^2 2p^4$

Mg^{2+}: $1s^2 2s^2 2p^6$ O^{2-}: $1s^2 2s^2 2p^6$

(c) Ne and Ne

PP 2.5 **(a)** Using the ionic radii data in Appendix 2, calculate the coulombic force of attraction between the Mg^{2+}–O^{2-} ion pair. **(b)** What is the repulsive force in this case? (See Sample Problems 2.5 and 2.6.)

PP 2.5 (a) From Appendix 2,

$$r_{Mg^{2+}} = 0.078\,nm$$

$$r_{O^{2-}} = 0.132\,nm$$

Then,

$$a_o = r_{Mg^{2+}} + r_{O^{2-}} = 0.078\,nm + 0.132\,nm = 0.210\,nm$$

$$F_c = -\frac{(9\times10^9\,V\cdot m/C)(+2)(0.16\times10^{-18}\,C)(-2)(0.16\times10^{-18}\,C)}{(0.210\times10^{-9}\,m)^2}$$

$$= 20.9\times10^{-9}\,N$$

(b) $F_R = -F_c = -20.9\times10^{-9}\,N$

PP 2.6 Calculate the minimum radius ratio for a coordination number of (**a**) 4 and (**b**) 6. (See Sample Problem 2.7.)

PP 2.6 (a)

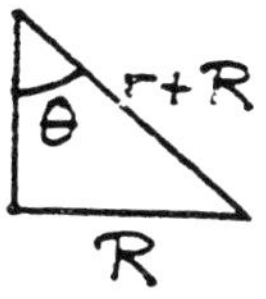

$\theta = \frac{109.5°}{2}$ (see Figure 2-19)

$\sin\left(\frac{109.5°}{2}\right) = \frac{R}{r+R}$

$0.8166r + 0.8166R = R$

$0.1834R = 0.8166r$

giving, finally

$\frac{r}{R} = 0.225$

(b)

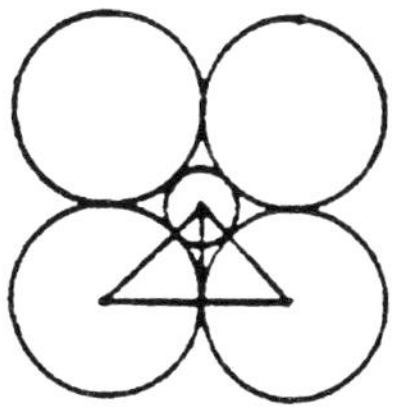

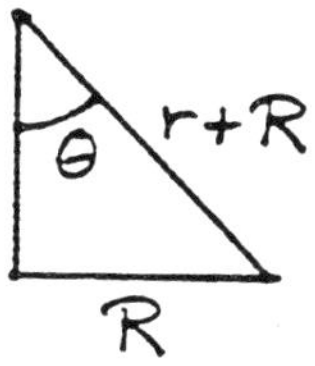

$\theta = 45°$

$\sin 45° = \frac{R}{r+R}$

$0.707r + 0.707R = R$

giving, finally

$\frac{r}{R} = 0.414$

PP 2.7 In the next chapter we shall see that MgO, CaO, FeO, and NiO all share the NaCl crystal structure. As a result, in each case the metal ions will have the same coordination number (6). The case of MgO and CaO is treated in Sample Problem 2.8. Use the radius ratio calculation to see if it estimates the CN = 6 for FeO and NiO.

PP 2.7 From Appendix 2,

$r_{Fe^{2+}} = 0.087\,nm$, $r_{Ni^{2+}} = 0.078\,nm$, $r_{O^{2-}} = 0.132\,nm$

For FeO,

$\frac{r}{R} = \frac{0.087\,nm}{0.132\,nm} = 0.66$ for which Table 2.1 gives

$\underline{\underline{CN = 6}}$

For NiO,

$\frac{r}{R} = \frac{0.078\,nm}{0.132\,nm} = 0.59$ giving $\underline{\underline{CN = 6}}$

2.13 Make an accurate plot of F_c versus a (comparable to Figure 2-6) for an $Mg^{2+} - O^{2-}$ pair. Consider the range of a from 0.2 to 0.7 nm.

2.13 $F_c = -\frac{(9 \times 10^9\,V\cdot m/C)(+2)(0.16 \times 10^{-18}\,C)(-2)(0.16 \times 10^{-18}\,C)}{a^2}$

a	F_c
$0.2 \times 10^{-9}\,m$	$23.0 \times 10^{-9}\,N$
$0.3 \times$ "	$10.2 \times$ "
$0.4 \times$ "	$5.76 \times$ "
$0.5 \times$ "	$3.69 \times$ "
$0.6 \times$ "	$2.56 \times$ "
$0.7 \times$ "	$1.88 \times$ "

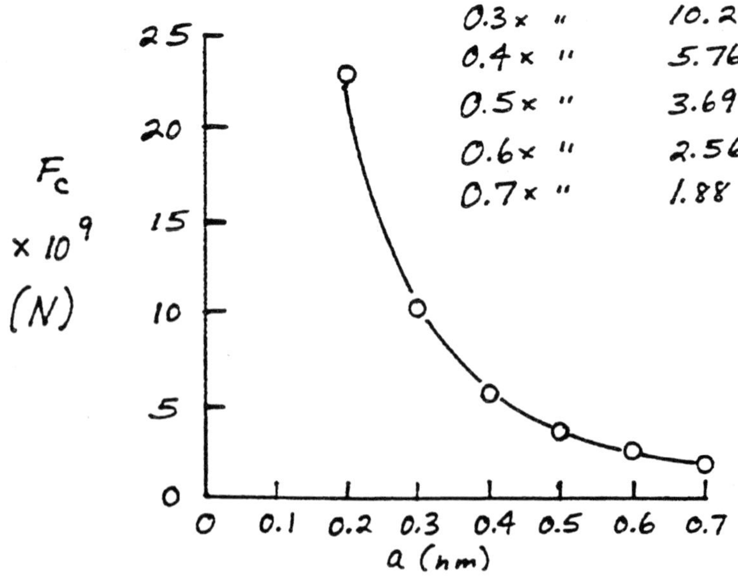

2.14 Make an accurate plot of F_c versus a for a $Na^{+}-O^{2-}$ pair.

2.14

$$F_c = -\frac{(9\times10^{9}\ V\cdot m/C)(+1)(0.16\times10^{-18}\ C)(-2)(0.16\times10^{-18}\ C)}{a^2}$$

a	F_c
0.2×10^{-9} m	11.5×10^{-9} N
$0.3\times$ "	$5.12\times$ "
$0.4\times$ "	$2.88\times$ "
$0.5\times$ "	$1.84\times$ "
$0.6\times$ "	$1.28\times$ "
$0.7\times$ "	$0.940\times$ "

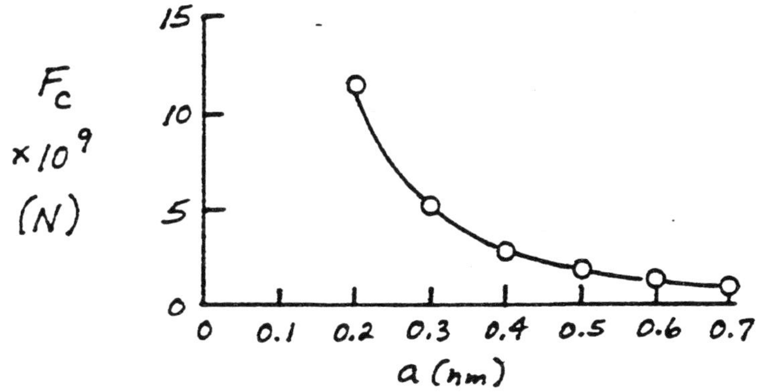

2.15 So far, we have concentrated on the coulombic force of attraction between ions. But like ions repel each other. A nearest-neighbor pair of Na^{+} ions in Figure 2-5 are separated by a distance of $\sqrt{2}a_0$, where a_0 is defined in Figure 2-7. Calculate the coulombic force of *repulsion* between such a pair of like ions.

2.15

$$F_c = -\frac{k_0(Z_1 q)(Z_2 q)}{a^2}$$

$$= -\frac{(9\times10^{9}\ V\cdot m/C)(+1)(0.16\times10^{-18}C)(+1)(0.16\times10^{-18}C)}{(2)(0.278\times10^{-9}\ m)^2}$$

$$= -1.49\times10^{-9}\ N$$

2.16 Calculate the coulombic force of attraction between Ca^{2+} and O^{2-} in CaO, which has the NaCl-type structure.

2.16 From Appendix 2,

$r_{Ca^{2+}} = 0.106\ nm$ & $r_{O^{2-}} = 0.132\ nm$

Then,

$$a_0 = r_{Ca^{2+}} + r_{O^{2-}} = 0.106\ nm + 0.132\ nm = 0.238\ nm$$

$$F_c = -\frac{(9\times10^9\ V\cdot m/C)(+2)(0.16\times10^{-18}\ C)(-2)(0.16\times10^{-18}\ C)}{(0.238\times10^{-9}\ m)^2}$$

$$= \underline{\underline{16.3\times10^{-9}\ N}}$$

2.17 Calculate the coulombic force of repulsion between nearest-neighbor Ca^{2+} ions in CaO. (Note Problems 2.15 and 2.16.)

2.17 As noted in Problem 2.15,

$$a = \sqrt{2}\,a_0$$

Using the calculation for Problem 2.16,

$$a = \sqrt{2}\,(0.238\ nm) = 0.337\ nm$$

Then,

$$F_c = -\frac{(9\times10^9\ V\cdot m/C)(+2)^2(0.16\times10^{-18}\ C)^2}{(0.337\times10^{-9}\ m)^2}$$

$$= \underline{\underline{-8.13\times10^{-9}\ N}}$$

2.18 Calculate the coulombic force of repulsion between nearest-neighbor O^{2-} ions in CaO. (Note Problems 2.15, 2.16, and 2.17.)

2.18 As in Problem 2.17,

$$a = \sqrt{2}\,a_0 = \sqrt{2}\,(0.238\ nm) = 0.337\ nm$$

And,

$$F_c = -\frac{(9\times10^9\ V\cdot m/C)(-2)^2(0.16\times10^{-18}\ C)^2}{(0.337\times10^{-9}\ m)^2}$$

$$= \underline{\underline{-8.13\times10^{-9}\ N}}$$

2.19 Calculate the coulombic force of repulsion between nearest-neighbor Ni^{2+} ions in NiO, which has the NaCl-type structure. (Note Problem 2.17.)

2.19

$$a = \sqrt{2}\,a_0 = \sqrt{2}\,(r_{Ni^{2+}} + r_{O^{2-}})$$

From Appendix 2,

$$a = \sqrt{2}\,(0.078\,nm + 0.132\,nm) = 0.297\,nm$$

Then,

$$F_c = -\frac{(9\times10^9\,V\cdot m/C)(+2)^2(0.16\times10^{-18}\,C)^2}{(0.297\times10^{-9}\,m)^2} = \underline{\underline{-10.4\times10^{-9}\,N}}$$

2.20 Calculate the coulombic force of repulsion between nearest-neighbor O^{2-} ions in NiO. (Note Problems 2.18 and 2.19.)

2.20

As in Problem 2.19,

$$a = \sqrt{2}\,(0.078\,nm + 0.132\,nm) = 0.297\,nm$$

and,

$$F_c = -\frac{(9\times10^9\,V\cdot m/C)(-2)^2(0.16\times10^{-18}\,C)^2}{(0.297\times10^{-9}\,m)^2} = \underline{\underline{-10.4\times10^{-9}\,N}}$$

2.21 SiO_2 is known as a "glass former" because of the tendency of SiO_4^{4-} tetrahedra (Figure 2-17) to link together in a noncrystalline network. Al_2O_3 is known as an intermediate glass former due to the ability of Al^{3+} to substitute for Si^{4+} in the glass network, although Al_2O_3 does not by itself tend to be noncrystalline. Discuss the substitution of Al^{3+} for Si^{4+} in terms of the radius ratio.

2.21

As discussed in Section 2.3, the radius ratio for $Si^{4+}-O^{2-}$ is:

$$r/R = \frac{0.039\,nm}{0.132\,nm} = 0.295,$$ well within the range for 4-fold coordination.

For $Al^{3+}-O^{2-}$, data in Appendix 2 gives:

$$r/R = \frac{0.057\,nm}{0.132\,nm} = 0.432,$$ just above the range for 4-fold coordination indicating that the role of Al_2O_3 as an intermediate glass former is reasonably consistent with this simple ionic calculation.

2.22 Repeat Problem 2.21 for TiO_2, which like Al_2O_3, is an intermediate glass former.

2.22 For $Ti^{4+}-O^{2-}$, Appendix 2 gives:

$$r/R = 0.064\,nm / 0.132\,nm = 0.485$$

As with Al^{3+}, Ti^{4+} gives a value just above the range for 4-fold coordination consistent with its intermediate role.

2.23 The coloration of glass by certain ions is often sensitive to the coordination of the cation by oxygen ions. For example, Co^{2+} gives a blue-purple color when in the fourfold coordination characteristic of the silica network (see Problem 2.21) and gives a pink color when in a sixfold coordination. Which color from Co^{2+} is predicted by the radius ratio?

2.23 Using the data from Appendix 2,

$$\frac{r_{Co^{2+}}}{r_{O^{2-}}} = \frac{0.082\,nm}{0.132\,nm} = 0.621$$

which is in the range for 6-fold coordination in Table **2.1**. Therefore, a <u>pink</u> color is indicated.

Note: The rich blue-purple color known as "cobalt blue" associated with 4-fold coordination of Co^{2+} is, then, determined by more than simple ionic considerations.

2.24 One of the first nonoxide materials to be produced as a glass was BeF_2. As such, it was found to be similar to SiO_2 in many ways. Calculate the radius ratio for Be^{2+} and F^- and comment.

2.24 For $Be^{2+}-F^-$, Appendix 2 gives:

$$r/R = 0.054\,\text{nm}/0.133\,\text{nm} = 0.406,$$ which is

in the range for 4-fold coordination. As a result, tetrahedrally-coordinated Be^{2+} leads to network formation similar to the case for Si^{4+} in SiO_2.

2.25 A common feature in the first generation of high-temperature ceramic superconductors is a Cu–O sheet that serves as a superconducting plane. Calculate the coulombic force of attraction between a Cu^{2+} and an O^{2-} within one of these sheets.

2.25 From Appendix 2,

$$r_{Cu^{2+}} = 0.072\,\text{nm} \;\&\; r_{O^{2-}} = 0.132\,\text{nm}$$

Then,

$$a_0 = r_{Cu^{2+}} + r_{O^{2-}} = 0.072\,\text{nm} + 0.132\,\text{nm} = 0.204\,\text{nm}$$

$$F_c = -\frac{(9\times10^9\,\text{V}\cdot\text{m}/\text{C})(+2)(0.16\times10^{-18}\,\text{C})(-2)(0.16\times10^{-18}\,\text{C})}{(0.204\times10^{-9}\,\text{m})^2}$$

$$= \underline{\underline{22.1\times10^{-9}\,\text{N}}}$$

2.26 In contrast to the calculation for the superconducting Cu–O sheets discussed in Problem 2.25, calculate the coulombic force of attraction between a Cu^{+} and an O^{2-}.

2.26 From Appendix 2,

$$r_{Cu^+} = 0.096\ nm \ \& \ r_{O^{2-}} = 0.132\ nm$$

Then,

$$a_0 = r_{Cu^+} + r_{O^{2-}} = 0.096\ nm + 0.132\ nm = 0.228\ nm$$

$$F_c = -\frac{(9\times10^9\ V\cdot m/C)(+1)(0.16\times10^{-18}\ C)(-2)(0.16\times10^{-18}\ C)}{(0.228\times10^{-9}\ m)^2}$$

$$= \underline{8.86\times10^{-9}\ N}$$

• 2.27 For an ionic crystal, such as NaCl, the net coulombic bonding force is a simple multiple of the force of attraction between an adjacent ion pair. To demonstrate this, consider the hypothetical, one-dimensional "crystal" shown:

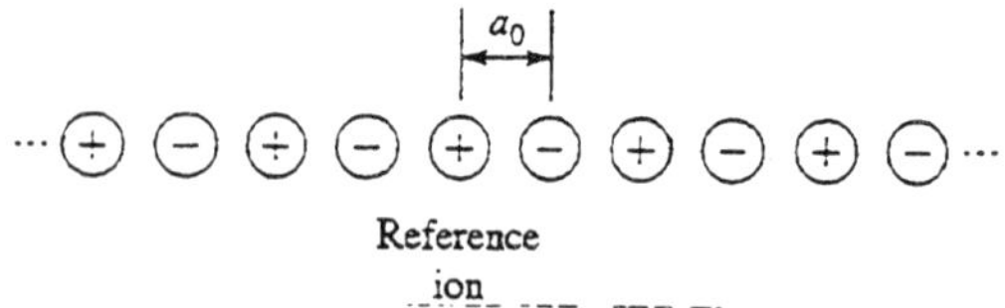

(a) Show that the net coulombic force of attraction between the reference ion and all other ions in the crystal is

$$F = AF_c$$

where F_c is the force of attraction between an adjacent ion pair (see Equation 2.1) and A is a series expansion.

(b) Determine the value of A.

2.27 (a) For the "crystal,"

$$F_{c,net} = F = -K\left(\frac{2}{a_0^2} - \frac{2}{(2a_0)^2} + \frac{2}{(3a_0)^2} - \frac{2}{(4a_0)^2} + \cdots\right)$$

$$= -\frac{2K}{a_0^2}\left(+1 - \frac{1}{4} + \frac{1}{9} - \frac{1}{16} + \cdots\right)$$

For an adjacent ion pair,

$$F_c = -\frac{K}{a_0^2}$$ (Of course, K for the adjacent pair is negative in sign.)

Or $$F = F_c\, 2\left(1 - \frac{1}{4} + \frac{1}{9} - \frac{1}{16} + \cdots\right)$$

$$= A\, F_c \quad \text{where } A = 2\left(1 - \frac{1}{4} + \frac{1}{9} - \frac{1}{16} + \cdots\right)$$

(b) One can evaluate A by carrying out the series a sufficiently large number of terms until the net value converges.

One can also note that

$$1 - \frac{1}{2^2} + \frac{1}{3^2} - \frac{1}{4^2} + \cdots = \frac{\pi^2}{12}$$

giving:

$$A = \frac{2\pi^2}{12} = \underline{\underline{1.645}}$$

2.28 In Problem 2.27, a value for A was calculated for the simple one-dimensional case. For the three-dimensional NaCl structure, A has been calculated to be 1.748. Calculate the net coulombic force of attraction, F, for this case.

2.28

$$F = A F_c$$

From Sample Problem 2.5,

$$F_c = 2.98 \times 10^{-9}\ N$$

$$\therefore\ F = (1.748)(2.98 \times 10^{-9}\ N) = \underline{\underline{5.21 \times 10^{-9}\ N}}$$

Section 2.3 – The Covalent Bond

PP 2.8 In Figure 2-14 we see the polymerization of polyethylene $-(C_2H_4)_n-$ illustrated. Sample Problem 2.9 illustrates polymerization for poly(vinyl chloride) $-(C_2H_3Cl)_n-$. Make a similar sketch to illustrate the polymerization of polypropylene $-(C_2H_3R)_n-$, where R is a CH$_3$ group.

[PP 2.8]

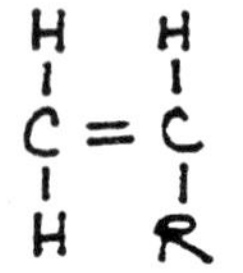

$$\begin{array}{c} H \quad H \\ | \quad\; | \\ C = C \\ | \quad\; | \\ H \quad R \end{array}$$ propylene molecule

$$\cdots-\underset{H}{\overset{H}{C}}-\underset{R}{\overset{H}{C}}-\underset{H}{\overset{H}{C}}-\underset{R}{\overset{H}{C}}-\underset{H}{\overset{H}{C}}-\underset{R}{\overset{H}{C}}-\underset{H}{\overset{H}{C}}-\underset{R}{\overset{H}{C}}-\cdots$$ polypropylene molecule

$\rightarrow|\;\;\;|\leftarrow$ — propylene mer

where R = CH$_3$

PP 2.9 Use a sketch to illustrate the polymerization of polystyrene $-(C_2H_3R)_n-$, where R is a benzene group, C$_6$H$_5$.

[PP 2.9]

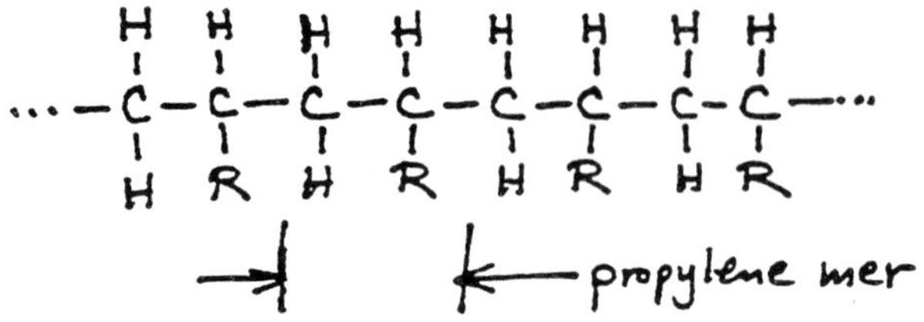

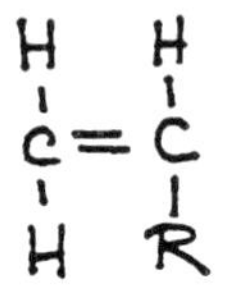

$$\begin{array}{c} H \quad H \\ | \quad\; | \\ C = C \\ | \quad\; | \\ H \quad R \end{array}$$ styrene molecule

$$\cdots-\underset{H}{\overset{H}{C}}-\underset{R}{\overset{H}{C}}-\underset{H}{\overset{H}{C}}-\underset{R}{\overset{H}{C}}-\underset{H}{\overset{H}{C}}-\underset{R}{\overset{H}{C}}-\underset{H}{\overset{H}{C}}-\underset{R}{\overset{H}{C}}-\cdots$$ polystyrene molecule

$\rightarrow|\;\;\;|\leftarrow$ — styrene mer

where R = C$_6$H$_5$

PP 2.10 Calculate the reaction energy for polymerization of (a) propylene (see Practice Problem 2.8) and (b) styrene (see Practice Problem 2.9).

[PP 2.10] (a) The backbone reaction is the same. Therefore, the calculation is the same:

$(740-680)$ kJ/mol = __60 kJ/mol__

(b) Again,

$(740-680)$ kJ/mol = __60 kJ/mol__

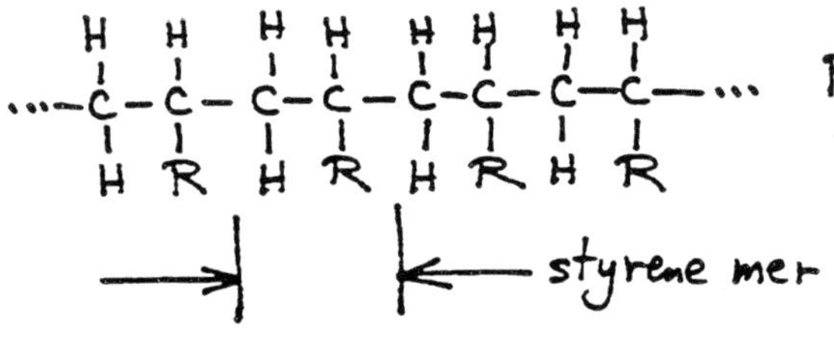

PP 2.11 The length of an average polyethylene molecule in a commercial clear plastic wrap is 0.2 μm. What is the average degree of polymerization (n) for this material? (See Sample Problem 2.11.)

PP 2.11 As illustrated in Sample Problem 2.11,
$$L = 2n\ell$$
or
$$n = \frac{L}{2\ell} = \frac{0.2 \times 10^{-6} \text{ m}}{2 \times 0.126 \times 10^{-9} \text{ m}} = \underline{\underline{794}}$$

2.29 Calculate the total reaction energy for polymerization required to produce the roll of clear plastic wrap described in Problem 2.4.

2.29 From Problem 2.4, we obtain:
4.60×10^{24} C atoms in the sheet of polyethylene.

From Sample Problem 2.10, we note there is a reaction energy of 60 kJ/mol (of double bonds).

Then, the total reaction energy for the polymer wrap is:

$$E_{reaction} = \frac{60 \text{ kJ}}{\text{mol bonds}} \times 4.60 \times 10^{24} \text{ atoms C} \times \frac{1 \text{ mol C atoms}}{0.6023 \times 10^{24} \text{ atoms C}} \times \frac{1 \text{ bond}}{2 \text{ atoms}}$$

$$= \underline{\underline{229 \text{ kJ}}}$$

2.30 Natural rubber is polyisoprene. The polymerization reaction can be illustrated as

$$n \begin{pmatrix} H & H & CH_3 & H \\ | & | & | & | \\ C=C-C=C \\ | & & & | \\ H & & & H \end{pmatrix} \rightarrow \begin{pmatrix} H & H & CH_3 & H \\ | & | & | & | \\ -C-C=C-C- \\ | & & & | \\ H & & & H \end{pmatrix}_n$$

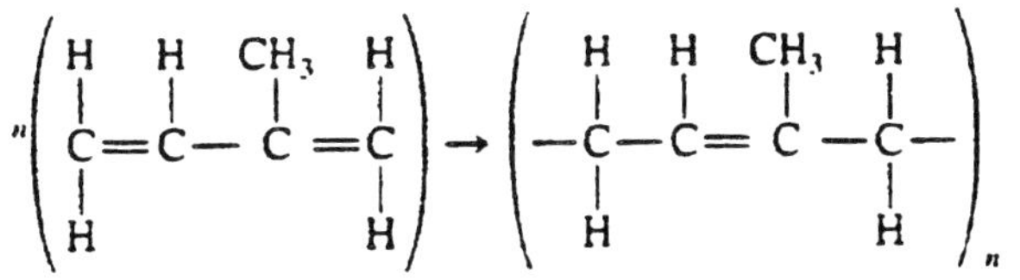

Calculate the reaction energy (per mole) for polymerization.

2.30 Although this reaction appears more complex, the net effect is (as in Sample Problem 2.10):
$$1 \, C=C \rightarrow 2 \, C-C$$
giving a reaction energy of:
$$(740 - 680) \text{ kJ/mol} = \underline{\underline{60 \text{ kJ/mol}}}$$

2.31 Neoprene is a synthetic rubber, polychloroprene, with a chemical structure similar to natural rubber (see Problem 2.30) except that it contains a Cl atom in place of the CH₃ group of the isoprene molecule. (a) Sketch the polymerization reaction for neoprene, and (b) calculate the reaction energy (per mole) for this polymerization. (c) Calculate the total energy released during the polymerization of 1 kg of chloroprene.

2.31 (a) Similar to the reaction in Problem 2.30:

$$n\begin{pmatrix} H & H & Cl & H \\ | & | & | & | \\ C=C-C=C \\ | & & & | \\ H & & & H \end{pmatrix} \rightarrow \left\{ \begin{array}{c} H & H & Cl & H \\ | & | & | & | \\ C-C=C-C \\ | & & & | \\ H & & & H \end{array} \right\}_n$$

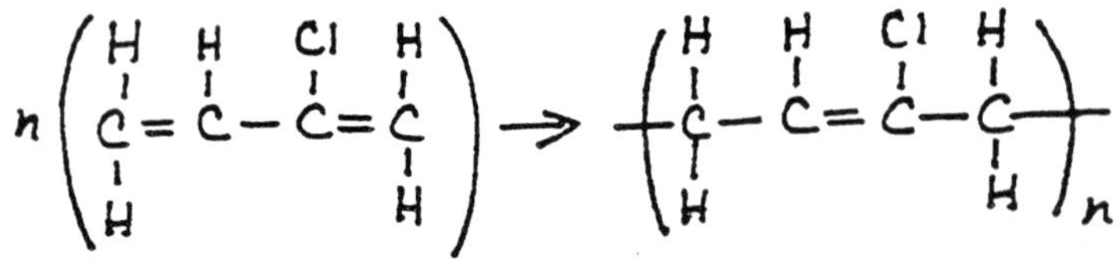

(b) Again (as in Sample Problem 2.10), the net reaction is: $1\ C=C \rightarrow 2\ C-C$

giving a reaction energy of:

$(740 - 680)\ kJ/mol = \underline{60\ kJ/mol}$

(c) The mer molecular weight is:

$(4 \times 12.01 + 5 \times 1.008 + 35.45)\ g = 88.53\ g$

As there is one bond reaction [as shown in part (b)] for each mer, we can write:

energy released $= 1\ kg \times \dfrac{1\ mol}{88.53\ g} \times \dfrac{1000\ g}{kg} \times 60\ kJ/mol = \underline{678\ kJ}$

2.32 Acetal polymers, which are widely used for engineering applications, can be represented by the following reaction, the polymerization of formaldehyde:

$$n\begin{pmatrix} H \\ \backslash \\ C=O \\ / \\ H \end{pmatrix} \longrightarrow \left\{ \begin{array}{c} H \\ | \\ C-O \\ | \\ H \end{array} \right\}_n$$

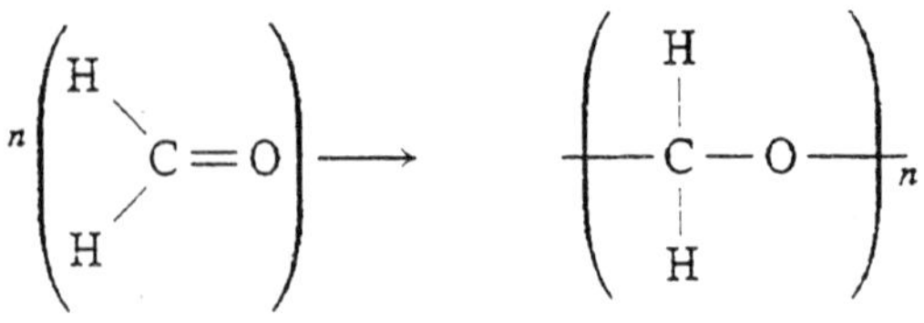

Calculate the reaction energy for this polymerization.

2.32 In this case, the net reaction is: $1\ C=O \rightarrow 2\ C-O$

giving a reaction energy of:

$2(360\ kJ/mol) - 535\ kJ/mol = \underline{185\ kJ/mol}$

2.33 The first step in the formation of phenolformaldehyde, a common phenolic polymer, is shown in Figure 9-6. Calculate the net reaction energy (per mole) for this step in the overall polymerization reaction.

2.33 In this case, the net reaction is:

$$1\ C{=}O + 2\ C{-}H \rightarrow 2\ C{-}C + 2\ O{-}H$$

giving a reaction energy of:

$$2(370\ kJ/mol) + 2(500\ kJ/mol) - 2(435\ kJ/mol) - 1(535\ kJ/mol) = 335\ kJ/mol$$

2.34 Calculate the molecular weight of a polyethylene molecule with $n = 500$.

2.34 Using the data of Appendix 1, we obtain

$$\text{mol. wt.} \left(C_2H_4\right)_{n=500} = 500[2(12.01) + 4(1.008)]\ \text{amu}$$

$$= 14{,}030\ \text{amu}$$

2.35 The monomer upon which a common acrylic polymer, polymethyl methacrylate, is based is given in Table 9.1. Calculate the molecular weight of a polymethyl methacrylate molecule with $n = 500$.

2.35 Using the chemical formula given by Table 9.1 and the data of Appendix 1, we obtain

$$\text{mol. wt.} \left(C_5H_8O_2\right)_{n=500}$$

$$= 500[5(12.01) + 8(1.008) + 2(16.00)]\ \text{amu}$$

$$= 50{,}060\ \text{amu}$$

2.36 Bone "cement" used by orthopedic surgeons to set artificial hip implants in place is methyl methacrylate polymerized during the surgery. The resulting polymer has a relatively wide range of molecular weights. Calculate the resulting range of molecular weights if $200 < n < 700$.

2.36 For: $\left(C_5H_8O_2\right)_{200<n<700}$

$$\text{mol. wt.} = 200\,[5(12.01)+8(1.008)+2(16.00)]\,\text{amu to } 700\,[\;"\;+\;"\;+\;"\;]\,\text{amu}$$

$$= \underline{\underline{20{,}020\ \text{amu to } 70{,}080\ \text{amu}}}$$

2.37 Orthopedic surgeons notice a substantial amount of heat evolution from polymethyl methacrylate bone cement during surgery. Calculate the reaction energy if a surgeon uses 15 gm of polymethyl methacrylate to set a given hip implant.

2.37 Note that 1 mol of polymethyl methacrylate contains 1 mol of C=C double bonds.

1 mol $C_5H_8O_2$ has $[5(12.01)+8(1.008)+2(16.00)]\,g = 100.1\,g$

As calculated numerous times in this section, the reaction energy for 1 C=C → 2 C–C is:

$$(740-680)\,kJ/mol = 60\,kJ/mol$$

Then, the total reaction energy the implant cement is:

$$E_{reaction} = \frac{60\,kJ}{\text{mol. bonds}} \times \frac{1\ \text{mol. bonds}}{100.1\,g} \times 15\,g = \underline{\underline{8.99\,kJ}}$$

2.38 The monomer for the common fluoroplastic, polytetrafluoroethylene, is

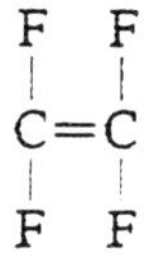

(a) Sketch the polymerization of polytetrafluoroethylene. (b) Calculate the reaction energy (per mole) for this polymerization. (c) Calculate the molecular weight of a molecule with $n = 500$.

2.38 (a)

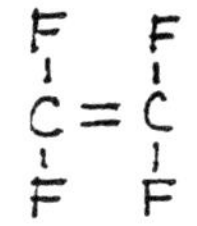

tetrafluoroethylene molecule

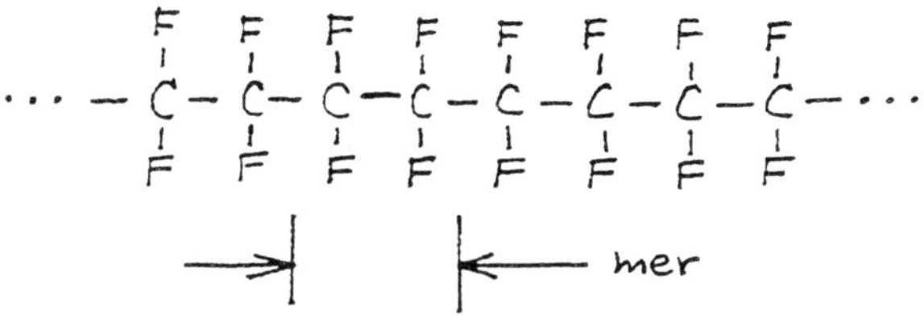

polytetrafluoroethylene molecule

(b) Reaction energy $= (740 - 680)$ kJ/mol
$= \underline{\underline{60 \text{ kJ/mol}}}$

(c) Using the data of Appendix 1,

mol. wt. $= 500[2(12.01) + 4(19.00)]$ amu

$= \underline{\underline{50{,}010 \text{ amu}}}$

2.39 Repeat Problem 2.38 for polyvinylidene fluoride, an ingredient in various commercial fluoroplastics, having the monomer:

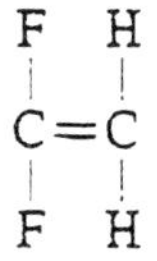

2.39 (a)

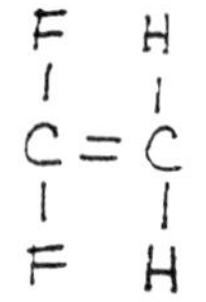

$$\begin{array}{c} F \\ | \\ C \\ | \\ F \end{array} = \begin{array}{c} H \\ | \\ C \\ | \\ H \end{array} \quad \text{vinylidene fluoride molecule}$$

$$\cdots - \underset{F}{\overset{F}{C}} - \underset{H}{\overset{H}{C}} - \underset{F}{\overset{F}{C}} - \underset{H}{\overset{H}{C}} - \underset{F}{\overset{F}{C}} - \underset{H}{\overset{H}{C}} - \underset{F}{\overset{F}{C}} - \underset{H}{\overset{H}{C}} - \cdots \quad \text{polyvinylidene fluoride molecule}$$

(b) Reaction energy = (740 − 680) kJ/mol
= **60 kJ/mol**

(c) Using the data of Appendix 1,

mol. wt. = 500 [2(12.01) + 2(1.008) + 2(19.00)] amu

= **32,020 amu**

2.40 Repeat Problem 2.38 for polyhexafluoropropylene, an ingredient in various commercial fluoroplastics, having the monomer:

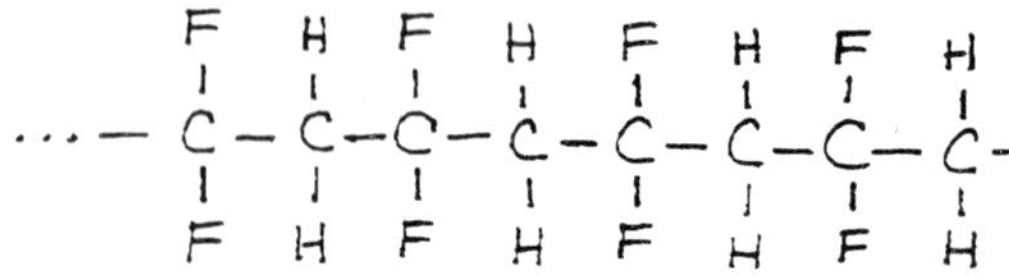

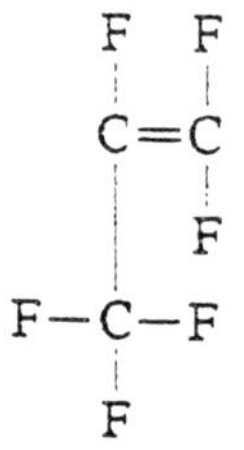

$$\begin{array}{c} F \quad F \\ | \quad | \\ C = C \\ | \quad | \\ \quad \; F \\ F-C-F \\ | \\ F \end{array}$$

2.40 (a)

$$\begin{array}{c} F \quad F \\ | \quad | \\ C = C \\ | \quad | \\ \quad \; F \\ F-C-F \\ | \\ F \end{array} \quad \text{hexafluoropropylene molecule & represent as:} \quad \begin{array}{c} F \quad F \\ | \quad | \\ C = C \\ | \quad | \\ R \quad F \end{array}$$

$$\cdots - \underset{R}{\overset{F}{C}} - \underset{F}{\overset{F}{C}} - \underset{R}{\overset{F}{C}} - \underset{F}{\overset{F}{C}} - \underset{R}{\overset{F}{C}} - \underset{F}{\overset{F}{C}} - \underset{R}{\overset{F}{C}} - \underset{F}{\overset{F}{C}} - \cdots \quad \text{polyhexafluoropropylene molecule}$$

(b) Reaction energy = (740−680) kJ/mol = **60 kJ/mol**

(c) Using Appendix 1,
mol. wt. = 500 [3(12.01) + 6(19.00)] amu = **75,020 amu**

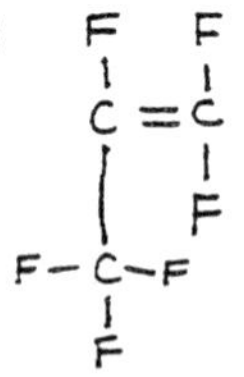

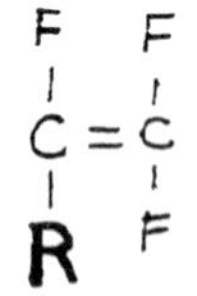

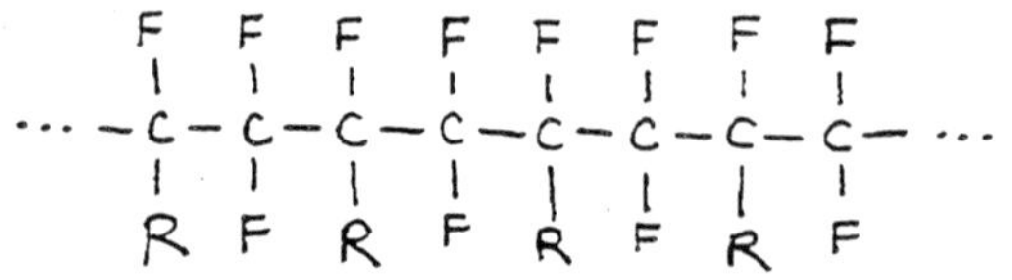

Section 2.4 – The Metallic Bond

PP 2.12 Discuss the low coordination number (= 4) for the diamond cubic structure found for some elemental solids, such as silicon. (See Sample Problem 2.12.)

PP 2.12 A greater degree of covalency in the Si–Si bond provides even stronger directionality and lower coordination number.

2.41 In Table 2.3, the heat of sublimation was used to indicate the magnitude of the energy of the metallic bond. A significant range of energy values is indicated by the data. The melting point data in Appendix 1 are another, more indirect indication of bond strength. Plot heat of sublimation versus melting point for the five metals of Table 2.3 and comment on the correlation.

2.41 Using Table **2.3** and Appendix 1, we obtain:

Atomic No.	Metal	$T_{m.pt.}$ (°C)	Heat of Subl. (kJ/mol)
12	Mg	649	148
13	Al	660	326
22	Ti	1660	473
26	Fe	1535	416
29	Cu	1083	338

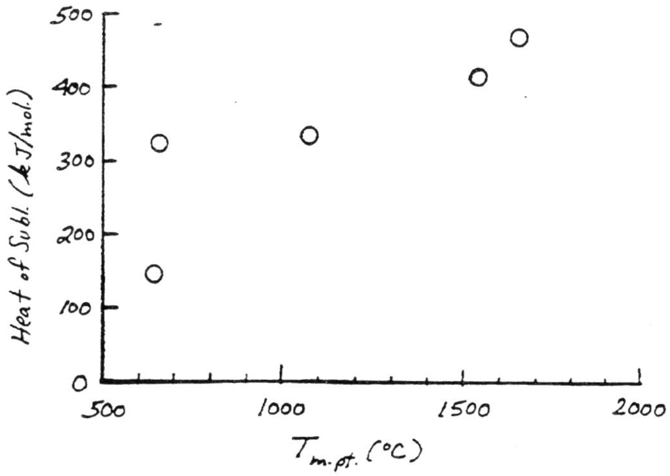

Comment: There is clearly a positive correlation between these two indicators of bond strength. However, the nature of the two processes (sublimation and melting) are sufficiently different to prevent a more precise relationship.

2.42 In order to explore a trend within the periodic table, plot the bond length of the group IIA metals (Be to Ba) as a function of atomic number. (Refer to Appendix 2 for necessary data.)

2.42 Using Figure 2-2 (the periodic table) with Appendix 2 we obtain:

IIA element	Atomic number	Atomic radius (r)	Bond length (= 2r)
Be	4	0.114 nm	0.228 nm
Mg	12	0.160 nm	0.320 nm
Ca	20	0.197 nm	0.394 nm
Sr	38	0.215 nm	0.430 nm
Ba	56	0.217 nm	0.434 nm

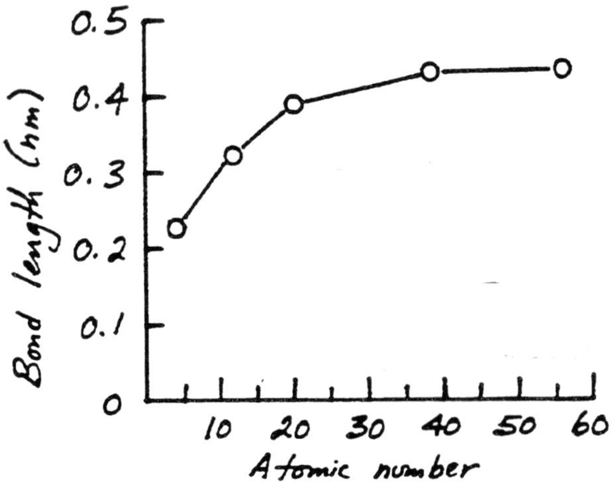

2.43 Superimpose on the plot generated for Problem 2.42 the metal–oxide bond lengths for the same range of elements.

2.43 Using Figure 2-2 and Appendix 2 gives us:

IIA element	Atomic number	Ionic radius + $r_{O^{2-}}$ (0.132 nm) =	Bond length
Be	4	0.054 nm	0.186 nm
Mg	12	0.078 nm	0.210 nm
Ca	20	0.106 nm	0.238 nm
Sr	38	0.127 nm	0.259 nm
Ba	56	0.143 nm	0.275 nm

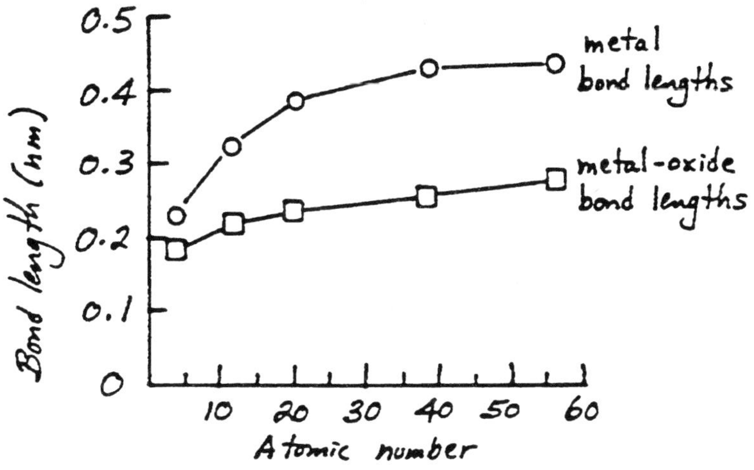

2.44 To explore another trend within the periodic table, plot the bond length of the metals in the row Na to Si as a function of atomic numbers. (For this purpose, Si is treated as a semimetal.)

2.44 Using Figure 2-2 and Appendix 2 gives us:

Element	Atomic Number	Atomic Radius (r)	Bond Length (=2r)
Na	11	0.186 nm	0.372 nm
Mg	12	0.160 nm	0.320 nm
Al	13	0.143 nm	0.286 nm
Si	14	0.117 nm	0.234 nm

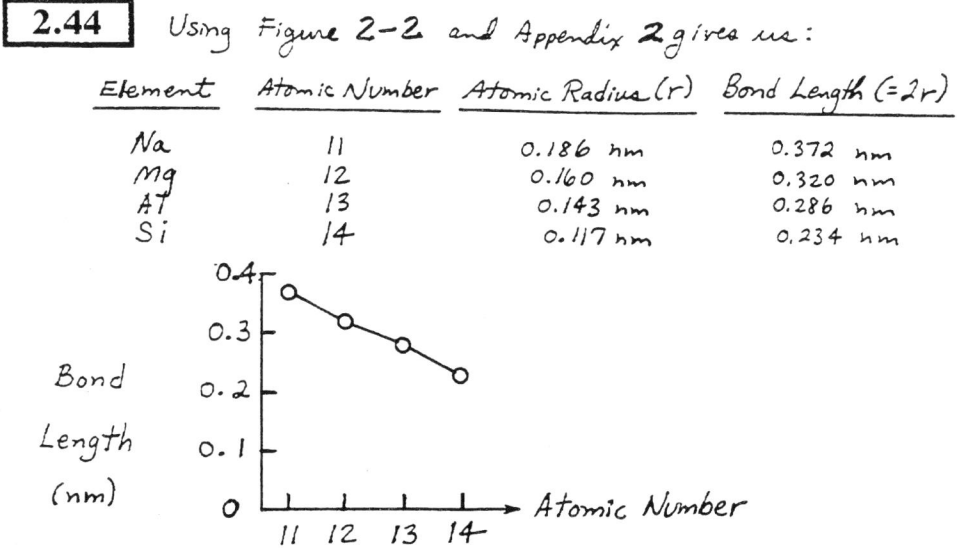

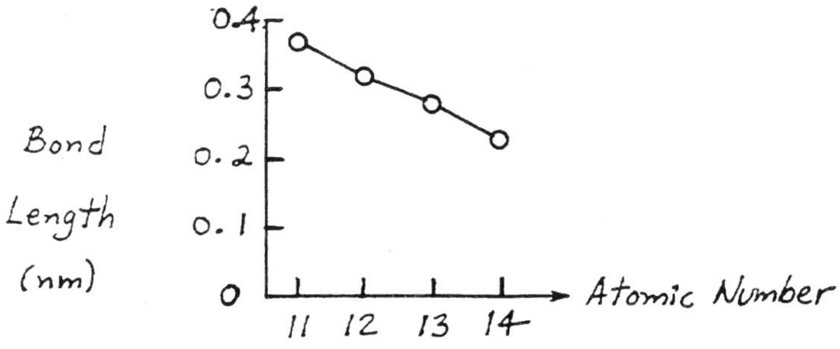

2.45 Superimpose on the plot generated for Problem 2.44 the metal–oxide bond lengths for the same range of elements.

2.45 Using Figure 2-2 and Appendix 2 gives:

Element	At. No.	Ionic Radius* + $r_{O^{2-}}$ (= 0.132 nm)	= Bond Length
Na	11	0.098 nm	0.230 nm
Mg	12	0.078 nm	0.210 nm
Al	13	0.057 nm	0.189 nm
Si	14	0.039 nm	0.171 nm

* using the most common valence, as noted on inside, back cover

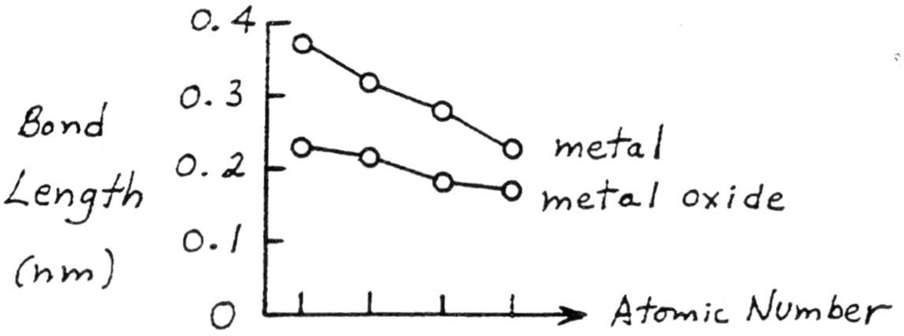

2.46 Plot the bond length of the metals in the long row of metallic elements (K to Ga).

2.46 Using Figure 2-2 and Appendix 2 gives us:

Element	Atomic number	Atomic radius (r)	Bond length (=2r)
K	19	0.231 nm	0.462 nm
Ca	20	0.197 nm	0.394 nm
Sc	21	0.160 nm	0.320 nm
Ti	22	0.147 nm	0.294 nm
V	23	0.132 nm	0.264 nm
Cr	24	0.125 nm	0.250 nm
Mn	25	0.112 nm	0.224 nm
Fe	26	0.124 nm	0.248 nm
Co	27	0.125 nm	0.250 nm
Ni	28	0.125 nm	0.250 nm
Cu	29	0.128 nm	0.256 nm
Zn	30	0.133 nm	0.266 nm
Ga	31	0.135 nm	0.270 nm

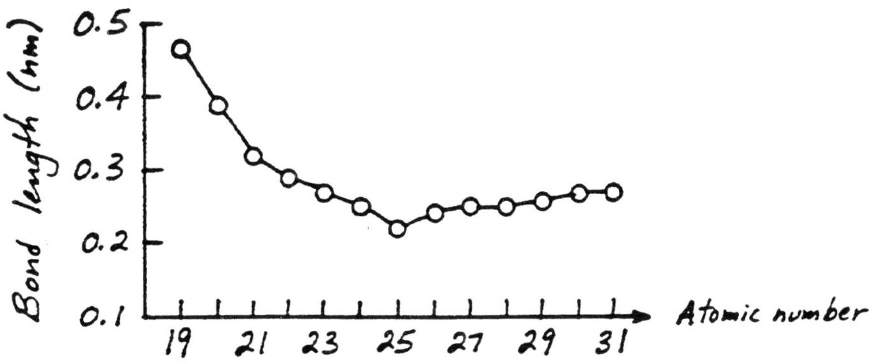

2.47 Superimpose on the plot generated for Problem 2.46 the metal–oxide bond lengths for the same range of elements.

2.47 Using Figure 2-2 and Appendix 2 gives:

Element	At. No.	Ionic radius* + $r_{O^{2-}}$ = (0.132 nm) =	Bond length
K	19	0.133 nm	0.265 nm
Ca	20	0.106 nm	0.238 nm
Sc	21	0.083 nm	0.215 nm
Ti	22	0.064 nm	0.196 nm
V	23	0.061 nm	0.193 nm
Cr	24	0.064 nm	0.196 nm
Mn	25	0.091 nm	0.223 nm
Fe	26	0.087 nm	0.219 nm
Co	27	0.082 nm	0.214 nm
Ni	28	0.078 nm	0.210 nm
Cu	29	0.096 nm	0.228 nm
Zn	30	0.083 nm	0.215 nm
Ga	31	0.062 nm	0.194 nm

*using the most common valence, as noted on inside, back cover

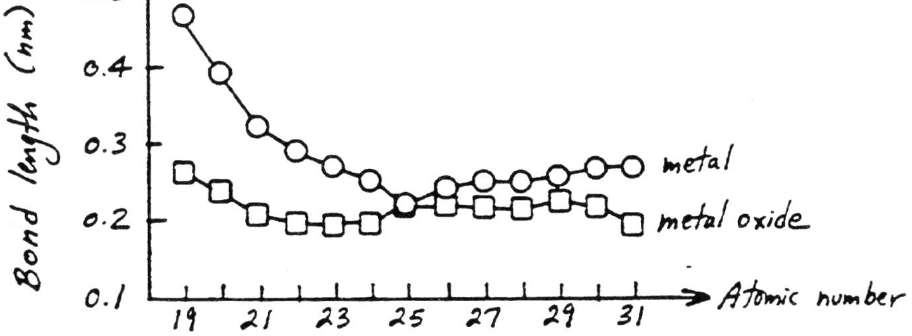

• 2.48 The heat of sublimation of a metal, introduced in Table 2.3, is related to the ionic bonding energy of a metallic compound discussed in Section 2.2. Specifically, these and related reaction energies are summarized in the Born–Haber cycle, illustrated below. For the simple example of NaCl

$$\text{Na (solid)} + \tfrac{1}{2}\text{Cl}_2\text{ (g)} \longrightarrow \text{Na (g)} + \text{Cl (g)}$$

$$\downarrow \Delta H_f^\circ \qquad\qquad \downarrow \qquad \downarrow$$

$$\text{NaCl (solid)} \longleftarrow \text{Na}^+\text{ (g)} + \text{Cl}^-\text{ (g)}$$

Given the heat of sublimation to be 100 kJ/mol for sodium, calculate the ionic bonding energy of sodium chloride. (Additional data: ionization energies for sodium and chlorine = 496 kJ/mol and −361 kJ/mol, respectively; dissociation energy for diatomic chlorine gas = 243 kJ/mol; heat of formation, ΔH_f°, of NaCl = −411 kJ/mol.)

2.48 Note that

$$\Delta H_f^\circ = \Delta E_{subl,\,Na} + \tfrac{1}{2}\Delta E_{dissoc,\,Cl_2} + \Delta E_{ion,\,Na} + \Delta E_{ion,\,Cl} + \Delta E_{ionic\ bonding,\,NaCl}$$

or,

$$\Delta E_{ionic\ bonding,\,NaCl} = \Delta H_f^\circ - \Delta E_{subl,\,Na} - \tfrac{1}{2}\Delta E_{dissoc,\,Cl_2} - \Delta E_{ion,\,Na} - \Delta E_{ion,\,Cl}$$

$$= \left(-411 - 100 - \frac{243}{2} - 496 + 361\right)\text{kJ/mol}$$

$$= \underline{\underline{-768\ \text{kJ/mol}}}$$

Section 2.5 – The Secondary, or van der Waals, Bond

PP 2.13 The bond energy and bond length for argon are calculated (assuming a "6–12" potential) in Sample Problem 2.13. Plot E as a function of a over the range 0.33 to 0.80 nm.

PP 2.13 From Sample Problem 2.13:

$$E_{bonding} = \left[-\frac{(10.37\times10^{-78}\ \text{J}\cdot\text{m}^6)}{a^6} + \frac{(16.16\times10^{-135}\ \text{J}\cdot\text{m}^{12})}{a^{12}}\right]$$

$$\times\ 0.6023\times10^{24}\ \text{mol}^{-1}$$

For the given range of a:

a	$E_{bonding}$
0.33×10^{-9} m	+0.999 kJ/mol
0.35 × "	− 0.517 "
0.382 × "	−0.999 " (from Sample Problem 2.5-1)
0.4 × "	−0.945 "
0.5 × "	−0.360 "
0.6 × "	−0.129 "
0.7 × "	−0.052 "
0.8 × "	−0.024 "

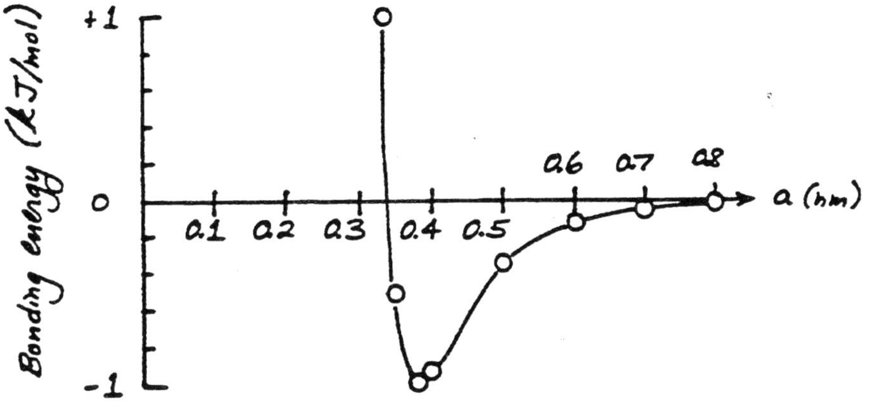

PP 2.14 Using the information from Sample Problem 2.13, plot the van der Waals bonding force curve for argon, that is, F versus a over the same range covered in Practice Problem 2.13.

PP 2.14

$$F = \frac{dE}{da} = \frac{6K_A}{a^7} - \frac{12K_R}{a^{13}}$$

$$F_{bonding} = \left[\frac{6\,(10.37\times10^{-78}\,J\cdot m^6)}{a^7} - \frac{12(16.16\times10^{-135}\,J\cdot m^{12})}{a^{13}}\right]$$

$$\times\ 0.6023\times10^{24}\,mol^{-1}$$

For the given range of a:

a	$F_{bonding}$
0.33×10^{-9} m	-124×10^{12} N/mol
0.35 × "	−40.5 × "
0.382 × "	0
0.4 × "	+5.47 × "
0.45 × "	+6.26 × "
0.5 × "	+3.84 × "
0.6 × "	+1.25 × "
0.7 × "	+0.44 × "
0.8 × "	+0.18 × "

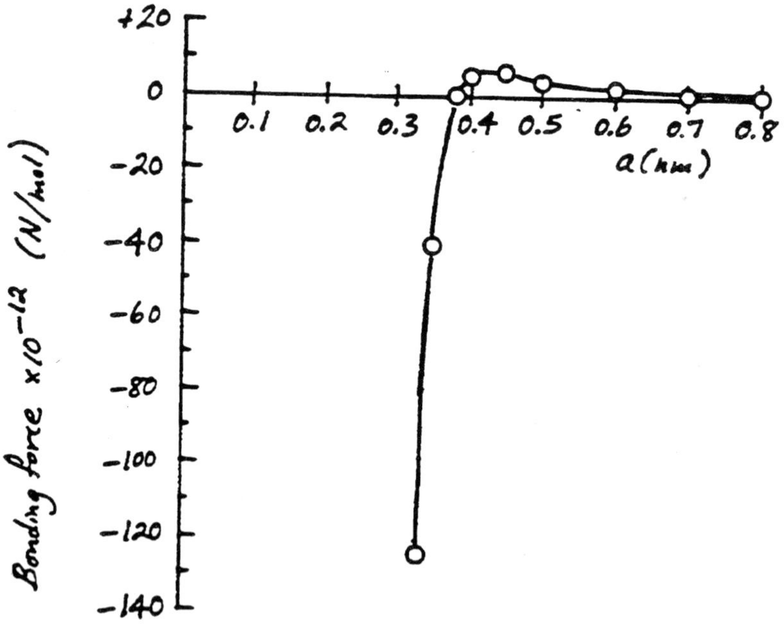

2.49 The secondary bonding of gas molecules to a solid surface is a common mechanism for measuring the surface area of porous materials. By lowering the temperature of a solid well below room temperature, a measured volume of the gas will condense to form a monolayer coating of molecules on the porous surface. For a 100 g sample of fused copper catalyst, a volume of 9 × 10^3 mm^3 of nitrogen (measured at standard temperature and pressure, 0°C and 1 atm) is required to form a monolayer upon condensation. Calculate the surface area of the catalyst in units of m^2/kg. (Take the area covered by a nitrogen molecule as 0.162 nm^2 and recall that, for an ideal gas, $pV = nRT$ where n is the number of moles of the gas.)

2.49 $pV = nRT$ or $n = \frac{pV}{RT} = \frac{(1\,atm)(9\times10^3\times10^{-9}\,m^3)}{(8.314\,J/K)(273\,K)}\,mol\,N_2$

$$\times \frac{1\,N/m^2}{9.869\times10^{-6}\,atm} \times 0.6023\times10^{24}\,\frac{molec.}{mol}$$

$$= 2.42\times10^{20}\ molec.\ N_2$$

$$\text{Area covered} = 0.162\times10^{-18}\,\frac{m^2}{molec.} \times 2.42\times10^{20}\,molec.$$

$$= 39.2\ m^2\ (\text{per } 100g\ Cu)$$

$$\text{or, } S = \frac{39.2\,m^2}{100\,g}\times\frac{1000\,g}{kg} = \underline{\underline{392\,m^2/kg}}$$

2.50 Repeat Problem 2.49 for a highly porous silica gel that has a volume of 1.16 × 10^7 mm^3 of N_2 gas (at STP) condensed to form a monolayer.

2.50 In this case,

$$n = \frac{pV}{RT} = \frac{(1\,atm)(1.16\times10^7\times10^{-9}\,m^3)(1\,N/m^2)(0.6023\times10^{24}\,molec./mol)}{(8.314\,J/K)(273\,K)(9.869\times10^{-6}\,atm)}$$

$$= 3.12\times10^{23}\ molec.\ N_2$$

$$\text{Area covered} = 0.162\times10^{-18}\,\frac{m^2}{molec.}\times 3.12\times10^{23}\,molec.$$

$$= 5.05\times10^4\ m^2\ (\text{per } 100g\ \text{silica gel})$$

$$\text{or, } S = \frac{5.05\times10^4\,m^2}{100\,g}\times\frac{1000\,g}{kg} = \underline{\underline{5.05\times10^5\,m^2/kg}}$$

2.51 Small-diameter noble gas atoms, such as helium, can dissolve in the relatively open network structure of silicate glasses. (See Figure 1-7b for a schematic of glass structure.) The secondary bonding of helium in vitreous silica is represented by a heat of solution, ΔH_s, of -3.96 kJ/mol. The relationship between solubility, S, and the heat of solution is

$$S = S_0 e^{-\Delta H_s/(RT)}$$

where S_0 is a constant, R the gas constant, and T the absolute temperature (in K). If the solubility of helium in vitreous silica is 5.51×10^{23} atoms/($m^3 \cdot$ atm) at 25°C, calculate the solubility at 200°C.

2.51 Using the given expression for solubility, we have

$$\frac{S_{200°C}}{S_{25°C}} = \frac{S_0\, e^{-\Delta H_s/R(200+273)K}}{S_0\, e^{-\Delta H_s/R(25+273)K}}$$

or

$$S_{200°C} = S_{25°C}\, e^{-\frac{\Delta H_s}{R}\left(\frac{1}{473\,K} - \frac{1}{298\,K}\right)}$$

$$= (5.51 \times 10^{23}\ \text{atoms}/(m^3 \cdot \text{atm}))$$

$$\times\, e^{-\frac{(-3{,}960\ J/mol)}{(8.314\ J/(mol \cdot K))}(-1.24 \times 10^{-3}\ K^{-1})}$$

$$= \underline{\underline{3.05 \times 10^{23}\ \text{atoms}/(m^3 \cdot \text{atm})}}$$

2.52 Due to its larger atomic diameter, neon has a higher heat of solution in vitreous silica than helium. If the heat of solution of neon in vitreous silica is -6.70 kJ/mol and the solubility at 25°C is 9.07×10^{23} atoms/($m^3 \cdot$ atm), calculate the solubility at 200°C. (See Problem 2.51.)

2.52 Using the solubility expression from Problem 2.51,

$$\frac{S_{200°C}}{S_{25°C}} = \frac{S_0\, e^{-\Delta H_s/R(200+273)K}}{S_0\, e^{-\Delta H_s/R(25+273)K}}$$

or

$$S_{200°C} = S_{25°C}\, e^{-\frac{\Delta H_s}{R}\left(\frac{1}{473\,K} - \frac{1}{298\,K}\right)}$$

$$= (9.07 \times 10^{23}\ \text{atoms}/(m^3 \cdot \text{atm}))$$

$$\times\, e^{-\frac{(-6{,}700\ J/mol)}{(8.314\ J/(mol \cdot K))}(-1.24 \times 10^{-3}\ K^{-1})}$$

$$= \underline{\underline{3.33 \times 10^{23}\ \text{atoms}/(m^3 \cdot \text{atm})}}$$

Section 3.1 – Seven Systems and Fourteen Lattices

PP 3.1 The note at the end of Sample Problem 3.1 comments that an area-centered square lattice can be resolved into a simple square lattice. Sketch this equivalence.

PP 3.1

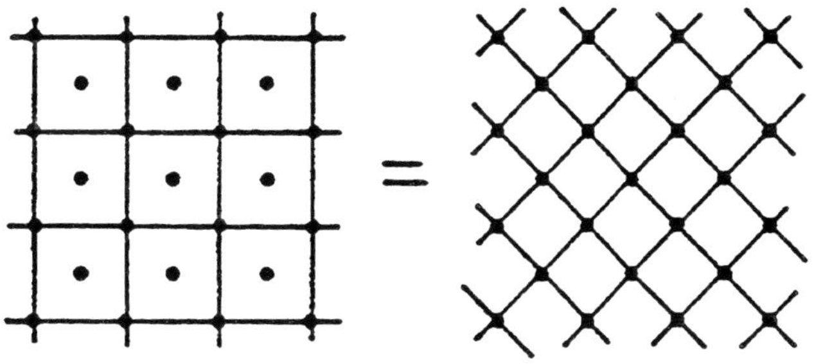

3.1 Why is the simple hexagon not a two-dimensional point lattice?

3.1 Consider a simple hexagonal lattice:

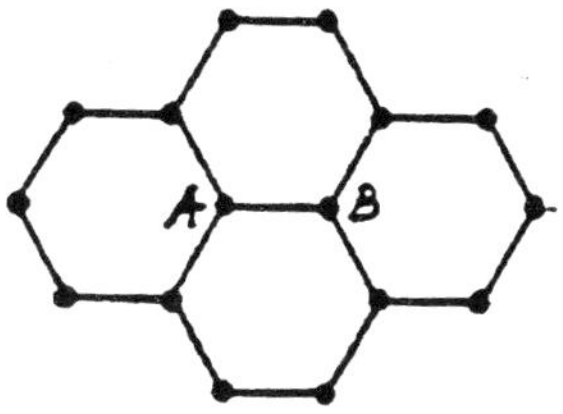

Points A and B do not have identical environments. (That for A is the mirror image of the environment for B.)

3.2 What would be an equivalent two-dimensional point lattice for the area-centered hexagon?

3.2

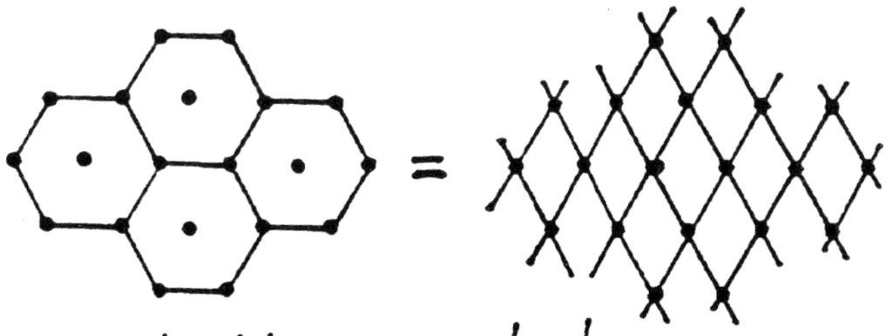

area-centered hexagon = rhombus

3.3 Why is there no base-centered cubic lattice in Table 3.2? (Use a sketch to answer.)

3.3 The base-centered cubic is equivalent to a simple tetragonal:

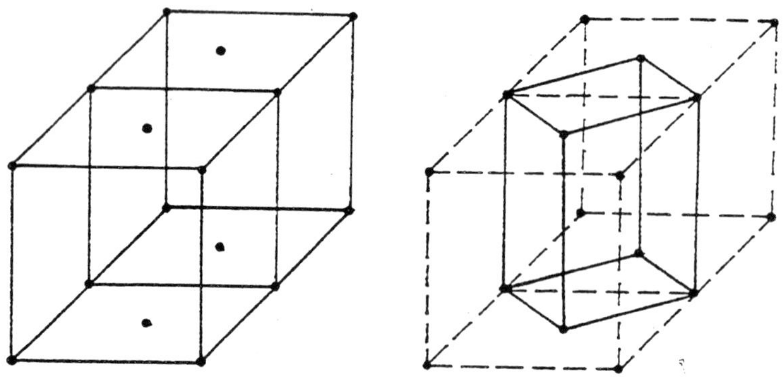

two adjacent base-centered unit cells

a simple tetragonal unit cell

• 3.4 (a) Which two-dimensional point lattice corresponds to the crystalline ceramic illustrated in Figure 1-7a? (b) Sketch the unit cell.

3.4 (a) Area-centered hexagon, as seen in (b):

(b)

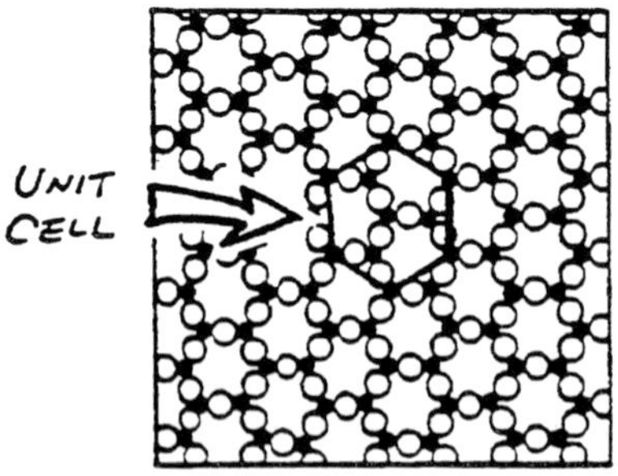

3.5 Under what conditions does the triclinic system reduce to the hexagonal system?

3.5 Noting the footnote for Table 3.1, we see that, when $a = b$, $\alpha = \beta$ ($= 90°$), and $\gamma = 120°$, the triclinic system becomes hexagonal.

3.6 Under what conditions does the monoclinic system reduce to the orthorhombic system?

3.6 Noting the footnote for Table 3.1, we see that, when $\beta = 90°$, the monoclinic system becomes orthorhombic.

Section 3.2 – Lattice Positions, Directions, and Planes

PP 3.2 From Table 3.2, list the lattice point positions for **(a)** the bcc Bravais lattice, **(b)** the body-centered tetragonal lattice, and **(c)** the body-centered orthorhombic lattice. (See Sample Problem 3.2.)

PP 3.2 (a) Corner positions: 000, 100, 010, 001, 110, 101, 011, 111

body-centered position: $\frac{1}{2}\frac{1}{2}\frac{1}{2}$

(b) Same

(c) same

PP 3.3 Use a sketch to determine which lattice points lie along the [111] direction in the **(a)** bcc, **(b)** body-centered tetragonal, and **(c)** body-centered orthorhombic unit cells of Table 3.2. (See Sample Problem 3.3.)

PP 3.3 (a)

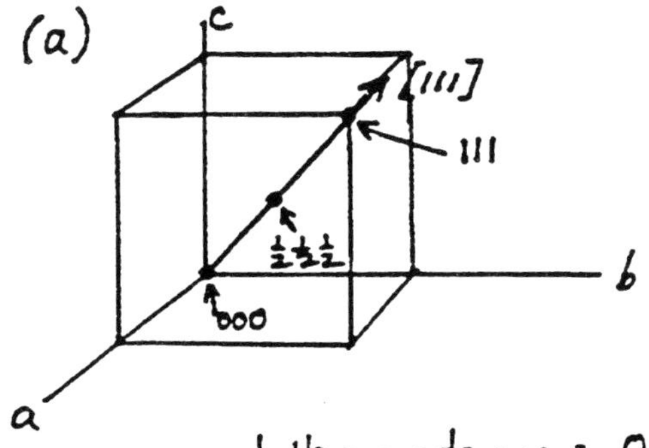

lattice points are: $\underline{\underline{000, \frac{1}{2}\frac{1}{2}\frac{1}{2}, 111}}$

(b) same

(c) same

PP 3.4 Sketch the 12 members of the ⟨110⟩ family determined in Sample Problem 3.4. (You may want to use more than one sketch.)

PP 3.4

In the a_2-a_3 plane:

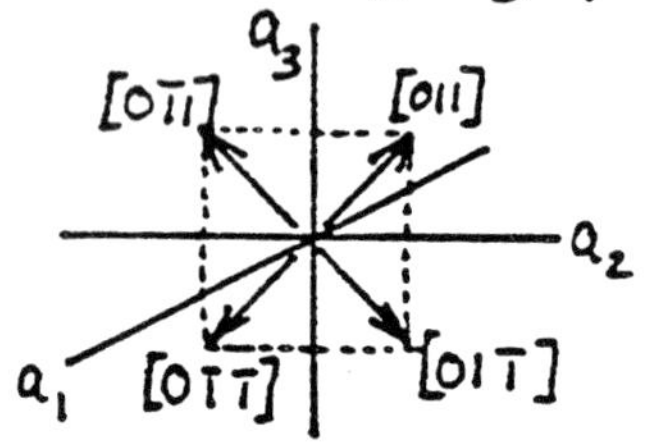

In the a_1-a_2 plane:

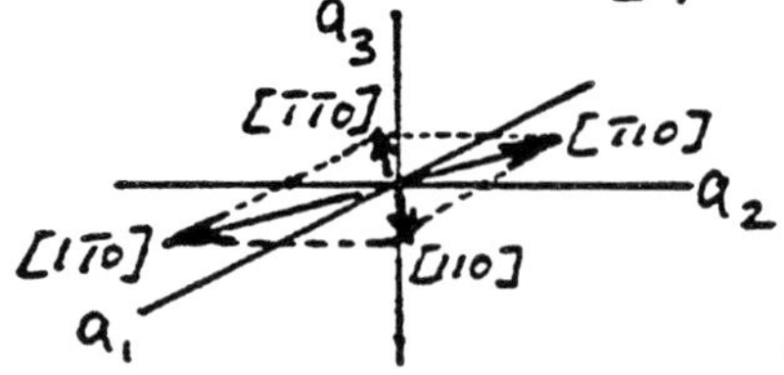

In the a_1-a_3 plane:

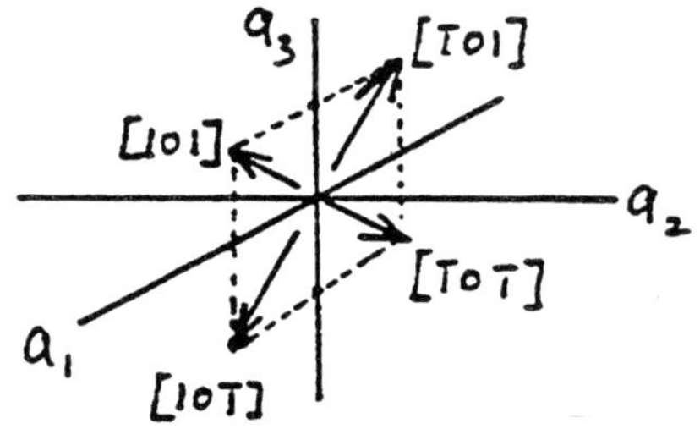

PP 3.5 (a) Determine the ⟨100⟩ family of directions in the cubic system, and (b) sketch the members of this family. (See Sample Problem 3.4 and Practice Problem 3.4.)

PP 3.5

(a) $\langle 100 \rangle = [100], [010], [001]$
$[\bar{1}00], [0\bar{1}0], [00\bar{1}]$

(b)

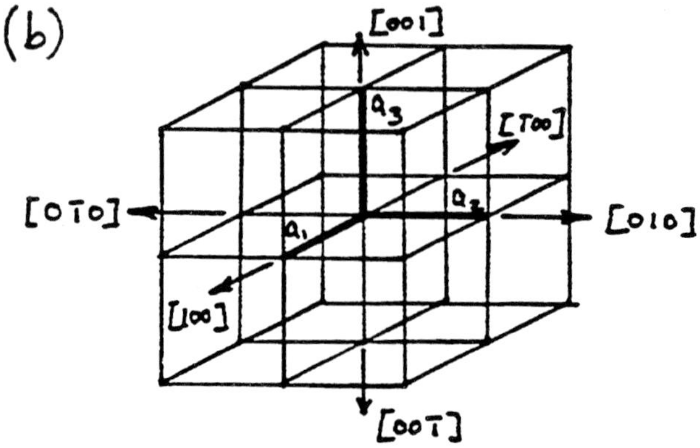

PP 3.6 Calculate the angles between **(a)** the [100] and [110] and **(b)** the [100] and [111] directions in the cubic system. (See Sample Problem 3.5.)

PP 3.6

(a) $\delta = \arccos\left(\frac{1+0+0}{\sqrt{1}\sqrt{2}}\right) = \arccos\frac{1}{\sqrt{2}} = \underline{\underline{45°}}$

(b) $\delta = \arccos\left(\frac{1+0+0}{\sqrt{1}\sqrt{3}}\right) = \arccos\frac{1}{\sqrt{3}} = \underline{\underline{54.7°}}$

PP 3.7 Sketch the $(3\bar{1}1)$ plane and its intercepts. (See Sample Problem 3.6 and Figure 3.8.)

PP 3.7

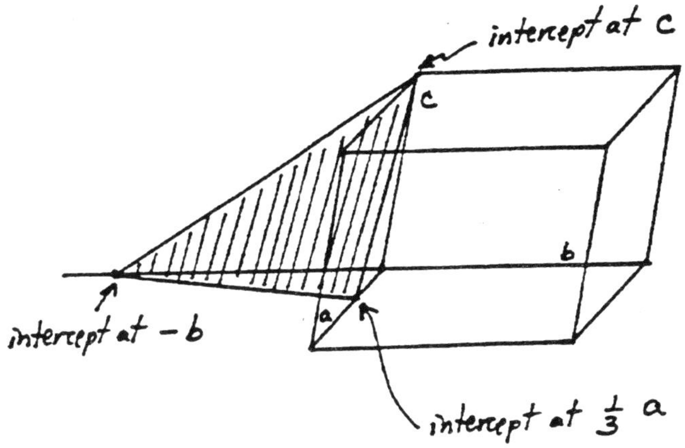

PP 3.8 Sketch the 12 members of the {110} family determined in Sample Problem 3.7. (To simplify matters, you will probably want to use more than one sketch.)

PP 3.8

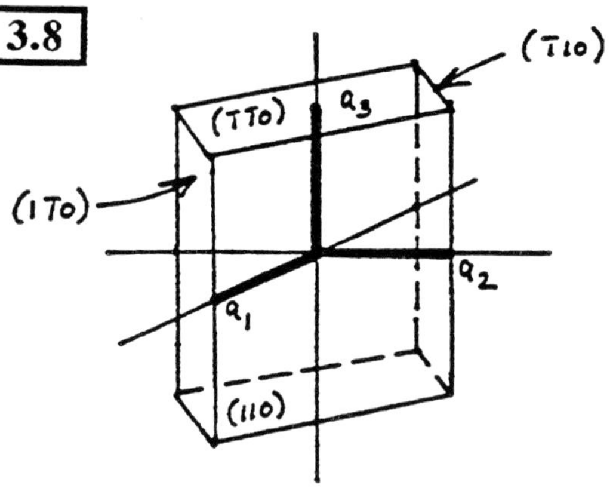

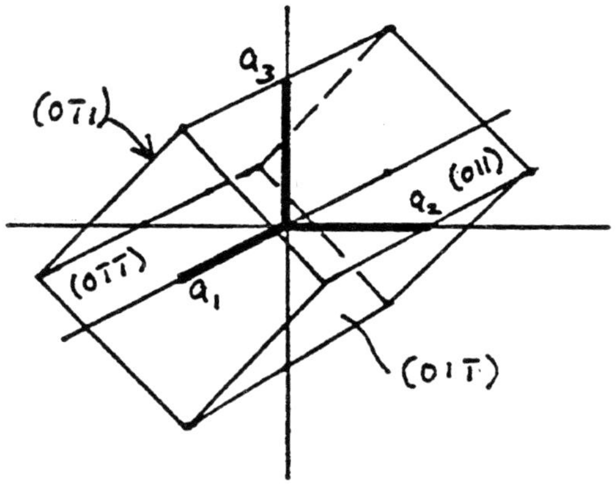

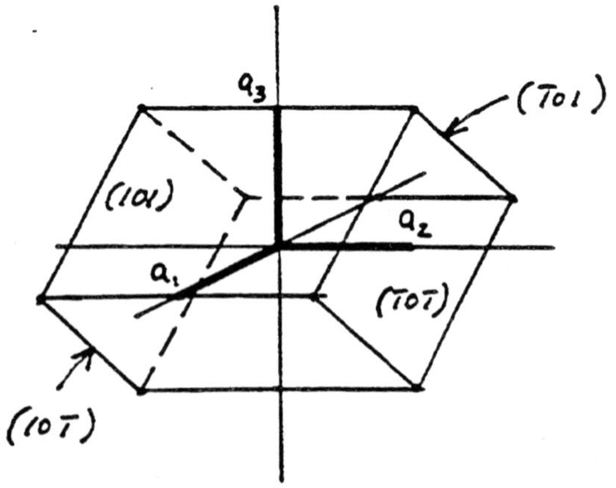

3.7 (a) Sketch, in a cubic unit cell, a [111] and a [112] lattice direction. (b) Use a trigonometric calculation to determine the angle between these two directions. (c) Use Equation 3.3 to determine the angle between these two directions.

3.7 (a)

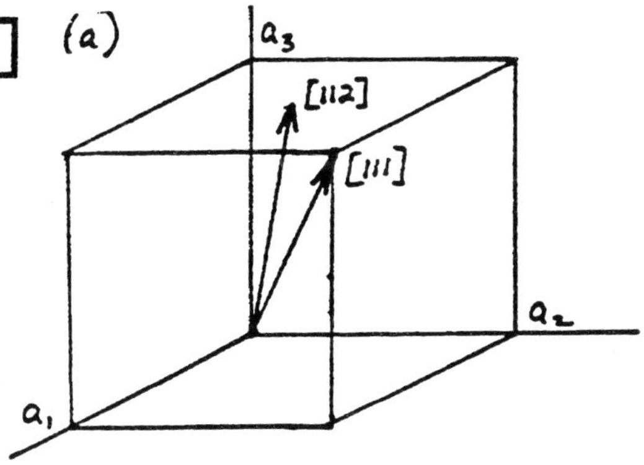

(b)

let $a = a_1 = a_2 = a_3$

$$\overline{[112]} = \sqrt{(\tfrac{1}{2}a)^2 + (\tfrac{1}{2}a)^2 + a^2}$$

$$= \sqrt{1.5a^2} = 1.225\,a$$

$$\overline{[111]} = \sqrt{3}\,a = 1.732a$$

$$x = \frac{\sqrt{2}}{2}a = 0.707a$$

In general,

$$x^2 = (\overline{[111]})^2 + (\overline{[112]})^2 - 2(\overline{[111]})(\overline{[112]})\cos\delta$$

or

$$\cos\delta = -\frac{x^2 - (\overline{[111]})^2 - (\overline{[112]})^2}{2(\overline{[111]})(\overline{[112]})}$$

$$= -\frac{0.5a^2 - 3a^2 - 1.5a^2}{2(\sqrt{3}a)(\sqrt{1.5}a)} = \frac{4}{4.243}$$

$$= 0.9428$$

$$\rightarrow \quad \delta = \underline{\underline{19.5°}}$$

(c)

$$\delta = \arccos\frac{1+1+2}{\sqrt{1+1+1}\sqrt{1+1+4}} =$$

$$= \arccos\frac{4}{\sqrt{3}\sqrt{6}} = \underline{\underline{19.5°}}$$

3.8 List the lattice point positions for (a) the base-centered orthorhombic lattice and (b) the triclinic lattice.

3.8 (a) Corner positions: 000, 100, 010, 001, 110, 101, 011, 111

base-centered positions: $\frac{1}{2}\frac{1}{2}0$, $\frac{1}{2}\frac{1}{2}1$

(b) Corner positions only: 000, 100, 010, 001
110, 101, 011, 111

3.9 (a) Sketch, in a cubic unit cell, [100] and [210] directions. (b) Use a trigonometric calculation to determine the angle between these directions. (c) Use Equation 3.3 to determine this angle.

3.9 (a)

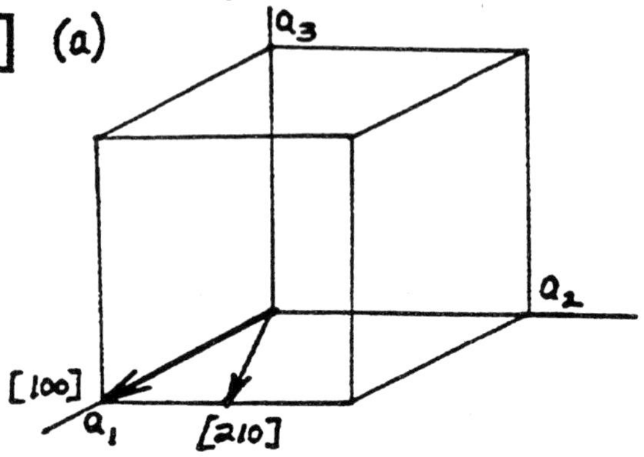

(b)

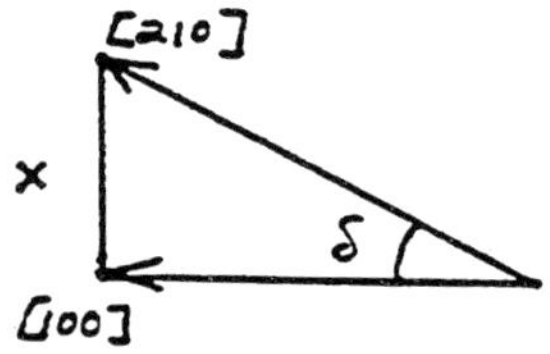

let $a = a_1 = a_2 = a_3$

$x = a/2$

$|[100]| = a$

$$\therefore \delta = \arctan \frac{a/2}{a}$$

$$= \arctan (1/2) = \underline{\underline{26.6°}}$$

(c) $$\delta = \arccos \frac{2+0+0}{\sqrt{1}\,\sqrt{4+1}}$$

$$= \arccos \frac{2}{\sqrt{5}} = \underline{\underline{26.6°}}$$

3.10 List the lattice point positions for (a) the body-centered orthorhombic lattice and (b) the base-centered monoclinic lattice.

3.10

(a) Corner positions : 000, 100, 010, 001, 110, 101, 011, 111

body-centered position: $\frac{1}{2}\frac{1}{2}\frac{1}{2}$

(b) corner positions: 000, 100, 010, 001, 110, 101, 011, 111

base-centered positions: $\frac{1}{2}\frac{1}{2}0$, $\frac{1}{2}\frac{1}{2}1$

• 3.11 What polyhedron is formed by "connecting the dots" between a corner atom in the fcc lattice and the 3 adjacent face-centered positions? Illustrate your answer with a sketch.

3.11

The figure is seen to be a <u>tetrahedron</u>:

• 3.12 Repeat Problem 3.11 for the 6 face-centered positions on the surface of the fcc unit cell.

3.12

The figure is seen to be an <u>octahedron</u>:

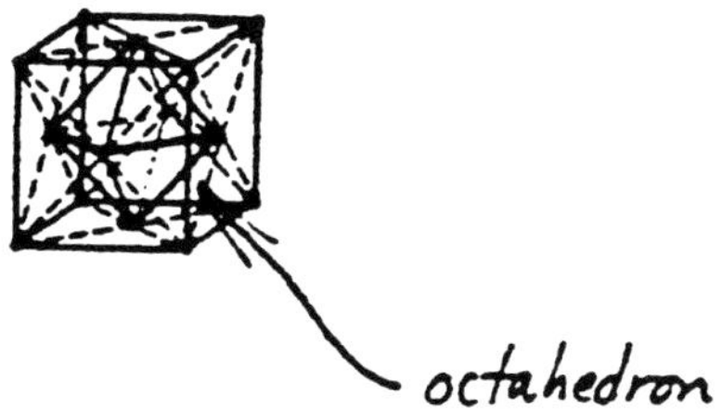

3.13 What $[hkl]$ direction connects the adjacent face-centered positions $\frac{1}{2}\,\frac{1}{2}\,0$ and $\frac{1}{2}\,0\,\frac{1}{2}$? Illustrate your answer with a sketch.

3.13 We can note that, if $\frac{1}{2}\,\frac{1}{2}\,0$ becomes the origin 000, then $\frac{1}{2}\,0\,\frac{1}{2}$ would become

$$\begin{array}{rccc} & \frac{1}{2} & 0 & \frac{1}{2} \\ - & \frac{1}{2} & \frac{1}{2} & 0 \\ \hline & 0 & -\frac{1}{2} & \frac{1}{2} \end{array}$$

Extending the line from the new "origin" through the new $0\,-\frac{1}{2}\,\frac{1}{2}$ position will lead to the position $1\,-1\,0$:

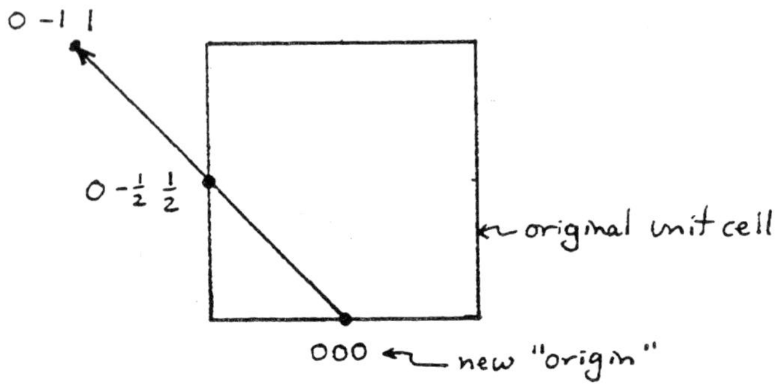

Therefore, it is the $[0\bar{1}1]$ direction.

3.14 A useful rule of thumb for the cubic system is that a given $[hkl]$ direction is the normal to the (hkl) plane. Using this rule and Equation 3.3, determine which members of the $\langle 110 \rangle$ family of directions lie within the (111) plane. (*Hint:* The dot product of two perpendicular vectors is zero.)

3.14 [111] is the plane normal to (111) and [111] will give a zero dot product with:

$[\bar{1}10]$, $[1\bar{1}0]$, $[\bar{1}01]$, $[10\bar{1}]$, $[01\bar{1}]$, and $[0\bar{1}1]$

3.15 Which members of the ⟨111⟩ family of directions lie within the (110) plane? (See the comments in Problem 3.14.)

3.15 [110] is the plane normal to (110) and [110] will give a zero dot product with:

$[\bar{1}11]$, $[1\bar{1}1]$, $[1\bar{1}\bar{1}]$, and $[\bar{1}1\bar{1}]$

3.16 Repeat Problem 3.14 for the $(\bar{1}11)$ plane.

3.16 $[\bar{1}11]$ is the plane normal to $(\bar{1}11)$ and $[\bar{1}11]$ will give a zero dot product with:

$[110]$, $[\bar{1}\bar{1}0]$, $[101]$, $[\bar{1}0\bar{1}]$, $[0\bar{1}1]$, and $[01\bar{1}]$

3.17 Repeat Problem 3.14 for the $(11\bar{1})$ plane.

3.17 $[11\bar{1}]$ is the plane normal to $(11\bar{1})$ and $[11\bar{1}]$ will give a zero dot product with:

$[\bar{1}10]$, $[1\bar{1}0]$, $[101]$, $[\bar{1}0\bar{1}]$, $[011]$, and $[0\bar{1}\bar{1}]$

3.18 Repeat Problem 3.15 for the (101) plane.

3.18 [101] is the plane normal to (101) and [101] will give a zero dot product with:

$[\bar{1}11]$, $[\bar{1}\bar{1}1]$, $[11\bar{1}]$, and $[1\bar{1}\bar{1}]$

3.19 Repeat Problem 3.15 for the $(10\bar{1})$ plane.

3.19

$[10\bar{1}]$ is the plane normal to $(10\bar{1})$
and $[10\bar{1}]$ will give a zero dot product with:

$[111]$, $[\bar{1}\bar{1}\bar{1}]$, $[1\bar{1}1]$, and $[\bar{1}1\bar{1}]$

3.20 Repeat Problem 3.15 for the $(\bar{1}01)$ plane.

3.20

$[\bar{1}01]$ is the plane normal to $(\bar{1}01)$
and $[\bar{1}01]$ will give a zero dot product with:

$[111]$, $[\bar{1}\bar{1}\bar{1}]$, $[1\bar{1}1]$, and $[\bar{1}1\bar{1}]$

3.21 Sketch the basal plane for the hexagonal unit cell having the Miller-Bravais indices (0001) (see Figure 3-9).

3.21

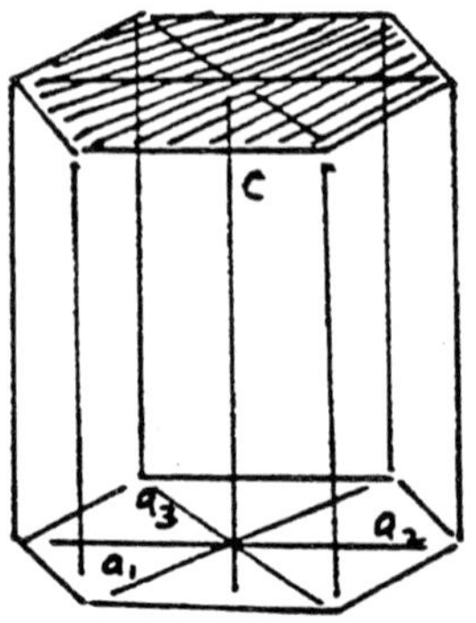

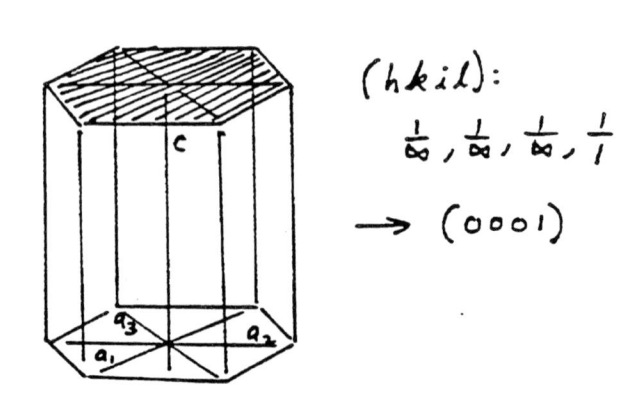

$(hkil)$:

$\frac{1}{\infty}, \frac{1}{\infty}, \frac{1}{\infty}, \frac{1}{1}$

$\rightarrow (0001)$

3.22 List the members of the family of prismatic planes for the hexagonal unit cell $\{01\bar{1}0\}$ (see Figure 3-9).

3.22 $(01\bar{1}0)$, $(0\bar{1}10)$, $(10\bar{1}0)$, $(\bar{1}010)$, $(1\bar{1}00)$, and $(\bar{1}100)$

3.23 The four-digit notation system (Miller-Bravais indices) introduced for planes in the hexagonal system can also be used for describing crystal directions. In a hexagonal unit cell sketch (a) the [0001] direction, and (b) the $[11\bar{2}0]$ direction.

3.23

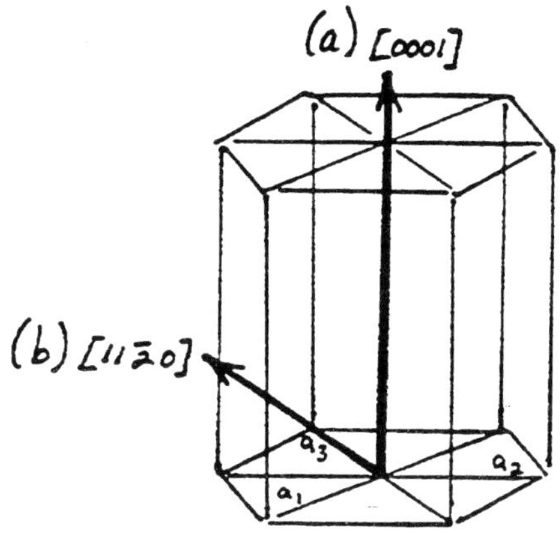

3.24 The family of directions described in Practice Problem 3.5 contains six members. The size of this family will be diminished for noncubic unit cells. List the members of the ⟨100⟩ family for (a) the tetragonal system, and (b) the orthorhombic system.

3.24 (a) Assuming $a = b \neq c$:

$[100]$, $[010]$, $[\bar{1}00]$, $[0\bar{1}0]$

(b) Now, $a \neq b \neq c$ giving us:

$[100]$, $[\bar{1}00]$

3.25 The comment in Problem 3.24 about families of directions also applies to families of planes. Figure 3-10 illustrates the six members of the {100} family of planes for the cubic system. List the members of the {100} family for (a) the tetragonal system, and (b) the orthorhombic system.

3.25 (a) Assuming $a = b \neq c$:

$(100), (010), (\bar{1}00), (0\bar{1}0)$

(b) Now, $a \neq b \neq c$ giving us:

$(100), (\bar{1}00)$

3.26 (a) List the first three lattice points (including the 000 point) lying on the [112] direction in the fcc lattice. (b) Illustrate your answer to part (a) with a sketch.

3.26 (a) 000, $\frac{1}{2}\frac{1}{2}1$, 112

(b)

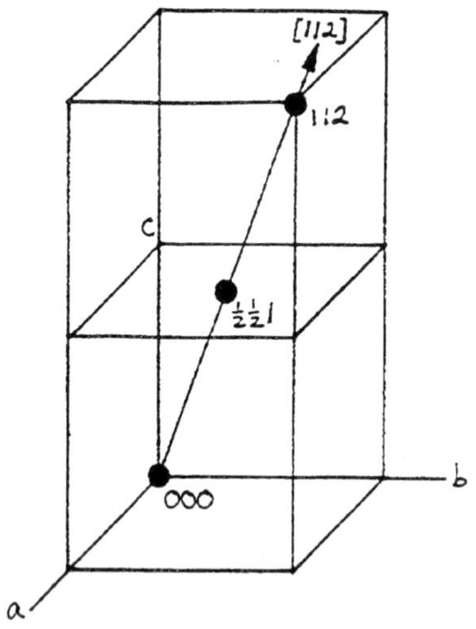

3.27 Repeat Problem 3.26 for the bcc lattice.

3.27 (a) 000, 112, 224

(b)

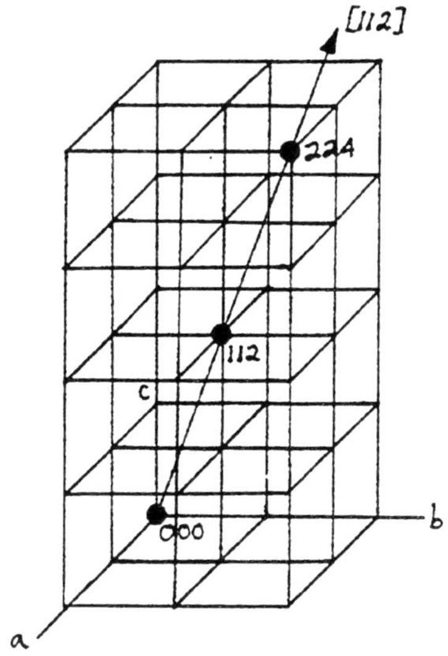

3.28 Repeat Problem 3.26 for the bct lattice.

3.28 (a) 000, 112, 224

(b)

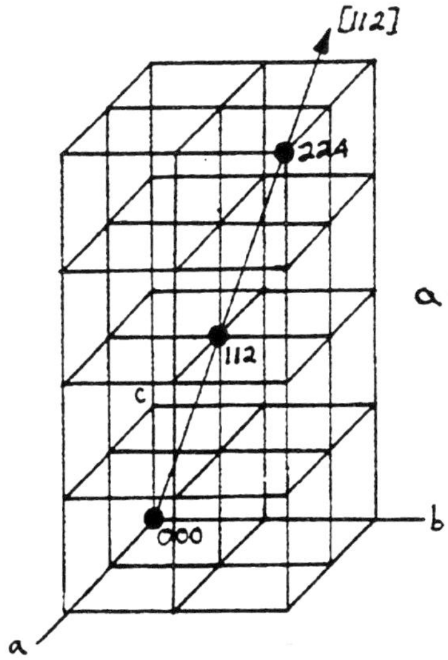

$a = a \neq c$

3.29 Repeat Problem 3.26 for the base-centered orthorhombic lattice.

3.29

(a) 000, $\frac{1}{2}\frac{1}{2}1$, 112

(b)

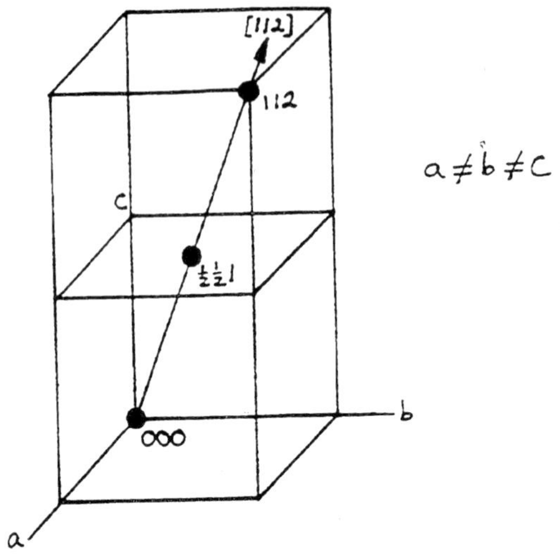

3.30 In the cubic system, which of the ⟨110⟩ family of directions represents the line of intersection between the (111) and (11$\bar{1}$) planes? (Note the comment in Problem 3.14.)

3.30

The line of intersection between the two planes lies within each plane and, therefore, must be perpendicular to the plane normal of each plane. In other words, the dot product of the direction along the line of intersection with the Miller indices of each plane must equal zero.

For (111) and (11$\bar{1}$), either [$\bar{1}$10] or [1$\bar{1}$0] meet these conditions.

3.31 Sketch the directions and planar intersection described in Problem 3.30.

3.31

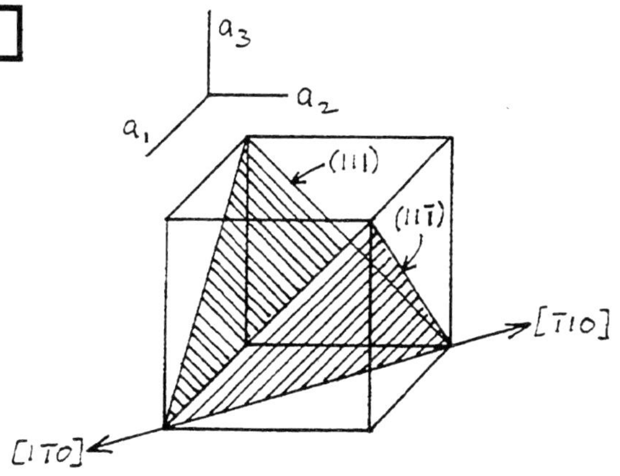

3.32 Sketch the members of the {100} family of planes in the triclinic system.

3.32 Analogous to Figure 3-10, only (100) and ($\bar{1}$00) are equivalent in this least symmetrical of the crystal systems.

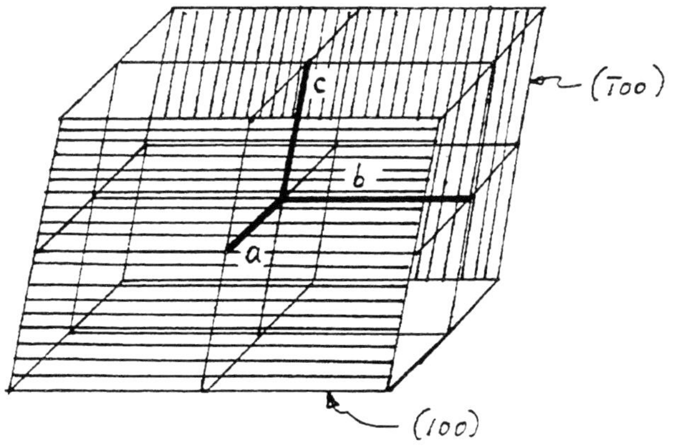

• 3.33 The first eight planes that give x-ray diffraction peaks for aluminum are indicated in Figure 3-40. Sketch each plane and its intercepts relative to a cubic unit cell. (To avoid confusion, use a separate sketch for each plane.)

3.33

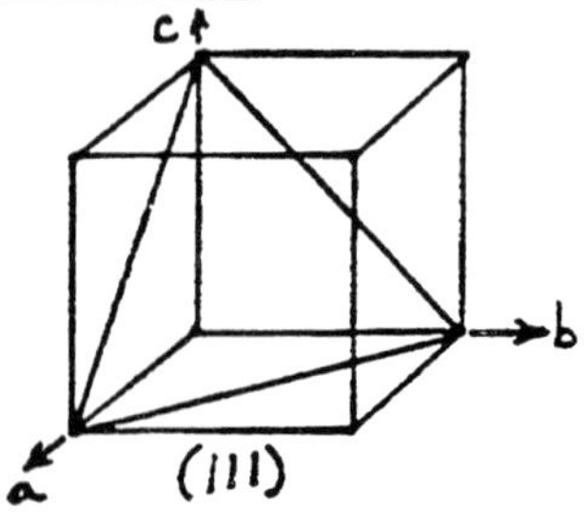

(111)

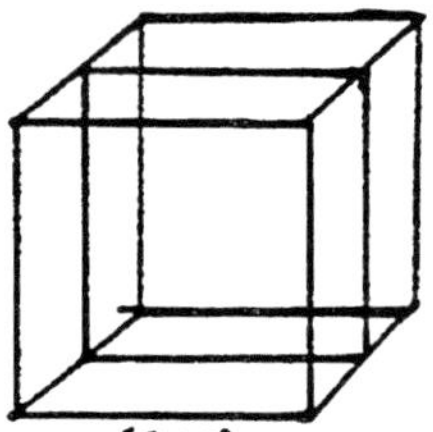

(200)

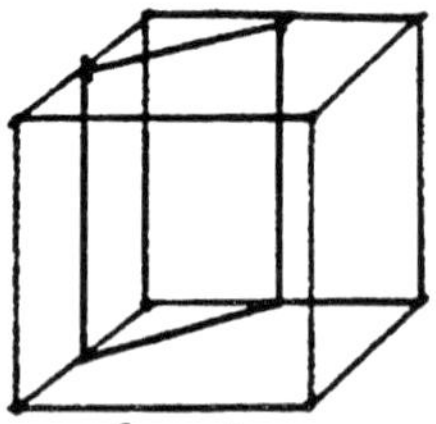

(220)

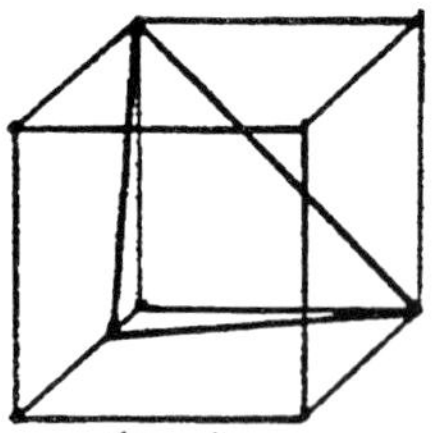

(311)

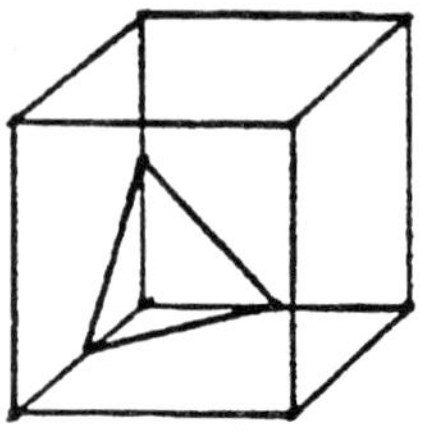

(222)

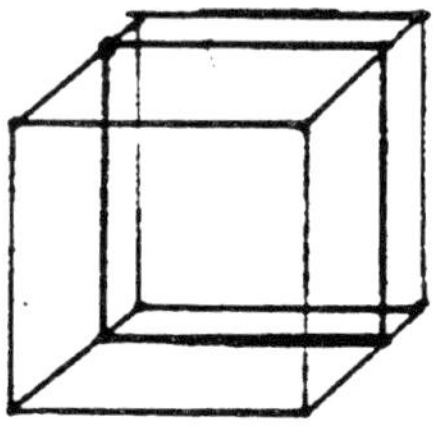

(400)

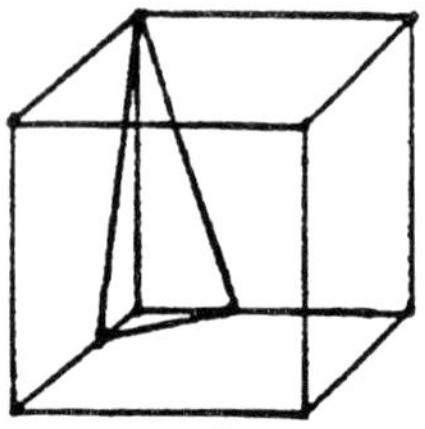

(331)

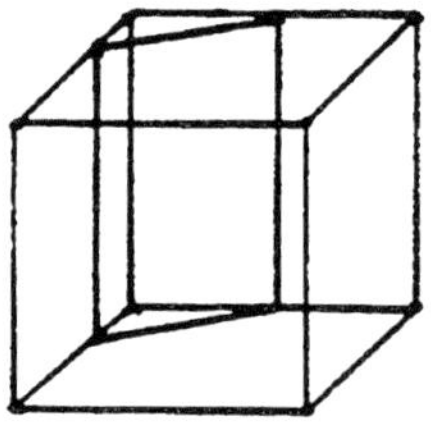

(420)

• 3.34 (a) List the ⟨112⟩ family of directions in the cubic system.
(b) Sketch this family. (You will want to use more than one sketch.)

3.34 (a) $\langle 112 \rangle = [112], [121], [211], [\bar{1}\bar{1}\bar{2}], [\bar{1}\bar{2}\bar{1}], [\bar{2}\bar{1}\bar{1}]$
$[11\bar{2}], [12\bar{1}], [21\bar{1}], [\bar{1}\bar{1}2], [\bar{1}\bar{2}1], [\bar{2}\bar{1}1]$
$[1\bar{1}2], [1\bar{2}1], [2\bar{1}1], [\bar{1}1\bar{2}], [\bar{1}2\bar{1}], [\bar{2}1\bar{1}]$
$[\bar{1}12], [\bar{1}21], [\bar{2}11], [1\bar{1}\bar{2}], [1\bar{2}\bar{1}], [2\bar{1}\bar{1}]$

(b) Consider the eight adjacent unit cells shown in Figure 3-7:

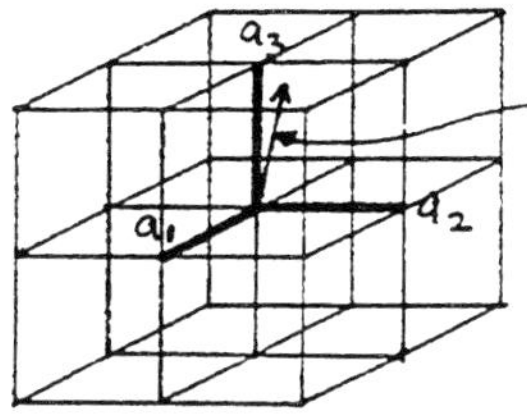

In all cases, a [112]-type direction goes from the origin to the center of an outside face on a given unit cell (example of [112] shown above). Then, the 24 directions can be shown as 8 clustered around $\pm a_1$, 8 clustered around $\pm a_2$, and 8 clustered around $\pm a_3$.

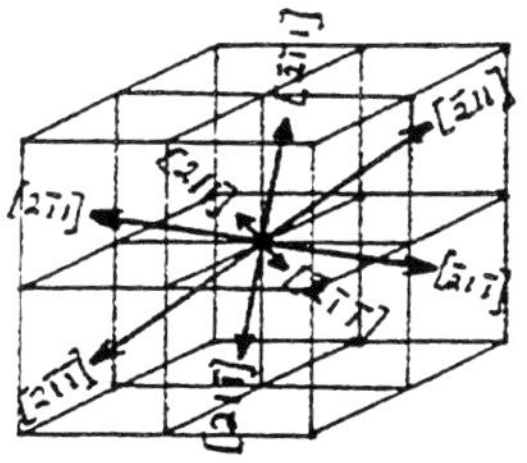

around a_1

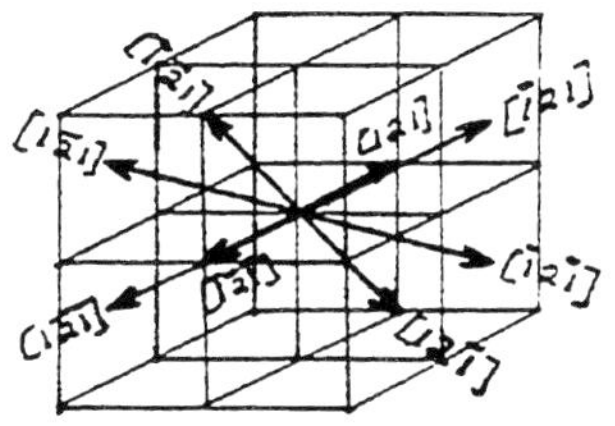

around a_2

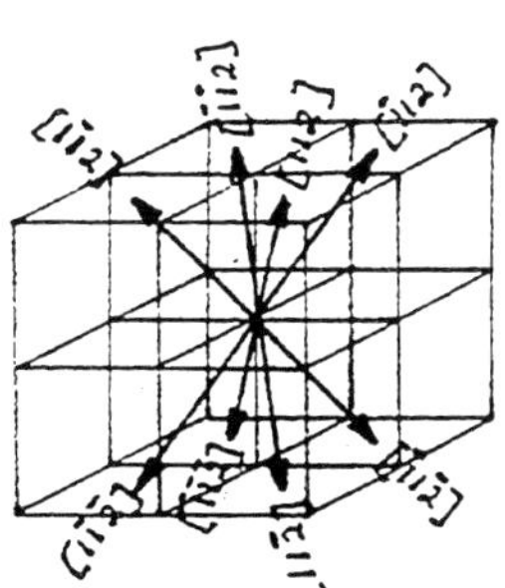

around a_3

Section 3.3 – Metal Structures

PP 3.9 Calculate the linear density of atoms along the [111] direction in **(a)** bcc iron and **(b)** fcc nickel. (See Sample Problem 3.8.)

PP 3.9

(a) For bcc Fe:

$$r^{-1} = \frac{1}{2r_{Fe}} = \frac{1}{2(0.124\text{ nm})} = \underline{\underline{4.03 \text{ atoms/nm}}}$$

(b) For fcc Ni:

$$a = \frac{4}{\sqrt{2}} r_{Ni} = \frac{4}{\sqrt{2}}(0.125\text{ nm}) = 0.354\text{ nm}$$

$$r^{-1} = \frac{1}{\sqrt{3}a} = \frac{1}{\sqrt{3}(0.354\text{ nm})} = \underline{\underline{1.63 \text{ atoms/nm}}}$$

PP 3.10 Calculate the planar density of atoms in the (111) plane of **(a)** bcc iron and **(b)** fcc nickel. (See Sample Problem 3.9.)

PP 3.10

(a) For bcc Fe:

$$a = \frac{4}{\sqrt{3}} r_{Fe} = \frac{4}{\sqrt{3}}(0.124\text{ nm}) = 0.286\text{ nm}$$

$$l = \sqrt{2}a = \sqrt{2}(0.286\text{ nm}) = 0.405\text{ nm}$$

$$A = \frac{\sqrt{3}}{4} l^2 = \frac{\sqrt{3}}{4}(0.405\text{ nm})^2 = 0.0710\text{ nm}^2$$

$$\text{atomic density} = \frac{0.5\text{ atom}}{A} = \frac{0.5\text{ atom}}{0.0710\text{ nm}^2}$$

$$= \underline{\underline{7.04 \text{ atoms/nm}^2}}$$

(b) For fcc Ni:

$$l = \sqrt{2}a$$

Using $a = 0.354\text{ nm}$ (from PP 3.9) gives:

$$l = \sqrt{2}(0.354\text{ nm}) = 0.500\text{ nm}$$

$$A = \frac{\sqrt{3}}{4} l^2 = \frac{\sqrt{3}}{4}(0.500\text{ nm})^2 = 0.108\text{ nm}^2$$

$$\text{atomic density} = \frac{2\text{ atoms}}{A} = \frac{2\text{ atoms}}{0.108\text{ nm}^2}$$

$$= \underline{\underline{18.5 \text{ atoms/nm}^2}}$$

PP 3.11 In Sample Problem 3.10 the relationship between lattice parameter, a, and atomic radius, r, for an fcc metal is found to be $a = (4/\sqrt{2})r$. Derive similar relationships for (**a**) a bcc metal and (**b**) an hcp metal.

PP 3.11 (a) The length, l, of a body diagonal is:

$$l = 4r = \sqrt{3}\,a$$

or

$$a = \frac{4}{\sqrt{3}}r$$

(b) By inspection of Figure 3–13, we see that:

$$a = 2r$$

(Note: We shall see in Problem 3.39 that $c = 1.633a$.)

PP 3.12 Calculate the density of α-Fe, which is a bcc metal. (*Caution:* A different relationship between lattice parameter, a, and atomic radius, r, applies to this different crystal structure. See Practice Problem 3.11.)

PP 3.12 From PP 3.11, $a = (4/\sqrt{3})r$.

For α-Fe, $a = (4/\sqrt{3})(0.124\ nm) = 0.286\ nm$

giving:

$$\rho = \frac{2\ atoms}{(0.286\ nm)^3} \times \frac{55.85\ g}{0.6023 \times 10^{24}\ atoms} \times \left(\frac{10^7\ nm}{cm}\right)^3$$

$$= 7.90\ g/cm^3$$

(Note final comments in PP 3.11.)

3.35 Calculate the density of Mg, an hcp metal. (Note Problem 3.39 for the ideal c/a ratio.)

3.35 $a = 2r$, $c = 1.633a$ (see Problem 3.39)

Consider the area of the base of the unit cell shown in Figure 3–13:

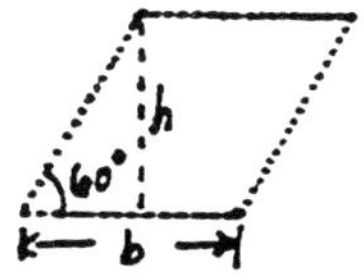

area = bh

$b = a$

$\sin 60° = \frac{h}{b} = \frac{h}{a} \rightarrow h = a \sin 60°$

or area $= a^2 \sin 60°$

Then, volume of unit cell = area × height

$$= (a^2 \sin 60°)(1.633a)$$

$$= 1.633 a^3 \sin 60°$$

$$= 1.414 a^3 = 1.414 \times 8 r_{Mg}^3$$

$$= 1.414 \times 8 (0.160\ nm)^3$$

$$= 0.04634\ nm^3$$

Finally giving:

$$\rho = \frac{2\ atoms}{0.04634\ nm^3} \times \frac{24.31\ g}{0.6023 \times 10^{24}\ atoms} \times \left(\frac{10^7\ nm}{cm}\right)^3$$

$$= \underline{\underline{1.74\ g/cm^3}}$$

3.36 Calculate the atomic packing factor (APF) of 0.68 for the bcc metal structure.

3.36

From PP 3.11, $a = (4/\sqrt{3})r$

$$V_{atom} = \tfrac{4}{3}\pi r^3$$

$$APF = \frac{2(\tfrac{4}{3}\pi r^3)}{[(4/\sqrt{3})r]^3} = \underline{\underline{0.680}}$$

3.37 Calculate the atomic packing factor (APF) of 0.74 for fcc metals.

3.37

From PP 3.11, $a = (4/\sqrt{2})r$

$$V_{atom} = \tfrac{4}{3}\pi r^3$$

$$APF = \frac{4(\tfrac{4}{3}\pi r^3)}{[(4/\sqrt{2})r]^3} = \underline{\underline{0.740}}$$

3.38 Calculate the atomic packing factor (APF) of 0.74 for hcp metals.

3.38

From Problems 3.35 and 3.39 $a = 2r$

$c = 1.633a$

and $V_{unit\ cell} = 1.633(2r)^3 \sin 60°$.

$$V_{atom} = \tfrac{4}{3}\pi r^3$$

$$APF = \frac{2(\tfrac{4}{3}\pi r^3)}{1.633(2r)^3 \sin 60°} = \underline{\underline{0.740}}$$

3.39 (a) Show that the c/a ratio (height of unit cell divided by its edge length) is 1.633 for the ideal hcp structure. (b) Comment on the fact that real hcp metals display c/a ratios varying from 1.58 (for Be) to 1.89 (for Cd).

3.39 (a) Note from Figure 3–13 that the central atom (at $\frac{2}{3}\frac{1}{3}\frac{1}{2}$) rests atop 3 basal plane atoms giving a tetrahedral configuration:

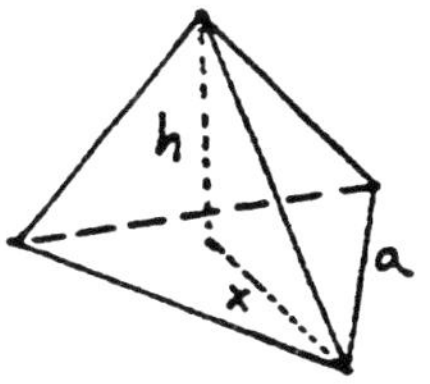

a = length of each side of tetrahedron

h = height of tetrahedron = $c/2$

In basal plane:

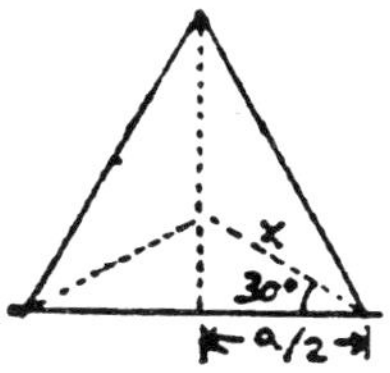

$$x = \frac{a/2}{\cos 30^\circ} = 0.5774a$$

$$h^2 = a^2 - x^2 = a^2 - (0.5774a)^2 = 0.6667a^2$$

$$h = 0.8165a$$

$$c = 2h = 1.633a$$

or $\underline{\underline{c/a = 1.633}}$

(b) Rather than perfect spheres, the atoms are effectively ellipsoids (due to some asymmetry in atomic bonding).

3.40 In Figures 3-11b and 3-12b we show atoms and fractional atoms making up a unit cell. An alternative convention is to describe the unit cell in terms of "equivalent points." For example, the two atoms in the bcc unit cell can be considered to be one corner atom at 000 and one body-centered atom at $\frac{1}{2}\frac{1}{2}\frac{1}{2}$. The one corner atom is equivalent to the eight $\frac{1}{8}$ atoms shown in Figure 3-11b. In a similar way, identify the four atoms associated with equivalent points in the fcc structure.

3.40 $\underline{\underline{000,\ \frac{1}{2}\frac{1}{2}0,\ \frac{1}{2}0\frac{1}{2},\ 0\frac{1}{2}\frac{1}{2}}}$

3.41 Identify the atoms associated with equivalent points in the hcp structure. (See Problem 3.40.)

3.41

$000, \frac{2}{3}\frac{1}{3}\frac{1}{2}$

3.42 Repeat Problem 3.41 for the body-centered orthorhombic lattice.

3.42

As unit cell dimensions are not involved, the body-centered orthorhombic result is the same as the body-centered cubic:

$000, \frac{1}{2}\frac{1}{2}\frac{1}{2}$

3.43 Repeat Problem 3.41 for the base-centered orthorhombic lattice.

3.43

$000, \frac{1}{2}\frac{1}{2}0$

3.44 Sketch the $[1\bar{1}0]$ direction within the (111) plane relative to an fcc unit cell. Include all atom center positions within the plane of interest.

3.44

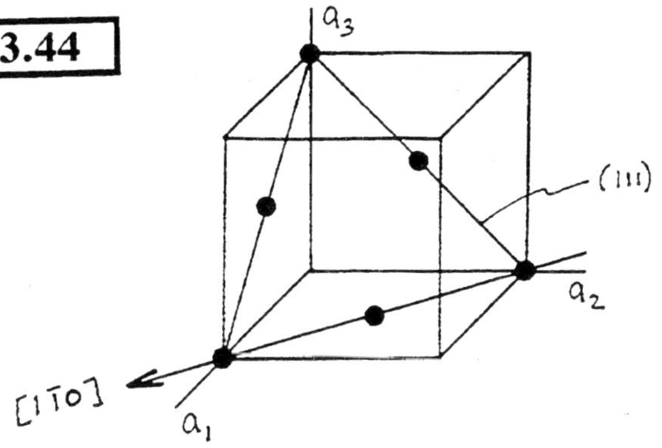

3.45 Sketch the $[1\bar{1}1]$ direction within the (110) plane relative to a bcc unit cell. Include all atom center positions within the plane of interest.

3.45

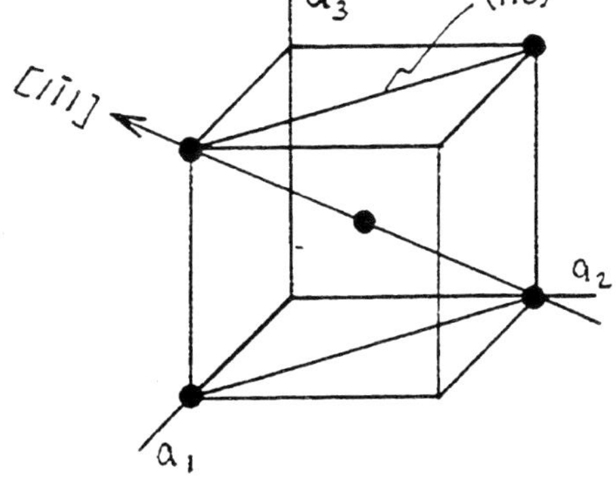

3.46 Sketch the $[11\bar{2}0]$ direction within the (0001) plane relative to an hcp unit cell. Include all atom center positions within the plane of interest.

3.46

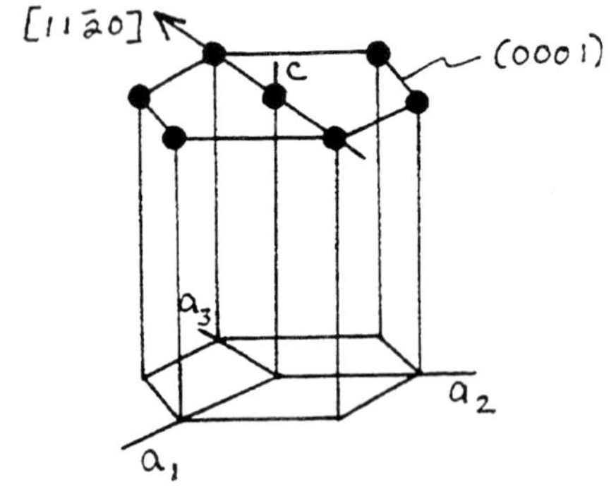

• 3.47 The $\frac{1}{4}\frac{1}{4}\frac{1}{4}$ position in the fcc structure is a "tetrahedral site," an interstice with fourfold atomic coordination. The $\frac{1}{2}\frac{1}{2}\frac{1}{2}$ position is an "octahedral site," an interstice with sixfold atomic coordination. How many tetrahedral and octahedral sites are there per fcc unit cell? Use a sketch to illustrate your answer.

3.47

By inspection of the unit cell, there are 8 of the $\frac{1}{4}\frac{1}{4}\frac{1}{4}$-type positions within the unit cell cube. There is only one $\frac{1}{2}\frac{1}{2}\frac{1}{2}$ position, but there are 12 sites with the equivalent 6-fold coordination at $\frac{1}{2}00$-type positions (midway along a unit cell edge). But, each of the edge sites is shared by four adjacent unit cells. Therefore, there are $1 + 12 \times \frac{1}{4} = 4$ octahedral sites.

In summary, there 8 tetrahedral sites and 4 octahedral sites per fcc unit cell.

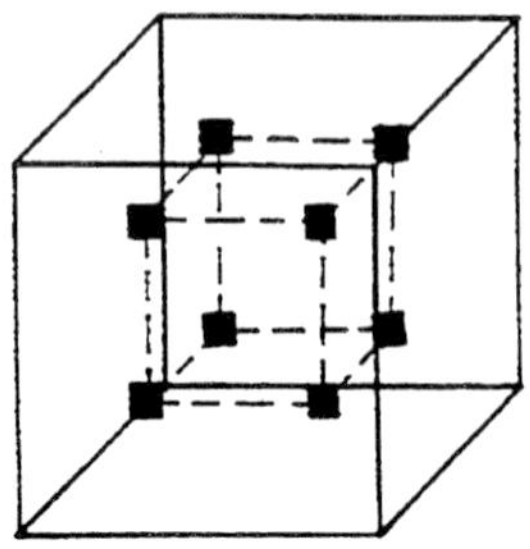

the 8 tetrahedral sites at $\frac{1}{4}\frac{1}{4}\frac{1}{4}$-type positions (forming an interior cube)

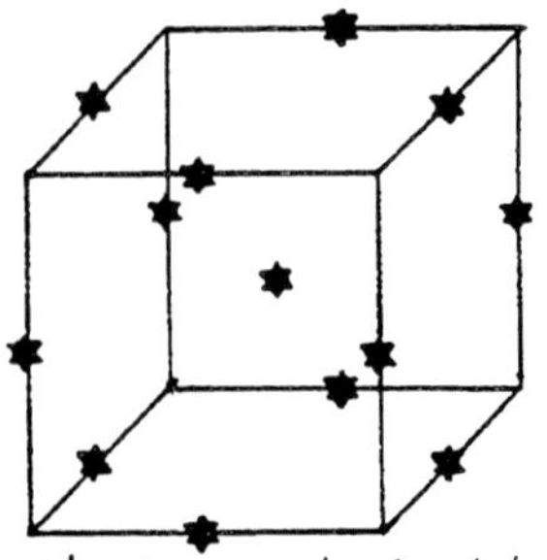

the body-centered octahedral site + the 12 "quarter" sites along each edge

• 3.48 The first eight planes that give x-ray diffraction peaks for aluminum are indicated in Figure 3-40. Sketch each plane relative to the fcc unit cell (Figure 3-12a) and emphasize atom positions within the planes. (Note Problem 3.33 and use a separate sketch for each plane.)

3.48

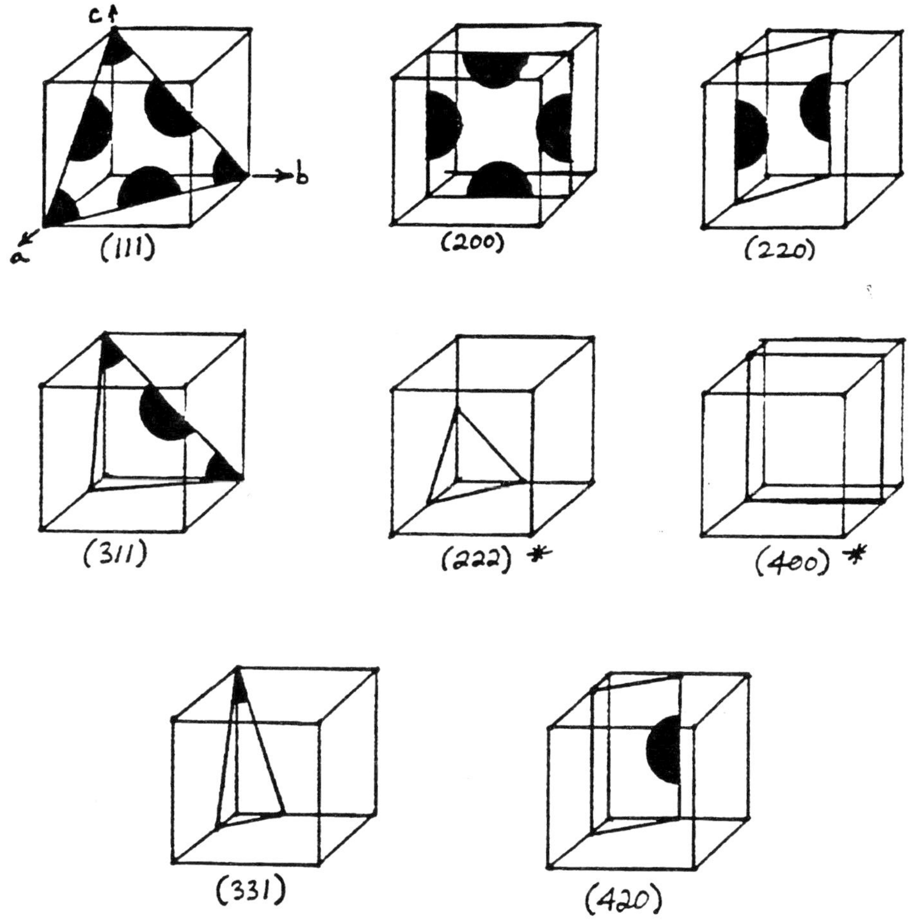

* Note absence of atoms. Diffraction occurs due to higher-order diffraction from parallel planes, e.g. (111) or (200)

Section 3.4 – Ceramic Structures

PP 3.13 Calculate the ionic packing factor of **(a)** CaO, **(b)** FeO, and **(c)** NiO. All of these compounds share the NaCl-type structure. **(d)** Is there a unique IPF value for the NaCl-type structure? Explain. (See Sample Problem 3.11.)

PP 3.13 (a) $a = 2r_{Ca^{2+}} + 2r_{O^{2-}} = 2(0.106\,nm) + 2(0.132\,nm)$

$= 0.476\,nm$

$V_{unit\ cell} = a^3 = (0.476\,nm)^3 = 0.108\,nm^3$

$V_{ions} = 4\left[\frac{4}{3}\pi r_{Ca^{2+}}^3 + \frac{4}{3}\pi r_{O^{2-}}^3\right]$

$= \frac{16\pi}{3}\left[(0.106\,nm)^3 + (0.132\,nm)^3\right] = 0.0585\,nm^3$

$IPF = \frac{0.0585\,nm^3}{0.108\,nm^3} = \underline{\underline{0.542}}$

(b) $a^3 = \left[2(0.087\,nm) + 2(0.132\,nm)\right]^3 = 0.0840\,nm^3$

$V_{ions} = \frac{16\pi}{3}\left[(0.087\,nm)^3 + (0.132\,nm)^3\right] = 0.0496\,nm^3$

$IPF = \frac{0.0496\,nm^3}{0.0840\,nm^3} = \underline{\underline{0.590}}$

(c) $a^3 = \left[2(0.078\,nm) + 2(0.132\,nm)\right]^3 = 0.0741\,nm^3$

$V_{ions} = \frac{16\pi}{3}\left[(0.078\,nm)^3 + (0.132\,nm)^3\right] = 0.0465\,nm^3$

$IPF = \frac{0.0465\,nm^3}{0.0741\,nm^3} = \underline{\underline{0.627}}$

(d) No. As seen by the calculations above, the IPF is a function of the radius ratio.

PP 3.14 Calculate the linear density of ions along the [111] direction for CaO. (See Sample Problem 3.12.)

PP 3.14

$a = 2r_{Ca^{2+}} + 2r_{O^{2-}} = 2(0.106\,nm) + 2(0.132\,nm)$

$= 0.476\,nm$

$l = \sqrt{3}\,a = \sqrt{3}(0.476\,nm) = 0.824\,nm$

giving us:

$(1\,Ca^{2+} + 1\,O^{2-})/0.824\,nm$

$= (1.21\,Ca^{2+} + 1.21\,O^{2-})/nm$

PP 3.15 Calculate the planar density of ions in the (111) plane for CaO. (See Sample Problem 3.13.)

PP 3.15

$a = 2r_{Ca^{2+}} + 2r_{O^{2-}} = 2(0.106\,nm) + 2(0.132\,nm) = 0.476\,nm$

$l = \sqrt{2}\,a = \sqrt{2}(0.476\,nm) = 0.673\,nm$

$A = \frac{1}{2}bh = \frac{\sqrt{3}}{4}(0.673\,nm)^2 = 0.196\,nm^2$

Ionic density $= 2\,ions/0.196\,nm^2$

$= 10.2\,(Ca^{2+}\text{ or }O^{2-})/nm^2$

PP 3.16 Calculate the density of CaO. (See Sample Problem 3.14.)

PP 3.16

$a = 2r_{Ca^{2+}} + 2r_{O^{2-}} = 2(0.106\,nm) + 2(0.132\,nm) = 0.476\,nm$

$V_{unit\,cell} = a^3 = (0.476\,nm)^3 = 0.108\,nm^3$

$$\rho = \frac{[4(40.08\,g) + 4(16.00\,g)]/(0.6023\times10^{24})}{0.108\,nm^3}\times\left(\frac{10^7\,nm}{cm}\right)^3$$

$= 3.45\,g/cm^3$

3.49 Calculate the ionic packing factor for UO_2, which has the CaF_2 structure (Figure 3-18).

3.49

$\frac{\sqrt{3}\,a}{4} = r_{U^{4+}} + r_{O^{2-}} = 0.105\,nm + 0.132\,nm$

giving $a = 0.548\,nm$

and $V_{unit\,cell} = a^3 = (0.548\,nm)^3 = 0.164\,nm^3$

$$V_{ions} = 4 \times \frac{4}{3}\pi r_{U^{4+}}^3 + 8 \times \frac{4}{3}\pi r_{O^{2-}}^3$$

$$= \frac{16\pi}{3}(0.105\,nm)^3 + \frac{32\pi}{3}(0.132\,nm)^3 = 0.0965\,nm^3$$

$$IPF = \frac{0.0965\,nm^3}{0.164\,nm^3} = \underline{\underline{0.588}}$$

3.50 Calculate the linear density of ions along the [111] direction in UO_2, which has the CaF_2 structure (Figure 3-18).

3.50 $\frac{\sqrt{3}a}{4} = r_{U^{4+}} + r_{O^{2-}}$

but $l = \sqrt{3}a = 4(r_{U^{4+}} + r_{O^{2-}}) = 4(0.105\,nm + 0.132\,nm)$

$= 0.948\,nm$

giving us:

$(1\,U^{4+} + 2\,O^{2-})/0.948\,nm$

$$= \underline{\underline{(1.05\,U^{4+} + 2.11\,O^{2-})/nm}}$$

3.51 In Section 3.4 the open nature of the CaF_2 structure was given credit for the ability of UO_2 to absorb He gas atoms and, thereby, resist swelling. Confirm that an He atom (diameter ≈ 0.2 nm) can fit in the center of the UO_2 unit cell (see Figure 3-18 for the CaF_2 structure).

3.51 body diagonal $= \sqrt{3}a = 4(r_{U^{4+}} + r_{O^{2-}}) = 4(0.105\,nm + 0.132\,nm)$

$= 0.948\,nm$

dia. of center opening $= \frac{1}{2}(\text{body diagonal}) - 2r_{O^{2-}}$

$= \frac{1}{2}(0.948\,nm) - 2(0.132\,nm)$

$= 0.21\,nm$

$\therefore$ diameter of opening in center of unit cell = 0.21 nm

3.52 Calculate the ionic packing factor for $CaTiO_3$, (Figure 3-22).

3.52 $a = 2r_{Ti^{4+}} + 2r_{O^{2-}} = 2(0.064\text{ nm}) + 2(0.132\text{ nm}) = 0.392\text{ nm}$

$$V_{unit\ cell} = a^3 = 0.0602\text{ nm}^3$$

$$V_{ions} = \frac{4}{3}\pi r^3_{Ca^{2+}} + \frac{4}{3}\pi r^3_{Ti^{4+}} + 3 \times \frac{4}{3}\pi r^3_{O^{2-}}$$

$$V_{ions} = \frac{4}{3}\pi\left[(0.106\text{ nm})^3 + (0.064\text{ nm})^3 + 3(0.132\text{ nm})^3\right]$$

$$= 0.0350\text{ nm}^3$$

$$IPF = \frac{0.0350\text{ nm}^3}{0.0602\text{ nm}^3} = \underline{\underline{0.581}}$$

Note: One may choose to solve this problem by taking $\sqrt{2}a = 2r_{Ca^{2+}} + 2r_{O^{2-}}$, giving $a = 0.337\text{nm}$. Unfortunately, this method yields an unrealistic IPF > 1.0. (This is a problem due to the imperfect values of ionic radii.)

3.53 Calculate the linear density of ions along the [111] direction in $CaTiO_3$ (Figure 3-22).

3.53 In observing Figure 3-22, we see that along a body diagonal there is 1 Ca^{2+} + 1 Ti^{4+}. Alternatively, we could define a unit cell with an origin centered on an O^{2-}. For that case, there is 1 O^{2-} per body diagonal.

$$a = 2r_{Ti^{4+}} + 2r_{O^{2-}} = 2(0.064\text{ nm}) + 2(0.132\text{ nm})$$

$$= 0.392\text{ nm}$$

$$l = \sqrt{3}a = \sqrt{3}(0.392\text{ nm}) = 0.679\text{ nm}$$

giving us:

$$(1\ Ca^{2+} + 1\ Ti^{4+})/0.679\text{ nm}$$

$$= \underline{\underline{(1.47\ Ca^{2+} + 1.47\ Ti^{4+})/\text{nm}}}$$

and: $1 O^{2-}/0.679\text{ nm} = \underline{\underline{1.47 O^{2-}/\text{nm}}}$

Note: If one chooses to solve this problem by taking $\sqrt{2}a = 2r_{Ca^{2+}} + 2r_{O^{2-}}$, one obtains $a = 0.337\text{ nm}$ and ionic densities of 1.72/nm rather than 1.47. (The discrepency is due to imperfect values of ionic radii.)

3.54 Show that the unit cell in Figure 3-24 gives the chemical formula $2(OH)_4Al_2Si_2O_5$.

3.54 Summing the five individual layers of ions gives:

$$6\,O + 4\,Si + 4\,O + 2\,OH + 4\,Al + 6\,OH$$

$$= 10\,O + 4\,Si + 8\,OH + 4\,Al$$

$$= 2(4\,OH + 2\,Al + 2\,Si + 5\,O)$$

which corresponds to:

$$2(OH)_4Al_2Si_2O_5$$

3.55 Identify the ions associated with equivalent points in the NaCl structure (Note Problem 3.40).

3.55 After Problem 3.40 and Figure 3–17,

equivalent points = Cl^- at 000, $\frac{1}{2}\frac{1}{2}0$, $\frac{1}{2}0\frac{1}{2}$, $0\frac{1}{2}\frac{1}{2}$

and Na^+ at $00\frac{1}{2}$, $\frac{1}{2}\frac{1}{2}\frac{1}{2}$, $\frac{1}{2}01$, $0\frac{1}{2}1$

3.56 Identify the ions associated with equivalent points in the $CaTiO_3$ structure. (Note Problem 3.40.)

3.56 Equivalent points from inspection of Figure 3–22:

Ca^{2+} at 000

Ti^{4+} at $\frac{1}{2}\frac{1}{2}\frac{1}{2}$

O^{2-} at $\frac{1}{2}\frac{1}{2}0$, $\frac{1}{2}0\frac{1}{2}$, $0\frac{1}{2}\frac{1}{2}$

3.57 Calculate the planar density of ions in the (111) plane of $CaTiO_3$.

3.57 Inspecting the unit cell of Figure 3-22, we note an alternating arrangement of ions for a (111) plane:

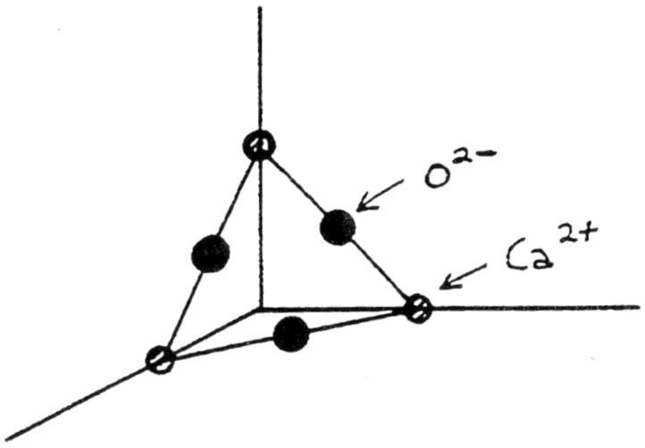

The length of each side of a (111) "triangle" is

$$l = \sqrt{2}\, a$$

where

$$a = 2r_{Ti^{4+}} + 2r_{O^{2-}} = 2(0.064\ nm) + 2(0.132\ nm)$$

$$= 0.392\ nm$$

giving $l = \sqrt{2}(0.392\ nm) = 0.554\ nm$

and a planar area:

$$A = \frac{1}{2}bh = \frac{1}{2}(0.554\ nm)\left(\frac{\sqrt{3}}{2}\,0.554\ nm\right) = 0.133\ nm^2$$

giving an

$$\text{ionic density} = \frac{3 \times \frac{1}{6}\ Ca^{2+} + 3 \times \frac{1}{2}\ O^{2-}}{0.133\ nm^2}$$

$$= (3.76\ Ca^{2+} + 11.3\ O^{2-})/nm^2$$

Note the comments for Problems **3.52 and 3.53** in regard to an alternate way to determine the (111) triangle side as $2r_{Ca^{2+}} + 2r_{O^{2-}}$.

3.58 Calculate the density of UO_2.

3.58 As determined in Problem 3.49,

$$\frac{\sqrt{3}\,a}{4} = r_{U^{4+}} + r_{O^{2-}} = 0.105\text{ nm} + 0.132\text{ nm}$$

giving $a = 0.548\text{ nm}$

and $V_{unit\,cell} = a^3 = (0.548\text{ nm})^3 = 0.164\text{ nm}^3$.

$$\rho = \frac{[4(238.03) + 8(16.00)]\text{ g}/0.6023\times10^{24}}{0.164\text{ nm}^3} \times \left(\frac{10^7\text{ nm}}{\text{cm}}\right)^3$$

$$= \underline{\underline{10.9\text{ g/cm}^3}}$$

3.59 Calculate the density of $CaTiO_3$.

3.59 $a = 2r_{Ti^{4+}} + 2r_{O^{2-}} = 2(0.064\text{ nm}) + 2(0.132\text{ nm})$

$= 0.392\text{ nm}$

$V_{unit\,cell} = a^3 = 0.0602\text{ nm}^3$

$$\rho = \frac{[(40.08\text{ g}) + (47.90\text{ g}) + 3(16.00\text{ g})]/(0.6023\times10^{24})}{0.0602\text{ nm}^3} \times \left(\frac{10^7\text{ nm}}{\text{cm}}\right)^3$$

$$= \underline{\underline{3.75\text{ g/cm}^3}}$$

Note the comments for Problems **3.52, 3.53, and 3.57** in regard to an alternate way to estimate the lattice parameter, a, viz. $\sqrt{2}\,a = 2r_{Ca^{2+}} + 2r_{O^{2-}}$.

• **3.60** (**a**) Derive a general relationship between the ionic packing factor of the NaCl-type structure and the radius ratio (r/R). (**b**) Over what r/R range is this relationship reasonable?

3.60 (a) $a = 2r + 2R$

$$V_{unit\ cell} = a^3 = (2r+2R)^3 = 2^3(r+R)^3 = 8(r+R)^3$$

$$V_{ions} = 4\left[\tfrac{4}{3}\pi r^3 + \tfrac{4}{3}\pi R^3\right] = \tfrac{16}{3}\pi(r^3+R^3)$$

$$IPF = \frac{\frac{16}{3}\pi(r^3+R^3)}{8(r+R)^3} = \frac{2}{3}\pi\frac{(r^3+R^3)}{(r+R)^3}$$

$$= \frac{2}{3}\pi\frac{R^3\left(\left[\frac{r}{R}\right]^3+1\right)}{r^3+3r^2R+3rR^2+R^3} = \frac{2}{3}\pi\frac{\left(\left[\frac{r}{R}\right]^3+1\right)}{\left(\left[\frac{r}{R}\right]^3+3\left[\frac{r}{R}\right]^2+3\left[\frac{r}{R}\right]+1\right)}$$

$$IPF = \frac{2}{3}\pi\frac{\left(\left[\frac{r}{R}\right]^3+1\right)}{\left(\left[\frac{r}{R}\right]+1\right)^3}$$

(b) Corresponding to the range for 6-fold coordination in Table 2.1:

$$0.414 \leq \frac{r}{R} < 0.732$$

• **3.61** Sketch the ion positions in a (111) plane through the cristobalite unit cell (Figure 3-19).

3.61

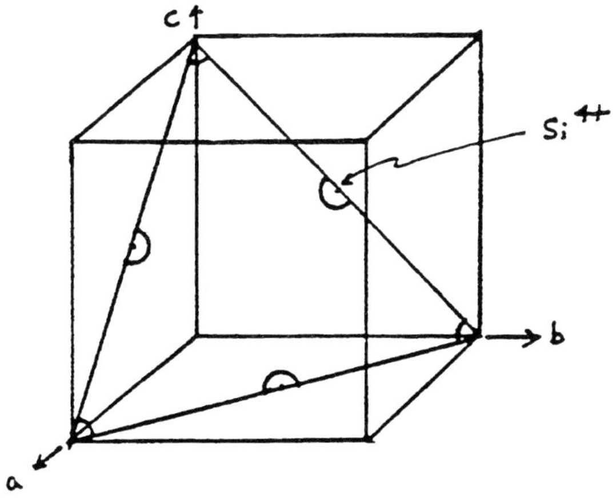

• **3.62** Sketch the ion positions in a (101) plane through the cristobalite unit cell (Figure 3-19).

3.62

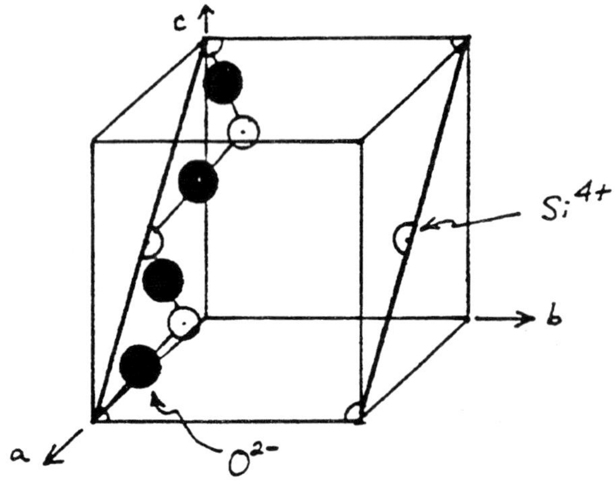

• **3.63** Calculate the ionic packing factor for cristobalite (Figure 3-19).

3.63 The key to the solution is to note that there are 4 - Si^{4+} ions in fcc positions and 4 - Si^{4+} in the interior ⇒ 8 Si^{4+} total.

Then, there are $2 \times 8 = 16$ O^{2-}.

The size of the unit cell is determined by the identity:

$$2(r_{Si^{4+}} + r_{O^{2-}}) = \tfrac{1}{4}\sqrt{3}\,a$$

or $$a = \frac{8}{\sqrt{3}}(r_{Si^{4+}} + r_{O^{2-}}).$$

Using Appendix 2 gives:

$$a = \frac{8}{\sqrt{3}}(0.039 + 0.132)\text{nm} = 0.7898\text{nm}$$

or

$$a^3 = 0.4927\text{nm}^3$$

Next,

$$V_{ions} = 8 \times \tfrac{4}{3}\pi(0.039\text{nm})^3 + 16 \times \tfrac{4}{3}\pi(0.132\text{nm})^3$$

$$= \frac{32}{3}\pi\left[(0.039\text{nm})^3 + 2(0.132\text{nm})^3\right]$$

$$= 0.1561\text{ nm}^3$$

$$\therefore \text{IPF} = \frac{0.1561\text{ nm}^3}{0.4927\text{ nm}^3} = \underline{\underline{0.317}}$$

• **3.64** Calculate the ionic packing factor for corundum (Figure 3-21).

3.64 Note that the corundum structure is essentially a set of close-packed O^{2-} layers with two interstitial Al^{3+} ions per every 3 O^{2-} ions.

Using a basis of 3 O^{2-} ions, the associated volume of corundum would determined by the APF of a hexagonal close-packed structure:

$$0.74 = \frac{3 \times V_{O^{2-}}}{V_{corundum}}$$

or $$V_{corundum} = \frac{3 \times \frac{4}{3}\pi(0.132\ nm)^3}{0.74} = 0.03906\ nm^3$$

The total volume of ions within that volume would be:

$$V_{ions} = 3 \times V_{O^{2-}} + 2 \times V_{Al^{3+}}$$

$$= 3 \times \tfrac{4}{3}\pi(0.132\ nm)^3 + 2 \times \tfrac{4}{3}\pi(0.057\ nm)^3$$

$$= 0.03045\ nm^3$$

Giving, $$IPF = \frac{0.03045\ nm^3}{0.03906\ nm^3} = \underline{\underline{0.78}}$$

<u>Note</u>: This value is greater than the APF for hcp metals because of the "extra" Al^{3+} ions in corundum.

Section 3.5 – Polymeric Structures

PP 3.17 How many unit cells are contained in 1 kg of commercial polyethylene that is 50 vol % crystalline (balance amorphous) and has an overall product density of 0.940 Mg/m^3? (See Sample Problem 3.15.)

PP 3.17 First, note that unit cells are present only in the crystalline portion.

$$V_{total} = \frac{mass}{density} = \frac{1\,kg \times (1\,Mg/1000\,kg)}{0.940\,Mg/m^3} = 1.06 \times 10^{-3}\,m^3$$

$$V_{crystalline} = 0.5\,(1.06 \times 10^{-3}\,m^3) = 0.532 \times 10^{-3}\,m^3$$

$$V_{unit\ cell} = 0.0933\,nm^3 \times \left(\frac{10^{-9}\,m}{nm}\right)^3 = 9.33 \times 10^{-29}\,m^3$$

giving us:

$$N_{unit\ cells} = \frac{0.532 \times 10^{-3}\,m^3}{9.33 \times 10^{-29}\,m^3/unit\ cell} = \underline{\underline{5.70 \times 10^{24}\ unit\ cells}}$$

3.65 Calculate the reaction energy involved in forming a single unit cell of polyethylene.

3.65 Following the methods of Section 2.3, we can note that for the C=C → 2 C–C reaction, Table 2.2 indicates the reaction energy to be:

$$2(370\,kJ/mol) - 680\,kJ/mol = 60\,kJ/mol.$$

As there are two C_2H_4 "molecules" associated with the unit cell (as shown in **Sample** Problem 3.**15**), the reaction energy/unit cell is:

$$E_{reac./u.c.} = \frac{2}{N_{AV}} \times 60{,}000\,J/mol$$

$$= \frac{2}{0.6023 \times 10^{24}\,mol^{-1}} \times 6 \times 10^4\,J \times \frac{6.242 \times 10^{18}\,eV}{J}$$

$$= \underline{\underline{1.24\,eV}}$$

3.66 How many unit cells are contained in the thickness of a 10-nm-thick polyethylene platelet (Figure 3-29)?

3.66 dimension of unit cell in c-direction = 0.255 nm (see Figure 3-28)

plate thickness is in c-direction (see Figure 3-29)

giving us:

$$N_{unit\ cells} = \frac{100\ nm}{0.255\ nm/unit\ cell} = \underline{\underline{392\ unit\ cells}}$$

3.67 Calculate the atomic packing factor for polyethylene.

3.67

$$V_{atoms} = 4 \times V_C + 8 \times V_H = 4 \times \frac{4}{3}\pi r_C^3 + 8 \times \frac{4}{3}\pi r_H^3$$

$$= 4 \times \frac{4}{3}\pi \left[(0.077\ nm)^3 + 2\,(0.046\ nm)^3\right]$$

$$= 0.0109\ nm^3$$

$$V_{unit\ cell} = 0.0933\ nm^3 \quad \text{(from Sample Problem 3.15)}$$

$$\therefore APF = \frac{0.0109\ nm^3}{0.0933\ nm^3} = \underline{\underline{0.12}}$$

Section 3.6 – Semiconductor Structures

PP 3.18 In Sample Problem 3.16 we find the atomic packing factor for silicon to be quite low compared to the common metal structures. Comment on the relationship between this characteristic and the nature of bonding in semiconductor silicon.

PP 3.18 Bonding in silicon is highly covalent. The associated directionality of the bonds dominates over efficient packing of spheres (a characteristic of non-directional, metallic bonding).

PP 3.19 Find the linear density of atoms along the [111] direction for germanium. (See Sample Problem 3.17.)

PP 3.19

$$\text{linear density} = \frac{2\ \text{atoms}}{8 r_{Ge}} = \frac{2\ \text{atoms}}{8(0.122\ \text{nm})} = \underline{\underline{2.05\ \frac{\text{atoms}}{\text{nm}}}}$$

PP 3.20 Find the planar density of atoms in the (111) plane for germanium. (See Sample Problem 3.18.)

PP 3.20

$$A = \frac{\sqrt{3}}{4}\left[\frac{\sqrt{2}}{\sqrt{3}}\,8\,(0.122\,\text{nm})\right]^2 = 0.275\ \text{nm}^2$$

$$\text{planar density} = \frac{2\ \text{atoms}}{0.275\ \text{nm}^2} = \underline{\underline{7.27\ \text{atoms}/\text{nm}^2}}$$

PP 3.21 Calculate the density of germanium, using data from Appendixes 1 and 2. (See Sample Problem 3.19.)

PP 3.21

$$\sqrt{3}\,a = 8 r_{Ge}$$

$$a = \frac{8}{\sqrt{3}} r_{Ge} = \frac{8}{\sqrt{3}}(0.122\,\text{nm}) = 0.563\ \text{nm}$$

$$a^3 = 0.179\ \text{nm}^3$$

$$\rho = \frac{8\ \text{atoms}}{0.179\ \text{nm}^3} \times \frac{72.59\ \text{g}}{0.6023 \times 10^{24}\ \text{atoms}} \times \left(\frac{10^7\ \text{nm}}{\text{cm}}\right)^3$$

$$= \underline{\underline{5.39\ \text{g}/\text{cm}^3}}$$

(Obviously, the comment at the bottom of Sample Problem 3.19 applies here also.)

3.68 Calculate the ionic packing factor for the zinc blende structure (Figure 3-32).

3.68

$$\frac{\sqrt{3}}{4}a = r_{Zn^{2+}} + r_{S^{2-}} = 0.083\ \text{nm} + 0.174\ \text{nm} = 0.257\ \text{nm}$$

$$a = \frac{4}{\sqrt{3}}(0.257\ \text{nm}) = 0.594\ \text{nm}$$

$$V_{\text{unit cell}} = a^3 = (0.594\ \text{nm})^3 = 0.209\ \text{nm}^3$$

$$V_{\text{ions}} = 4 \times \frac{4\pi}{3} r^3_{Zn^{2+}} + 4 \times \frac{4\pi}{3} r^3_{S^{2-}}$$

$$= \frac{16\pi}{3}\left[(0.083\,\text{nm})^3 + (0.174\,\text{nm})^3\right] = 0.0978\ \text{nm}^3$$

$$IPF = \frac{0.0978\ \text{nm}^3}{0.209\ \text{nm}^3} = \underline{\underline{0.468}}$$

3.69 Calculate the linear density of ions along the [111] direction in zinc blende (Figure 3-32).

3.69 There is 1 Zn^{2+} + 1 S^{2-} per body diagonal.

$\sqrt{3}a = 4(r_{Zn^{2+}} + r_{S^{2-}}) = 4[(0.083 nm) + (0.174 nm)]$

$= 1.028 nm$

giving us:

$(1\ Zn^{2+} + 1\ S^{2-})/1.028 nm$

$= \underline{(0.973\ Zn^{2+} + 0.973\ S^{2-})/nm}$

3.70 Calculate the planar density of ions along the (111) plane in zinc blende (Figure 3-32).

3.70 The (111) plane through Figure 3-32(a) strikes an fcc array of Zn^{2+} ions, but a (111) plane with origin at $\frac{1}{4}\frac{1}{4}\frac{1}{4}$ will have a similar array of S^{2-} ions.

As in Sample Problem 3.13, $l = \sqrt{2}a$

$a = \frac{4}{\sqrt{3}}(r_{Zn^{2+}} + r_{S^{2-}}) = \frac{4}{\sqrt{3}}(0.083 nm + 0.174 nm)$

$= 0.594 nm$

$l = \sqrt{2}a = \sqrt{2}(0.594 nm) = 0.839 nm$

$\text{area} = \frac{\sqrt{3}}{4}(0.839 nm)^2 = 0.305 nm^2$

Ionic density $= 2 \text{ ions}/0.305 nm^2$

$= \underline{6.56\ (Zn^{2+} \text{ or } S^{2-})/nm^2}$

3.71 Calculate the density of zinc blende, using data from Appendixes 1 and 2.

3.71 $a = \frac{4}{\sqrt{3}}(r_{Zn^{2+}} + r_{S^{2-}}) = \frac{4}{\sqrt{3}}(0.083 nm + 0.174 nm) = 0.594 nm$

$a^3 = 0.209 nm^3$

$$\rho = \frac{[4(65.38 g) + 4(32.06 g)]/(0.6023 \times 10^{24})}{0.209 nm^3} \times \left(\frac{10^7 nm}{cm}\right)^3$$

$= \underline{3.09 g/cm^3}$

3.72 Identify the ions associated with equivalent points in the diamond cubic structure. (Note Problem 3.40.)

3.72

The equivalent points are the fcc points identified in Problem 3.40 plus four more points offset by a $\frac{1}{4}\frac{1}{4}\frac{1}{4}$ translation from those fcc points:

$$000, \tfrac{1}{2}\tfrac{1}{2}0, \tfrac{1}{2}0\tfrac{1}{2}, 0\tfrac{1}{2}\tfrac{1}{2}, \tfrac{1}{4}\tfrac{1}{4}\tfrac{1}{4}, \tfrac{3}{4}\tfrac{3}{4}\tfrac{1}{4}, \tfrac{3}{4}\tfrac{1}{4}\tfrac{3}{4}, \tfrac{1}{4}\tfrac{3}{4}\tfrac{3}{4}$$

3.73 Identify the ions associated with equivalent points in the zinc blende structure. (Note Problem 3.40.)

3.73

This answer is similar to that for Problem 3.72 except that the four fcc points indicated by Figure 3-32 are occupied by Zn^{2+} ions and the four offset points are occupied by S^{2-} ions:

Zn^{2+} at $000, \tfrac{1}{2}\tfrac{1}{2}0, \tfrac{1}{2}0\tfrac{1}{2}, 0\tfrac{1}{2}\tfrac{1}{2}$ and
S^{2-} at $\tfrac{1}{4}\tfrac{1}{4}\tfrac{1}{4}, \tfrac{3}{4}\tfrac{3}{4}\tfrac{1}{4}, \tfrac{3}{4}\tfrac{1}{4}\tfrac{3}{4}, \tfrac{1}{4}\tfrac{3}{4}\tfrac{3}{4}$

3.74 Identify the ions associated with equivalent points in the wurtzite structure. (Note Problem 3.40.)

3.74

Inspection of Figure 3-33 indicates

Zn^{2+} at $000, \tfrac{1}{3}\tfrac{2}{3}\tfrac{1}{2}$ and

S^{2-} at $0,0,0.375; \tfrac{1}{3}, \tfrac{2}{3}, 0.875$

• 3.75 (a) Derive a general relationship between the ionic packing factor of the zinc blende structure and the radius ratio (r/R). (b) What is a primary limitation of such IPF calculations for these compound semiconductors?

3.75 (a) Similar to the procedure used in Problem **3.60** and recalling the specific solution for Problem 3.68,

$$\frac{\sqrt{3}}{4}a = r + R \quad \text{or} \quad a = \frac{4}{\sqrt{3}}(r+R)$$

$$V_{unit\,cell} = a^3 = \frac{64}{3\sqrt{3}}(r+R)^3$$

$$V_{ions} = 4\left[\frac{4}{3}\pi r^3 + \frac{4}{3}\pi R^3\right] = \frac{16}{3}\pi(r^3+R^3)$$

$$IPF = \frac{\frac{16}{3}\pi(r^3+R^3)}{\frac{64}{3\sqrt{3}}(r+R)^3}$$

$$= \frac{\sqrt{3}\,\pi}{4}\,\frac{(r^3+R^3)}{(r+R)^3}$$

$$IPF = \frac{\sqrt{3}\,\pi}{4}\,\frac{\left(\left[\frac{r}{R}\right]^3+1\right)}{\left(\left[\frac{r}{R}\right]+1\right)^3}$$

(b) The compound semiconductors have a high degree of covalent bonding, so the ionic radii in Appendix 2 are highly approximate. (The "ionic/covalent" radii will tend to be intermediate between the "atomic" and "ionic" radii of Appendix 2.)

• 3.76 Calculate the ionic packing factor for the wurtzite structure (Figure 3-33).

3.76

To determine the size of the unit cell, note that the height determined by $r_{Zn^{2+}} + r_{S^{2-}} = (0.875-0.500)c$

$$= 0.375c$$

$$\text{or } c = \frac{r_{Zn^{2+}} + r_{S^{2-}}}{0.375}$$

$$= \frac{0.083\,nm + 0.174\,nm}{0.375}$$

$$= 0.6853\,nm$$

And, a = edge of a ZnS_4^{6-} tetrahedron

$$= \sqrt{\frac{8}{3}}\left(r_{Zn^{2+}} + r_{S^{2-}}\right)$$

$$= \sqrt{\frac{8}{3}}\left[(0.083\,nm) + (0.174\,nm)\right] = 0.4197\,nm$$

The volume of the unit cell is

$$V_{uc} = ca^2 \sin 60°$$

$$= (0.6853\,nm)(0.4197\,nm)^2 \sin 60°$$

$$= 0.1045\,nm^3$$

$$V_{ions} = 2\left(\frac{4}{3}\pi r_{Zn^{2+}}^3 + \frac{4}{3}\pi r_{S^{2-}}^3\right)$$

$$= \frac{8}{3}\pi\left[(0.083\,nm)^3 + (0.174\,nm)^3\right] = 0.04892\,nm^3$$

$$\therefore IPF = \frac{0.04892\,nm^3}{0.1045\,nm^3} = \underline{\underline{0.468}}$$

• 3.77 Calculate the density of wurtzite, using data from Appendixes 1 and 2.

3.77 Using the calculations for Problem 3.76,

$$V_{unit\ cell} = 0.1045\ nm^3$$

Then,

$$\rho = \frac{[2(65.38g) + 2(32.06g)]/(0.6023 \times 10^{24})}{0.1045\ nm^3} \times \left(\frac{10^7\ nm}{cm}\right)^3$$

$$= \underline{\underline{3.10\ g/cm^3}}$$

Section 3.7 – X-ray Diffraction

PP 3.22 In Sample Problem 3.20 we characterize the geometry for diffraction by (111) planes in MgO. Suppose the crystal is tilted slightly so that the (111) diffraction spot is shifted to a position 0.5 cm from the film center. What wavelength (λ) would produce first-order diffraction in this case?

PP 3.22 $\phi = \arctan\left(\frac{0.5\ cm}{3\ cm}\right) = 9.46°$

$$2\theta = 180° - \phi = 180° - 9.46° = 170.5°$$

$$\theta = 85.3°$$

$$n\lambda = 2d \sin\theta$$

or

$$\lambda = \frac{2d}{n} \sin\theta$$

noting the d calculated in Sample Problem 3.20:

$$\lambda = \left(\frac{2 \times 0.242\ nm}{1}\right) \sin(85.3°) = \underline{\underline{0.483\ nm}}$$

PP 3.23 The diffraction angles for the first three peaks in Figure 3-40 are calculated in Sample Problem 3.21. Calculate the diffraction angles for the remainder of the peaks in Figure 3-40.

PP 3.23

$$d_{311} = \frac{0.404\,nm}{\sqrt{3^2+1+1}} = \frac{0.404\,nm}{\sqrt{11}} = 0.122\,nm$$

$$d_{222} = \frac{0.404\,nm}{\sqrt{2^2+2^2+2^2}} = \frac{0.404\,nm}{\sqrt{12}} = 0.117\,nm$$

$$d_{400} = \frac{0.404\,nm}{\sqrt{4^2+0+0}} = \frac{0.404\,nm}{4} = 0.101\,nm$$

$$d_{331} = \frac{0.404\,nm}{\sqrt{3^2+3^2+1}} = \frac{0.404\,nm}{\sqrt{19}} = 0.0927\,nm$$

$$d_{420} = \frac{0.404\,nm}{\sqrt{4^2+2^2+0}} = \frac{0.404\,nm}{\sqrt{20}} = 0.0903\,nm$$

$$\theta_{311} = \arcsin \frac{0.1542\,nm}{2 \times 0.122\,nm} = 39.3° \quad \text{or} \quad (2\theta)_{311} = 78.5°$$

$$\theta_{222} = \arcsin \frac{0.1542\,nm}{2 \times 0.117\,nm} = 41.4° \quad \text{or} \quad (2\theta)_{222} = 82.8°$$

$$\theta_{400} = \arcsin \frac{0.1542\,nm}{2 \times 0.101\,nm} = 49.8° \quad \text{or} \quad (2\theta)_{400} = 99.5°$$

$$\theta_{331} = \arcsin \frac{0.1542\,nm}{2 \times 0.0927\,nm} = 56.3° \quad \text{or} \quad (2\theta)_{331} = 113°$$

$$\theta_{420} = \arcsin \frac{0.1542\,nm}{2 \times 0.0903\,nm} = 58.6° \quad \text{or} \quad (2\theta)_{420} = 117°$$

3.78 The diffraction peaks labeled in Figure 3-40 correspond to the reflection rules for an fcc metal (h, k, l unmixed, as shown in Table 3.3). What would be the (hkl) indices for the three lowest diffraction angle peaks for a bcc metal?

3.78 First, we must satisfy the condition:

$h+k+l$ = even number

Second, we can combine Equations 3.5 and 3.6:

$$\sin\theta = \frac{n\lambda}{2a}\sqrt{h^2+k^2+l^2}$$

It is apparent that the smallest values of θ correspond to the smallest values of $h^2+k^2+l^2$

By inspection:

(hkl):	(110)	(200)	(211)
$h+k+l$:	2	2	4
$h^2+k^2+l^2$:	2	4	6

3.79 Using the result of Problem 3.78, calculate the diffraction angles (2θ) for the first three peaks in the diffraction pattern of α-Fe powder using CuK_α-radiation ($\lambda = 0.1542$ nm).

3.79 For α-Fe, $\sqrt{3}\,a = 4r_{Fe}$ or $a = (4/\sqrt{3})r_{Fe}$

$$= (4/\sqrt{3})(0.124\,nm)$$
$$= 0.286\,nm$$

$$d_{110} = \frac{0.286\,nm}{\sqrt{1+1+0}} = \frac{0.286\,nm}{\sqrt{2}} = 0.202\,nm$$

$$d_{200} = \frac{0.286\,nm}{\sqrt{2^2+0+0}} = \frac{0.286\,nm}{2} = 0.143\,nm$$

$$d_{211} = \frac{0.286\,nm}{\sqrt{2^2+1+1}} = \frac{0.286\,nm}{\sqrt{6}} = 0.117\,nm$$

$$\theta_{110} = \arcsin\frac{0.1542\,nm}{2\times 0.202\,nm} = 22.4° \text{ or } (2\theta)_{110} = 44.8°$$

$$\theta_{200} = \arcsin\frac{0.1542\,nm}{2\times 0.143\,nm} = 32.6° \text{ or } (2\theta)_{200} = 65.3°$$

$$\theta_{211} = \arcsin\frac{0.1542\,nm}{2\times 0.117\,nm} = 41.3° \text{ or } (2\theta)_{211} = 82.5°$$

3.80 Repeat Problem 3.79 using CrK_α-radiation ($\lambda = 0.2291$ nm).

3.80 The d values were calculated for Problem 3.79:

$$\theta_{110} = \arcsin\frac{0.2291\,nm}{2\times 0.202\,nm} = 34.5° \text{ or } (2\theta)_{110} = 68.9°$$

$$\theta_{200} = \arcsin\frac{0.2291\,nm}{2\times 0.143\,nm} = 53.1° \text{ or } (2\theta)_{200} = 106°$$

$$\theta_{211} = \arcsin\frac{0.2291\,nm}{2\times 0.117\,nm} = 78.5° \text{ or } (2\theta)_{211} = 157°$$

3.81 Repeat Problem 3.78 for the *next* three lowest diffraction angle peaks for a bcc metal.

3.81

By inspection:

(hkl) :	(110)	(200)	(211)	(220)	(310)	(222)
h+k+l :	2	2	4	4	4	6
$h^2+k^2+l^2$:	2	4	6	8	10	12

3.82 Repeat Problem 3.79 for the *next* three lowest diffraction angle peaks for α-Fe powder using CuK_α-radiation.

3.82

From Problem 3.79, $a = 0.286$ nm

$$d_{220} = \frac{0.286\text{nm}}{\sqrt{2^2+2^2+0}} = \frac{0.286\text{nm}}{\sqrt{8}} = 0.101$$

$$d_{310} = \frac{0.286\text{nm}}{\sqrt{3^2+1+0}} = \frac{0.286\text{ nm}}{\sqrt{10}} = 0.0904$$

$$d_{222} = \frac{0.286\text{ nm}}{\sqrt{2^2+2^2+2^2}} = \frac{0.286\text{ nm}}{\sqrt{12}} = 0.0826$$

$$\theta_{220} = \arcsin\frac{0.1542\text{ nm}}{2\times 0.101\text{ nm}} = 49.7° \quad \text{or} \quad (2\theta)_{220} = 99.4°$$

$$\theta_{310} = \arcsin\frac{0.1542\text{ nm}}{2\times 0.0904\text{nm}} = 58.5° \quad \text{or} \quad (2\theta)_{310} = 117.0°$$

$$\theta_{222} = \arcsin\frac{0.1542\text{ nm}}{2\times 0.0826\text{nm}} = 69.0° \quad \text{or} \quad (2\theta)_{222} = 138.1°$$

3.83 Assuming the relative peak heights would be the same for given (*hkl*) planes, sketch a diffraction pattern similar to Figure 3-40 for copper powder using CuK$_\alpha$-radiation. Cover the range of $20° < 2\theta < 90°$.

3.83 To obtain diffraction angles, we follow the method for Sample Problem 3.21 and Practice Problem 3.23.

For fcc Cu, using Appendix 2,

$$a = \frac{4}{\sqrt{2}} r_{Cu} = \frac{4}{\sqrt{2}}(0.128\text{ nm}) = 0.362\text{ nm}$$

$$d_{111} = \frac{0.362\text{ nm}}{\sqrt{3}} = 0.209\text{ nm} \rightarrow \theta_{111} = \arcsin\left(\frac{0.1542\text{ nm}}{2 \times 0.209\text{ nm}}\right) = 21.6°$$

$$\text{or } (2\theta)_{111} = 43.3°$$

$$d_{200} = \frac{0.362\text{ nm}}{2} = 0.181\text{ nm} \rightarrow \theta_{200} = \arcsin\left(\frac{0.1542\text{ nm}}{2 \times 0.181\text{ nm}}\right) = 25.2°$$

$$\text{or } (2\theta)_{200} = 50.4°$$

$$d_{220} = \frac{0.362\text{ nm}}{\sqrt{8}} = 0.128\text{ nm} \rightarrow \theta_{220} = \arcsin\left(\frac{0.1542\text{ nm}}{2 \times 0.128\text{ nm}}\right) = 37.0°$$

$$\text{or } (2\theta)_{220} = 74.1°$$

$$d_{311} = \frac{0.362\text{ nm}}{\sqrt{11}} = 0.109\text{ nm} \rightarrow \theta_{311} = \arcsin\left(\frac{0.1542\text{ nm}}{2 \times 0.109\text{ nm}}\right) = 44.9°$$

$$\text{or } (2\theta)_{311} = 89.9°$$

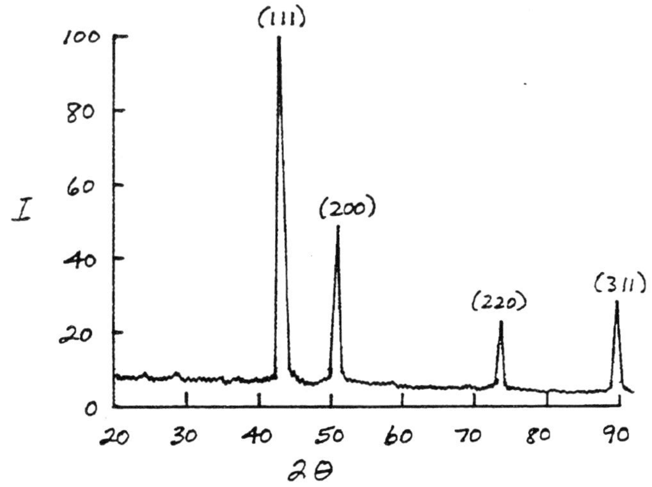

3.84 Repeat Problem 3.83 for lead powder.

3.84 For fcc Pb, $a = \frac{4}{\sqrt{2}} r_{Pb} = \frac{4}{\sqrt{2}}(0.175\,nm) = 0.495\,nm$

$$d_{111} = \frac{0.495\,nm}{\sqrt{3}} = 0.286\,nm \rightarrow \theta_{111} = \sin^{-1}\left(\frac{0.1542\,nm}{2 \times 0.286\,nm}\right) = 15.7° \rightarrow (2\theta)_{111} = 31.3°$$

$$d_{200} = \frac{0.495\,nm}{2} = 0.247\,nm \rightarrow \theta_{200} = \sin^{-1}\left(\frac{0.1542\,nm}{2 \times 0.247\,nm}\right) = 18.2° \rightarrow (2\theta)_{200} = 36.3°$$

$$d_{220} = \frac{0.495\,nm}{\sqrt{8}} = 0.175\,nm \rightarrow \theta_{220} = \sin^{-1}\left(\frac{0.1542\,nm}{2 \times 0.175\,nm}\right) = 26.1° \rightarrow (2\theta)_{220} = 52.3°$$

$$d_{311} = \frac{0.495\,nm}{\sqrt{11}} = 0.149\,nm \rightarrow \theta_{311} = \sin^{-1}\left(\frac{0.1542\,nm}{2 \times 0.149\,nm}\right) = 31.1° \rightarrow (2\theta)_{311} = 62.2°$$

$$d_{222} = \frac{0.495\,nm}{\sqrt{12}} = 0.143\,nm \rightarrow \theta_{222} = \sin^{-1}\left(\frac{0.1542\,nm}{2 \times 0.143\,nm}\right) = 32.7° \rightarrow (2\theta)_{222} = 65.3°$$

$$d_{400} = \frac{0.495\,nm}{4} = 0.124\,nm \rightarrow \theta_{400} = \sin^{-1}\left(\frac{0.1542\,nm}{2 \times 0.124\,nm}\right) = 38.5° \rightarrow (2\theta)_{400} = 77.1°$$

$$d_{331} = \frac{0.495\,nm}{\sqrt{19}} = 0.114\,nm \rightarrow \theta_{331} = \sin^{-1}\left(\frac{0.1542\,nm}{2 \times 0.114\,nm}\right) = 42.8° \rightarrow (2\theta)_{331} = 85.5°$$

$$d_{420} = \frac{0.495\,nm}{\sqrt{20}} = 0.111\,nm \rightarrow \theta_{420} = \sin^{-1}\left(\frac{0.1542\,nm}{2 \times 0.111\,nm}\right) = 44.2° \rightarrow (2\theta)_{420} = 88.3°$$

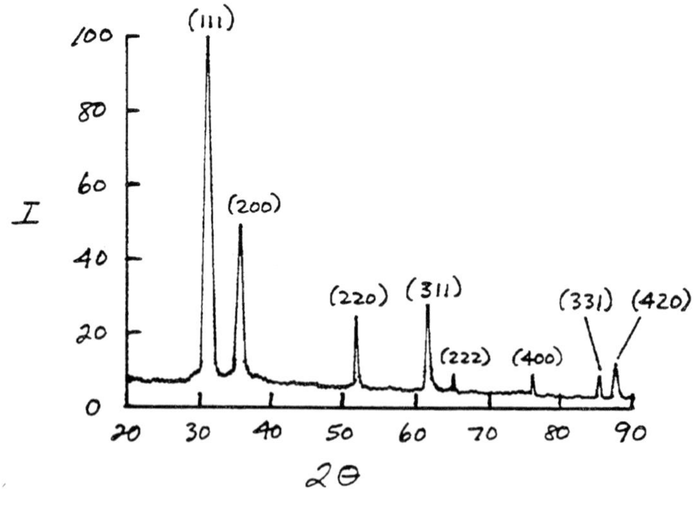

• **3.85** What would be the (hkl) indices for the three lowest diffraction angle peaks for an hcp metal?

3.85 The desired indices must provide a maximum value of d_{hkl} from Equation 3.7 along with the satisfaction of the reflection rules for the hcp structure.

Inspection of Equation 3.7 indicates that the maximum d occurs for the minimum value of

$$s = \frac{4}{3}(h^2 + hk + k^2) + l^2(a^2/c^2)$$

Using the ideal c/a ratio of 1.633*, the three lowest values of s consistent with the reflection rules are:

(hkl)	$s = 4/3(h^2+hk+k^2)+l^2(a^2/c^2)$
100	1.33
002	1.50
101	1.71

*Note Problem 3.39

• **3.86** Using the result of Problem 3.85, calculate the diffraction angles (2θ) for the first three peaks in the diffraction pattern of magnesium powder using CuK_α-radiation (λ = 0.1542 nm). Note that the c/a ratio for Mg is 1.62.

3.86 Using Equation 3.7 and Bragg's law, and noting in Figure 3-13 that $a = 2r_{Mg} = 2(0.160\text{ nm}) = 0.320\text{ nm}$,

$$d_{100} = \frac{0.320\text{ nm}}{\sqrt{4/3}} = 0.2771\text{ nm} \qquad \theta_{100} = \arcsin\left(\frac{0.1542\text{ nm}}{2\times 0.2771\text{ nm}}\right) = 16.2° \quad \text{or} \quad \underline{\underline{(2\theta)_{100} = 32.3°}}$$

$$d_{002} = \frac{0.320\text{ nm}}{\sqrt{4(1.62)^{-2}}} = 0.2592\text{ nm} \qquad \theta_{002} = \arcsin\left(\frac{0.1542\text{ nm}}{2\times 0.2592\text{ nm}}\right) = 17.3° \quad \text{or} \quad \underline{\underline{(2\theta)_{002} = 34.6°}}$$

$$d_{101} = \frac{0.320\text{ nm}}{\sqrt{4/3 + (1.62)^{-1}}} = 0.2444\text{ nm} \qquad \theta_{101} = \arcsin\left(\frac{0.1542\text{ nm}}{2\times 0.2444\text{ nm}}\right) = 18.4° \quad \text{or} \quad \underline{\underline{(2\theta)_{101} = 36.8°}}$$

3.87 Calculate the first six diffraction peak positions for MgO powder using CuK$_\alpha$-radiation. (This ceramic structure based on the fcc lattice shares the reflection rules of the fcc metals.)

3.87

The first six (hkl) values for this fcc ceramic will be the same as for the fcc metal in Figure 3-**40**, i.e.,

(111), (200), (220), (311), (222), and (400).

The lattice parameter for MgO was calculated in Sample Problem **3.11**, i.e. $a = 0.420$ nm

$$d_{111} = \frac{0.420\text{nm}}{\sqrt{3}} = 0.242\text{nm} \rightarrow \theta_{111} = \sin^{-1}\left(\frac{0.1542\text{nm}}{2\times 0.242\text{nm}}\right) = 18.5° \rightarrow \underline{\underline{(2\theta)_{111} = 37.1°}}$$

$$d_{200} = \frac{0.420\text{nm}}{\sqrt{4}} = 0.210\text{nm} \rightarrow \theta_{200} = \sin^{-1}\left(\frac{0.1542\text{nm}}{2\times 0.210\text{nm}}\right) = 21.5° \rightarrow \underline{\underline{(2\theta)_{200} = 43.1°}}$$

$$d_{220} = \frac{0.420\text{nm}}{\sqrt{8}} = 0.148\text{nm} \rightarrow \theta_{220} = \sin^{-1}\left(\frac{0.1542\text{nm}}{2\times 0.148\text{nm}}\right) = 31.3° \rightarrow \underline{\underline{(2\theta)_{220} = 62.6°}}$$

$$d_{311} = \frac{0.420\text{nm}}{\sqrt{11}} = 0.127\text{nm} \rightarrow \theta_{311} = \sin^{-1}\left(\frac{0.1542\text{nm}}{2\times 0.127\text{nm}}\right) = 37.5° \rightarrow \underline{\underline{(2\theta)_{311} = 75.0°}}$$

$$d_{222} = \frac{0.420\text{nm}}{\sqrt{12}} = 0.121\text{nm} \rightarrow \theta_{222} = \sin^{-1}\left(\frac{0.1542\text{nm}}{2\times 0.121\text{nm}}\right) = 39.5° \rightarrow \underline{\underline{(2\theta)_{222} = 79.0°}}$$

$$d_{400} = \frac{0.420\text{nm}}{\sqrt{16}} = 0.105\text{nm} \rightarrow \theta_{400} = \sin^{-1}\left(\frac{0.1542\text{nm}}{2\times 0.105\text{nm}}\right) = 47.2° \rightarrow \underline{\underline{(2\theta)_{400} = 94.5°}}$$

3.88 Repeat Problem 3.87 for CrK$_\alpha$-radiation ($\lambda = 0.2291$ nm).

3.88

Using the calculations from Problem **3.87**,

$$\theta_{111} = \sin^{-1}\left(\frac{0.2291\text{nm}}{2\times 0.242\text{nm}}\right) = 28.3° \rightarrow \underline{\underline{(2\theta)_{111} = 56.5°}}$$

$$\theta_{200} = \sin^{-1}\left(\frac{0.2291\text{nm}}{2\times 0.210\text{nm}}\right) = 33.1° \rightarrow \underline{\underline{(2\theta)_{200} = 66.1°}}$$

$$\theta_{220} = \sin^{-1}\left(\frac{0.2291\text{nm}}{2\times 0.148\text{nm}}\right) = 50.7° \rightarrow \underline{\underline{(2\theta)_{220} = 101°}}$$

$$\theta_{311} = \sin^{-1}\left(\frac{0.2291\text{nm}}{2\times 0.127\text{nm}}\right) = 64.4° \rightarrow \underline{\underline{(2\theta)_{311} = 129°}}$$

$$\theta_{222} = \sin^{-1}\left(\frac{0.2291\text{nm}}{2\times 0.121\text{nm}}\right) = 71.2° \rightarrow \underline{\underline{(2\theta)_{222} = 142°}}$$

$$\theta_{400} = \sin^{-1}\left(\frac{0.2291\text{nm}}{2\times 0.105\text{nm}}\right) =$$ NO SOLUTION (THERE ARE NO DIFFRACTION PEAKS FOR PLANES WITH THIS d-SPACING OR SMALLER USING Cr-K$_\alpha$.)

• **3.89** The first three diffraction peaks of a metal powder are $2\theta = 44.4°, 64.6°$, and $81.7°$ using CuK_α-radiation. Is the metal Cr, Ni, Ag, or W?

3.89 Noting Sample Problem 3.21 and Problem 3.78, we can recall that the first three peaks for an fcc metal are the (111), (200), and (220) and the first three peaks for a bcc metal are (110), (200), and (211).

Appendix 1 indicates that Ni and Ag are fcc and Cr and W are bcc.

Combining Equations 3.5 and 3.6 gives

$$\sin\theta = \frac{\lambda}{2a}\sqrt{h^2+k^2+l^2}$$

or $\sin\theta \propto \sqrt{h^2+k^2+l^2}$

For the given data,

$$\sin\theta_1 = \sin\frac{44.4°}{2} = 0.3778 \text{ and } \frac{\sin\theta_1}{\sin\theta_1} = 1.000$$

$$\sin\theta_2 = \sin\frac{64.6°}{2} = 0.5344 \text{ and } \frac{\sin\theta_2}{\sin\theta_1} = 1.415$$

$$\sin\theta_3 = \sin\frac{81.7°}{2} = 0.6541 \text{ and } \frac{\sin\theta_3}{\sin\theta_1} = 1.731$$

For an fcc metal,

$$\frac{\sqrt{h_2^2+k_2^2+l_2^2}}{\sqrt{h_1^2+k_1^2+l_1^2}} = \frac{\sqrt{2^2+0^2+0^2}}{\sqrt{1^2+1^2+1^2}} = 1.155$$

$$\frac{\sqrt{h_3^2+k_3^2+l_3^2}}{\sqrt{h_1^2+k_1^2+l_1^2}} = \frac{\sqrt{2^2+2^2+0^2}}{\sqrt{1^2+1^2+1^2}} = 1.633$$

For a bcc metal,

$$\frac{\sqrt{h_2^2+k_2^2+l_2^2}}{\sqrt{h_1^2+k_1^2+l_1^2}} = \frac{\sqrt{2^2+0^2+0^2}}{\sqrt{1^2+1^2+0^2}} = 1.414$$

$$\frac{\sqrt{h_3^2+k_3^2+l_3^2}}{\sqrt{h_1^2+k_1^2+l_1^2}} = \frac{\sqrt{2^2+1^2+1^2}}{\sqrt{1^2+1^2+0^2}} = 1.732$$

The ratio of $\sin\theta$ corresponds to the ratio of $\sqrt{h^2+k^2+l^2}$ for a bcc metal.

∴ The metal is either Cr or W.

Using data from Appendix 2,

$$a_{Cr} = \frac{4}{\sqrt{3}} r_{Cr} = \frac{4}{\sqrt{3}}(0.125\text{ nm}) = 0.289\text{ nm}$$

$$\& \; a_W = \frac{4}{\sqrt{3}} r_W = \frac{4}{\sqrt{3}}(0.137\text{ nm}) = 0.316\text{ nm}$$

Then

$$d_{110,Cr} = \frac{0.289\text{ nm}}{\sqrt{2}} = 0.204\text{ nm} \quad \& \quad d_{110,W} = \frac{0.316\text{ nm}}{\sqrt{2}} = 0.224\text{ nm}$$

and

$$2\theta_{110,Cr} = 2\arcsin\left(\frac{0.1542\text{ nm}}{2\times 0.204\text{ nm}}\right) = 44.4°$$

$$\& \; 2\theta_{110,W} = 2\arcsin\left(\frac{0.1542\text{ nm}}{2\times 0.224\text{ nm}}\right) = 40.3°$$

∴ The diffraction data indicate that the metal is Cr.

Section 4.1 – The Solid Solution - Chemical Imperfection

PP 4.1 Copper and nickel (which are completely soluble in each other) satisfy the first Hume-Rothery rule of solid solubility, as shown in Sample Problem 4.1. Aluminum and silicon are soluble in each other to only a limited degree. Do they satisfy the first Hume-Rothery rule?

PP 4.1

$r_{Al} = 0.143\,nm$, $r_{Si} = 0.117\,nm$

$$\frac{0.143\,nm - 0.117\,nm}{0.143\,nm} \times 100\% = 18.2\%$$

therefore, no (% radius difference > 15%)

PP 4.2 The interstitial site for dissolving a carbon atom in α-Fe was shown in Figure 4-4. Sample Problem 4.2 shows that a carbon atom is more than four times too large for the site and, consequently, carbon solubility in α-Fe is quite low. Consider now the case for interstitial solution of carbon in the high-temperature (fcc) structure of γ-Fe. The largest interstitial site for a carbon atom is a $\frac{1}{2}01$ type. (a) Sketch this interstitial solution in a manner similar to Figure 4-4. (b) Determine by how much the C atom in γ-Fe is oversize. (Note that the atomic radius for fcc iron is 0.127 nm.)

PP 4.2 (a)

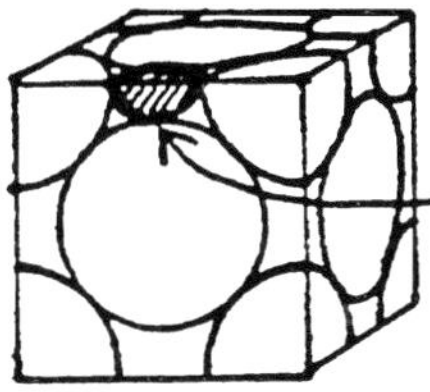

(b) $a = \frac{4}{\sqrt{2}} r_{Fe} = 2r_{Fe} + 2r_{interstitial}$

$$2r_{interstitial} = \frac{4}{\sqrt{2}} r_{Fe} - 2r_{Fe}$$

$$r_{interstitial} = \sqrt{2}\, r_{Fe} - r_{Fe} = 0.414\, r_{Fe} = 0.414(0.127\,nm)$$

$$= 0.0526\,nm$$

or

$$\frac{r_{carbon}}{r_{interstitial}} = \frac{0.077\,nm}{0.0526\,nm} = 1.46$$

Therefore, the carbon atom is roughly 50% too large.

Note: This is an important problem. It shows that the individual interstices in γ-Fe are larger than those in α-Fe, even though the overall APF for γ-Fe is greater.

4.1 In Chapter 5 we shall find a phase diagram for the Al–Cu system that indicates that these two metals do not form a complete solid solution. Which of the Hume-Rothery rules can you identify for Al–Cu that are violated? (We do not have electronegativity data, so rule 3 cannot be tested.)

4.1

Rule #1: $r_{Al} = 0.143\ nm$, $r_{Cu} = 0.128\ nm$

$$\% \text{ difference} = \frac{0.143-0.128}{0.128} \times 100\% = 11.7\% \quad (<15\%)$$

Rule #2: both Al and Cu are fcc

Rule #4: Al has valence of 3+ and Cu has valence of 1+ (or 2+).

Therefore, rule number 4 (valences are different)

4.2 For the Al–Mg system with a phase diagram in Chapter 5 showing incomplete solid solution, which of the Hume-Rothery rules are violated? (See Problem 4.1.)

4.2

Rule # 1: $r_{Al} = 0.143\ nm$, $r_{Mg} = 0.160\ nm$

$$\% \text{ difference} = \frac{0.160-0.143}{0.143} \times 100\% = 11.9\% \quad (<15\%)$$

Rule #2: Al is fcc, Mg is hcp

Rule #4: Al has valence of 3+ and Mg has valence of 2+.

Therefore, rule number 2 (different crystal structures) and rule number 4 (different valences)

4.3 For the Cu–Zn system with a phase diagram in Chapter 5 showing incomplete solid solution, which of the Hume-Rothery rules are violated? (See Problem 4.1.)

4.3

Rule #1: $r_{Cu} = 0.128\ nm$, $r_{Zn} = 0.133\ nm$

$$\% \text{ difference} = \frac{0.133-0.128}{0.128} \times 100\% = 3.9\% \quad (<15\%)$$

Rule #2: Cu is fcc, Zn is hcp

Rule # 4: Cu has valence of 2+ (although 1+ is also stable) and Zn has a valence of 2+

Therefore, rule number 2 (different crystal structures) and possibly rule number 4 (same valences shown in Appendix 2, although Cu^{1+} is also stable)

4.4 For the Pb–Sn system with a phase diagram in Chapter 5 showing incomplete solid solution, which of the Hume-Rothery rules are violated? (See Problem 4.1.)

4.4

Rule #1: $r_{Pb} = 0.175\ nm$, $r_{Sn} = 0.158\ nm$

$$\%\,difference = \frac{0.175 - 0.158}{0.158} \times 100\,\% = 10.8\%\ (<15\%)$$

Rule #2: Pb is fcc, Sn is bct

Rule #4: Pb has a valence of 4+ and 2+ while Sn has a valence of 4+

Therefore, rule number 2 (different crystal structures) and rule number 4 (different valences)

4.5 Sketch the pattern of atoms in the (111) plane of the ordered $AuCu_3$ alloy shown in Figure 4-3. (Show an area at least five atoms wide by five atoms high.)

4.5

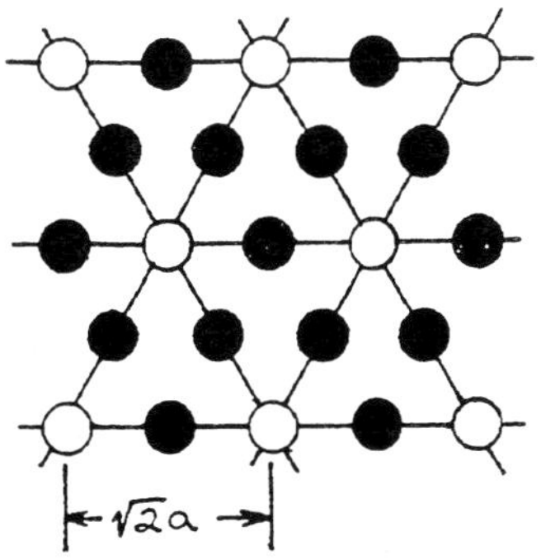

4.6 Sketch the pattern of atoms in the (110) plane of the ordered $AuCu_3$ alloy shown in Figure 4-3. (Show an area at least five atoms wide by five atoms high.)

4.6

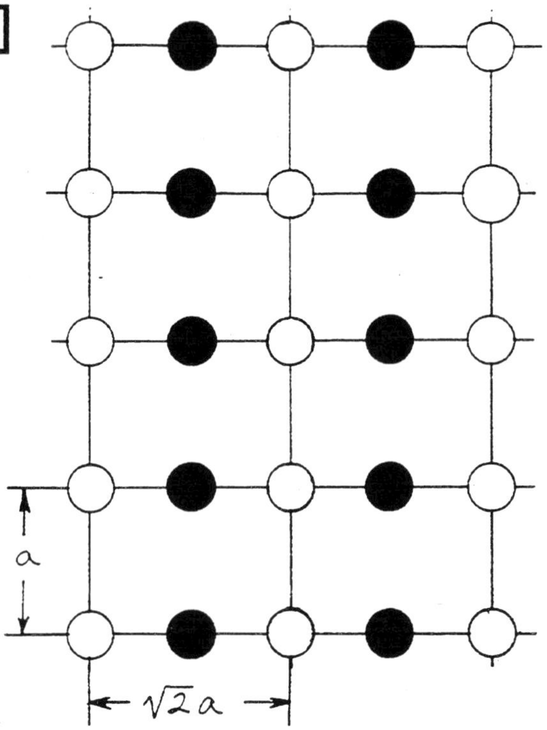

4.7 Sketch the pattern of atoms in the (200) plane of the ordered $AuCu_3$ alloy shown in Figure 4-3. (Show an area at least five atoms wide by five atoms high.)

4.7

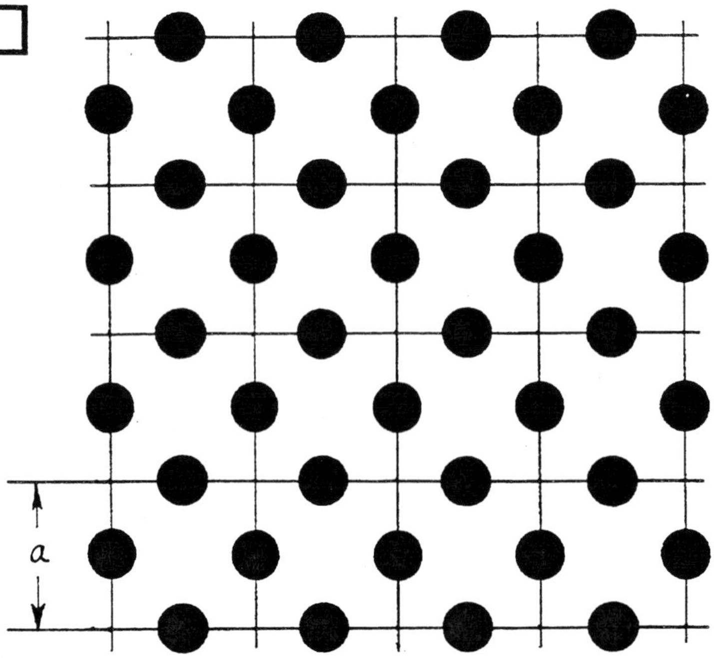

4.8 What are the equivalent points for ordered $AuCu_3$ (Figure 4-3)? Note Problem 3.40.

4.8

Au at 000

Cu at $\frac{1}{2}\frac{1}{2}0, \frac{1}{2}0\frac{1}{2}, 0\frac{1}{2}\frac{1}{2}$

4.9 Although the Hume-Rothery rules apply strictly only to metals, the concept of similarity of cations corresponds to the complete solubility of NiO in MgO (Figure 4-5). Calculate the percent difference between cation sizes in this case.

4.9

$r_{Ni^{2+}} = 0.078\ nm,\ r_{Mg^{2+}} = 0.078\ nm$

$$\%\ \text{difference} = \frac{0.078 - 0.078}{0.078} \times 100\ \% = \underline{0\ \%}$$

4.10 Calculate the percent difference between cation sizes for Al_2O_3 in MgO (Figure 4-6), a system that does not exhibit complete solid solubility.

4.10

$r_{Al^{3+}} = 0.057\ nm,\ r_{Mg^{2+}} = 0.078\ nm$

$$\%\ \text{difference} = \frac{0.078 - 0.057}{0.057} \times 100\ \% = \underline{\underline{36.9\ \%}}$$

4.11 Calculate the number of Mg^{2+} vacancies produced by the solubility of 1 mol of Al_2O_3 in 99 mol of MgO (see Figure 4-6).

4.11

1 mol Al_2O_3 in 99 mol MgO →

99 mol O sites (from MgO)
+ 3 mol O sites (from Al_2O_3)
102 mol O sites in solid solution

99 mol cation sites (from MgO)
+ 2 mol cation sites (from Al_2O_3)
101 mol cation sites in solid solution

→ 102 mol − 101 mol = 1 mol Mg^{2+} vacancies

$= \underline{\underline{0.6023 \times 10^{24}\ \text{vacancies}}}$

4.12 Calculate the number of Fe^{2+} vacancies in 1 mol of $Fe_{0.95}O$ (see Figure 4-7).

4.12

$1\ \text{mol}\ Fe_{0.95}O \rightarrow 1\ \text{mol O sites}$
$+ 0.95\ \text{mol Fe sites}$
$+ 0.05\ \text{mol}\ Fe^{2+}\ \text{vacancies}$

or $0.05 \times 0.6023 \times 10^{24}$ vacancies

$= \underline{\underline{3.01 \times 10^{22}\ \text{vacancies}}}$

4.13 In Part III of the text we shall be especially interested in "doped" semiconductors, in which small levels of impurities are added to an essentially pure semiconductor in order to produce desirable electrical properties. For silicon with 5×10^{21} aluminum atoms per cubic meter in solid solution, calculate (a) the atomic percent of aluminum atoms and (b) the weight percent of aluminum atoms.

4.13

(a) number Si atoms in $1\ m^3$ (undoped):

$$\rho = \frac{2.33\ Mg}{m^3} \times \frac{1\ \text{mol}}{28.09\ g} \times \frac{10^6\ g}{Mg} \times \frac{0.6023 \times 10^{24}\ \text{atoms}}{\text{mol}}$$

$= 5.00 \times 10^{28}$ atoms

$$\text{at.\% Al} = \frac{5 \times 10^{21}}{(5.00 \times 10^{28} + 5 \times 10^{21})} \times 100\% = \underline{\underline{10.0 \times 10^{-6}\ \text{at.\%}}}$$

(b) $$m_{Al} = 5 \times 10^{21}\ \text{atoms} \times \frac{26.98\ g}{0.6023 \times 10^{24}\ \text{atoms}} = 0.224\ g$$

$$m_{Si} = 5 \times 10^{28}\ \text{atoms} \times \frac{28.09\ g}{0.6023 \times 10^{24}\ \text{atoms}} = 2.33 \times 10^6\ g$$

$$\text{wt.\% Al} = \frac{0.224}{(2.33 \times 10^6 + 0.224)} \times 100\% = \underline{\underline{9.60 \times 10^{-6}\ \text{wt.\%}}}$$

4.14 For 5×10^{21} aluminum atoms/m^3 in solid solution in germanium, calculate (a) the atomic percent of aluminum atoms and (b) the weight percent of aluminum atoms.

4.14

(a) number Ge atoms in $1\ m^3$ (undoped):

$$\rho = \frac{5.32\ Mg}{m^3} \times \frac{1\ mol}{72.59\ g} \times \frac{10^6\ g}{Mg} \times \frac{0.6023 \times 10^{24}\ atoms}{mol}$$

$$= 4.41 \times 10^{28}\ atoms$$

$$at.\%\ Al = \frac{5 \times 10^{21}}{(4.41 \times 10^{28} + 5 \times 10^{21})} \times 100\% = \underline{\underline{11.3 \times 10^{-6}\ at.\%}}$$

(b) as in Problem 4.13,

$$m_{Al} = 0.224\ g$$

$$m_{Ge} = 4.41 \times 10^{28}\ atoms \times \frac{72.59\ g}{0.6023 \times 10^{24}\ atoms} = 5.32 \times 10^6\ g$$

$$\therefore wt.\%\ Al = \frac{0.224\ g}{(5.32 \times 10^6 + 0.224)\ g} \times 100\% = \underline{\underline{4.21 \times 10^{-6}\ wt.\%}}$$

4.15 For 5×10^{21} phosphorus atoms/m^3 in solid solution in silicon, calculate (a) the atomic percent of phosphorous atoms and (b) the weight percent of phosphorous atoms.

4.15

(a) as in Problem 4.13,

no. Si atoms in $1\ m^3$ (undoped) $= 5.00 \times 10^{28}$ atoms

$$at\%\ P = \frac{5 \times 10^{21}}{(5.00 \times 10^{28} + 5 \times 10^{21})} \times 100\% = \underline{\underline{10.0 \times 10^{-6}\ at.\%}}$$

(b) $$m_P = 5 \times 10^{21}\ atoms \times \frac{30.97\ g}{0.6023 \times 10^{24}\ atoms} = 0.257\ g$$

as in Problem 4.13,

$$m_{Si} = 2.33 \times 10^6\ g$$

$$\therefore wt.\%\ P = \frac{0.257\ g}{(2.33 \times 10^6 + 0.257\ g)} \times 100\% = \underline{\underline{11.0 \times 10^{-6}\ wt.\%}}$$

4.16 One way to determine a structural defect model (such as Figure 4-6 for a solid solution of Al_2O_3 in MgO) is to make careful density measurements. What would be the percent change in density for a 5 at % solution of Al_2O_3 in MgO (compared to pure, defect-free MgO)?

4.16 For 5 mol Al_2O_3 and 95 mol MgO, we have

95 mol O sites	(from MgO)	
+ 15 " " "	(" Al_2O_3)	
110 " " "	in solid solution	

95 mol cation sites	(from MgO)	
+ 10 mol " "	(" Al_2O_3)	
105 " " "	in solid solution	

→ 110 − 105 = 5 mol. Mg^{2+} vacancies

We can assume that the volume occupied by a given number of oxygens would be the same in either a pure material or the solid solution. Then, the density of either material would be proportional to the molecular weight of each system, i.e.,

$$\rho_{MgO} \propto 110 \times \text{at.wt. Mg} + 110 \times \text{at.wt. O}$$
$$= (110)(24.31\text{ amu}) + (110)(16.00\text{ amu}) = 4434.1\text{ amu}$$

and,

$$\rho_{soln} \propto 95 \times \text{at. wt. Mg} + 10 \times \text{at.wt. Al} + 110 \times \text{at.wt. O}$$
$$= (95)(24.31\text{ amu}) + (10)(26.98\text{ amu}) + (110)(16.00\text{ amu})$$
$$= 4339.3\text{ amu}$$

or,

$$\%\text{ change} = \frac{4339.3\text{ amu} - 4434.1\text{ amu}}{4434.1\text{ amu}} \times 100\% = \underline{\underline{-2.14\%}}$$

Section 4.2 – Point Defects - Zero-Dimensional Imperfections

PP 4.3 Calculate the fraction of aluminum lattice sites vacant at (a) 500°C, (b) 200° C, and (c) room temperature (25°C). (See Sample Problem 4.3.)

PP 4.3

(a) At 500°C (= 773 K),

$$\frac{n_v}{n_{sites}} = Ce^{-E_v/kT} = (11.2)e^{\frac{-(0.76\,eV)}{(86.2\times10^{-6}\,eV/K)(773\,K)}}$$

$$= \underline{\underline{1.25\times10^{-4}}}$$

(b) At 200°C (= 473 K),

$$\frac{n_v}{n_{sites}} = (11.2)e^{-\frac{(0.76\,eV)}{(86.2\times10^{-6}\,eV/K)(473\,K)}} = \underline{\underline{9.00\times10^{-8}}}$$

(c) At 25°C (=298 K),

$$\frac{n_v}{n_{sites}} = (11.2)\,e^{-\frac{(0.76\,eV)}{(86.2\times10^{-6}\,eV/K)(298\,K)}} = \underline{\underline{1.59\times10^{-12}}}$$

4.17 Verify that the data represented by Figure 4-13b correspond to an energy of formation of 0.76 eV for a defect in aluminum.

4.17

Extending the straight line plot to the edges of the figure gives a value of $\ln(n_v/n_{sites}) = -6.4$ at $1/T = 1.0\times10^{-3}\,K^{-1}$ and a value of $\ln(n_v/n_{sites}) = -10.85$ at $1/T = 1.5\times10^{-3}\,K^{-1}$.

Using Equation 4.7,

$$\ln(n_v/n_{sites}) = \ln C - \frac{E_v}{kT}$$

or $$-6.4 = \ln C - \frac{E_v}{k}(1.0\times10^{-3}\,K^{-1})$$

$$-\left[\;-10.85 = \ln C - \frac{E_v}{k}(1.5\times10^{-3}\,K^{-1})\right]$$

$$4.45 = \frac{E_v}{k}(1.5-1.0)\times10^{-3}\,K^{-1}$$

or $$E_v = \frac{4.45}{0.5\times10^{-3}\,K^{-1}} \times 86.20\times10^{-6}\,eV/K$$

$= 0.767\,eV$, which is in good agreement with 0.76 eV given the graphical measurements required

4.18 What type of crystal direction corresponds to the movement of interstitial carbon in α-Fe between equivalent ($\frac{1}{2}0\frac{1}{2}$-type) interstitial positions? Illustrate your answer with a sketch.

4.18 [110]-type direction

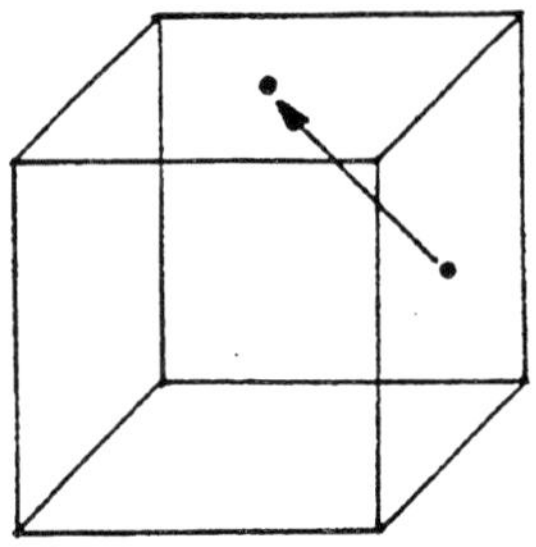

4.19 Repeat Problem 4.18 for the movement between equivalent interstices in γ-Fe. (Note Practice Problem 4.2.)

4.19 [110]-type direction

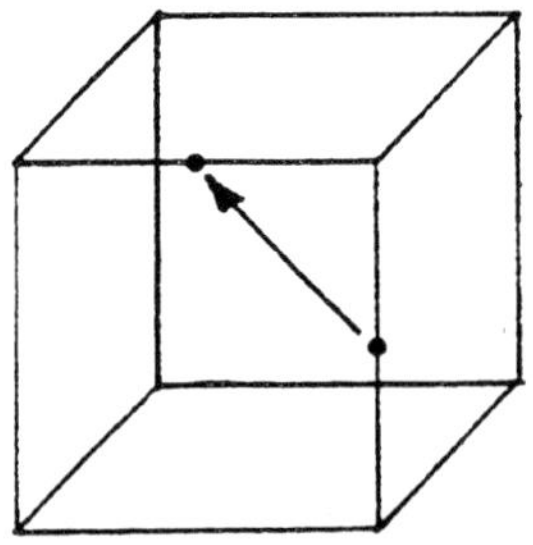

4.20 What crystallographic positions and directions are indicated by the migration shown in Figure 4-14? [Assume the atoms are in a (100) plane of an fcc metal.]

4.20 000-type position and [110]-type direction

Section 4.3 – Point Defects and Solid-State Diffusion

PP 4.4 Suppose that the carbon concentration gradient described in Sample Problem 4.4 occurred at 1100°C rather than 1000°C. Calculate the carbon atom flux for this case.

PP 4.4 The only difference for the current calculation is the diffusivity:

$$D_{C \text{ in } \gamma\text{-Fe}, 1100°C} = D_0 e^{-Q/RT}$$

$$= (20 \times 10^{-6} m^2/s) e^{-(142{,}000 \text{ J/mol})/(8.314 \text{ J/mol}\cdot K)(1373 K)}$$

$$= 7.92 \times 10^{-11} m^2/s$$

giving us:

$$J_x \cong -D \frac{\Delta c}{\Delta x} = -(7.92 \times 10^{-11} m^2/s)(-8.23 \times 10^{29} \text{ atoms}/m^4)$$

$$= \underline{\underline{6.52 \times 10^{19} \text{ atoms}/(m^2 \cdot s)}}$$

PP 4.5 In Sample Problem 4.5 the time to generate a given carbon concentration profile is calculated using the error function table. The carbon content at the surface was 1.0 wt % and at 1 mm from the surface was 0.6 wt %. For this diffusion time, what is the carbon content at a distance (**a**) 0.5 mm from the surface and (**b**) 2 mm from the surface?

PP 4.5 (a)

$$z = \frac{x}{2\sqrt{Dt}} = \frac{5 \times 10^{-4} m}{2\sqrt{(2.98 \times 10^{-11} m^2/s)(3.68 \times 10^4 s)}} = 0.2387$$

Interpolating from Table 4.1 gives:

$$\frac{0.25 - 0.2387}{0.25 - 0.20} = \frac{0.2763 - \text{erf} z}{0.2763 - 0.2227}$$

or

$$\text{erf} z = 0.2642$$

giving

$$\frac{C_x - C_0}{C_s - C_0} = 1 - \text{erf} z$$

or

$$\frac{C_x - 0.2 \text{ wt.\%}}{(1.0 - 0.2) \text{ wt.\%}} = 1 - 0.2642$$

giving

$$C_x = \underline{\underline{0.79 \text{ wt.\%}}}$$

(b) $z = \frac{2 \times 10^{-3}\,m}{2\sqrt{(2.98 \times 10^{-11}\,m^2/s)(3.68 \times 10^4\,s)}} = 0.9549$

& interpolating from Table 4.1 →

$$\frac{1.00 - 0.9549}{1.00 - 0.95} = \frac{0.8427 - \mathrm{erf}\,z}{0.8427 - 0.8209}$$

or $\mathrm{erf}\,z = 0.8230$

giving

$$\frac{C_x - 0.2\,wt\%}{(1.0 - 0.2)\,wt.\%} = 1 - 0.8230$$

or

$$C_x = \underline{\underline{0.34\,wt.\%}}$$

PP 4.6 Repeat Practice Problem 4.5 using the graphical method of Sample Problem 4.6.

PP 4.6

(a) For Dt fixed, we need only look for reducing x by one-half, i.e. go from $x/\sqrt{Dt} = 0.95$ to $x/\sqrt{Dt} = 0.475$.

Figure 4-19 gives for the latter case:

$$\frac{C - C_0}{C_s - C_0} \cong 0.733 \quad \text{or} \quad \frac{C - 0.2}{1.0 - 0.2} = 0.733$$

giving $C = \underline{\underline{0.79\ wt.\%\ C}}$

(b) For $x/\sqrt{Dt} = 2(0.95) = 1.90$, $\frac{C - C_0}{C_s - C_0} \cong 0.175$

or $\frac{C - 0.2}{1.0 - 0.2} = 0.175$ giving $C = \underline{\underline{0.34\ wt.\%\ C}}$

PP 4.7 In Sample Problem 4.7 a carburization temperature is calculated for a given carbon concentration profile. Calculate the carburization temperature if the given profile were obtained in 8 hours rather than 10 hours, as originally stated.

PP 4.7

In this case,

$$D = \frac{(0.75\times10^{-3}\,m)^2}{(0.95)^2(8\times3.6\times10^3\,s)} = 2.16\times10^{-11}\,m^2/s$$

$$= (20\times10^{-6}\,m^2/s)\,e^{-(142{,}000\,J/mol)/[8.314\,J/(mol\cdot K)]T}$$

giving $T = 1243\,K = \underline{\underline{970\,°C}}$

4.21 Verify that the data represented by Figure 4-21 correspond to an activation energy of 122,000 J/mol for the diffusion of carbon in α-iron.

4.21

Extending the straight line plot to the top and side of the figure gives a value of $D = 10^{-8}\,m^2/s$ at $1/T = 0.676\times10^{-3}\,K^{-1}$ and a value of $\log_{10} D = -13.27$ at $1/T = 1.5\times10^{-3}\,K^{-1}$

From Equation 4.13

$$D = D_0 e^{-Q/RT}$$

or

$$\ln D = \ln D_0 - \frac{Q}{R}\cdot\frac{1}{T}$$

or

$$2.303\,(-8.0) = \ln D_0 - \frac{Q}{R}(0.676\times10^{-3}\,K^{-1})$$

$$-\left[\,2.303\,(-13.27) = \ln D_0 - \frac{Q}{R}(1.5\times10^{-3}\,K^{-1})\,\right]$$

$$12.14 = \frac{Q}{R}(1.5-0.676)\times10^{-3}\,K^{-1}$$

or

$$Q = \frac{12.14}{0.824\times10^{-3}\,K^{-1}}\times 8.314\,J/(mol\cdot K)$$

$$= 122{,}000\,J/mol, \text{ as desired}$$

4.22 Carburization was introduced in Sample Problem 4.4. The decarburization of a steel can also be described by using the error function. Starting with Equation 4.11 and taking $c_s = 0$, derive an expression to describe the concentration profile of carbon as it diffuses out of a steel with initial concentration, c_0. (This situation can be produced by placing the steel in a vacuum at elevated temperature.)

4.22

$$\frac{c_x - c_0}{0 - c_0} = 1 - \text{erf}\left(\frac{x}{2\sqrt{Dt}}\right)$$

or

$$c_x - c_0 = -c_0 + c_0 \text{erf}\left(\frac{x}{2\sqrt{Dt}}\right)$$

giving,

$$\frac{c_x}{c_0} = \text{erf}\left(\frac{x}{2\sqrt{Dt}}\right)$$

4.23 Using the decarburization expression derived in Problem 4.22, plot the concentration profile of carbon within 1 mm of the carbon-free surface after 1 hour in vacuum at 1000°C. Take the initial carbon content of the steel to be 0.3 wt %.

4.23

From Problem 4.22,

$$c_x = c_0 \text{erf}\left(\frac{x}{2\sqrt{Dt}}\right)$$

Using the given data, D from Sample Problem 4.4 and representative x values:

x (mm)	c_x (wt.% C)
0	0
0.2	0.10
0.4	0.18
0.6	0.24
0.8	0.27
1.0	0.29

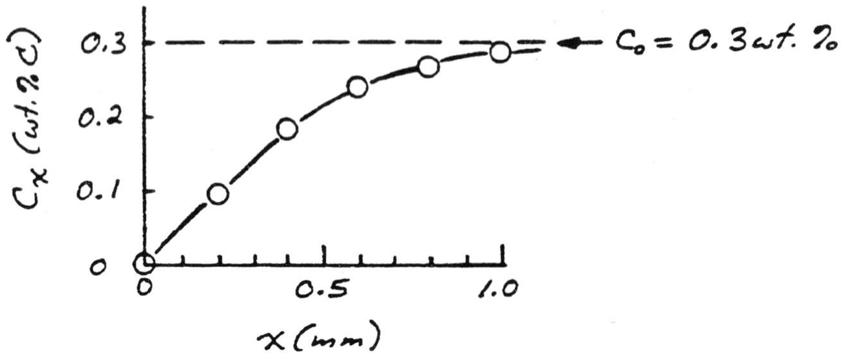

4.24 A "diffusion couple" is formed when two different materials are allowed to interdiffuse at elevated temperature. For a block of pure metal A adjacent to a block of pure metal B, the concentration profile of A (in at %) after interdiffusion is given by

$$c_x = 50\left[1 - \text{erf}\left(\frac{x}{2\sqrt{Dt}}\right)\right]$$

where x is measured from the original interface. For a diffusion couple with $D = 10^{-14}$ m^2/s, plot the concentration profile of metal A over a range of 20 μm on either side of the original interface ($x = 0$) after a time of 1 hour. [Note that erf $(-z) = -\text{erf}(z)$.]

4.24

Using the given data and representative x values:

x (μm)	c_x (at.% A) $= 50\left[1-\text{erf}\left(\frac{x}{2\sqrt{Dt}}\right)\right]$
+20	1
+15	4
+10	12
+5	29
0	50
−5	72
−10	87
−15	96
−20	99

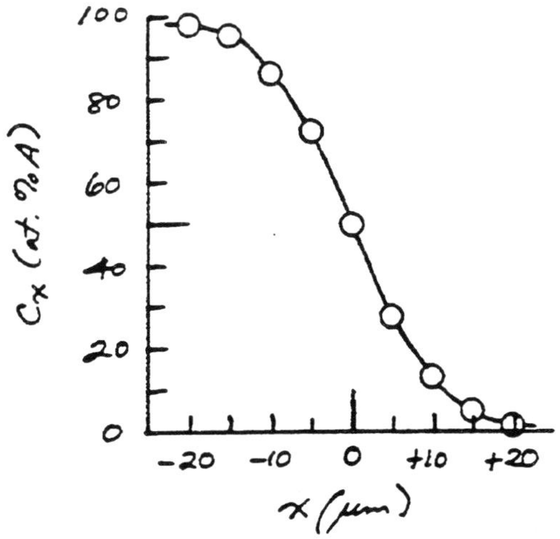

4.25 Use the information from Problem 4.24 to plot the progress of interdiffusion of two metals, X and Y, with $D = 10^{-12}$ m^2/s. Plot the concentration profile of metal X over a range of 300 μm on either side of the original interface after times of 1, 2, and 3 hr.

4.25 Using the given data and representative x values:

x (μm)	C_x (at. % X) $= 50\left[1-\text{erf}\left(\frac{x}{2\sqrt{Dt}}\right)\right]$		
	$t = 1$ hr	2 hr	3 hr
+300	<0.24	0.65	2.1
+250	<0.24	1.9	4.5
+200	0.93	4.8	8.7
+150	3.9	11	15
+100	12	20	25
+50	28	34	37
0	50	50	50
−50	72	66	63
−100	88	80	75
−150	96	89	85
−200	99	95	91
−250	>99	98	95
−300	>99	99	98

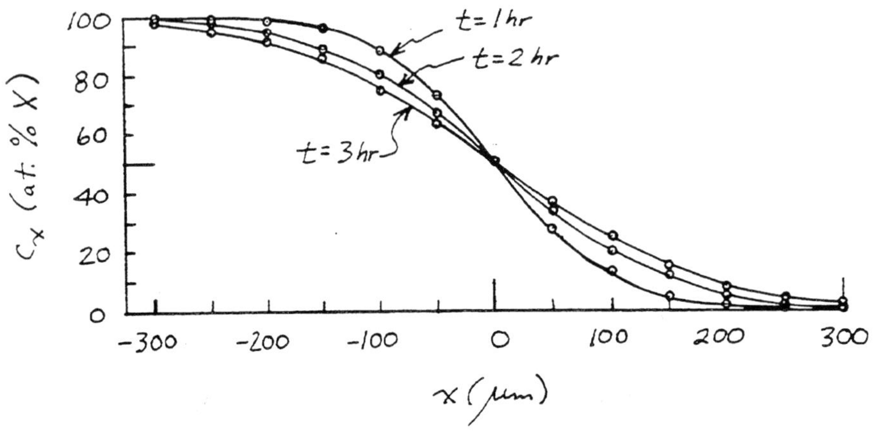

4.26 Using the results of Problem 4.24 and assuming that profile occurred at a temperature of 1000°C, superimpose the concentration profile of metal A for the same diffusion couple for 1 hour but heated at 1200°C at which $D = 10^{-13}$ m^2/s.

4.26 Recalculating C_x for $D = 10^{-13}$ m^2/s at the same x values:

x (μm)	C_x (at. % A) $= 50\left[1-\text{erf}\left(\frac{x}{2\sqrt{Dt}}\right)\right]$
+20	23
+15	29
+10	35
+5	43
0	50
−5	57
−10	65
−15	71
−20	77

/

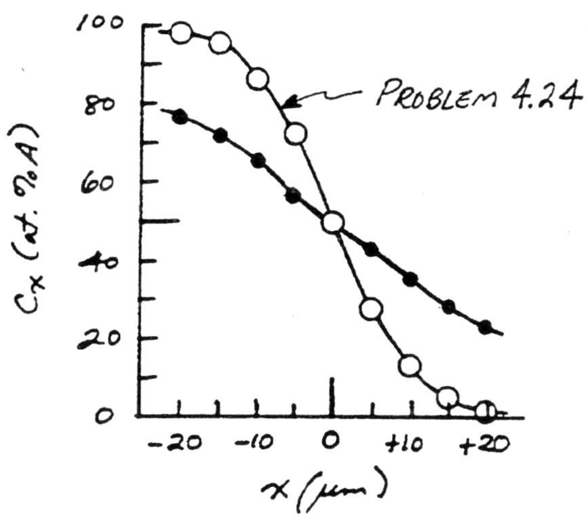

4.27 Given the information in Problems 4.24 and 4.26, calculate the activation energy for the interdiffusion of metals A and B.

4.27

$$\frac{D_{1200^\circ C}}{D_{1000^\circ C}} = \frac{D_0 e^{-Q/RT_2}}{D_0 e^{-Q/RT_1}} = e^{-Q/R\left(\frac{1}{T_2} - \frac{1}{T_1}\right)}$$

$$= \frac{10^{-13}}{10^{-14}} = 10$$

then, by re-arranging:

$$Q = \frac{-R \ln 10}{\left(1/1473\,K - 1/1273\,K\right)} = \frac{-[8.314\,J/(mol \cdot K)] \ln 10}{\left(1/1473\,K - 1/1273\,K\right)}$$

$$= 179 \times 10^3\,J/mol = 179\,kJ/mol$$

4.28 Use the result of Problem 4.27 to calculate the diffusion coefficient for the interdiffusion of metals A and B at 1400°C.

4.28

$$D = D_0 e^{-Q/RT} \quad \text{or} \quad D_0 = D e^{+Q/RT}$$

Using $Q = 179$ kJ/mol from Problem 4.27,

$$D_0 = D_{1000°C}\, e^{+(179{,}000 \text{ J/mol})/[8.314 \text{ J/(mol}\cdot\text{K)}]1273 \text{ K}}$$

$$= 10^{-14} \text{ m}^2\text{/s}\; e^{+(179{,}000/8.314 \times 1273)} = 2.317 \times 10^{-7} \text{ m}^2\text{/s}$$

$$\therefore\; D_{1400°C} = 2.317 \times 10^{-7} \text{ m}^2\text{/s}\; e^{-(179{,}000/8.314 \times 1673)}$$

$$= \underline{\underline{5.77 \times 10^{-13} \text{ m}^2\text{/s}}}$$

4.29 Using data in Table 4.2, calculate the self-diffusivity for iron in bcc iron at 900°C.

4.29

$$D_{Fe,\,bcc,\,900°C} = (200 \times 10^{-6} \text{ m}^2\text{/s})\, e^{-\frac{(240 \times 10^3 \text{ J/mol})}{[8.314 \text{ J/(mol}\cdot\text{K)}](1173 \text{ K})}}$$

$$= \underline{\underline{4.10 \times 10^{-15} \text{ m}^2\text{/s}}}$$

4.30 Using data in Table 4.2, calculate the self-diffusivity for iron in fcc iron at 1000°C.

4.30

$$D_{Fe,\,fcc,\,1000°C} = (22 \times 10^{-6} \text{ m}^2\text{/s})\, e^{-\frac{(268 \times 10^3 \text{ J/mol})}{[8.314 \text{ J/(mol}\cdot\text{K)}](1273 \text{ K})}}$$

$$= \underline{\underline{2.21 \times 10^{-16} \text{ m}^2\text{/s}}}$$

4.31 Using data in Table 4.2, calculate the self-diffusivity for copper in copper at 1000°C.

4.31

$$D_{Cu,\,1000°C} = \left(20\times10^{-6}\,m^2/s\right)e^{-\frac{(197\times10^3\,J/mol)}{[8.314\,J/(mol\cdot K)](1273\,K)}}$$

$$= \underline{\underline{1.65\times10^{-13}\,m^2/s}}$$

4.32 The diffusivity of copper in a commercial brass alloy is 10^{-20} m^2/s at 400°C. The activation energy for diffusion of copper in this system is 195 kJ/mol. Calculate the diffusivity at 600°C.

4.32

$$D = D_0 e^{-Q/RT}$$

$$D_0 = D e^{+Q/RT}$$

$$= \left(10^{-20}\,m^2/s\right)e^{+\frac{(195\times10^3\,J/mol)}{[8.314\,J/(mol\cdot K)](673\,K)}} = 1.37\times10^{-5}\,m^2/s$$

$$D_{600°C} = \left(1.37\times10^{-5}\,m^2/s\right)e^{-\frac{(195\times10^3\,J/mol)}{[8.314\,J/(mol\cdot K)](873\,K)}}$$

$$= \underline{\underline{2.93\times10^{-17}\,m^2/s}}$$

4.33 The diffusion coefficient of nickel in an austenitic (fcc structure) stainless steel is 10^{-22} m^2/s at 500°C and 10^{-15} m^2/s at 1000°C. Calculate the activation energy for the diffusion of nickel in this alloy over this temperature range.

4.33

$$\frac{D_{1000°C}}{D_{500°C}} = \frac{D_0 e^{-Q/RT_2}}{D_0 e^{-Q/RT_1}} = e^{-\frac{Q}{R}\left(\frac{1}{T_2}-\frac{1}{T_1}\right)}$$

$$= \frac{10^{-15}}{10^{-22}} = 10^7$$

or

$$-\frac{Q}{R}\left(\frac{1}{1273\,K} - \frac{1}{773\,K}\right) = \ln 10^7$$

giving

$$Q = \frac{-R\ln 10^7}{\left(1/1273\,K - 1/773\,K\right)} = -\frac{[8.314\,J/(mol\cdot K)]\ln 10^7}{\left(1/1273\,K - 1/773\,K\right)}$$

$$= 264\times10^3\,J/mol = \underline{\underline{264\,kJ/mol}}$$

• 4.34 Show that the relationship between vacancy concentration and fractional dimension changes for the case shown in Figure 4-13 is approximately

$$\frac{n_v}{n_{\text{sites}}} = 3\left(\frac{\Delta L}{L} - \frac{\Delta a}{a}\right)$$

[Note that $(1+x)^3 \simeq 1+3x$ for small x.]

4.34 Consider the volume of a cube of defect-free crystal with edge dimension ℓ:

$$V_1 = \ell^3$$

For the same mass of material (at the same temperature), a concentration of vacancies will produce a larger cube:

$$V_2 = (\ell + \Delta\ell)^3 = \left[\ell\left(1 + \frac{\Delta\ell}{\ell}\right)\right]^3 = \ell^3\left(1 + \frac{\Delta\ell}{\ell}\right)^3$$

As $(1+x)^3 \cong 1+3x$ for small x,

$$V_2 \cong \ell^3\left(1 + 3\frac{\Delta\ell}{\ell}\right)$$

Then, $\Delta V = V_2 - V_1 = \ell^3 + \ell^3\left(3\frac{\Delta\ell}{\ell}\right) - \ell^3 = \ell^3\left(3\frac{\Delta\ell}{\ell}\right)$

or, $\dfrac{\Delta V}{V} = 3\dfrac{\Delta\ell}{\ell}$

This volume fraction is, then, equal to the fraction of vacant sites:

$$\frac{n_v}{n_{\text{sites}}} = \frac{\Delta V}{V} = 3\frac{\Delta\ell}{\ell}$$

But, by definition, the value of $\Delta\ell/\ell$ at a given temperature is the overall thermal expansion term less the inherent lattice expansion:

$$\frac{\Delta\ell}{\ell} = \frac{\Delta L}{L} - \frac{\Delta a}{a}$$

giving, finally:

$$\frac{n_v}{n_{\text{sites}}} = 3\left(\frac{\Delta L}{L} - \frac{\Delta a}{a}\right)$$

• 4.35 A popular use of diffusion data in materials science is to identify mechanisms for certain phenomena. This is done by comparison of activation energies. For example, consider the oxidation of an aluminum alloy. The rate-controlling mechanism is the diffusion of ions through an Al_2O_3 surface layer. This means that the rate of growth of the oxide layer thickness is directly proportional to a diffusion coefficient. We can specify whether oxidation is controlled by Al^{3+} diffusion or O^{2-} diffusion by comparing the activation energy for oxidation with the activation energies of the two species as given in Table 4.3. Given that the rate constant for oxide growth is 4.00×10^{-8} kg/($m^4 \cdot s$) at 500°C and 1.97×10^{-4} kg/($m^4 \cdot s$) at 600°C, determine if the oxidation process is Al^{3+} diffusion controlled or O^{2-} diffusion controlled.

4.35

$$\frac{k_{600°C}}{k_{500°C}} = \frac{C e^{-Q/RT_2}}{C e^{-Q/RT_1}} = e^{-\frac{Q}{R}\left(\frac{1}{T_2}-\frac{1}{T_1}\right)}$$

$$= \frac{1.97\times10^{-4}}{4.00\times10^{-8}} = 4.93\times10^{3}$$

$$Q = -\frac{[8.314\ J/(mol\cdot K)]\ \ln(4.93\times10^{3})}{(1/873 - 1/773)\ K^{-1}} = 477\ kJ/mol$$

Therefore, Al^{3+} diffusion controlled ($Q = 477$ kJ/mol)

4.36 *Diffusion length*, λ, is a popular term in characterizing the production of semiconductors by the controlled diffusion of impurities into a high-purity material. The value of λ is taken as $2\sqrt{Dt}$ where λ represents the extent of diffusion for an impurity with a diffusion coefficient, D, over a period of time, t. Calculate the diffusion length for B in Ge for a total diffusion time of 30 minutes at a temperature of (**a**) 800°C and (**b**) 900°C.

4.36 $\lambda = 2\sqrt{Dt}$

Using Table 4.3, we see that for B in Ge:

$$D = D_0 e^{-Q/RT} = (1.1\times10^{3}\, m^2/s)\, e^{-439{,}000\ J/mol / RT}$$

(a) At 800°C,

$$D = (1.1\times10^{3}\, m^2/s)\, e^{-(439{,}000\ J/mol)/[8.314\ J/(mol\cdot K)]1073K}$$

$$= 4.67\times10^{-19}\ m^2/s$$

$$\therefore \lambda = 2\sqrt{Dt} = 2\sqrt{(4.67\times10^{-19}\ m^2/s)(30\times60\ s)}$$

$$= \underline{\underline{5.80\times10^{-8}\ m}}$$

(b) At 900°C,

$$D = (1.1\times10^{3}\, m^2/s)\, e^{-(439{,}000\ J/mol)/[8.314\ J/(mol\cdot K)]1173K}$$

$$= 3.10\times10^{-17}\ m^2/s$$

$$\therefore \lambda = 2\sqrt{Dt} = 2\sqrt{(3.10\times10^{-17}\ m^2/s)(30\times60\ s)}$$

$$= \underline{\underline{4.72\times10^{-7}\ m}}$$

Section 4.4 – Linear Defects, or Dislocations - One-Dimensional Imperfections

PP 4.8 Calculate the magnitude of the Burgers vector for an hcp metal, Mg. (See Sample Problem 4.8.)

PP 4.8 For Mg, an hcp metal, again $|\vec{b}| = d_{Mg} = 2R_{Mg}$

$$= 2(0.160\ nm) = \underline{\underline{0.320\ nm}}$$

PP 4.9 Repeat Sample Problem 4.9, assuming that the two directions are 45° rather than 60° and 40°.

PP 4.9 $\tau = \sigma \cos\lambda \cos\phi = (0.690\ MPa)\cos(45°)\cos(45°)$

$$= \underline{\underline{0.345\ MPa\ (50.0\ psi)}}$$

4.37 The energy necessary to generate a dislocation is proportional to the square of the length of the Burgers vector, $|\mathbf{b}|^2$. This means that the most stable (lowest energy) dislocations have the minimum length, $|\mathbf{b}|$. This is justification for the concept shown in Figure 4-33. For the bcc metal structure, calculate (relative to $E_{\mathbf{b}=[111]}$) the dislocation energies for (a) $E_{\mathbf{b}=[110]}$ and (b) $E_{\mathbf{b}=[100]}$.

4.37 (a) For a bcc metal, $\sqrt{3}a = 4R$ or $a = (4/\sqrt{3})R$

$$|b|_{110} = \sqrt{2}a = \sqrt{2}(4/\sqrt{3})R$$

But, $2R = |b|_{111}$

Therefore,

$$|b|_{110} = \left(\sqrt{\tfrac{2}{3}}\right)2|b|_{111} \quad \text{or} \quad \frac{|b|_{110}}{|b|_{111}} = 2\sqrt{\tfrac{2}{3}}$$

And, finally:

$$\left(\frac{|b|_{110}}{|b|_{111}}\right)^2 = 4 \times \frac{2}{3} = \underline{\underline{2.67}}$$

(b) $$|b|_{100} = a = \frac{4R}{\sqrt{3}} = \frac{2|b|_{111}}{\sqrt{3}}$$

$$\text{or} \quad \frac{|b|_{100}}{|b|_{111}} = \frac{2}{\sqrt{3}} \quad \text{giving} \quad \left(\frac{|b|_{100}}{|b|_{111}}\right)^2 = \frac{4}{3} = \underline{\underline{1.33}}$$

4.38 The comments in Problem 4.37 also apply for the fcc metal structure. Calculate (relative to $E_{\mathbf{b}=[110]}$) the dislocation energies for (a) $E_{\mathbf{b}=[111]}$ and (b) $E_{\mathbf{b}=[100]}$.

4.38 (a) For an fcc metal, $\sqrt{2}a = 4R$ or $a = (4/\sqrt{2})R$

$$|b|_{111} = \sqrt{3}a = \sqrt{3}(4/\sqrt{2})R = \left(\sqrt{\tfrac{3}{2}}\right)2|b|_{110}$$

$$\text{or} \quad \frac{|b|_{111}}{|b|_{110}} = 2\sqrt{\tfrac{3}{2}} \quad \text{giving} \quad \left(\frac{|b|_{111}}{|b|_{110}}\right)^2 = 4 \times \frac{3}{2} = \underline{\underline{6.00}}$$

(b) $$|b|_{100} = a = \frac{4R}{\sqrt{2}} = \frac{2|b|_{110}}{\sqrt{2}}$$

$$\text{or} \quad \frac{|b|_{100}}{|b|_{110}} = \frac{2}{\sqrt{2}} = \sqrt{2} \quad \text{giving} \quad \left(\frac{|b|_{100}}{|b|_{110}}\right)^2 = \underline{\underline{2.00}}$$

4.39 The comments in Problem 4.37 also apply for the hcp metal structure. Calculate (relative to $E_{b=[11\bar{2}0]}$) the dislocation energies for (a) $E_{b=[1\bar{1}00]}$ and (b) $E_{b=[0001]}$.

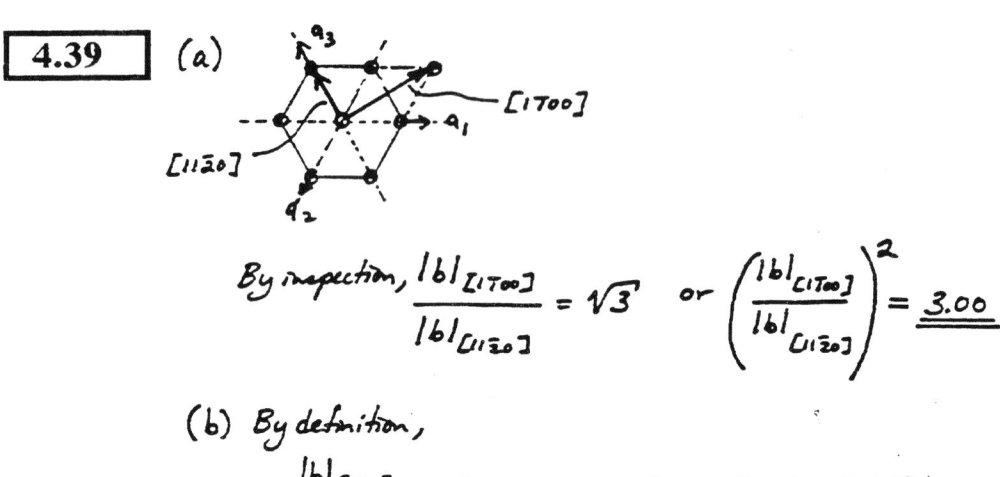

4.39 (a)

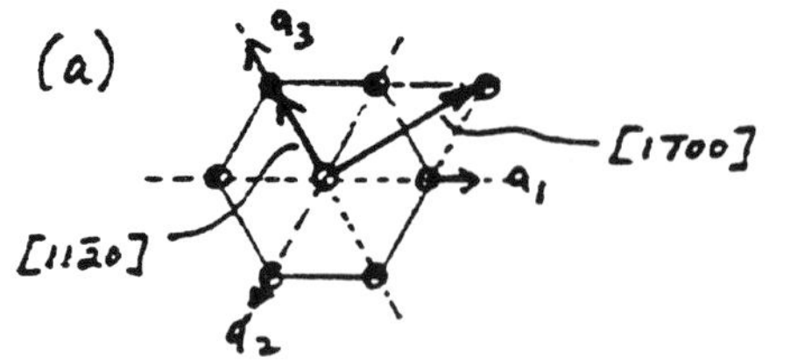

By inspection, $\frac{|b|_{[1\bar{1}00]}}{|b|_{[11\bar{2}0]}} = \sqrt{3}$ or $\left(\frac{|b|_{[1\bar{1}00]}}{|b|_{[11\bar{2}0]}}\right)^2 = \underline{\underline{3.00}}$

(b) By definition,

$$\frac{|b|_{[0001]}}{|b|_{[11\bar{2}0]}} = \frac{c}{a} = 1.633 \quad \text{(see Problem 3.39)}$$

giving:

$$\left(\frac{|b|_{[0001]}}{|b|_{[11\bar{2}0]}}\right)^2 = (1.633)^2 = \underline{\underline{2.67}}$$

• 4.40 Figure 4-34 lists the slip systems for an fcc and an hcp metal. For each case, this represents all unique combinations of close-packed planes and close-packed directions (contained within the close-packed planes). Make a similar list for the 12 slip systems in the bcc structure (see Table 4.4). (*A few important hints:* It will help to first verify the list for the fcc metal. Note that each slip system involves a plane $(h_1k_1l_1)$ and a direction $[h_2k_2l_2]$ whose indices give a dot product of zero (i.e., $h_1h_2 + k_1k_2 + l_1l_2 = 0$). Further, all members of the $\{hkl\}$ family of planes are not listed. Because a stress involves simultaneous force application in two antiparallel directions, only nonparallel planes need to be listed. Similarly, antiparallel crystal directions are redundant. You may want to review Problems 3.14 to 3.16.)

4.40 The slip system is $\{110\}\langle\bar{1}11\rangle$

Non parallel members of $\{110\}$ include:

$(110), (101), (011), (\bar{1}10), (\bar{1}01), (0\bar{1}1)$

[Note: It is acceptable to substitute other members of the $\{110\}$ family. For example, we could list (101) or $(\bar{1}0\bar{1})$ which are parallel. In general, the hkl indices of parallel planes are negative multiples of each other.]

For each plane, there are four members of the $\langle\bar{1}11\rangle$ family which will give a dot product of zero with the $\{110\}$ indices. Only two are not antiparallel. Again, the hkl indices of antiparallel directions are negative multiples of each other.

A resulting list is:

$(110)[\bar{1}11]$, $(110)[1\bar{1}1]$, $(101)[\bar{1}11]$, $(101)[11\bar{1}]$

$(011)[1\bar{1}1]$, $(011)[11\bar{1}]$, $(\bar{1}10)[111]$, $(\bar{1}10)[11\bar{1}]$

$(\bar{1}01)[111]$, $(\bar{1}01)[\bar{1}\bar{1}1]$, $(0\bar{1}1)[111]$, $(0\bar{1}1)[\bar{1}11]$

[Again, note: This list represents 12 of 48 possible $\{110\}\langle\bar{1}11\rangle$ combinations. Which set of twelve nonparallel planes and non-antiparallel directions listed is arbitrary.]

4.41 A crystalline grain of aluminum in a metal plate is oriented so that a tensile load is oriented along the [111] crystal direction. If the applied stress is 0.5 MPa (72.5 psi), what will be the resolved shear stress, τ, along the [101] direction within the $(11\bar{1})$ plane? (Review the comments in Problem 3.14.)

4.41

F along [111] with $\sigma = 0.5$ MPa

λ: angle between [101] and [111]

$$\cos\lambda = \frac{1+0+1}{\sqrt{2}\sqrt{3}}$$

ϕ: angle between [111] and $[11\bar{1}]$

$$\cos\phi = \frac{1+1-1}{\sqrt{3}\sqrt{3}}$$

$$\tau = \sigma\cos\lambda\cos\phi = (0.5\text{ MPa})\left(\frac{2}{\sqrt{2}\sqrt{3}}\right)\left(\frac{1}{3}\right) = \underline{0.136\text{ MPa}}$$

4.42 In Problem 4.41, what tensile stress is required to produce a critical resolved shear stress, τ_c, of 0.242 MPa?

4.42

$$\sigma_c = \frac{\tau_c}{\cos\lambda \cos\phi} = \frac{0.242\text{ MPa}}{(2/\sqrt{6})(1/3)}$$

$$= \underline{\underline{0.889\text{ MPa}}}$$

4.43 A crystalline grain of iron in a metal plate is oriented so that a tensile load is oriented along the [110] crystal direction. If the applied stress is 50 MPa (7.25×10^3 psi), what will be the resolved shear stress, τ, along the $[11\bar{1}]$ direction within the (101) plane? (Review the comments in Problem 3.14.)

4.43 F along [110] with $\sigma = 50\text{MPa}$

$$\cos\lambda = \frac{1+1+0}{\sqrt{2}\sqrt{3}}, \quad \cos\phi = \frac{1+0+0}{\sqrt{2}\sqrt{2}}$$

$$\tau = \sigma\cos\lambda\cos\phi = (50\text{MPa})\left(\frac{2}{\sqrt{6}}\right)\left(\frac{1}{2}\right) = \underline{\underline{20.4\text{MPa}}}$$

4.44 In Problem 4.43, what tensile stress is required to produce a critical resolved shear stress, τ_c, of 31.1 MPa?

4.44

$$\sigma_c = \frac{\tau_c}{\cos\lambda\cos\phi} = \frac{31.1\text{MPa}}{(2/\sqrt{6})(1/2)} = \underline{\underline{76.2\text{MPa}}}$$

• **4.45** Consider the slip systems for aluminum shown in Figure 4-34. For an applied tensile stress in the [111] direction, which slip system(s) would be most likely to operate?

4.45 $\tau \propto \cos\lambda\cos\phi$

For each angle, the cosine is determined by the dot product of [111] with {111} (for ϕ) or $\langle\bar{1}10\rangle$ (for λ).

The "most likely" slip systems are those with maximum τ. For the twelve systems in Fig. 4-34, $|\cos\lambda\cos\phi| = 2$ or 0. The six systems for which $|\cos\lambda\cos\phi| = 2$ are:

$\underline{\underline{(1\bar{1}1)[\bar{1}10], (\bar{1}11)[\bar{1}10], (11\bar{1})[\bar{1}01], (\bar{1}11)[\bar{1}01], (11\bar{1})[0\bar{1}1], (1\bar{1}1)[0\bar{1}1]}}$

4.46 Sketch the atomic arrangement and Burgers vector orientations in the slip plane of a bcc metal. (Note the shaded area of Table 4.4.)

4.46 For example, two of the 12 systems would be:

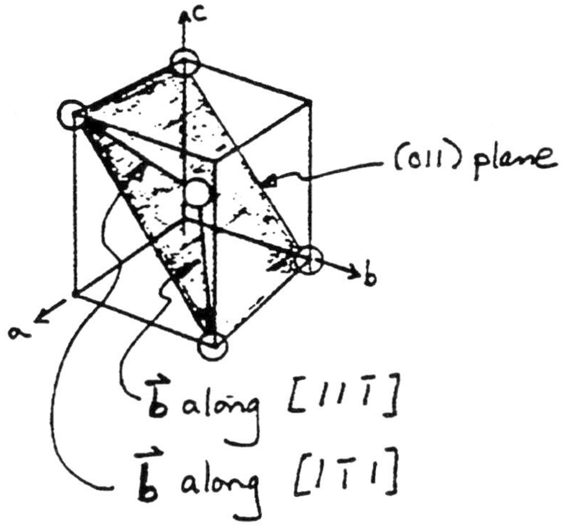

4.47 Sketch the atomic arrangement and Burgers vector orientation in the slip plane of an fcc metal. (Note the shaded area of Table 4.4.)

4.47 For example, three of the 12 systems would be:

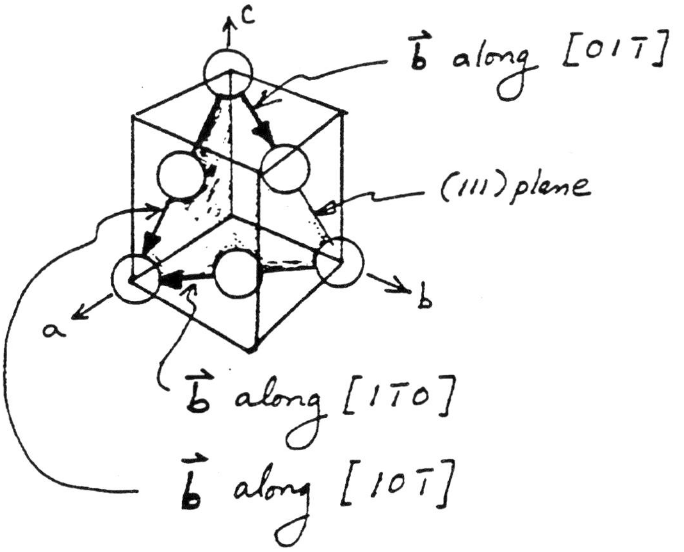

4.48 Sketch the atomic arrangement and Burgers vector orientation in the slip plane of an hcp metal. (Note the shaded area of Table 4.4.)

4.48 The three systems would be:

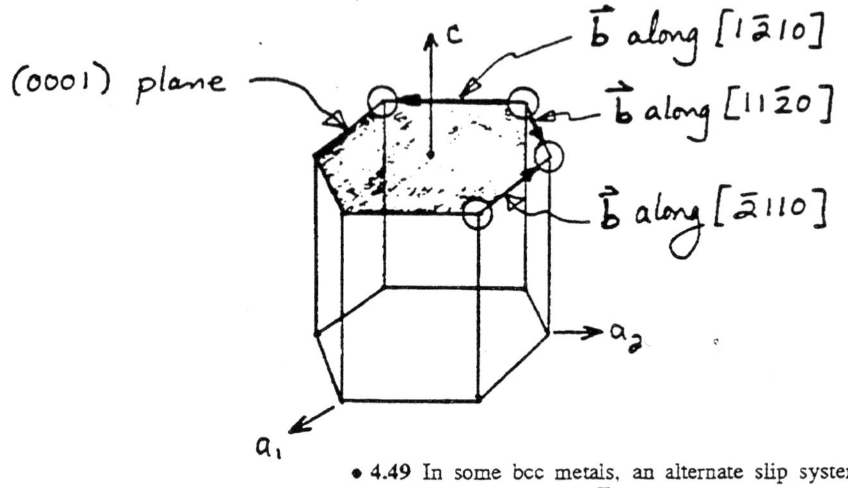

• 4.49 In some bcc metals, an alternate slip system operates, namely, the $\{211\}\langle\bar{1}11\rangle$. This system has the same Burgers vector but a lower-density slip plane, as compared to the slip system in Table 4.4. Sketch the unit cell geometry for this alternate slip system in the manner used in Table 4.4.

4.49

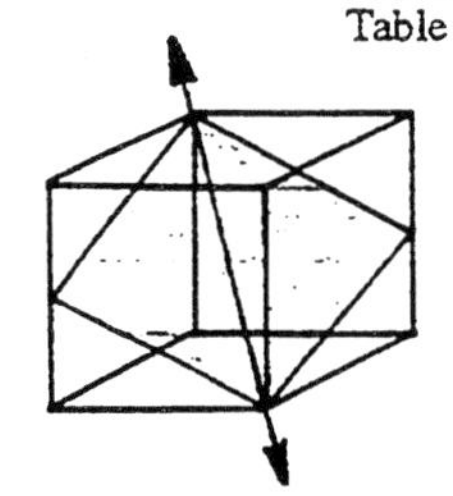

A single combination of $\langle\bar{1}11\rangle$-type direction (body diagonal) and $\{211\}$-type plane

• 4.50 Identify the 12 individual slip systems for the alternate system given for bcc metals in Problem 4.49. (Recall the comments in Problem 4.40.)

4.50 First, one can identify the 12 non-parallel planes as:

$(211), (121), (112), (\bar{2}11), (1\bar{2}1), (11\bar{2})$
$(2\bar{1}1), (\bar{1}21), (\bar{1}12), (21\bar{1}), (12\bar{1}), (1\bar{1}2)$

Then, we find a single direction to give a dot product of zero with each of the above planes:

$(211)[\bar{1}11]$, $(121)[1\bar{1}1]$, $(112)[11\bar{1}]$

$(\bar{2}11)[111]$, $(1\bar{2}1)[111]$, $(11\bar{2})[111]$

$(2\bar{1}1)[11\bar{1}]$, $(\bar{1}21)[11\bar{1}]$, $(\bar{1}12)[1\bar{1}1]$

$(21\bar{1})[1\bar{1}1]$, $(12\bar{1})[\bar{1}11]$, $(1\bar{1}2)[\bar{1}11]$

• **4.51** Sketch the atomic arrangement and Burgers vector orientation in a (211) slip plane of a bcc metal. (Note Problems 4.49 and 4.50.)

4.51 For example,

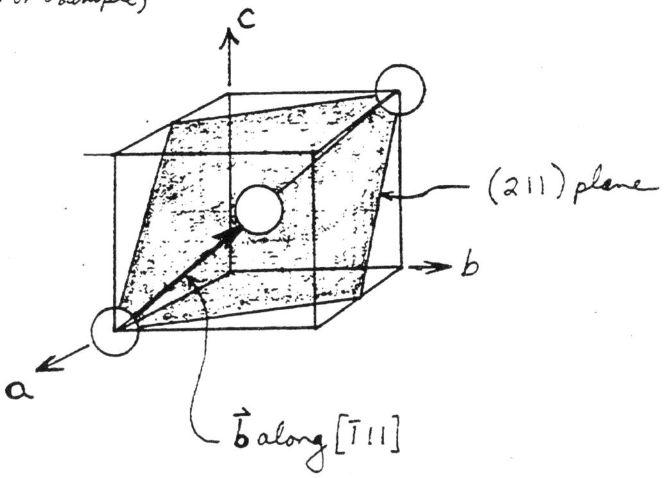

• **4.52** Figure 4-29 illustrates how a Burgers vector can be broken up into partials. The Burgers vector for an fcc metal can be broken up into two partials. (a) Sketch the partials relative to the full dislocation, and (b) identify the magnitude and crystallographic orientation of each partial.

4.52 (a)

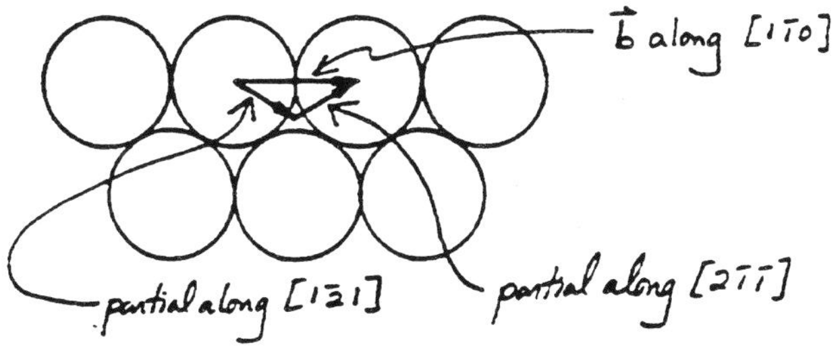

(b) By inspection, the angle between [1$\bar{1}$0] and [1$\bar{2}$1] or [2$\bar{1}\bar{1}$] is 30°, i.e.

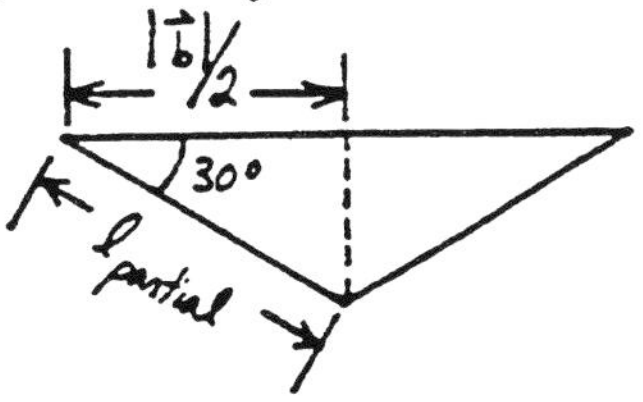

or,

$$l_{partial} = \frac{0.5\,|\vec{b}|}{\cos 30°} = \underline{\underline{0.577\,|\vec{b}|}}.$$

Section 4.5 – Planar Defects - Two-Dimensional Imperfections

PP 4.10 In Sample Problem 4.10 we find the separation distance between dislocations for a 2° tilt boundary in aluminum. Repeat this calculation for **(a)** $\theta = 1°$ and **(b)** $\theta = 5°$. **(c)** Plot the overall trend of D versus θ over the range $\theta = 0$ to $5°$.

PP 4.10

(a) $$D = \frac{|b|}{\theta} = \frac{0.286\text{ nm}}{1° \times (1\text{ rad}/57.3°)} = \underline{\underline{16.4\text{ nm}}}$$

(b) $$D = \frac{|b|}{\theta} = \frac{0.286\text{ nm}}{5° \times (1\text{ rad}/57.3°)} = \underline{\underline{3.28\text{ nm}}}$$

(c)

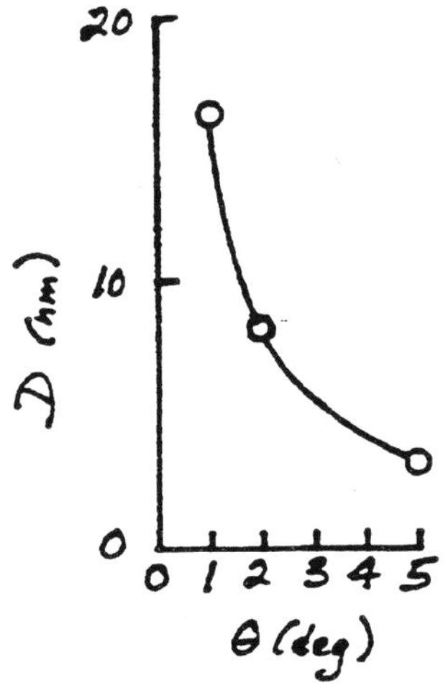

PP 4.11 Figure 4-45 gives a sample calculation of grain-size number, G. Sample Problem 4.11 recalculates G assuming a magnification of 300× rather than 100×. Repeat this process, assuming that the micrograph in Figure 4-45 is at 50× rather than 100×.

PP 4.11

$$A_{100\times} = 3.98\text{ in}^2 \times \left(\frac{100}{50}\right)^2 = 15.92\text{ in}^2$$

giving a grain density of

$$N = \frac{32\text{ grains}}{15.92\text{ in}^2} = 2.01\text{ grains/in}^2$$

$$N = 2^{G-1} \quad \text{or} \quad \ln N = (G-1)\ln 2$$

so that $$G = \frac{\ln N}{\ln 2} + 1 = \frac{\ln(2.01)}{\ln 2} + 1 = 2.01$$

or $\underline{\underline{G \cong 2}}$

4.53 Determine the grain-size number, G, for the microstructure shown in Figure 4-41. (Keep in mind that the precise answer will depend on your choice of an area of sampling.)

4.53 As in Figure 4–45 a $2\frac{1}{4}$" diameter circle in the middle of Figure 4–41 gives:

77 grains in the field of view + 39 cut by the circumference.

$77 + \frac{39}{2} = 96.5$

$$N = \frac{96.5 \text{ grains}}{\pi(2.25/2)^2 \text{ in}^2} = 24.3 \text{ grains/in}^2$$

$$G = \frac{\ln N}{\ln 2} + 1 = \frac{\ln(24.3)}{\ln 2} + 1 = \underline{\underline{5.6}}$$

4.54 Calculate the grain-size number for the microstructures in Figures 1-21a and 1-22a given that the magnifications are 160× and 330×, respectively.

4.54 As in Figure 4-45, a $2\frac{1}{4}$" diameter circle in the middle of Figure 1-21a gives:

3 grains in field of view + 8 cut by circumference

$3 + \frac{8}{2} = 7$

$$N = \frac{7 \text{ grains}}{\pi(2.25/2)^2 \text{ in}^2 \times (100/160)^2} = 4.51 \text{ grains/in}^2$$

$$G = \frac{\ln N}{\ln 2} + 1 = \frac{\ln(4.51)}{\ln 2} + 1 = \underline{\underline{3.2}} \text{ (for Fig. 1-21a)}$$

A $2\frac{1}{4}$" diameter circle in the middle of Fig. 1-22a gives:

33 full grains + 19 partial grains

$33 + \frac{19}{2} = 42.5$

$$N = \frac{42.5 \text{ grains}}{\pi(2.25/2)^2 \text{ in}^2 \times (100/330)^2} = 116 \text{ grains/in}^2$$

$$G = \frac{\ln N}{\ln 2} + 1 = \frac{\ln(116)}{\ln 2} + 1 = \underline{\underline{7.9}} \text{ (for Fig. 1-22a)}$$

Note: These results are quantitative indicators of the substantially finer grain structure associated with the translucent ceramic.

4.55 Using Equation 4.17, estimate the average grain diameter of Figures 1-21a and 1-22a using "random lines" cutting across the diagonal of each figure from its lower left corner to its upper right corner. (See Problem 4.54 for magnifications.)

4.55

For Figure 1-21a, the "random" diagonal gives

$$n_L = 6/109\ \text{mm} = 0.0550\ \text{mm}^{-1}$$

or

$$d = \frac{1.5}{(0.0550\ \text{mm}^{-1})(160)} = 0.170\ \text{mm} \times \frac{1000\ \mu\text{m}}{\text{mm}} = \underline{\underline{170\ \mu\text{m}}}$$

For Figure 1-22a, the "random" diagonal gives

$$n_L = 12/109\ \text{mm} = 0.110\ \text{mm}^{-1}$$

or

$$d = \frac{1.5}{(0.110\ \text{mm}^{-1})(330)} = 0.0413\ \text{mm} \times \frac{1000\ \mu\text{m}}{\text{mm}} = \underline{\underline{41.3\ \mu\text{m}}}$$

• 4.56 Note in Figure 4-44 that the crystalline regions in the fcc structure are represented by a repetitive polyhedra structure. This is an alternative to our usual unit cell configuration. In other words, the fcc structure can be equally represented by a space-filling stacking of regular polyhedra (tetrahedra and octahedra in a ratio of 2:1). (a) Sketch a typical tetrahedron (four-sided figure) on a perspective sketch such as Figure 3-12a. (b) Similarly, show a typical octahedron (eight-sided figure). (Note also Problem 3.47.)

4.56

(a)

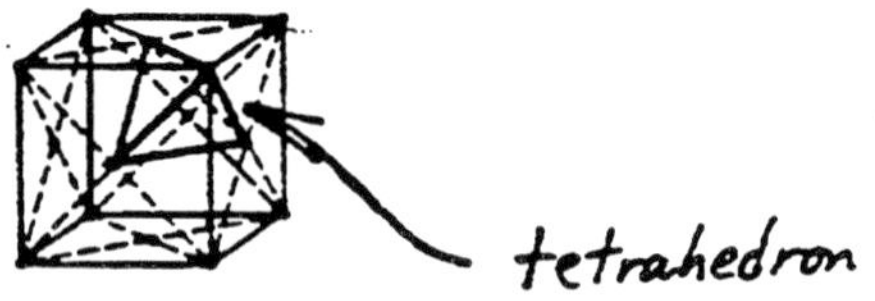

(b)

4.57 As implied in the text, demonstrate that the tilt angle for the $\Sigma 5$ boundary is defined by $\theta = 2\ \tan^{-1}(1/3)$. (**HINT:** Rotate two overlapping square lattices by 36.9° about a given common point and note the direction corresponding to one-half the rotation angle.)

4.57

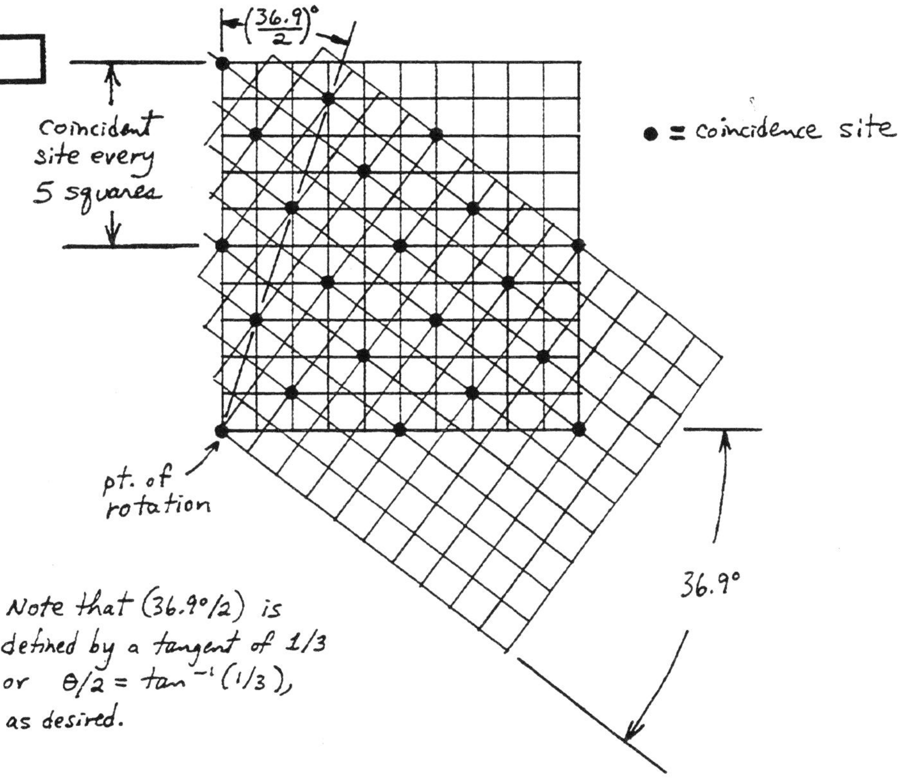

4.58 Show that the tilt angle for the $\Sigma 13$ grain boundary is defined by $\theta = 2\ \tan^{-1}(1/5) = 22.6°$. (Note Problem 4.57.)

4.58

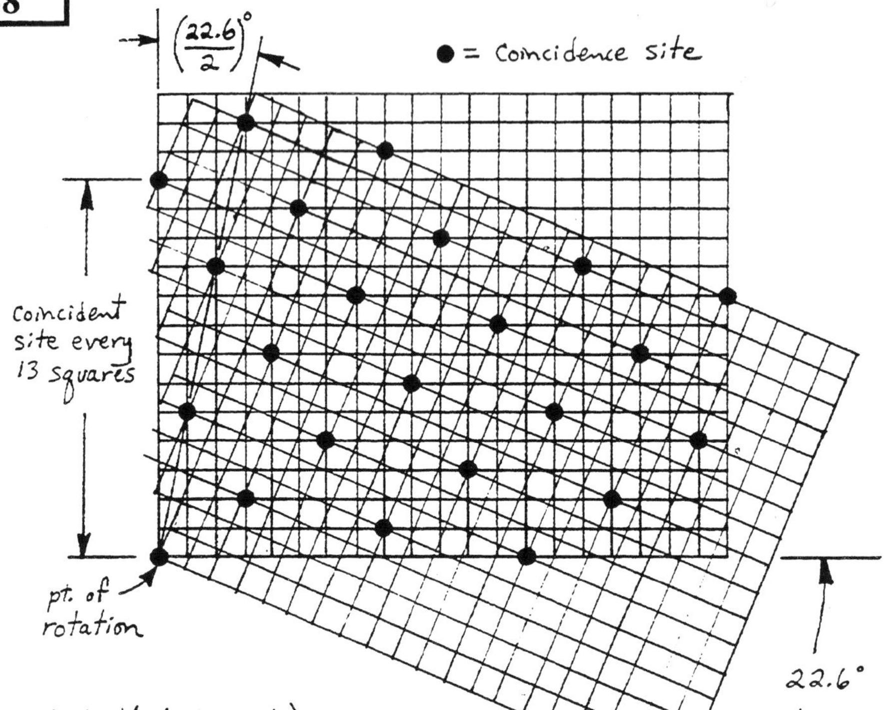

Note that (22.6°/2) is defined by a tangent of 1/5 or $\theta/2 = \tan^{-1}(1/5)$, as desired.

Section 4.6 – Noncrystalline Solids - Three-Dimensional Imperfection

PP 4.12 Estimate the atomic packing factor of amorphous silicon if its density is reduced by 1% relative to the crystalline state. (Review Sample Problem 3.16.)

PP 4.12 From Appendix 1,

$$\rho_{crystal} = 2.33\ Mg/m^3$$

$$\rho_{amor} = 0.99(2.33\ Mg/m^3) = 2.307\ Mg/m^3$$

Therefore, $(APF)_{amor} = (APF)_{crystal} \times 0.99$

Using the result of Sample Problem 3.16,

$$(APF)_{amor} = 0.340 \times 0.99 = \underline{\underline{0.337}}$$

4.59 Figure 4-46b is a useful schematic for simple B_2O_3 glass, composed of rings of BO_3^{3-} triangles. To appreciate the openness of this glass structure, calculate the size of the interstice (i.e., largest inscribed circle) of a regular six-membered ring of BO_3^{3-} triangles.

4.59 Using the ionic radii of Appendix 2:

$r_{B^{3+}} = 0.02\,nm$ & $r_{O^{2-}} = 0.132\,nm$,

we can draw a six-membered ring of BO_3^{3-} "triangles" to scale:

(Note that the B^{3+} radius is at the minimum radius ratio for 3-fold coordination, as given by Table 2.1, so that the O^{2-} ions, in effect are just touching each other as well as the central B^{3+} ion.)

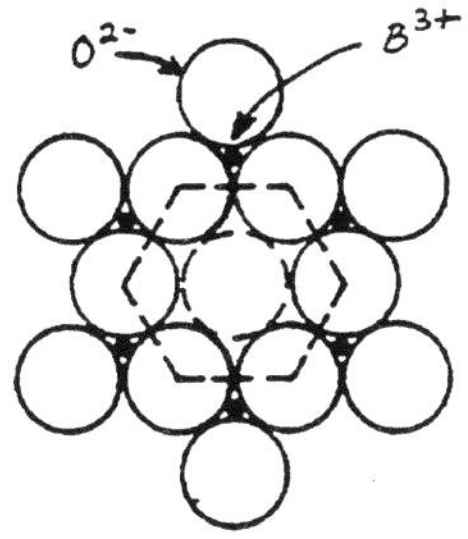

Because of the close-packed nature of the O^{2-} ions about the interstice, the largest inscribed circle (dashed) is equal in size to the O^{2-} ion.

$\therefore$ interstitial size $= d_{O^{2-}} = 2\,r_{O^{2-}} = 2(0.132\,nm)$
$= \underline{\underline{0.264\,nm}}$

4.60 In amorphous silicates, a useful indication of the lack of crystallinity is the "ring statistics." For the schematic illustration in Figure 4-46b, plot a histogram of the n-membered rings of O^{2-} ions, where n = number of O^{2-} ions in a loop surrounding an open interstice in the network structure. [*Note:* In Figure 4-46a, all rings are six-membered ($n = 6$).] (HINT: Ignore incomplete rings at the edge of the illustration.)

4.60 By observation, we see there are:

12 5-membered rings
12 6- " "
9 7- " "
2 8- " "

giving:

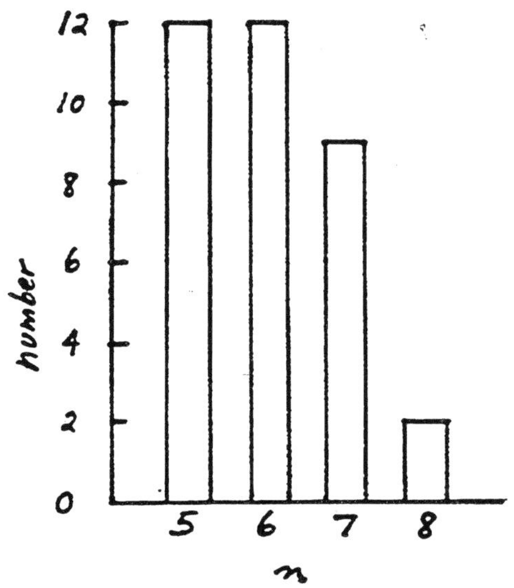

• **4.61** In Problem 4.56 a tetrahedron and octahedron were identified as the appropriate polyhedra to define an fcc structure. For the hcp structure, the tetrahedron and octahedron are also the appropriate polyhedra. **(a)** Sketch a typical tetrahedron on a perspective sketch such as Figure 3-13a. **(b)** Similarly, show a typical octahedron. (Of course, we are dealing with a crystalline solid in this example. But as Figure 4-47 shows, the noncrystalline, amorphous metal has a range of such polyhedra that fill space.)

4.61 (a)

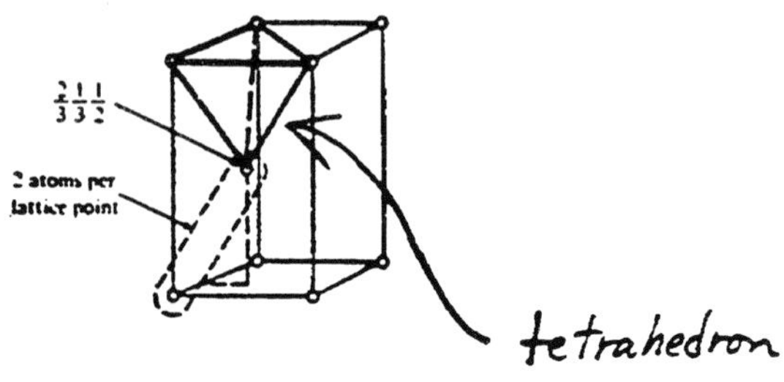

(b)

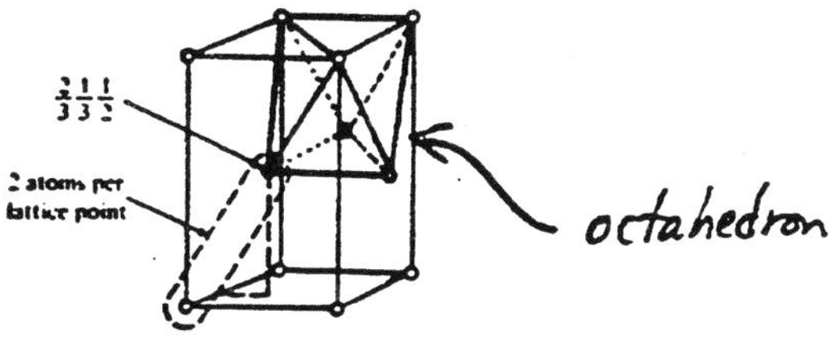

• **4.62** There are several polyhedra that can occur at grain boundaries, as discussed relative to Figure 4-44. The tetrahedron and octahedron treated in Problems 4.56 and 4.61 are the simplest. The next simplest is the pentagonal bipyramid, which consists of 10 equilateral triangle faces. Sketch this polyhedron as accurately as you can.

4.62

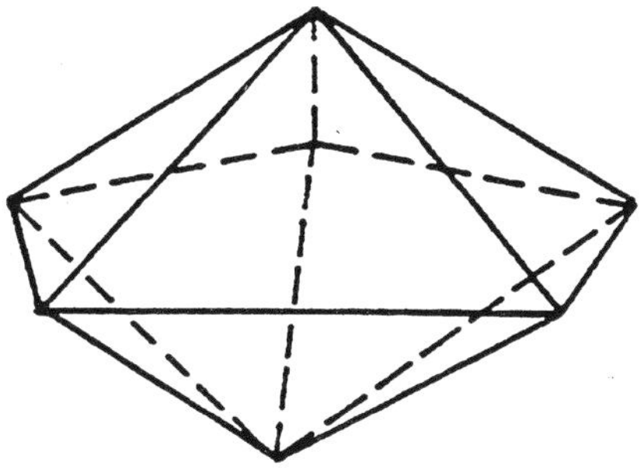

4.63 Sketch a few adjacent CaO_6 octahedra in the pattern shown in Figure 4-49. Indicate both the nearest neighbor $Ca^{2+} - Ca^{2+}$ distance, R_1, and the *next*-nearest neighbor $Ca^{2+} - Ca^{2+}$ distance, R_2.

4.63

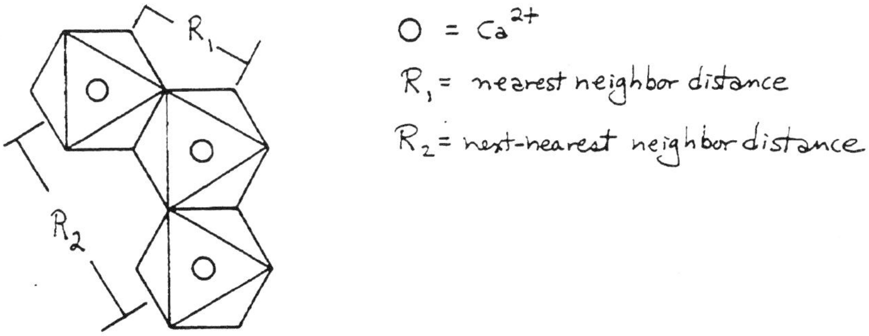

4.64 Diffraction measurements on the CaO-SiO_2 glass represented by Figure 4-49 show that the nearest neighbor $Ca^{2+} - Ca^{2+}$ distance, R_1, is 0.375 nm. What would be the next-nearest neighbor $Ca^{2+} - Ca^{2+}$ distance, R_2? (Note the results of Problem 4.63.)

4.64 By inspection of the result for Problem 4.63:

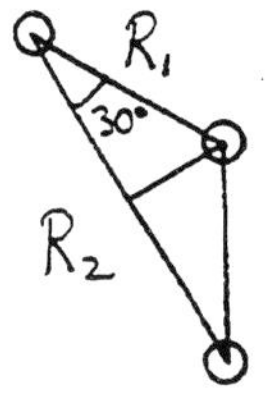

Note that

$$\cos 30° = \frac{\sqrt{3}}{2} = \frac{(R_2/2)}{R_1}$$

$$\therefore\ R_2 = \sqrt{3}\,R_1 = \sqrt{3}\,(0.375\text{ nm})$$

$$= \underline{\underline{0.650\text{ nm}}}$$

Section 4.7 – Recent Structural Discoveries

PP 4.13

In Sample Problem 4.13, we show one way in which the golden ratio, ϕ, appears in a Penrose tiling. To demonstrate that this tiling is isotropic, show that the decorative pentagons in Figure 4-53 are equally probable to be oriented "up" or "down."

PP 4.13 By inspection of Figure 4-53, we find 20 pentagons oriented "up" and 21 pentagons oriented "down," giving a ratio:

$$\frac{\text{"up"}}{\text{"down"}} = \frac{20}{21} = 0.95 \approx 1:1.$$

PP 4.14 In Sample Problem 4.14, we calculate the lower density associated with a larger segment of a fractal solid. Similarly, calculate the density associated with a smaller segment of radius 0.5 mm.

PP 4.14 Using the constant determined in Sample Problem 4.14, we have:

$$\rho_{r=0.5\,mm} = (6.32\times10^{-3}\,Mg/m^{2.5})(5\times10^{-4}\,m)^{-0.5}$$

$$= \underline{\underline{0.283\,Mg/m^3}}$$

4.65 Show, by using a graph, that the ratio of two consecutive terms in the Fibonacci series (1, 2, 3, 5, 8, . . .) approaches the golden ratio, ϕ.

4.65 We can extend the series (by adding the two previous terms) and monitor the ratio of the $(n+1)$th/nth terms:

nth term	$\frac{(n+1)\text{th}}{n\text{th}}$ ratio
1	2.000
2	1.500
3	1.667
5	1.600
8	1.625
13	1.615
21	1.619
34	1.618
55	1.618
89	

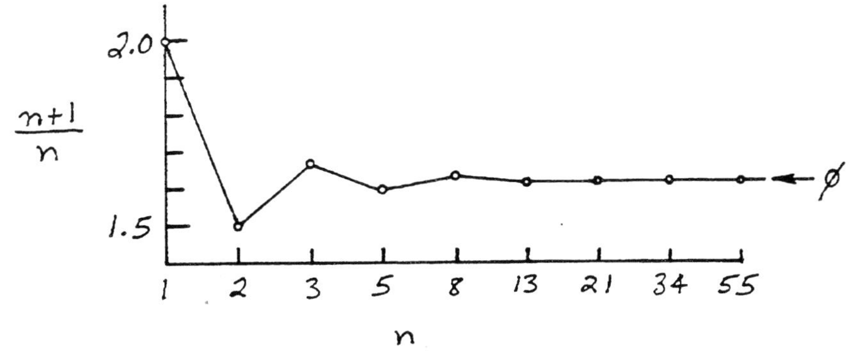

4.66 The Fibonacci series can also be used to describe crack branching in a material undergoing failure. Starting with a single, initial crack, sketch a branching pattern that follows the Fibonacci series. How many subcracks appear after five branching steps?

4.66

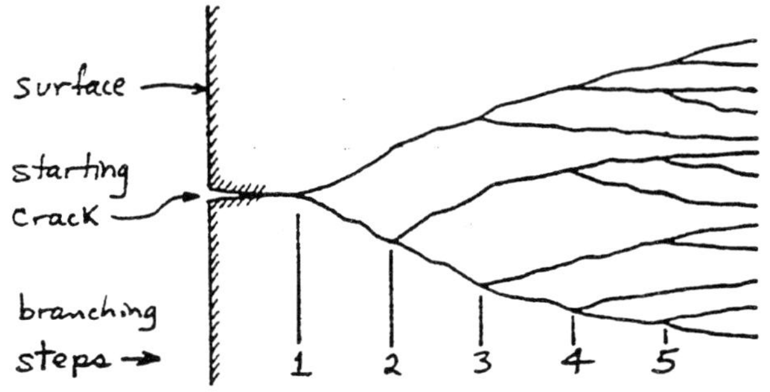

no. cracks → 1 2 3 5 8 13 (= no. after 5 steps)

4.67 In Problems 4.56, 4.61, and 4.62, we looked at the tetrahedron, the octahedron, and the pentagonal bipyramid as polyhedra with equilateral triangle faces. These are three of the family of polyhedra that can occur in metallic grain boundaries (Figure 4-44) or in amorphous metals (Figure 4-47). The icosahedron shown in Figure 4-55 is the largest such polyhedron. The interstice in the center of the icosahedron can *almost* accommodate another metal atom. (The next largest polyhedron is large enough to accommodate another atom and, consequently, the interstice gets filled in.) If the distance from the center of an icosahedron to the outermost vertex is 0.95 times the edge length of a triangular face, how large is the interstice (described as an inscribed sphere) compared to the atomic diameter?

4.67 Because the "triangular face" is formed by 3 adjacent metal atoms, the edge length is equal to the atomic diameter, d.

The center-to-vertex dist. = radius atom + radius interstice

or

$$0.95\,d = 0.50\,d + r_{int}$$

giving

$$r_{int} = 0.45\,d$$

or $d_{int} = 2(0.45\,d) = 0.90\,d$ or $\frac{d_{int}}{d} = 0.90$

4.68 Beyond determining the general icosahedral nature of quasicrystals, materials scientists are concerned with the location of specific atoms within the structure. One model for Al_6Mn is the Mackay icosahedron, in which there is an inner shell in the form of an icosahedron (with an Al atom at each vertex) and an outer shell icosahedron (with an Mn atom above each of the inner shell Al atoms and an Al atom in the center of each of the edges between Mn atoms). How many total atoms appear in this overall cluster?

4.68

Inspection of Figure 4-55 indicates that an icosahedron has 12 vertices and 30 edges. In the inner shell, then, there are 12 Al atoms. Correspondingly, there will be 12 Mn atoms in the outer shell. In addition, there will be 30 Al atoms along the edges of the outer shell for a total of:

$$12 + 12 + 30 = \underline{\underline{54 \text{ atoms}}}$$

• **4.69** The C_{60} buckyball introduced in Figure 3-27a is related to quasicrystals in an interesting way. The soccer ball geometry of C_{60} can be described as a truncated icosahedron. Illustrate this on a sketch of Figure 4-55.

4.69

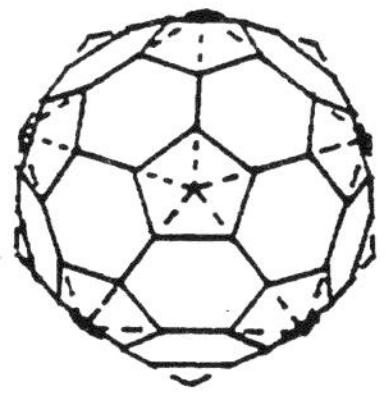

← For example, taking Figure 4-55a and cutting each edge in 3 equal parts gives a "buckyball." The middle part of each edge becomes the edge of a hexagon or a pentagon.

Each triangular face in the icosahedron becomes a hexagon in the buckyball, and each corner in the icosahedron is "truncated" and becomes a pentagon.

4.70 Given the relationship identified in Problem 4.69, describe how the golden ratio appears in the geometry of the C_{60} molecule.

4.70

Comparable to Figure 4-52a, the golden ratio, ϕ, appears in the pentagonal faces:

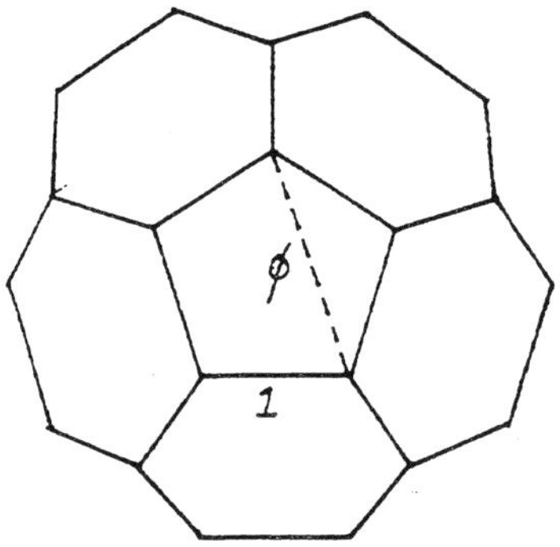

4.71 A given aerogel sample (with fractal dimension = 2.52) is homogeneous for spherical volumes with radii greater than 7.5 mm. If the homogeneous density is 0.150 Mg/m^3, what is the density of a 1-mm-radius spherical volume of this material?

4.71

Solving for the constant in Equation 4.20 using the data at the homogeneous limit:

$$C = \rho r^{3-d}$$

$$= (0.150 \, Mg/m^3)(7.5 \times 10^{-3} \, m)^{3-2.52}$$

$$= 0.0143 \, Mg/m^{2.52}$$

Then,

$$\rho_{1mm=r} = (0.0143 \, Mg/m^{2.52})(1 \times 10^{-3} \, m)^{2.52-3}$$

$$= \underline{\underline{0.395 \, Mg/m^3}}$$

4.72 Show how the fracture pattern described in Problem 4.66 can be fractal in nature.

4.72 The systematic nature of generating increasingly fine crack branches can exhibit self-similarity. For example, note similar patterns in the two dashed areas below:

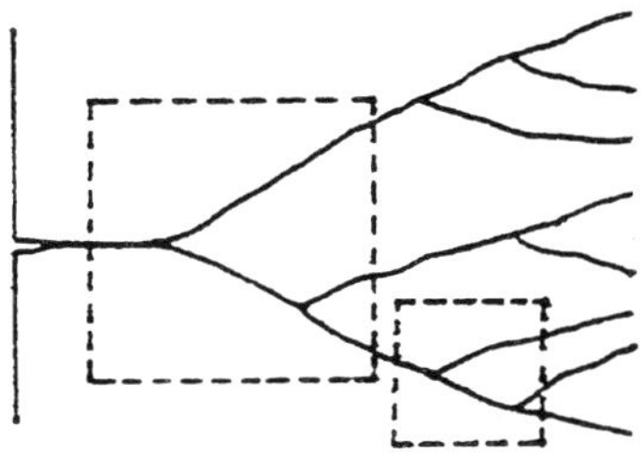

Section 4.8 – Electron Microscopy

PP 4.15 In Sample Problem 4.15 we calculate the diffraction angle (2θ) for 100-keV electrons diffracted from (111) planes in aluminum. What would be the diffraction angle from **(a)** the (200) planes and **(b)** the (220) planes?

PP 4.15 (a) $d_{200} = \dfrac{0.404\,nm}{\sqrt{2^2+0+0}} = 0.202\,nm$

$\theta = \arcsin\left(\dfrac{3.7\times10^{-3}\,nm}{2\times0.202\,nm}\right) = 0.525°$ or $(2\theta)_{200} = \underline{\underline{1.05°}}$

(b) $d_{220} = \dfrac{0.404\,nm}{\sqrt{2^2+2^2+0}} = 0.143\,nm$

$\theta = \arcsin\left(\dfrac{3.7\times10^{-3}\,nm}{2\times0.143\,nm}\right) = 0.742°$ or $(2\theta)_{220} = \underline{\underline{1.48°}}$

4.73 Suppose that the electron microscope in Figure 4-58c is used to make a simple diffraction spot pattern (rather than a magnified microstructural image). That is done by turning off the electromagnetic magnifying lenses. (Figure 4-50 was produced in this way.) The result is analogous to the Laue x-ray experiment described in Section 3.7 but with very small 2θ values. If the aluminum specimen described in Sample Problem 4.15 and Practice Problem 4.15 is 1 m from the photographic plate, (a) how far is the (111) diffraction spot from the direct (undiffracted) beam? Repeat part (a) for (b) the (200) spot and (c) the (220) spot.

4.73

(a) $r = (1\,m)\tan(2\theta)_{111}$

Using result of Sample Problem 4.15,

$r = (1\,m)\tan(0.906°) = 1.58\times10^{-2}\,m = \underline{\underline{15.8\,mm}}$

(b) Using result of PP 4.15a,

$r = (1\,m)\tan(1.05°) = 1.83\times10^{-2}\,m = \underline{\underline{18.3\,mm}}$

(c) Using result of PP 4.15b,

$r = (1\,m)\tan(1.48°) = 2.58\times10^{-2}\,m = \underline{\underline{25.8\,mm}}$

4.74 Repeat Problem 4.73 for (a) the (110) spot, (b) the (200) spot, and (c) the (211) spot produced by replacing the aluminum specimen with one composed of α-iron.

4.74

$a = (4/\sqrt{3})\,r_{Fe}$, as shown for Problem 3.79

$= (4/\sqrt{3})(0.124\,nm)$

$= 0.286\,nm$

then, $d_{110} = \dfrac{0.286\,nm}{\sqrt{1+1+0}} = 0.202\,nm$

$d_{200} = \dfrac{0.286\,nm}{\sqrt{2^2+0+0}} = 0.143\,nm$

$d_{211} = \dfrac{0.286\,nm}{\sqrt{2^2+1+1}} = 0.117\,nm$

giving,

$\theta_{110} = \arcsin\left(\dfrac{3.7\times10^{-3}\,nm}{2\times0.202\,nm}\right) = 0.525°$ or $(2\theta)_{110} = 1.05°$

$\theta_{200} =$ " $\left(\dfrac{3.7\times10^{-3}\,nm}{2\times0.143\,nm}\right) = 0.741°$ or $(2\theta)_{200} = 1.48°$

$\theta_{211} =$ " $\left(\dfrac{3.7\times10^{-3}\,nm}{2\times0.117\,nm}\right) = 0.906°$ or $(2\theta)_{211} = 1.81°$

for which,

$r_{110} = (1\,m)\tan(1.05°) = 1.83\times10^{-3}\,m = \underline{\underline{18.3\,mm}}$ (a)

$r_{200} = (1\,m)\tan(1.48°) = 2.59\times10^{-3}\,m = \underline{\underline{25.9\,mm}}$ (b)

$r_{211} = (1\,m)\tan(1.81°) = 3.16\times10^{-3}\,m = \underline{\underline{31.6\,mm}}$ (c)

4.75 A transmission electron microscope is used to produce a diffraction ring pattern for a thin, polycrystalline sample of copper. The (111) ring is 12 mm from the center of the film (corresponding to the undiffracted, transmitted beam). How far would the (200) ring be from the film center?

4.75 For the small angles involved,

$\theta \approx \sin\theta \approx \tan\theta$ which is proportional to the ring distance (from the film center) $= x$

Bragg's law gives us

$$\sin\theta = \frac{n\lambda}{2d} = \frac{n\lambda}{2a}\sqrt{h^2+k^2+l^2}$$

or

$$x \propto \sqrt{h^2+k^2+l^2}$$

$$\therefore \frac{x_{111}}{x_{200}} = \frac{\sqrt{1+1+1}}{\sqrt{4+0+0}}$$

or

$$x_{200} = \frac{2}{\sqrt{3}} \times 12\,mm = \underline{13.9\,mm}$$

4.76 The microchemical analysis discussed relative to Figure 4-65 is based on x-rays of characteristic wavelengths. As will be discussed in Section 8.5 on optical properties, a specific wavelength x-ray is equivalent to a photon of specific energy. Characteristic x-ray photons are produced by an electron transition between two energy levels in a given atom. For tin, the electron energy levels are

Electron Shell	Electron Energy
K	−29,199 eV
L	−3,929 eV
M	−709 eV

Which electron transition produces the characteristic K_α photon with energy of 25,270 eV?

4.76 As $-3{,}929\,eV - (-29{,}199\,eV) = 25{,}270\,eV$, the transition is <u>L to K</u>.

4.77 Repeat Problem 4.76, calculating the electron transition for lead in which a characteristic L_α photon with energy of 10,553 eV is used for the microanalysis. Relevant data are

Electron Shell	Electron Energy
K	−88,018 eV
L	−13,773 eV
M	−3,220 eV

4.77 As $-3{,}220\,eV - (-13{,}773\,eV) = 10{,}553\,eV$, the transition is <u>M to L</u>.

4.78 (a) Given only the data in Problems 4.76 and 4.77, determine whether a 28,490-eV characteristic x-ray photon would be produced by tin or by lead. (b) Which electron transition produces the characteristic photon in (a)?

4.78 (a) Inspection of the tin data shows:

$$-709\text{ eV} - (-29{,}199\text{ eV}) = 28{,}490\text{ eV}$$

(b) The transition is M to K.

(The resulting photon is referred to as the K_β, with β referring to the M shell being second closest to the K shell.)

Section 5.1 – The Phase Rule

PP 5. 1 Calculate the degrees of freedom at a constant pressure of 1 atm for (a) a single-phase solid solution of Sn dissolved in the solvent Pb, (b) pure Pb below its melting point, and (c) pure Pb at its melting point. (See Sample Problem 5.1.)

PP 5.1

(a) $F = C - P + 1 = 2 - 1 + 1 = \underline{\underline{2}}$

(b) $F = 1 - 1 + 1 = \underline{\underline{1}}$

(c) $F = 1 - 2 + 1 = \underline{\underline{0}}$

5.1 Apply the Gibbs phase rule to the various points in the one-component H_2O phase diagram (Figure 5-3).

5.1 In this case, pressure is not fixed, and we must use Equation 5.1 $(F = C - P + 2)$.

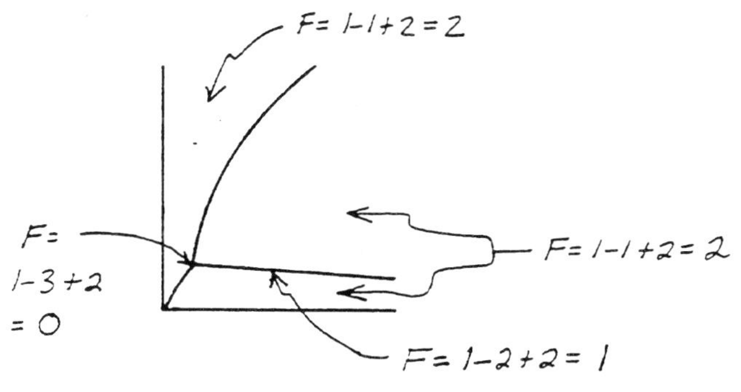

5.2 Apply the Gibbs phase rule to the various points in the one-component iron phase diagram (Figure 5-4).

5.2 In this case, pressure is not fixed, and we must use Equation 5.1 $(F=C-P+2)$.

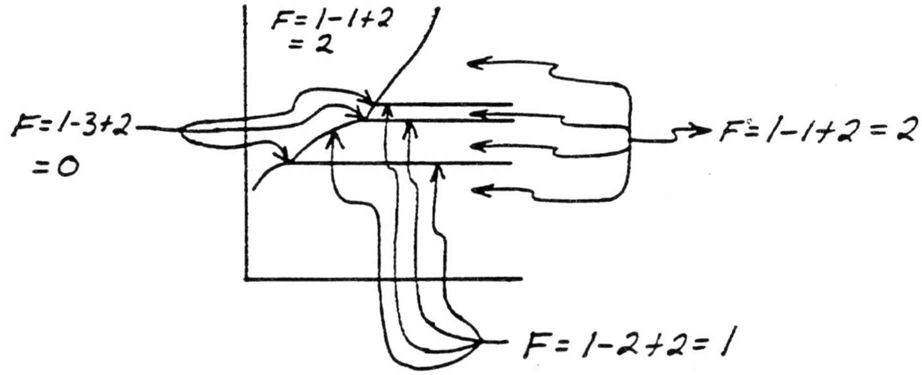

5.3 Calculate the degrees of freedom for a 50:50 copper–nickel alloy at (a) 1400°C where it exists as a single, liquid phase, (b) 1300°C where it exists as a two-phase mixture of liquid and solid solutions, and (c) 1200°C where it exists as a single, solid-solution phase. Assume a constant pressure of 1 atm above the alloy in each case.

5.3 (a) $F=C-P+1$

$= 2-1+1 = \underline{\underline{2}}$

(b) $F=2-2+1=\underline{\underline{1}}$

(c) $F=2-1+1=\underline{\underline{2}}$

5.4 In Figure 5-7, the Gibbs phase rule was applied to a hypothetical phase diagram. In a similar way, apply the phase rule to a sketch of the Pb–Sn phase diagram (Figure 5-38).

5.4

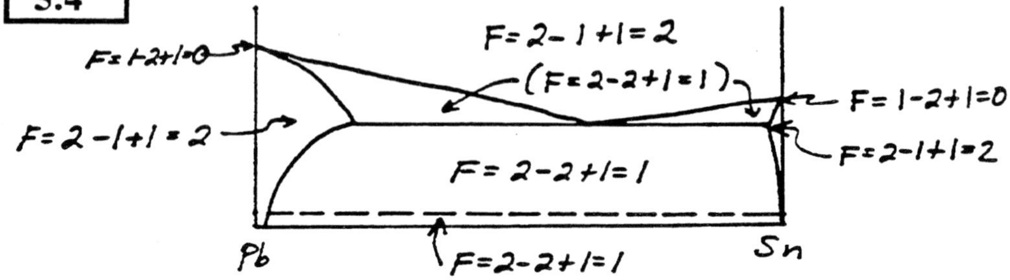

5.5 Apply the Gibbs phase rule to a sketch of the MgO–Al_2O_3 phase diagram (Figure 5-40).

5.5

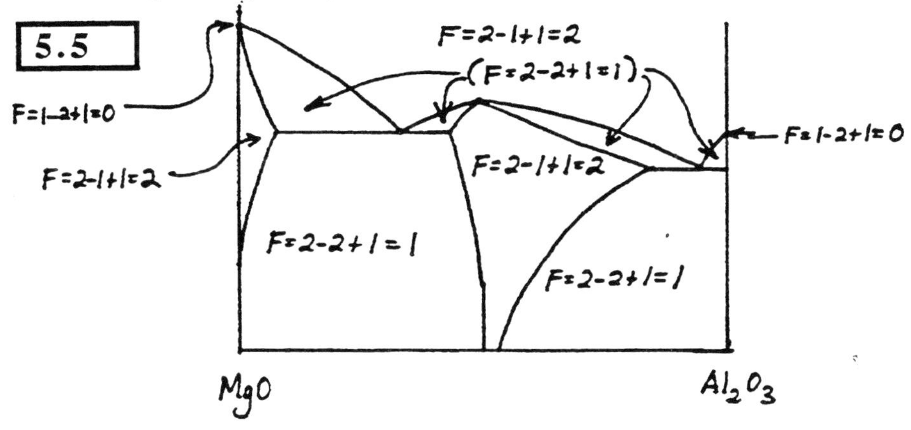

5.6 Apply the Gibbs phase rule to the various points in the Al_2O_3–SiO_2 phase diagram (Figure 5-39).

5.6

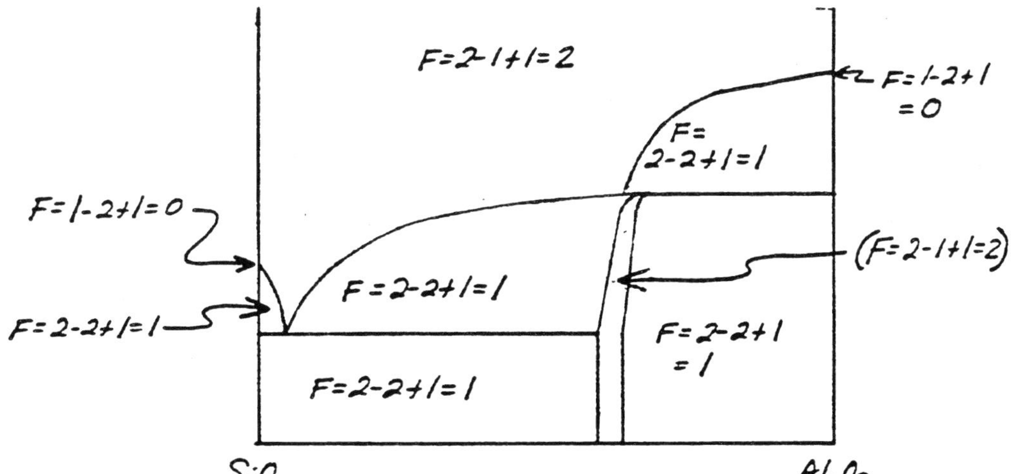

Section 5.2 – The Phase Diagram

PP 5.2 Qualitatively describe the microstructural development that will occur upon slow cooling of a melt of an alloy of equal parts of A_2B and AB.

PP 5.2

The first solid to precipitate is β. At the peritectic temperature, the remaining liquid solidifies leaving a two-phase microstructure of solid solutions β and γ.

5.7 Describe qualitatively the microstructural development that will occur upon slow cooling of a melt of equal parts (by weight) of copper and nickel (see Figure 5-36).

5.7

The first solid to precipitate is a solid solution, α, near 1315°C. By 1275°C, the complete system has solidified as α.

5.8 Describe qualitatively the microstructural development that will occur upon slow cooling of a melt composed of 50 wt % Al and 50 wt % Si (see Figure 5-33).

5.8

The first solid to precipitate is a solid solution, β, near 1045°C. At the eutectic temperature (577°C), the remaining liquid solidifies leaving a two-phase microstructure of solid solutions α and β.

5.9 Describe qualitatively the microstructural development that will occur upon slow cooling of a melt composed of 87.4 wt % Al and 12.6 wt % Si.

5.9

The system remains 100% liquid until reaching the eutectic temperature (577°C), at which the entire system solidifies leaving a two-phase microstructure of solid solutions α and β.

5.10 Describe qualitatively the microstructural development that will occur upon slow cooling of an alloy with equal parts (by weight) of aluminum and θ phase (Al_2Cu) (see Figure 5-34).

5.10

The first solid to precipitate is κ. At the eutectic temperature (548.2°C) the remaining liquid solidifies leaving a two-phase microstructure of solid solutions θ and κ.

5.11 Describe qualitatively the microstructural development that will occur upon slow cooling of a melt composed of (a) 20 wt % Mg, 80 wt % Al and (b) 80 wt % Mg, 20 wt % Al (see Figure 5-35).

5.11

(a) The first solid to precipitate is α. At the eutectic temperature (450°C), the remaining liquid solidifies leaving a two-phase microstructure of solid solutions α and β.

(b) The first solid to precipitate is δ. At the eutectic temperature (437°C), the remaining liquid solidifies leaving a two-phase microstructure of solid solutions γ' and δ.

5.12 Describe qualitatively the microstructural development during the slow cooling of a 30:70 brass (Cu with 30 wt % Zn). See Figure 5-37 for the Cu–Zn phase diagram.

5.12

The first solid to precipitate is α (near 950°C). By 920°C, the complete system has solidified as α.

5.13 Repeat Problem 5.12 for a 35:65 brass.

5.13 The first solid to precipitate is α (near 910°C). The remaining liquid solidifies at 903°C leaving a two-phase microstructure of solid solutions α and β. By 750°C, the complete system is solid solution α. That situation remains until approximately 220°C, at which β again precipitates.

5.14 Describe qualitatively the microstructural development during the slow cooling of a melt composed of (**a**) 30 wt % Pb–70 wt % Sn, (**b**) 40 wt % Pb–60 wt % Sn, and (**c**) 50 wt % Pb–50 wt % Sn (see Figure 5-38).

5.14 (a) The first solid to precipitate is solid solution β near 200°C. At the eutectic temperature (577°C), the remaining liquid solidifies leaving a two-phase microstructure of solid solutions α and β.

(b) The first solid to precipitate is solid solution α near 188°C. At the eutectic temperature (577°C), the remaining liquid solidifies leaving a two-phase microstructure of solid solutions α and β.

(c) The first solid to precipitate is solid solution α near 210°C. At the eutectic temperature (577°C), the remaining liquid solidifies leaving a two-phase microstructure of solid solutions α and β.

5.15 Repeat Problem 5.14 for a melt composed of 38.1 wt % Pb–61.9 wt % Sn.

5.15

The system remains 100% liquid until reaching the eutectic temperature (183°C), at which the entire system solidifies leaving a two-phase microstructure of solid solutions α and β.

5.16 Describe qualitatively the microstructural development during the slow cooling of (a) a 50 mol % Al_2O_3–50 mol % SiO_2 ceramic and (b) a 70 mol % Al_2O_3–30 mol % SiO_2 ceramic (see Figure 5-39).

5.16

(a) The first solid to precipitate is mullite solid solution. At the eutectic temperature (1587°C), the remaining liquid solidifies leaving a two-phase microstructure of mullite and SiO_2 (cristobalite).

(b) The first solid to precipitate is Al_2O_3. At the peritectic temperature (1828°C), the remaining liquid solidifies leaving a two-phase microstructure of Al_2O_3 and mullite solid solution.

Section 5.3 – The Lever Rule

PP 5.3 Suppose the alloy in Sample Problem 5.3 is reheated to a temperature at which the liquid composition is 48 wt % B and the solid-solution composition is 90 wt % B. Calculate the amount of each phase.

PP 5.3

$$m_L = \frac{90-50}{90-48}(1\,kg) = \underline{\underline{952\,g}}$$

$$m_{SS} = \frac{50-48}{90-48}(1\,kg) = \underline{\underline{48\,g}}$$

5.17 Calculate the amount of each phase present in 1 kg of a 50 wt % Ni–50 wt % Cu alloy at (a) 1400°C, (b) 1300°C, and (c) 1200°C (see Figure 5-36).

5.17 (a) In the single (L) phase region:

$$\underline{\underline{m_L = 1\,kg}},\ \underline{\underline{m_\alpha = 0\,kg}}$$

(b)

$$m_L = \frac{x_\alpha - x}{x_\alpha - x_L}(1\,kg) = \frac{58-50}{58-46}(1\,kg)$$

$$= 0.667\,kg = \underline{\underline{667\,g}}$$

$$m_\alpha = \frac{x - x_L}{x_\alpha - x_L}(1\,kg) = \frac{50-46}{58-46}(1\,kg)$$

$$= 0.333\,kg = \underline{\underline{333\,g}}$$

(c) In the single (α) phase region:

$$\underline{\underline{m_L = 0\,kg}},\ \underline{\underline{m_\alpha = 1\,kg}}$$

5.18 Calculate the amount of each phase present in 1 kg of a 50 wt % Pb–50 wt % Sn solder alloy at (a) 300°C, (b) 200°C, (c) 100°C, and (d) 0°C (see Figure 5-38).

5.18 (a) 1 kg L

(b) $$m_L = \frac{x - x_{\alpha\text{-Pb}}}{x_L - x_{\alpha\text{-Pb}}}(1\,kg) = \frac{50-18}{54-18}(1\,kg)$$

$$= 0.889\ kg = 889\ g$$

$$m_{\alpha\text{-Pb}} = \frac{x_L - x}{x_L - x_{\alpha\text{-Pb}}}(1\,kg) = \frac{54-50}{54-18}(1\,kg)$$

$$= 0.111\ kg = 111\ g$$

(c) $$m_{\alpha\text{-Pb}} = \frac{x_{\beta\text{-Sn}} - x}{x_{\beta\text{-Sn}} - x_{\alpha\text{-Pb}}}(1\,kg) = \frac{99-50}{99-5}(1\,kg)$$

$$= 0.521\ kg = 521\ g$$

$$m_{\beta\text{-Sn}} = \frac{x - x_{\alpha\text{-Pb}}}{x_{\beta\text{-Sn}} - x_{\alpha\text{-Pb}}}(1\,kg) = \frac{50-5}{99-5}(1\,kg)$$

$$= 0.479\ kg = 479\ g$$

(d) $$m_{\alpha\text{-Pb}} = \frac{x_{\alpha\text{-Sn}} - x}{x_{\alpha\text{-Sn}} - x_{\alpha\text{-Pb}}}(1\,kg) = \frac{100-50}{100-1}(1\,kg)$$

$$= 0.505\ kg = 505\ g$$

$$m_{\alpha\text{-Sn}} = \frac{x - x_{\alpha\text{-Pb}}}{x_{\alpha\text{-Sn}} - x_{\alpha\text{-Pb}}}(1\,kg) = \frac{50-1}{100-1}(1\,kg)$$

$$= 0.495\ kg = 495\ g$$

5.19 Repeat Problem 5.18 for a 60 wt % Pb–40 wt % Sn solder alloy.

5.19

(a) 1 kg L

(b) $$m_L = \frac{x - x_{\alpha\text{-}Pb}}{x_L - x_{\alpha\text{-}Pb}}(1\,kg) = \frac{40-18}{54-18}(1\,kg)$$

$$= 0.611\,kg = 611\,g$$

$$m_{\alpha\text{-}Pb} = \frac{x_L - x}{x_L - x_{\alpha\text{-}Pb}}(1\,kg) = \frac{54-40}{54-18}(1\,kg)$$

$$= 0.389\,kg = 389\,g$$

(c) $$m_{\alpha\text{-}Pb} = \frac{x_{\beta\text{-}Sn} - x}{x_{\beta\text{-}Sn} - x_{\alpha\text{-}Pb}}(1\,kg) = \frac{99-40}{99-5}(1\,kg)$$

$$= 0.628\,kg = 628\,g$$

$$m_{\beta\text{-}Sn} = \frac{x - x_{\alpha\text{-}Pb}}{x_{\beta\text{-}Sn} - x_{\alpha\text{-}Pb}}(1\,kg) = \frac{40-5}{99-5}(1\,kg)$$

$$= 0.372\,kg = 372\,g$$

(d) $$m_{\alpha\text{-}Pb} = \frac{x_{\alpha\text{-}Sn} - x}{x_{\alpha\text{-}Sn} - x_{\alpha\text{-}Pb}}(1\,kg) = \frac{100-40}{100-1}(1\,kg)$$

$$= 0.606\,kg = 606\,g$$

$$m_{\alpha\text{-}Sn} = \frac{x - x_{\alpha\text{-}Pb}}{x_{\alpha\text{-}Sn} - x_{\alpha\text{-}Pb}}(1\,kg) = \frac{40-1}{100-1}(1\,kg)$$

$$= 0.394\,kg = 394\,g$$

5.20 Repeat Problem 5.18 for an 80 wt % Pb–20 wt % Sn solder alloy.

5.20

(a) 1 kg L

(b) $m_L = \dfrac{x - x_{\alpha\text{-Pb}}}{x_L - x_{\alpha\text{-Pb}}}(1\ kg) = \dfrac{20-18}{54-18}(1\ kg)$

$= 0.056\ kg = 56\ g$

$m_{\alpha\text{-Pb}} = \dfrac{x_L - x}{x_L - x_{\alpha\text{-Pb}}}(1\ kg) = \dfrac{54-20}{54-36}(1\ kg)$

$= 0.944\ kg = 944\ g$

(c) $m_{\alpha\text{-Pb}} = \dfrac{x_{\beta\text{-Sn}} - x}{x_{\beta\text{-Sn}} - x_{\alpha\text{-Pb}}}(1\ kg) = \dfrac{99-20}{99-5}(1\ kg)$

$= 0.840\ kg = 840\ g$

$m_{\beta\text{-Sn}} = \dfrac{x - x_{\alpha\text{-Pb}}}{x_{\beta\text{-Sn}} - x_{\alpha\text{-Pb}}}(1\ kg) = \dfrac{20-5}{99-5}(1\ kg)$

$= 0.160\ kg = 160\ g$

(d) $m_{\alpha\text{-Pb}} = \dfrac{x_{\alpha\text{-Sn}} - x}{x_{\alpha\text{-Sn}} - x_{\alpha\text{-Pb}}}(1\ kg) = \dfrac{100-20}{100-1}(1\ kg)$

$= 0.808\ kg = 808\ g$

$m_{\alpha\text{-Sn}} = \dfrac{x - x_{\alpha\text{-Pb}}}{x_{\alpha\text{-Sn}} - x_{\alpha\text{-Pb}}}(1\ kg) = \dfrac{20-1}{100-1}(1\ kg)$

$= 0.192\ kg = 192\ g$

5.21 Calculate the amount of each phase present in 50 kg of a brass with composition 35 wt % Zn–65 wt % Cu at (a) 1000°C, (b) 900°C, (c) 800°C, (d) 700°C, (e) 100°C, and (f) 0°C (see Figure 5-37).

5.21

(a) $m_L = 1.0 \times 50\,kg = \underline{\underline{50\,kg}}$

(b) $m_\alpha = \frac{36.8-35}{36.8-32.5}(50\,kg) = \underline{\underline{20.9\,kg}}$

$m_\beta = \frac{35-32.5}{36.8-325}(50\,kg) = \underline{\underline{29.1\,kg}}$

(c) $x_\alpha \approx 34.1$ and $x_\beta \approx 39.6$ giving:

$m_\alpha = \frac{39.6-35}{39.6-34.1}(50\,kg) = \underline{\underline{41.8\,kg}}$

$m_\beta = \frac{35-34.1}{39.6-34.1}(50\,kg) = \underline{\underline{8.2\,kg}}$

(d) $m_\alpha = 1.0 \times 50\,kg = \underline{\underline{50\,kg}}$

(e) $x_\alpha \approx 32.1$ and $x_\beta \approx 46.5$ giving:

$m_\alpha = \frac{46.5-35}{46.5-32.1}(50\,kg) = \underline{\underline{39.9\,kg}}$

$m_\beta = \frac{35-32.1}{46.5-32.1}(50\,kg) = \underline{\underline{10.1\,kg}}$

(f) $x_\alpha \approx 29.2$ and $x_\beta \approx 46.5$ giving:

$m_\alpha = \frac{46.5-35}{46.5-29.2}(50\,kg) = \underline{\underline{33.2\,kg}}$

$m_\beta = \frac{35-29.2}{46.5-29.2}(50\,kg) = \underline{\underline{16.8\,kg}}$

5.22 Calculate the amount of each phase present in a 1-kg alumina refractory with composition 70 mol % Al_2O_3–30 mol % SiO_2 at (a) 2000°C, (b) 1900°C, and (c) 1800°C (see Figure 5-39).

5.22 (a) $m_L = 1.0 \times 1\,kg = \underline{\underline{1\,kg}}$

(b) In order to calculate the mass of each phase, it will be necessary to first convert the liquid composition to weight % using data from Appendix 1:

$x_L \approx 66$ mol.% Al_2O_3

$$\text{wt.\% } Al_2O_3 = \frac{66[2(26.98)+3(16.00)]}{66[2(26.98)+3(16.00)]+34[28.09+2(16.00)]} \times 100\%$$

$$= 76.7\%$$

We also need to know the overall wt.%:

$$x = 70 \text{ mol\%}$$

$$= \frac{70[2(26.98)+3(16.00)]}{70[2(26.98)+3(16.00)]+30[28.09+2(16.00)]} \times 100 \text{ wt.\%}$$

$$= 79.8\%$$

$$m_L = \frac{100-79.8}{100-76.7}(1\,kg) = 0.867\,kg = \underline{\underline{867\,g}}$$

$$m_{Al_2O_3} = \frac{79.8-76.7}{100-76.7}(1\,kg) = 0.133\,kg = \underline{\underline{133\,g}}$$

(c) The mullite composition is $\approx$ **66** mol.% Al_2O_3 or

$$x = \frac{66[2(26.98)+3(16.00)]}{66[2(26.98)+3(16.00)]+34[28.09+2(16.00)]} \times 100 \text{ wt.\%}$$

$$= 76.7\%$$

$$m_{mullite} = \frac{100-79.8}{100-76.7}(1\,kg) = 0.867\,kg = \underline{\underline{867\,g}}$$

$$m_{Al_2O_3} = \frac{79.8-76.7}{100-76.7}(1\,kg) = 0.133\,kg = \underline{\underline{133\,g}}$$

5.23 Some aluminum from a "metallization" layer on a solid-state electronic device has diffused into the silicon substrate. Near the surface, the silicon has an overall concentration of 1.0 wt % Al. In this region, what percentage of the microstructure would be composed of α-phase precipitates, assuming equilibrium? (See Figure 5-33 and assume the phase boundaries at 300°C will be essentially unchanged to room temperature.)

5.23

$$\text{wt.\% } \alpha \cong \frac{100-99}{100-0} \times 100\% = \underline{1\%}$$

5.24 In a test laboratory, quantitative x-ray diffraction determines that a refractory brick has 25 wt % alumina phase and 75 wt % mullite solid solution. What is the overall SiO_2 content (in wt %) of this material? (See Figure 5-39.)

5.24

Using a basis of 1 kg of the refractory,

$1 \text{ kg} \rightarrow 250 \text{ g } Al_2O_3 + 750 \text{ g mullite (ss)}$

The Al_2O_3 corresponds to:

$$\frac{250 \text{ g}}{(2 \times 26.98 + 3 \times 16.00) \text{ g/mol}} = 2.45 \text{ mol}$$

The mullite corresponds to:

$$\frac{750 \text{ g}}{0.64(2 \times 26.98 + 3 \times 16.00) \text{ g/mol} + 0.36(28.09 + 2 \times 16.00) \text{ g/mol}}$$

$$= 8.63 \text{ mol}$$

$\therefore$ mol. frac. Al_2O_3 in microstructure $= \dfrac{2.45 \text{ mol}}{(2.45 + 8.63) \text{ mol}}$

$= 0.221$

$\therefore$ overall composition is:

$64 \text{ mol\%} + 0.221(100 - 64) \text{ mol\%} = 72.0 \text{ mol\% } Al_2O_3$

or mol % $SiO_2 = 100 \text{ mol\%} - 72.0 \text{ mol\%} = 28.0 \text{ mol\%}$

For a basis of 1 mol $\rightarrow$

$0.280(28.09 + 32.00) \text{ g } SiO_2 + 0.720(2 \times 26.98 + 48.00) \text{ g } Al_2O_3$

$= 16.8 \text{ g } SiO_2 + 73.4 \text{ g } Al_2O_3$

$$\therefore \text{wt.\% } SiO_2 = \frac{16.8 \text{ g}}{16.8 \text{ g} + 73.4 \text{ g}} \times 100\% = \underline{18.6 \text{ wt.\% } SiO_2}$$

5.25 An important structural ceramic is partially stabilized zirconia (PSZ), which has a composition lying in the two-phase ZrO_2-cubic ZrO_2(ss) region. Use Figure 5-42 to calculate the amount of each phase present in a 10 mol % CaO PSZ at 500°C.

5.25

$$\text{mol. frac. cubic } ZrO_2(ss) = \frac{x - x_{mono\,ZrO_2}}{x_{cub\,ZrO_2} - x_{mono\,ZrO_2}} = \frac{10 - 2.1}{15.1 - 2.1} = \underline{\underline{0.608}}$$

$$\text{mol. frac. monoclinic } ZrO_2(ss) = \frac{x_{cub\,ZrO_2} - x}{x_{cub\,ZrO_2} - x_{mono\,ZrO_2}} = \frac{15.1 - 10}{15.1 - 2.1} = \underline{\underline{0.392}}$$

• 5.26 In a materials laboratory experiment, a student sketches a microstructure observed under an optical microscope. The sketch appears as

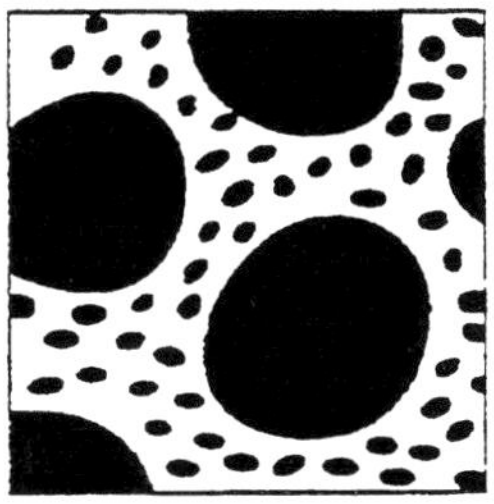

The phase diagram for this alloy system is

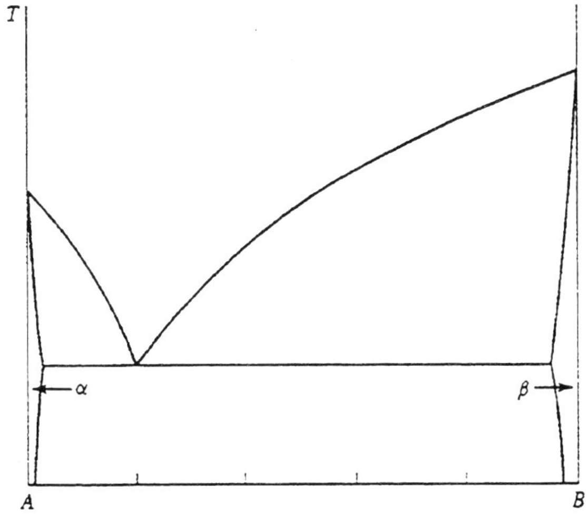

Determine (a) whether the black regions in the sketch represent α or β phase and (b) the approximate alloy composition.

5.26

(a) Because the eutectic region involves small black spots in a matrix of white <u>and</u> the phase diagram shows the eutectic is predominantly α, we conclude that <u>the black region is β.</u>

(b) Taking the overall areas of proeutectic and eutectic regions as approximately equal, the overall composition would be ≈ midway between the eutectic composition and the β phase solubility limit, i.e. <u>≈ 60% B</u>

Section 5.4 – Microstructural Development During Slow Cooling

PP 5.4 In Sample Problem 5.4, we calculate microstructural information about the β phase for the 70 wt % B alloy in Figure 5-23. In a similar way, calculate **(a)** the amount of α phase at T_3 for 1 kg of a 50 wt % B alloy and **(b)** the weight fraction of this α phase at T_3, which is proeutectic. (See also Figure 5-24.)

PP 5.4

(a) At T_3, $m_\alpha = \frac{90-50}{90-30}(1 kg) = 0.667 kg = \underline{\underline{667 g}}$

(b) At T_2, $m_\alpha = \frac{60-50}{60-30}(1 kg) = 0.333 kg = 333 g$

Fraction proeutectic $\alpha = \frac{333 g}{667 g} = \underline{\underline{0.50}}$

5.27 Calculate (**a**) the weight fraction of the α phase that is proeutectic in a 10 wt % Si–90 wt % Al alloy at 576°C and (**b**) the weight fraction of the β phase that is proeutectic in a 20 wt % Si–80 wt % Al alloy at 576°C (see Figure 5-33).

5.27

(a) At 576°C, $m_\alpha = \frac{100-10}{100-1.6}(m_{total}) = 0.915\, m_{total}$

At 578°C, $m_\alpha = \frac{12.6-10}{12.6-1.6}(m_{total}) = 0.236\, m_{total}$

Fraction proeutectic $\alpha = \frac{0.236\, m_{total}}{0.915\, m_{total}} = \underline{\underline{0.258}}$

(b) At 576°C, $m_\beta = \frac{20-1.6}{100-1.6}(m_{total}) = 0.187\, m_{total}$

At 578°C, $m_\beta = \frac{20-12.6}{100-12.6}(m_{total}) = 0.0847\, m_{total}$

Fraction proeutectic $\beta = \frac{0.0847\, m_{total}}{0.187\, m_{total}} = \underline{\underline{0.453}}$

5.28 Plot the weight percent of phases present as a function of temperature for a 10 wt % Si–90 wt % Al alloy slowly cooled from 700 to 300°C (see Figure 5-33).

5.28

Using results from Problem 5.27:

At 578°C, $\alpha = 23.6$ wt.% ($\therefore L = 100 - 23.6 = 76.4$ wt.%)
At 576°C, $\alpha = 91.5$ wt.% ($\therefore \beta = 100 - 91.5 = 8.5$ wt.%)

Also note that:

α precipitation begins just below 600°C
amount of β increases by a minor degree below 576°C

Allowing an overall plot:

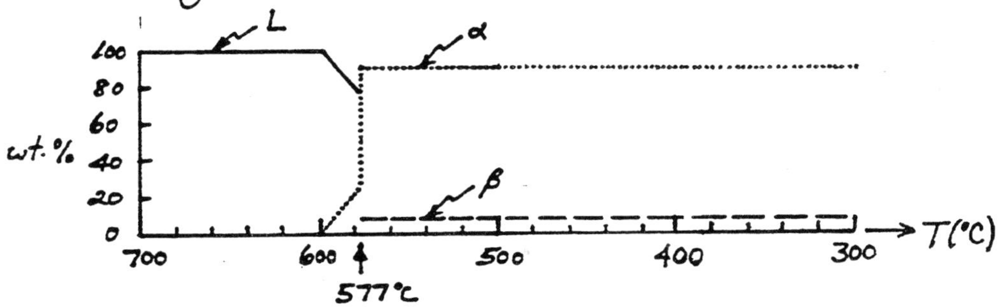

5.29 Plot the weight percent of phases present as a function of temperature for a 20 wt % Si–80 wt % Al alloy slowly cooled from 800 to 300°C (see Figure 5-33).

5.29

Using results from Problem 5.27:

At 578°C, $\beta = 8.47$ wt.% ($\therefore L = 100 - 8.47 = 91.53$ wt.%)
At 576°C, $\beta = 18.7$ wt.% ($\therefore \alpha = 100 - 18.7 = 81.3$ wt.%)

Also note that:

β precipitation begins at around 685°C
amount of β decreases by a minor degree below 576°C

Allowing an overall plot:

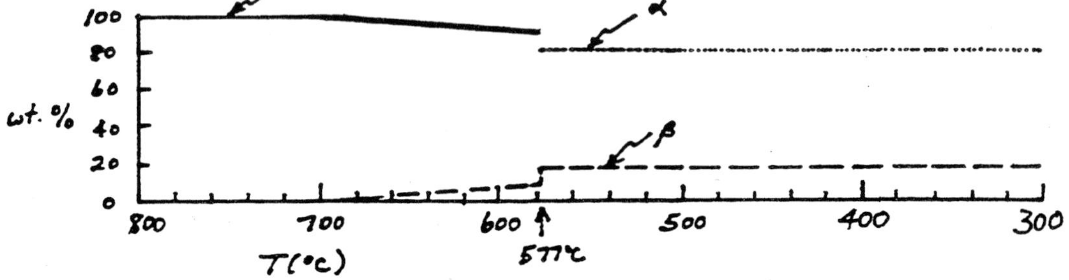

5.30 Calculate the *weight* fraction of mullite that is proeutectic in a slowly cooled 20 mol % Al_2O_3–80 mol % SiO_2 refractory cooled to room temperature (see Figure 5-39).

5.30

At 1588°C, frac mullite given by

$$\text{frac.}_{\text{mull.}} = \frac{x - x_{\text{eutectic}}}{x_{\text{mull}} - x_{\text{eutectic}}}$$

Converting mol. scale to wt. scale:

$$x = 20 \text{ mol.\% } Al_2O_3$$

or

$$= \frac{20 \text{ mol. } Al_2O_3}{20 \text{ mol. } Al_2O_3 + 80 \text{ mol. } SiO_2}$$

$$= \frac{20[2(26.98)+3(16.00)]}{20[2(26.98)+3(16.00)]+80[28.09+2(16.00)]}$$

$$= 0.298 \text{ (wt. fraction)}$$

Similarly,

$$x_{\text{eutectic}} = 4 \text{ mol.\% } Al_2O_3$$

or

$$= \frac{4[2(26.98)+3(16.00)]}{4[2(26.98+3(16.00)]+96[28.09+2(16.00)]} = 0.066$$

and

$$x_{\text{mull}} = 60 \text{ mol \% } Al_2O_3$$

or

$$= \frac{60[2(26.98)+3(16.00)]}{60[2(26.98)+3(16.00)]+40\cdot[28.09+2(16.00)]} = 0.718$$

∴ at 1588°C,

$$\text{wt. frac. mullite} = \frac{0.298-0.066}{0.718-0.066} = 0.356$$

Similarly, at 1586°C

$$\text{wt. frac. mullite} = \frac{0.298-0}{0.718-0} = 0.415$$

$$\therefore \text{Fraction proeutectic mullite} = \frac{0.356}{0.415} = \underline{\underline{0.858}}$$

5.31 Microstructural analysis of a slowly cooled Al–Si alloy indicates there is a 5 *volume* % silicon-rich proeutectic phase. Calculate the overall alloy composition (in weight percent) (see Figure 5-33).

5.31

To simplify, take the two phases as pure Al and Si at room temperature.

Consider a basis of 100 cc of alloy →

5 cc Si + 95 cc Al

Then,

$$m_{Si} = 5\,cc \times 2.33\,g/cc = 11.65\,g$$

and

$$m_{Al} = 95\,cc \times 2.70\,g/cc = 256.5\,g$$

$$\therefore \text{wt. \% Si} = \frac{11.65\,g}{(256.5 + 11.65)\,g} \times 100\% = \underline{\underline{4.34\,\%}}$$

5.32 Repeat Problem 5.31 for a 10 volume % silicon-rich proeutectic phase.

5.32

Again, simplifying by considering the two phases as pure Al and Si at room temperature:

100 cc alloy → 10 cc Si + 90 cc Al

Then,

$$m_{Si} = 10\,cc \times 2.33\,g/cc = 23.3\,g$$

and $m_{Al} = 90\,cc \times 2.70\,g/cc = 243\,g$

$$\therefore \text{wt. \% Si} = \frac{23.3\,g}{(243 + 23.3)\,g} \times 100\% = \underline{\underline{8.75\,\%}}$$

Section 5.5 – Some Important Binary Diagrams

PP 5.5 In Sample Problem 5.5, we found the amount of each phase in a eutectoid steel at room temperature. Repeat this calculation for the hypereutectoid steel (1.13 wt % C) illustrated in Figure 5-29.

PP 5.5

$$m_\alpha = \frac{6.69 - 1.13}{6.69 - 0}(1\,kg) = 0.831\,kg = \underline{\underline{831\,g}}$$

$$m_{Fe_3C} = \frac{1.13 - 0}{6.69 - 0}(1\,kg) = 0.169\,kg = \underline{\underline{169\,g}}$$

PP 5.6 Calculate the amount of proeutectoid cementite at the grain boundaries in 1 kg of the 1.13 wt % C hypereutectoid steel illustrated in Figure 5-29. (See Sample Problem 5.6.)

PP 5.6

In effect, we need to calculate the equilibrium amount of cementite at 728°C.

$$m_{Fe_3C} = \frac{1.13 - 0.77}{6.69 - 0.77}(1\,kg) = 0.0608\,kg = \underline{\underline{60.8\,g}}$$

PP 5.7 In Sample Problem 5.7, the amount of carbon in 1 kg of a 3 wt % C gray iron is calculated at two temperatures. Plot the amount as a function of temperature over the entire temperature range of 1135°C to room temperature.

PP 5.7

From Sample Problem 5.7,

at 1153°C, $m_C = 9.40\,g$

at room temperature, $m_C = 30.0\,g$ (this amount applies up to essentially 737°C)

At 739°C,

$$m_C = \frac{3.00 - 0.68}{100 - 0.68}(1\,kg) = 23.4\,g$$

The resulting plot is:

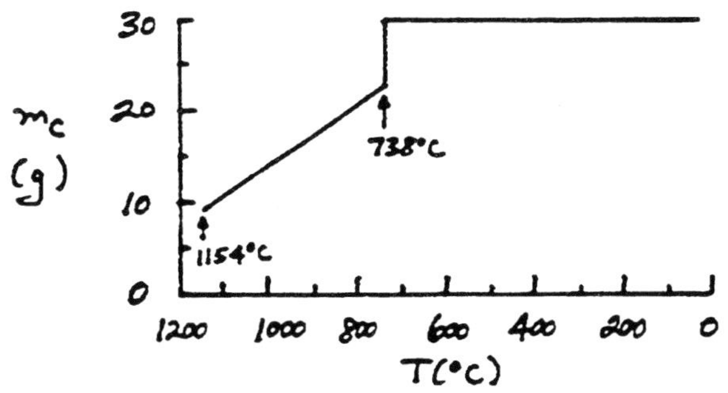

PP 5.8 In Sample Problem 5.8, we monitor the microstructural development for 1 kg of a 10 wt % Si–90 wt % Al alloy. Repeat this problem for a 20 wt % Si–80 wt % Al alloy.

PP 5.8

(a) $\approx 680°C$

(b) Solid solution β with a composition of ≈ 100 wt. % Si

(c) At the eutectic temperature, 577°C

(d) At 578°C,

$$m_\beta = \frac{20-12.6}{100-12.6}(1\,kg) = 0.0847\,kg = 84.7\,g$$

(e) At 576°C,

$$m_\alpha = \frac{100-20}{100-1.6}(1\,kg) = 0.813\,kg = 813\,g$$

$$m_\beta = 1000\,g - 813\,g = 187\,g$$

From (d), $m_{\text{proeutectic}\,\beta} = 85\,g$

Therefore, $m_{\text{eutectic}\,\beta} = m_{\beta,\text{total}} - m_{\text{proeutectic}\,\beta}$

$$= 187\,g - 85\,g = 102\,g$$

Finally,

Si in eutectic $\alpha = (0.0160)(813\,g) = 13.0\,g$

Si in eutectic $\beta = (1.00)(102\,g) = 102\,g$

Si in proeutectic $\beta = (1.00)(85\,g) = 85\,g$

PP 5.9

In Sample Problem 5.9, we calculate the weight percent of θ phase at room temperature in a 95.5 Al–4.5 Cu alloy. Plot the weight percent of θ (as a function of temperature) that would occur upon slow cooling over a temperature range of 548°C to room temperature.

PP 5.9

Sample Problem 5.9 gives the result near room temperature:

≈ 8.49 wt.% θ.

Cooling from 548°C, the first θ precipitation will occur at $\approx$ ~~480~~ °C

At 500°C,

wt.% θ = 0 % by definition

At 400°C,

$$\text{wt.\% } \theta = \frac{4.5-2.5}{53-2.5} \times 100\% = 4.0\%$$

At 300°C,

$$\text{wt.\% } \theta = \frac{4.5-1.0}{53-1.0} \times 100\% = 6.7\%$$

Resulting plot:

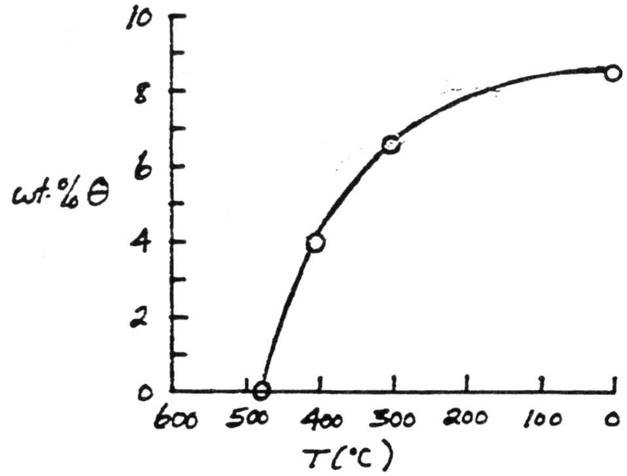

PP 5.10 Calculate microstructures for (a) a 40:60 Pb-Sn solder and (b) a 60:40 Pb-Sn solder at 200°C and 100°C. (See Sample Problem 5.10.)

PP 5.10 (a) For 200°C,

(i) liquid only

(ii) L is 60 wt.% Sn

(iii) wt.% L = 100%

For 100°C,

(i) α and β

(ii) α is ≈ 5 wt.% Sn and β is ≈ 99 wt.% Sn

(iii) $\text{wt.\% } \alpha = \frac{99-60}{99-5} \times 100\% = 41.5\%$

$\text{wt.\% } \beta = \frac{60-5}{99-5} \times 100\% = 58.5\%$

(b) For 200°C,

(i) α and liquid

(ii) α is ≈ 18 wt.% Sn and L is ≈ 54 wt.% Sn

(iii) $\text{wt.\% } \alpha = \frac{54-40}{54-18} \times 100\% = 38.9\%$

$\text{wt.\% } L = \frac{40-18}{54-18} \times 100\% = 61.1\%$

For 100°C,

(i) α and β

(ii) α is ≈ 5 wt.% Sn and β is ≈ 99 wt.% Sn

(iii) $\text{wt.\% } \alpha = \frac{99-40}{99-5} \times 100\% = 62.8\%$

$\text{wt.\% } \beta = \frac{40-5}{99-5} \times 100\% = 37.2\%$

PP 5.11 In the note at the end of Sample Problem 5.11, the point is made that the results can be easily converted to weight percent. Make these conversions.

PP 5.11 The results were: (i) SiO_2 + mullite

(ii) SiO_2 : 0% Al_2O_3

mullite: 60 mol.% Al_2O_3

(iii) 44.5 mol % SiO_2 + 55.5 mol.% mullite

For (i), there is no need for conversion.

For (ii), 0 mol.% Al_2O_3 = $\underline{0 \text{ wt.\% } Al_2O_3}$

On the basis of 100 moles of Al_2O_3 + SiO_2 :

$$m_{60\,mol.\,Al_2O_3} = 60\,[2(26.98) + 3(16.00)]\,\text{amu} = 6118\ \text{amu}$$

$$m_{40\,mol.\,SiO_2} = 40\,[(28.09) + 2(16.00)]\,\text{amu} = 2404\ \text{amu}$$

$$\underline{\text{wt.\% } Al_2O_3} = \frac{6118\ \text{amu}}{6118\ \text{amu} + 2404\ \text{amu}} \times 100\% = \underline{71.8\%}$$

For (iii), on the basis of 100 moles of SiO_2 + mullite:

$$m_{44.5\,moles\,SiO_2} = 44.5[(28.09) + 2(16.00)]\,\text{amu} = 2674\ \text{amu}$$

$$m_{55.5\,moles\,mullite} = 55.5 \times \frac{1}{5}\,[3(2)(26.98) + 3(3)(16.00) + 2(28.09) + 2(2)(16.00)]\,\text{amu} = 4729\ \text{amu} \quad *$$

$$\underline{\text{wt.\% } SiO_2} = \frac{2674\ \text{amu}}{(2674 + 4729)\,\text{amu}} \times 100\% = \underline{36.1\%}$$

$$\underline{\text{wt.\% mullite}} = \frac{4729\ \text{amu}}{(2674 + 4729)\,\text{amu}} \times 100\% = \underline{63.9\%}$$

* Note that the mol.wt. of mullite is normalized by a factor of 1/5 because 1 mullite formula consists of 5 moles of the components ($3Al_2O_3 + 2SiO_2$). The composition axis is normalized to $\Sigma(Al_2O_3 + SiO_2) = 1$ mole.

PP 5.12 In Sample Problem 5.12, the phase distribution in a partially stabilized zirconia is calculated. Repeat this calculation for a zirconia with 5 wt % CaO.

PP 5.12

Noting that 5 wt.% CaO ≈ 10 mol.% CaO:

$$\text{mol.\% monoclinic} = \frac{15-10}{15-2} \times 100\% =$$

$$= \underline{\underline{38.5 \text{ mol.\%}}}$$

$$\text{mol.\% cubic} = \frac{10-2}{15-2} \times 100\% =$$

$$= \underline{\underline{61.5 \text{ mol.\%}}}$$

5.33 Calculate the amount of proeutectic γ that has formed at 1149°C in the slow cooling of the 3.0 wt % C white cast iron illustrated in Figure 5-27. Assume a total of 100 kg of cast iron.

5.33

$$m_\gamma = \frac{4.30-3.00}{4.30-2.11}(100\,kg) = \underline{\underline{59.4\,kg}}$$

5.34 Plot the weight percent of phases present as a function of temperature for the 3.0 wt % C white cast iron illustrated in Figure 5-27 slowly cooled from 1400 to 0°C.

5.34

From Problem 5.33,

$\gamma = 59.4$ wt.% ($\xi\ L = 100 - 59.4 = 40.6$ wt.%) at 1149°C

At 1147°C,

$$\%\gamma = \frac{6.69-3.0}{6.69-2.11} \times 100\% = 80.6\text{ wt.\%}$$

and $\% Fe_3C = 100 - 80.6 = 19.4$ wt.%

Also note that γ precipitation begins just above 1300°C.

At 728°C,

$$\%\gamma = \frac{6.69-3.0}{6.69-0.77} \times 100\% = 62.3\text{ wt.\%}$$

and $\% Fe_3C = 100 - 62.3 = 37.7$ wt.%

At 726 °C,

$$\% \alpha = \frac{6.69 - 3.0}{6.69 - 0.02} \times 100\% = 55.3 \text{ wt.\%}$$

and $\% Fe_3C = 100 - 55.3 = 44.7$ wt.%
(with these percentages essentially unchanged upon cooling).

The resulting plot is:

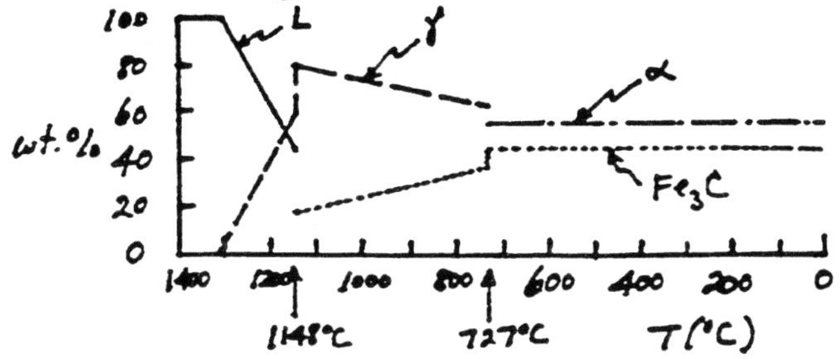

5.35 Plot the weight percent of phases present as a function of temperature from 1000 to 0°C for the 0.77 wt % C eutectoid steel illustrated in Figure 5-28.

5.35 From Sample Problem 5.5, at room temp.:

$\alpha = 88.5$ wt.% & $Fe_3C = 11.5$ wt.%
(This distribution is effectively unchanged from 726°C to 0°C.

By definition, there is 100% γ from 1000°C down to 728°C, allowing an overall plot:

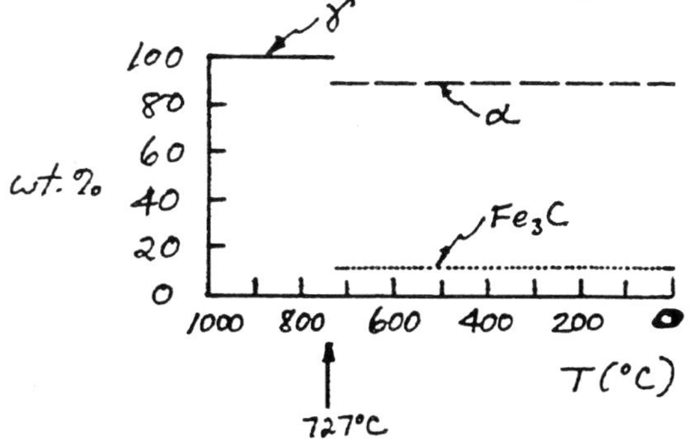

5.36 Plot the weight percent of phases present as a function of temperature from 1000 to 0°C for the 1.13 wt % C hypereutectoid steel illustrated in Figure 5-29.

5.36 From PP 5.6,

$Fe_3C = 6.1$ wt.% ($\&\ \gamma = 100 - 6.1 = 93.9$ wt.%) at 728°C

Also note that Fe_3C precipitation begins at $\approx 890°C$.

At 726°C,

$$\% \, Fe_3C = \frac{1.13 - 0.02}{6.69 - 0.02} \times 100\% = 16.6 \text{ wt.\%}$$

meaning that there will be $100 - 16.6 = 83.4$ wt.% α (and these percentages will not change significantly upon cooling).

Allowing an overall plot

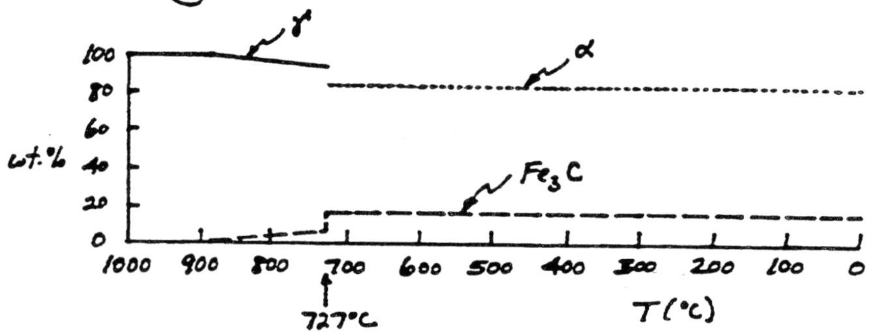

5.37 Calculate the amount of proeutectoid α present at the grain boundaries in 1 kg of a common 1020 structural steel (0.20 wt % C).

5.37 In effect, we need to calculate the equilibrium amount of ferrite at 728°C.

$$m_\alpha = \frac{0.77 - 0.20}{0.77 - 0.02}(1 \text{ kg}) = 0.76 \text{ kg}$$

$$= \underline{\underline{760 \text{ g}}}$$

5.38 Repeat Problem 5.37 for a 1040 structural steel (0.40 wt % C).

5.38

Again, calculating the amount of ferrite at 728°C:

$$m_\alpha = \frac{0.77-0.40}{0.77-0.02}(1\ kg) = 0.49\ kg$$

$$= \underline{\underline{490\ g}}$$

5.39 Plot the weight percent of phases present as a function of temperature from 1000 to 0°C for a common 1020 structural steel (0.20 wt % C).

5.39

Key temperatures and corresponding compositions include:

1000°C : 100 % γ

820°C : initial ppt. of α

728°C : $\%\ \gamma = \frac{0.20-0.02}{0.77-0.02} \times 100\ \% = 24\%$

$\%\ \alpha = 100\% - 24\% = 76\%$

726°C : $\%\ Fe_3C = \frac{0.20-0.02}{6.69-0.02} \times 100\ \% = 2.7\%$

$\%\ \alpha = 100\ \% - 2.7\% = 97.3\%$

Below 680°C : $\%\ Fe_3C = \frac{0.20-0}{6.69-0} \times 100\ \% = 3.0\%$

$\%\ \alpha = 100\% - 3.0\% = 97.0\%$

Giving an overall plot

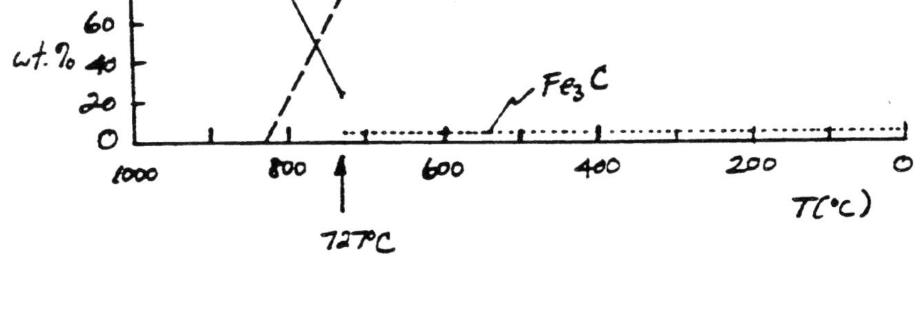

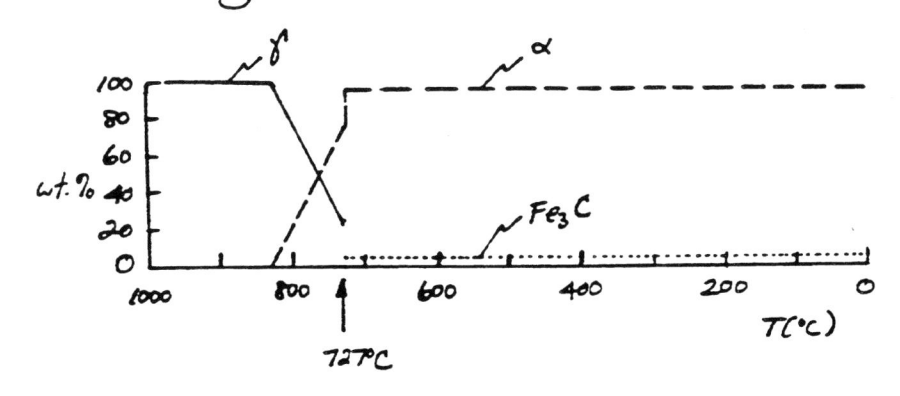

5.40 Repeat Problem 5.39 for a 1040 structural steel (0.40 wt % C).

5.40 Key temperatures and corresponding compositions include:

1000°C: 100% γ

780°C: initial ppt. of α

728°C: $\% \gamma = \frac{0.40 - 0.02}{0.77 - 0.02} \times 100\% = 51\%$

$\% \alpha = 100\% - 51\% = 49\%$

726°C: $\% Fe_3C = \frac{0.40 - 0.02}{6.69 - 0.02} \times 100\% = 5.7\%$

$\% \alpha = 100\% - 5.7\% = 94.3\%$

Below 680°C: $\% Fe_3C = \frac{0.40 - 0}{6.69 - 0} \times 100\% = 6.0\%$

$\% \alpha = 100\% - 6.0\% = 94.0\%$

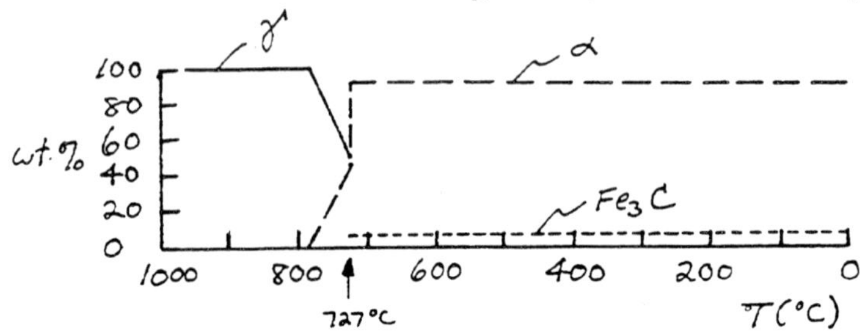

5.41 Plot the weight percent of phases present as a function of temperature from 1000 to 0°C for the 0.50 wt % C hypoeutectoid steel illustrated in Figure 5-30.

5.41 From Sample Problem 5.6, at 728°C:

$\alpha = 36.0$ wt.% ($\&\ \gamma = 100 - 36.0 = 64.0$ wt.%)

Also note that α precipitation begins at $\approx 760°C$.

At 726°C,

$$\%\alpha = \frac{6.69 - 0.50}{6.69 - 0.02} \times 100\% = 92.8\%$$

and, then, $\% Fe_3C = 100 - 92.8 = 7.2\%$
with these values not changing significantly upon cooling.

The resulting plot is:

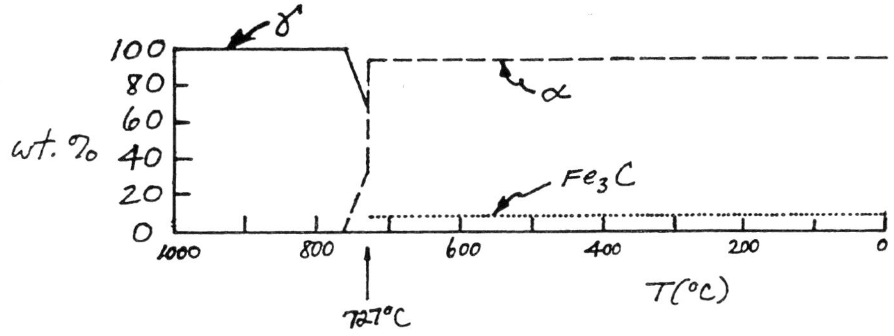

5.42 Plot the weight percent of phases present as a function of temperature from 1400 to 0°C for a white cast iron with an overall composition of 2.5 wt % C.

5.42 Key temperatures and compositions include:

1400–1360°C: 100% L

1149°C : $\frac{2.5-2.11}{4.30-2.11} \times 100\% = 18\%\,L$ ($\therefore$ 100 − 18 = 82% γ)

1147°C : $\frac{2.5-2.11}{6.69-2.11} \times 100\% = 8.5\%\,Fe_3C$ ($\therefore$ 100 − 8.5 = 91.5% γ)

728°C : $\frac{2.5-0.77}{6.69-0.77} \times 100\% = 29\%\,Fe_3C$ ($\therefore$ 100 − 29 = 71% γ)

726°C : $\frac{2.5-0.02}{6.69-0.02} \times 100\% = 37\%\,Fe_3C$ ($\therefore$ 100 − 37 = 63% α)

Giving a plot:

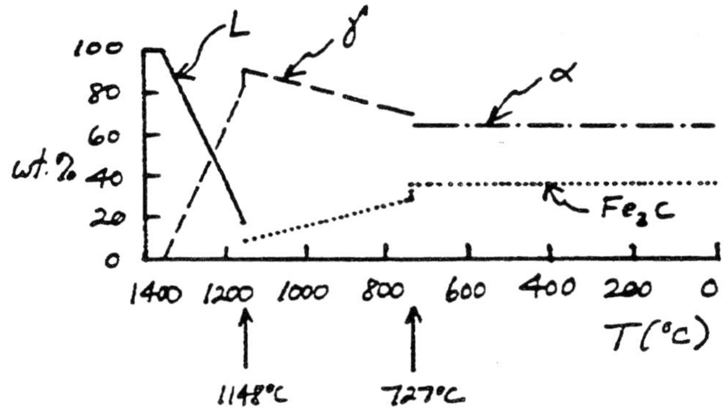

5.43 Plot the weight percent of all phases present as a function of temperature from 1400 to 0°C for a gray cast iron with an overall composition of 3.0 wt % C.

5.43

Key temperatures and compositions include:

1400–1300°C: 100% L

1155°C: $\frac{3.0-2.08}{4.26-2.08} \times 100\% = 42\%\ L$ ($\therefore$ 100 − 42 = 58% γ)

1153°C: $\frac{3.0-2.08}{100-2.08} \times 100\% = 0.9\%\ C$ ($\therefore$ 100 − 0.9 = 99.1% γ)

739°C: $\frac{3.0-0.68}{100-0.68} \times 100\% = 2.3\%\ C$ ($\therefore$ 100 − 2.3 = 97.7% γ)

737°C: $\frac{3.0-0.02}{100-0.02} \times 100\% = 3.0\%\ C$ ($\therefore$ 100 − 3 = 97% α)

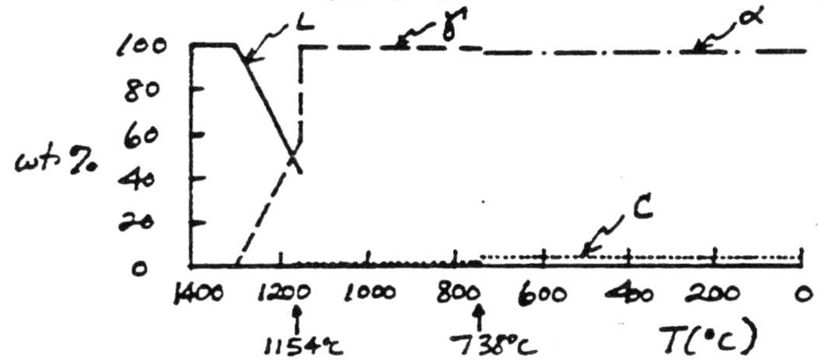

5.44 Repeat Problem 5.43 for a gray cast iron with an overall composition of 2.5 wt % C.

5.44

Key temperatures and compositions include:

1400–1350°C: 100% L

1155°C: $\frac{2.5-2.08}{4.26-2.08} \times 100\% = 19\%\ L$ ($\therefore$ 100 − 19 = 81% γ)

1153°C: $\frac{2.5-2.08}{100-2.08} \times 100\% = 0.4\%\ C$ ($\therefore$ 100 − 0.4 = 99.6% γ)

739°C: $\frac{2.5-0.68}{100-0.68} \times 100\% = 1.8\%\ C$ ($\therefore$ 100 − 1.8 = 98.2% γ)

737°C: $\frac{2.5-0.02}{100-0.02} \times 100\% = 2.5\%$ ($\therefore$ 100 − 2.5 = 97.5% α)

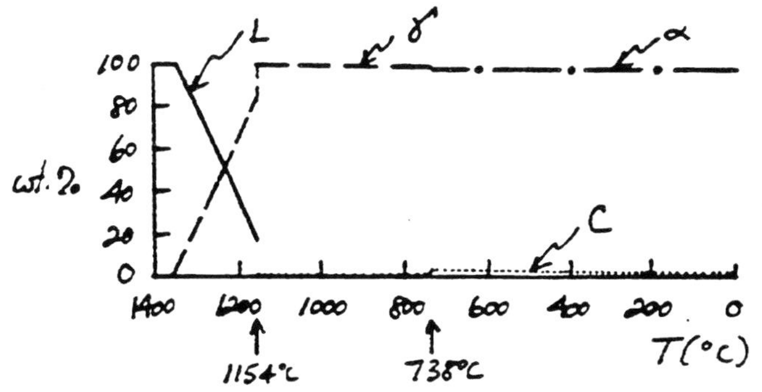

5.45 In comparing the equilibrium schematic microstructure in Figure 5-32 with the actual, room temperature microstructure shown in Figure 7-1b, it is apparent that metastable pearlite can form at the eutectoid temperature (due to insufficient time for the more stable, but slower, graphite formation). Assuming that Figures 5-31 and 5-32 are accurate for 100 kg of a gray cast iron (3.0 wt % C) down to 738°C but that pearlite forms upon cooling through the eutectoid temperature, calculate the amount of pearlite to be expected in the room temperature microstructure.

5.45 The amount of pearlite is equal to the amount of γ at 739°C:

$$m_\gamma = \frac{100-3.0}{100-0.68}(100\,kg) = \underline{\underline{97.7\,kg}}$$

Note: This is an overestimate as not all γ transforms.

5.46 For the assumptions in Problem 5.45, calculate the amount of flake graphite in the room temperature microstructure.

5.46 The flake graphite would be the C present at 739°C:

$$m_C = \frac{3.0-0.68}{100-0.68}(100\,kg) = \underline{\underline{2.3\,kg}}$$

5.47 Plot the weight percent of phases present as a function of temperature from 800 to 300°C for a 95 Al–5 Cu alloy.

5.47 Key temperatures and compositions include:

800 – 640°C: 100 % L

640 – 570°C: $\kappa + L$

570 – 500°C: 100 % κ

at 500°C: $\% \theta = 0\%$ by definition

$\% \kappa = 100\%$ by definition

at 400°C: $\% \theta = \frac{5-2.5}{53-2.5} \times 100\% = 5.0\%$

$\% \kappa = 100\% - 5.0\% = 95.0\%$

at 300°C: $\% \theta = \frac{5-1}{53-1} \times 100\% = 7.7\%$

$\% \kappa = 100\% - 7.7\% = 92.3\%$

Giving the resulting plot:

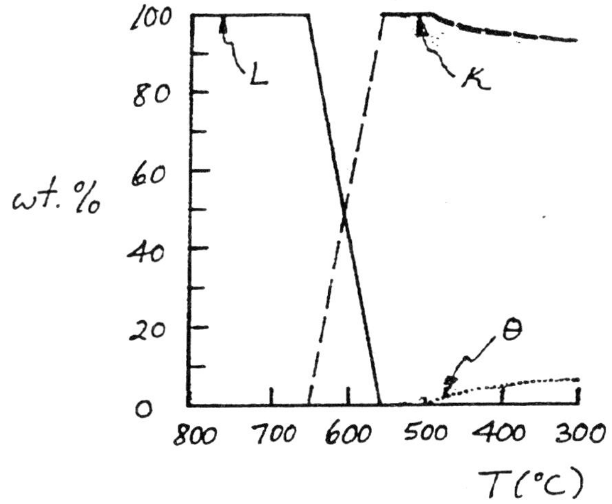

5.48 Consider 1 kg of a brass with composition 35 wt % Zn–65 wt % Cu (see Figure 5-37). (a) Upon cooling, at what temperature would the first solid appear? (b) What is the first solid phase to appear, and what is its composition? (c) At what temperature will the alloy completely solidify? (d) Over what temperature range will the microstructure be completely in the α-phase?

5.48

(a) ≈ 910°C

(b) α, ≈ 30 wt.% Zn

(c) 903°C

(d) From ≈ 750°C to ≈ 220°C.

5.49 Repeat Problem 5.48 for 1 kg of a brass with composition 30 wt % Zn–70 wt % Cu.

5.49

(a) ≈ 950°C

(b) Solid solution α with a composition of ≈ 26 wt.% Zn

(c) ≈ 920°C

(d) From ≈ 920°C to ≈ 40°C

5.50 Plot the weight percent of phases present as a function of temperature from 1000 to 0°C for a 35 wt % Zn–65 wt % Cu brass.

5.50

Key temperatures and compositions include:

1000–910°C: 100% L

904°C: $\frac{35-32.5}{37.5-32.5} \times 100\% = 50\%\ L$ (& 100−50 = 50% α)

902°C: $\frac{35-32.5}{36.8-32.5} \times 100\% = 58\%\ \beta$ (& 100−58 = 42% α)

750–220°C: 100% α

0°C: $\frac{35-29}{47-29} \times 100\% = 33\%\ \beta'$ (& 100−33 = 67% α)

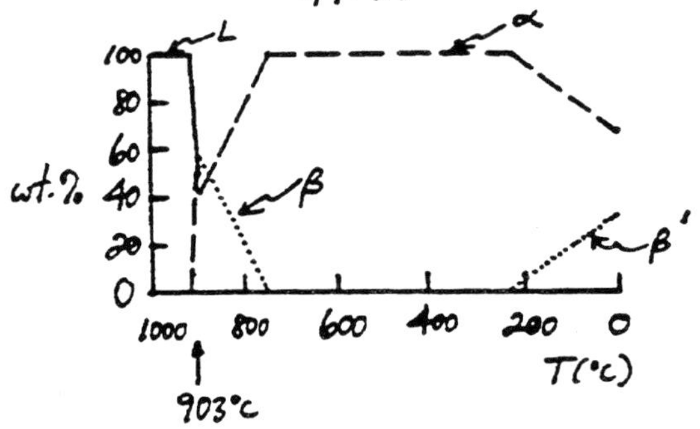

5.51 Repeat Problem 5.50 for a 30 wt % Zn–70 wt % Cu brass.

5.51

Key temperatures and compositions include:

1000–950°C: 100% L

920–40°C: 100% α

@ 0°C: $\frac{30-29}{47-29} \times 100\% = 6\%\ \beta'$

(& 100−6 = 94% α)

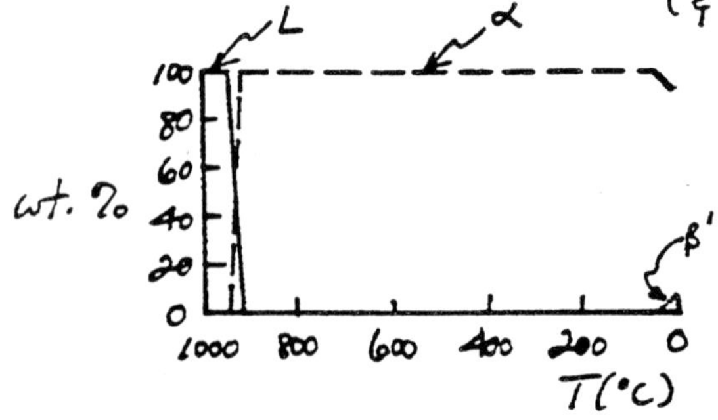

5.52 Repeat Problem 5.50 for 1 kg of brass with a composition of 15 wt % Zn–85 wt % Cu.

5.52 (a) ≈ 1030°C

(b) α, ≈ 12 wt. % Zn

(c) ≈ 1012°C

(d) from ≈ 1012°C to below 0°C

5.53 For a 15 wt % Zn–85 wt % Cu brass, plot the weight percent of phases present as a function of temperature from 1100°C to 0°C.

5.53 Key temperatures and compositions include:

1100 – 1030°C : 100% L

1012 – 0°C : 100% α

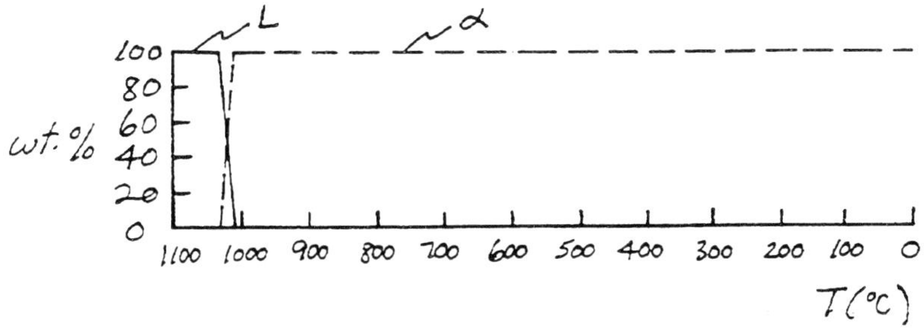

5.54 Calculate the amount of β phase that would precipitate from 1 kg of 95 wt % Al–5 wt % Mg alloy slowly cooled to 100°C.

5.54

$$\text{amt. } \beta = \frac{x - x_\alpha}{x_\beta - x_\alpha} \times 1\,kg$$

$$= \frac{5 - 2}{35.5 - 2} \times 1000\,g = \underline{\underline{89.5\,g}}$$

5.55 Identify the composition ranges in the Al–Mg system (Figure 5-35) for which precipitation of the type illustrated in Sample Problem 5.9 can occur (i.e., a second phase can precipitate from a single-phase microstructure upon cooling).

5.55

≈ 1 to 17.1 wt.% Mg, (extrapolating α range to room temp.)

≈ 42 to ≈ 57 wt.% Mg,

≈ 57 to 59.8 wt.% Mg,

87.4 to ≈ 99 wt.% Mg (extrapolating β range to room temp.)

5.56 is on page 173 (next page)

5.57 is on page 174 (page after next)

5.58 A solder batch is made by melting together 64 g of a 40:60 Pb–Sn alloy with 53 g of a 60:40 Pb–Sn alloy. Calculate the amounts of α and β phase that would be present in the overall alloy, assuming it is slowly cooled to room temperature, 25°C.

5.58

Total mass Pb: $0.4 \times 64\,g + 0.6 \times 53\,g = 57.4\,g$

" " Sn: $0.6 \times 64\,g + 0.4 \times 53\,g = 59.6\,g$

$$\therefore \text{wt.\% Sn} = \frac{59.6\,g}{59.6\,g + 57.4\,g} \times 100\% = 50.9\%$$

$$\therefore \text{amount } \beta = \frac{50.9 - 1.5}{99.8 - 1.5} m_{total} = 0.503\, m_{total}$$

$$\text{or } \alpha = (1.000 - 0.503)\, m_{total} = 0.497\, m_{total}$$

giving:

$$\alpha = 0.497\,(64 + 53)\,g = \underline{\underline{58.2\,g\ \alpha}}$$

$$\& \ \beta = 0.503\,(64 + 53)\,g = \underline{\underline{58.8\,g\ \beta}}$$

5.56 Plot the weight percent of phases present as a function of temperature from 700 to 100°C for a 90 Al–10 Mg alloy.

5.56 Key temperatures and compositions include:

700 – 610°C: 100% L

610 – 520°C: α + L

520 – 360°C: 100% α

at 300°C:

$$\% \alpha = \frac{36.1 - 10}{36.1 - 7} \times 100\% = 89.7\%$$

$$\% \beta = 100 - 89.7 = 10.3\%$$

at 200°C:

$$\% \alpha = \frac{36.1 - 10}{36.1 - 3} \times 100\% = 78.9\%$$

$$\% \beta = 100 - 78.9 = 21.1\%$$

at 100°C:

$$\% \alpha = \frac{36.1 - 10}{36.1 - 2} \times 100\% = 76.5\%$$

$$\% \beta = 100 - 76.5 = 23.5\%$$

Giving the resulting plot:

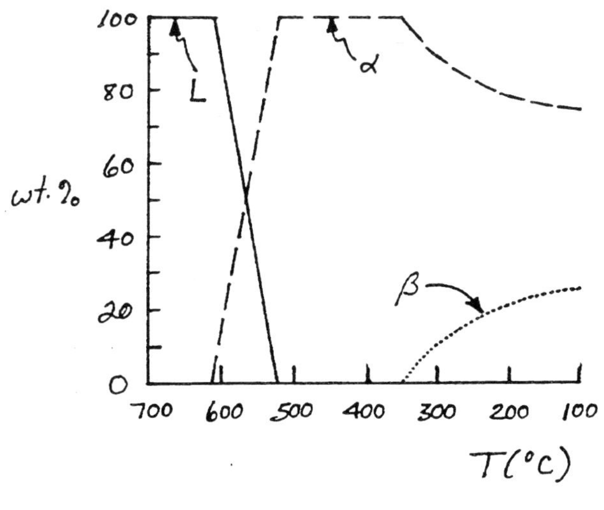

5.57 The ideal stoichiometry of the γ phase in the Al–Mg system is $Al_{12}Mg_{17}$. (a) What is the atomic percentage of excess Al in the most aluminum-rich γ composition at 450°C? (b) What is the atomic percentage of excess Mg in the most magnesium-rich γ composition at 437°C?

5.57

The ideal stoichiometry is:

$$\text{at. \% Al} = \frac{12}{12+17} \times 100\% = 41.4\%$$

$$\text{at. \% Mg} = \frac{17}{12+17} \times 100\% = 58.6\%$$

(a) For γ at 450°C: 42 wt. % Mg:

100 g alloy → 42 g Mg + 58 g Al

$$= \frac{42\text{ g Mg}}{24.31\text{ g Mg}}\text{ g·atom Mg} + \frac{58\text{ g Al}}{26.98\text{ g Al}}\text{ g·atom Al}$$

$$= 1.728\text{ g·atom Mg} + 2.150\text{ g·atom Al}$$

$$\rightarrow \frac{2.150}{2.150+1.728} \times 100\% = 55.4\text{ at. \% Al}$$

$$\text{\% excess} = \frac{55.4 - 41.4}{41.4} \times 100\% = \underline{\underline{33.8\%}}$$

(b) For γ at 437°C: 59.8 wt. % Mg

100 g alloy → 59.8 g Mg + 40.2 g Al

$$= \frac{59.8\text{ g Mg}}{24.31\text{ g Mg}}\text{ g·atom Mg} + \frac{40.2\text{ g Al}}{26.98\text{ g Al}}\text{ g·atom Al}$$

$$= 2.460\text{ g·atom Mg} + 1.490\text{ g·atom Al}$$

$$\rightarrow \frac{2.460}{2.460+1.490} \times 100\% = 62.3\text{ at. \% Mg}$$

$$\text{\% excess} = \frac{62.3 - 58.6}{58.6} \times 100\% = \underline{\underline{6.31\%}}$$

5.58 is on page 172 (two pages back)

5.59 Plot the weight percent of phases present as a function of temperature from 400 to 0°C for a slowly cooled 50:50 Pb–Sn solder.

5.59

Key temperatures and compositions include:

400 – 210°C : 100% L

At 184°C:

$$\% \alpha = \frac{61.9-50}{61.9-19} \times 100\% = 27.7\%$$

$$\% L = 100 - 27.7 = 72.3\%$$

At 182°C:

$$\% \alpha = \frac{97.5-50}{97.5-19} \times 100\% = 60.5\%$$

$$\% \beta = 100 - 60.5 = 39.5\%$$

At 100°C (from Sample Problem 5.10)

52.1% α and 47.9% β

At 0°C (from Problem 5.18)

50.5% α and 49.5% β

Giving a plot:

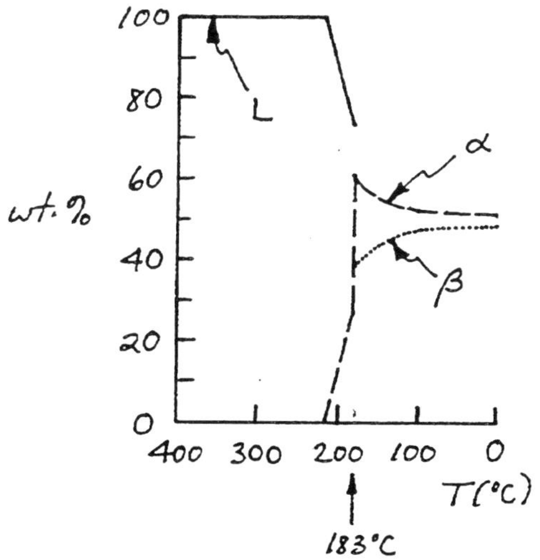

• 5.60 Suppose that you have a crucible containing 1 kg of an alloy of composition 90 wt % Sn–10 wt % Pb at a temperature of 184°C. How much Sn would you have to add to the crucible to completely solidify the alloy *without* changing the system temperature?

5.60

1 kg alloy → 900 g Sn + 100 g Pb

Complete solidification occurs (at 184°C) at 97.5% Sn

or $$\frac{900 + x}{1000 + x} = 0.975$$

giving $x = 3000$ g = 3 kg

5.61 Determine the phases present, their compositions, and their amounts (below the eutectic temperature) for a refractory made from equal molar fractions of kaolinite and mullite ($3Al_2O_3 \cdot 2SiO_2$).

5.61

1 mol. kaolinite → 1 mol. Al_2O_3 + 2 mol. SiO_2

1 mol. mullite → 3 mol. Al_2O_3 + 2 mol. SiO_2

4 mol. Al_2O_3 4 mol. SiO_2

The result is an overall composition of 50 mol.% Al_2O_3 which gives:

$$\text{mol. \% } SiO_2 = \frac{60-50}{60-0} \times 100\% = 16.7\%$$

$$\text{mol. \% mullite} = \frac{50-0}{60-0} \times 100\% = 83.3\%$$

5.62 Repeat Problem 5.61 for a refractory made from equal molar fractions of kaolinite and silica (SiO_2).

5.62

1 mol. kaolinite → 1 mol. Al_2O_3 + 2 mol. SiO_2

1 mol. SiO_2 →   1 mol. SiO_2

1 mol. Al_2O_3 3 mol. SiO_2

The overall composition is $\frac{1}{1+3} \times 100\% = 25$ mol. % Al_2O_3

which gives:

$$\text{mol. \% } SiO_2 = \frac{60-25}{60-0} \times 100\% = 58.3\%$$

$$\text{mol. \% mullite} = \frac{25-0}{60-0} \times 100\% = 41.7\%$$

• **5.63** Given that you have supplies of kaolinite, silica, and mullite as raw materials, calculate a batch of composition (in weight percent) using kaolinite plus *either* silica *or* mullite necessary to produce a final microstructure that is equimolar in silica and mullite.

5.63

By inspection of Figure 5-39, it is apparent that silica rather than mullite must be added to kaolinite (because kaolinite is $33\frac{1}{3}$ mol. % Al_2O_3 which is more than ½ the composition of mullite ss). Again, inspecting Figure 5-39 indicates that the desired, overall composition should be midway between pure SiO_2 (0% Al_2O_3) and mullite ss (60 mol % Al_2O_3) = 30 mol % Al_2O_3

Consider basis of 100 mole final product (30 mol % Al_2O_3)

$$= 30\text{ mol. } Al_2O_3 + 70\text{ mol. } SiO_2$$

All Al_2O_3 comes from the kaolinite. Specifically the 30 mol Al_2O_3 would come from 30 mol of kaolinite ($Al_2O_3 \cdot 2SiO_2 \cdot 2H_2O$), which obviously yields 30 mol SiO_2. To obtain an overall composition of 30 mol % Al_2O_3 in the final product, then, we must add x mol SiO_2:

$$\frac{30\text{ mol. } Al_2O_3}{30\text{ mol } Al_2O_3 + 60\text{ mol } SiO_2 + x\text{ mol. } SiO_2} = 0.30$$

or $x = 10$.

∴ overall batch is 10 mol. SiO_2 + 30 mol. kaolinite:

$$m_{SiO_2} = 10[28.09 + 2(16.00)]\text{ g} = 0.60\text{ kg}$$

$$m_{kaolinite} = 30[2(26.98) + 2(28.09) + 9(16.00) + 4(1.008)]\text{ g}$$

$$= 7.75\text{ kg}$$

$$\therefore \text{wt. \% kaolinite} = \frac{7.75\text{ kg}}{7.75 + 0.60\text{ kg}} \times 100\% = \underline{\underline{92.8\%}}$$

$$\& \text{ wt. \% } SiO_2 = 100\% - 92.8\% = \underline{\underline{7.2\%}}$$

5.64 Calculate the phases present, their compositions, and their amounts (in weight percent) for the microstructure at 1000°C for (a) a spinel ($MgO \cdot Al_2O_3$) refractory with 1 wt % excess MgO (i.e., 1 g MgO per 99 g $MgO \cdot Al_2O_3$), and (b) a spinel refractory with 1 wt % excess Al_2O_3.

5.64 (a) 1g MgO + 99g $MgO \cdot Al_2O_3$

$$\text{wt. frac. MgO in } MgO\cdot Al_2O_3 = \frac{(24.31+16.00)\text{ amu}}{[(24.31+16.00)+(2\times26.98+3\times16)]\text{ amu}}$$

$$= 0.283$$

$$m_{MgO \text{ in spinel}} = (0.283)(99\text{g}) = 28.05\text{g}$$

$$m_{Al_2O_3 \text{ in spinel}} = (1-0.283)(99\text{g}) = 70.95\text{g}$$

giving

$$\text{mole MgO} = (1\text{g} + 28.05\text{g}) \times 1\text{mol}/(24.31+16.00)\text{g}$$

$$= 0.7206 \text{ mole}$$

$$\text{mole } Al_2O_3 = 70.95\text{g} \times 1\text{mol}/(2\times26.98+3\times16.00)\text{g}$$

$$= 0.6959 \text{ mole}$$

$$\text{mol. \% } Al_2O_3 = \frac{0.6959 \text{ mol}}{(0.6959+0.7206)\text{mol}} \times 100\% = 49.1\%$$

Therefore, phases present are <u>periclase (ss) + spinel (ss)</u>

<u>Periclase is ≈100 mol % MgO</u>

<u>Spinel is ≈ 50 mol. % Al_2O_3</u>

By definition, <u>periclase is 1 wt.%</u> and <u>spinel is 99 wt.%.</u>

(b)

$$\text{mole MgO} = 28.05\text{g} \times 1\text{mol}/(24.31+16.00)\text{g}$$

$$= 0.6959 \text{ mole}$$

$$\text{mole } Al_2O_3 = (70.95\text{g} + 1\text{g}) \times 1\text{mol}/(2\times26.98+3\times16.00)\text{g}$$

$$= 0.7057 \text{ mole}$$

$$\text{mol. \% } Al_2O_3 = \frac{0.7057 \text{ mole}}{(0.6959+0.7057)\text{mole}} \times 100\% = 50.3\%$$

Therefore, phase present is <u>spinel (ss)</u> with composition <u>50.3 mol. % Al_2O_3</u>.

The microstructure is <u>100 wt.% spinel (ss)</u>.

5.65 Plot the phases present (in mole percent) as a function of temperature for the heating of a refractory with the composition 60 mol % Al_2O_3–40 mol % MgO from 1000 to 2500°C.

5.65 Key temperatures and compositions include:

1000°C:

$$\% Al_2O_3 = \frac{60-53}{100-53} \times 100\% = 14.9\%$$

$$\% \text{ spinel} = 100 - 14.9 = 85.1\%$$

1420 – 2030°C: 100% spinel

2070 – 2500°C: 100% liquid

Giving the plot:

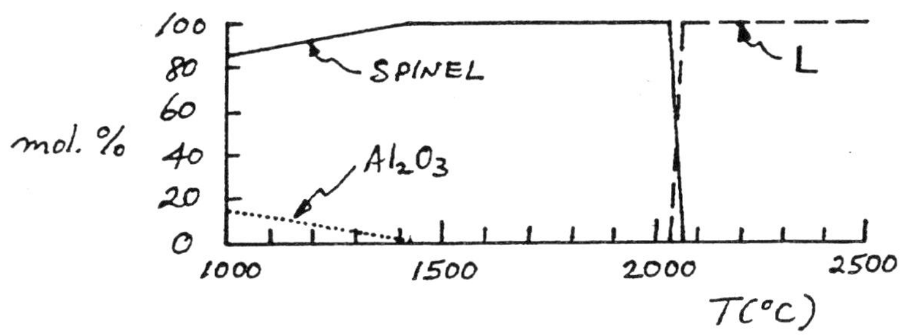

5.66 is on page 180 (next page)

5.67 A partially stabilized zirconia (for a novel structural application) is desired to have an equimolar microstructure of tetragonal and cubic zirconia at an operating temperature of 1250°C. Calculate the proper CaO content (in weight percent) for this structural ceramic.

5.67 Inspection of Figure 5-42 indicates that midpoint of the line at 1250°C falls at ≈14 mol. % CaO. Then,

$$\text{wt. \% CaO} = \frac{14 \text{ mol CaO} \times (40.08+16.00)\text{ g/mol}}{14(40.08+16.00)\text{g} + 86 \text{ mol } ZrO_2 (91.22+32.00)\frac{\text{g}}{\text{mol}}} \times 100\%$$

$$= \underline{\underline{6.9 \text{ wt. \%}}}$$

5.66 Plot the phases present (in mole percent) as a function of temperature for the heating of a partially stabilized zirconia with 10 mol % CaO from room temperature to 2800°C.

5.66

Note that: monoclinic + cubic from 0°C to ≈900°C
monoclinic + tetragonal from ≈900°C to ≈950°C
tetragonal only from ≈950°C to ≈2000°C
tetragonal + cubic from ≈2000°C to ≈2600°C
tetragonal + liquid from ≈2600°C to ≈2700°C
liquid only above ≈2700°C

Key Calculations:

At room temperature: see PP 5.12

At just below ≈900°C:

$$\text{mol. \% cubic} = \frac{10-3}{15-3} \times 100\% = 58\%$$

$$\text{mol. \% monoclinic} = 100 - 58 = 42\%$$

At just above ≈ 900°C:

$$\text{mol. \% tetragonal} = \frac{10-3}{13-3} \times 100\% = 70\%$$

$$\text{mol. \% monoclinic} = 100 - 70 = 30\%$$

At just below ≈2600°C:

$$\text{mol. \% cubic} = \frac{10-7}{13-7} \times 100\% = 50\%$$

$$\text{mol. \% tetragonal} = 100 - 50 = 50\%$$

At just above ≈2600°C:

$$\text{mol. \% tetragonal} = \frac{20-10}{20-7} \times 100\% = 77\%$$

$$\text{mol. \% liquid} = 100 - 77 = 23\%$$

Resulting plot:

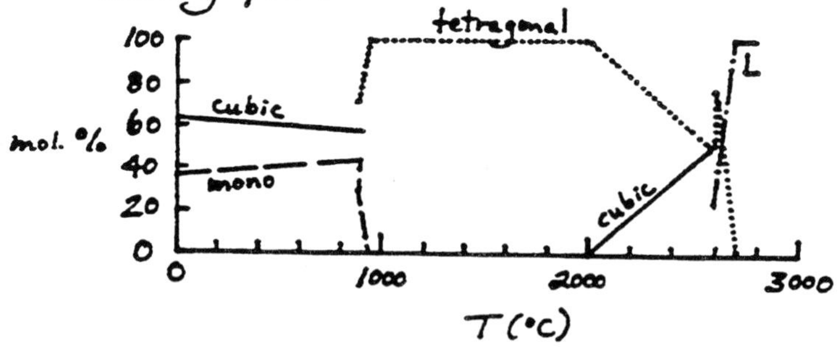

5.67 is on page 179 (previous page)

5.68 Repeat Problem 5.67 for a microstructure with equal weight fractions of tetragonal and cubic zirconia.

5.68 Inspection of Figure 5-42 indicates that the limit of tetragonal solid solution is ≈ 12 mol % and the limit of cubic solid solution is ≈ 15 mol %.

Then,

$$(\text{wt. \% CaO})_{tet} = \frac{12 \text{ mol CaO} \times (40.08+16.00)\text{ g/mol}}{12\,(40.08+16.00)\text{ g} + 88 \text{ mol } ZrO_2\,(91.22+32.00)\text{ g/mol}} \times 100\,\% = 5.8 \text{ wt.\%}$$

$$(\text{wt. \% CaO})_{cub} = \frac{15\,(40.08+16.00)\text{ g}}{15\,(40.08+16.00)\text{ g} + 85\,(91.22+32.00)\text{ g}} \times 100\,\% = 7.4 \text{ wt.\%}$$

∴ 50:50 microstructure by wt. lies midway between these values:

$$\frac{5.8 \text{ wt.\%} + 7.4 \text{ wt.\%}}{2} = \underline{\underline{6.6 \text{ wt.\%}}}$$

Section 6.1 – Time - The Third Dimension

PP 6.1 In Sample Problem 6.1, the activation energy for crystal growth in a copper alloy is calculated. Using that result, calculate the temperature at which the growth rate would have dropped three orders of magnitude relative to the rate at 900°C.

PP 6.1

$$\frac{\dot{G}_{900°C}}{\dot{G}_T} = e^{-[(181{,}000\ J/mol)/8.314\ J/(mol\cdot K)][\frac{1}{1173} - \frac{1}{T}]} = 10^3$$

or,

$$\ln 10^3 = -\left(\frac{181{,}000}{8.314}\right)\left(\frac{1}{1173} - \frac{1}{T}\right)$$

giving,

$$T = 855\,K = \underline{582°C}$$

6.1 For an aluminum alloy, the activation energy for crystal growth is 120 kJ/mol. By what factor would the rate of crystal growth change by dropping the alloy temperature from 500°C to room temperature (25°C)?

6.1

$$\frac{\dot{G}_{500°C}}{\dot{G}_{25°C}} = e^{-[(120{,}000\ J/mol)/8.314\ J/(mol\cdot K)][\frac{1}{773} - \frac{1}{298}]K^{-1}}$$

$$= \underline{8.43 \times 10^{12}}$$

6.2 Repeat Problem 6.1 for a copper alloy for which the activation energy for crystal growth is 195 kJ/mol.

6.2

$$\frac{\dot{G}_{500°C}}{\dot{G}_{25°C}} = e^{-[(195{,}000\ J/mol)/8.314\ J/(mol\cdot K)][\frac{1}{773} - \frac{1}{298}]K^{-1}}$$

$$= \underline{1.01 \times 10^{21}}$$

6.3 Although Section 6.1 concentrates on crystal nucleation and growth from a liquid, similar kinetics laws apply to solid state transformations. For example, Equation 6.1 can be used to describe the rate of β precipitation upon cooling supersaturated α phase in a 10 wt % Sn-90 wt % Pb alloy. Given precipitation rates of $3.77 \times 10^3\ s^{-1}$ and $1.40 \times 10^3\ s^{-1}$ at 20°C and 0°C, respectively, calculate the activation energy for this process.

6.3

$$\frac{(\text{Rate})_{20^\circ C}}{(\text{Rate})_{0^\circ C}} = \frac{C e^{-Q/R(293\,K)}}{C e^{-Q/R(273\,K)}}$$

$$\frac{3.77 \times 10^3\ s^{-1}}{1.40 \times 10^3\ s^{-1}} = e^{-\frac{Q}{[8.314\ J/(mol \cdot K)]}\left(\frac{1}{293\,K} - \frac{1}{273\,K}\right)}$$

or

$$Q = \frac{-[8.314\ J/(mol \cdot K)] \ln\left(\frac{3.77}{1.40}\right)}{\left(\frac{1}{293\,K} - \frac{1}{273\,K}\right)}$$

$$= 32.9 \times 10^3\ J/mol = \underline{\underline{32.9\ kJ/mol}}$$

6.4 Use the result of Problem 6.3 to calculate the precipitation rate at room temperature, 25°C.

6.4

$$(\text{Rate})_{25^\circ C} = (\text{Rate})_{0^\circ C}\, e^{-\frac{[32.9 \times 10^3\ J/mol]}{[8.314\ J/(mol \cdot K)]}\left(\frac{1}{298\,K} - \frac{1}{273\,K}\right)}$$

$$= (1.40 \times 10^3\ s^{-1})(3.37) = \underline{\underline{4.72 \times 10^3\ s^{-1}}}$$

6.5 The classical theory of nucleation is based on an energy balance between the nucleus and its surrounding liquid. The key principle is that a small cluster of atoms (the nucleus) will be stable only if further growth reduces the net energy of the system. Taking the nucleus in Figure 6-2a as spherical, the energy balance can be illustrated as follows:

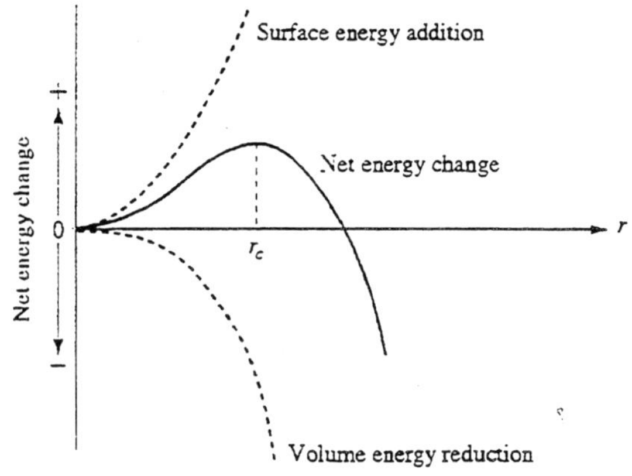

That is, a nucleus will be stable if its radius, r, is greater than a critical value, r_c. Derive an expression for r_c as a function of σ, the surface energy per unit area of the nucleus, and ΔG_v, the volume energy reduction per unit area. Recall that the area of a sphere is $4\pi r^2$, and its volume is $(\frac{4}{3})\pi r^3$.*

*As with other general discussions in this book dealing with stability and energy, a more rigorous treatment would introduce the thermodynamic free energy concept. An introduction to this is available in the thermodynamics chapter of the *Instructor's Manual for Introduction to Materials Science for Engineers*.

6.5

$$E_{total} = 4\pi r^2 \sigma + \frac{4}{3}\pi r^3 \Delta G_V$$

$$\frac{\partial E_{total}}{\partial r} = 8\pi r \sigma + 4\pi r^2 \Delta G_V$$

At r_c,

$$\left(\frac{\partial E_{total}}{\partial r}\right)_{r_c} = 0 = 8\pi r_c \sigma + 4\pi r_c^2 \Delta G_V$$

or

$$r_c = -\frac{2\sigma}{\Delta G_V}$$

• **6.6** The work of formation, W, for a stable nucleus is the maximum value of the net energy change (occurring at r_c) in the figure in Problem 6.5. Derive an expression for W in terms of σ and ΔG_v (defined in Problem 6.5).

6.6

Building on the derivation of Problem 6.5,

$$W = 4\pi\left(-\frac{2\sigma}{\Delta G_V}\right)^2\sigma + \frac{4}{3}\pi\left(-\frac{2\sigma}{\Delta G_V}\right)^3\Delta G_V$$

$$= 16\pi\frac{\sigma^3}{(\Delta G_V)^2} - \frac{32}{3}\pi\frac{\sigma^3}{(\Delta G_V)^2}$$

$$= \frac{16}{3}\pi\frac{\sigma^3}{(\Delta G_V)^2}$$

• **6.7** A theoretical expression for the rate of pearlite growth from austenite is

$$\dot{R} = Ce^{-Q/RT}(T_E - T)^2$$

where C is a constant, Q the activation energy for carbon diffusion in austenite, R the gas constant, and T an absolute temperature below the equilibrium transformation temperature, T_E. Derive an expression for the temperature, T_M, corresponding to the maximum growth rate, that is, the "knee" of the transformation curve.

6.7

$$\dot{R} = Ce^{-Q/RT}(T_E - T)^2$$

At the transformation "knee,"

$$\left(\frac{d\dot{R}}{dT}\right)_{T_{max}} = 0 = C\left[e^{-Q/RT_m} \times 2(T_E - T_m)(-1) + (T_E - T_m)^2 e^{-Q/RT_m}\left(+\frac{Q}{RT_m^2}\right)\right]$$

or

$$2 = (T_E - T_m)\frac{Q}{RT^2}$$

or

$$\frac{2R}{Q}T_m^2 = T_E - T_m$$

or

$$\frac{2R}{Q}T_m^2 + T_m - T_E = 0$$

with the solution:

$$T_m = \frac{-1 \pm \sqrt{1 - 4\cdot\frac{2R}{Q}(-T_E)}}{4R/Q}$$

6.8 Use the result of Problem 6.7 to calculate T_M (in °C). (Recall that the activation energy is given in Chapter 4 and the transformation temperature is given in Chapter 5.)

6.8 Substituting appropriate data and constants:

$$T_m = \frac{-1 \pm \sqrt{1 - 4 \cdot \frac{2(8.314\ J/mol\cdot K)}{142\ kJ/mol}(-1000\ K)}}{4(8.314\ J/mol\cdot K)/(142\ kJ/mol)}$$

Note: only + gives physically meaningful solution

&, finally, $T_m = 904\ K = \underline{\underline{631\ °C}}$

Section 6.2 – The TTT Diagram

PP 6.2 In Sample Problem 6.2, we use Figure 6-6 to determine the time for 50% transformation to pearlite and bainite at 600 and 300°C, respectively. Repeat these calculations for **(a)** 1% transformation and **(b)** 99% transformation.

PP 6.2
(a) ≈ 1 s (at 600°C) ≈ 80 s (at 300°C)
(b) ≈ 7 s (at 600°C) ≈ 1500 s = 25 m (at 300°C)

PP 6.3 A detailed thermal history is outlined in Sample Problem 6.3. Answer all of the questions in that problem if only one change is made in the history; namely, step (i) is an instantaneous quench to 400°C (not 500°C).

PP 6.3
(a) ≈10% fine pearlite + 90% γ

(b) ≈10% fine pearlite + 90% bainite

(c) ≈10% fine pearlite + 90% martensite (including a small amount of retained γ)

(d)

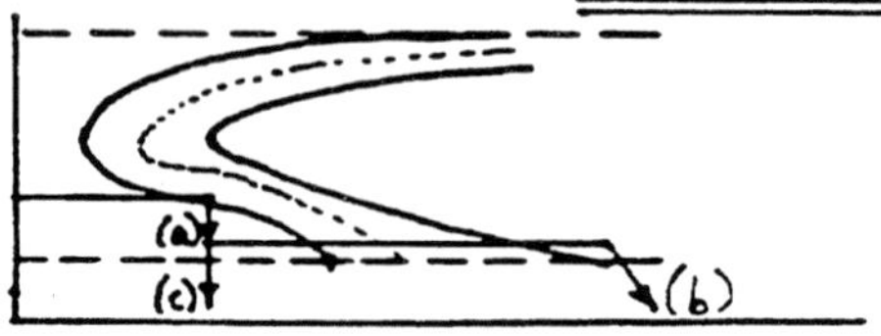

PP 6.4 In Sample Problem 6.4, we estimate quench rates necessary to retain austenite below the pearlite "knee." What would be the percentage of martensite formed in each of the alloys if these quenches were continued to 200°C?

PP 6.4

Figure 6-14 indicates the percentage of martensite formed will be $> 90\%$ for 0.5 wt.% C.

Figure 6-10 gives $\approx 20\%$ for 0.77 wt. % C.

Figure 6-13 gives 0% for 1.13 wt. % C.

PP 6.5 The time necessary for austempering is calculated for three alloys in Sample Problem 6.5. In order to do martempering (Figure 6-17), it is necessary to cool the alloy before bainite formation begins. How long can the alloy be held 5° above M_s before bainite formation begins in **(a)** 0.5 wt % C steel, **(b)** 0.77 wt % C steel, and **(c)** 1.13 wt % C steel?

PP 6.5

(a) ≈ 15 s (as indicated by Figure 6-14)

(b) $\approx 150\text{ s} \times 1\text{ min}/60\text{ s} = 2\frac{1}{2}$ min (from Figure 6-10)

(c) ≈ 1 hour (from Figure 6-13)

6.9 (a) A 1050 steel (iron with 0.5 wt % C) is rapidly quenched to 330°C, held for 10 minutes, and then cooled to room temperature. What is the resulting microstructure? (b) What is a name for this heat treatment?

6.9

(a) 100% bainite (by inspection of Figure 6-14)

(b) Austempering (see Figure 6-18)

6.10 (a) A eutectoid steel is (i) quenched instantaneously to 500°C, (ii) held for 5 seconds, (iii) quenched instantaneously to room temperature, (iv) reheated to 300°C for 1 hour, and (v) cooled to room temperature. What is the final microstructure? (b) A carbon steel with 1.13 wt % C is given exactly the same heat treatment described in part (a). What is the resulting microstructure in this case?

6.10

(a) This is comparable to the tempering history of Figure 6-15 except that some initial fine pearlite formation is allowed at 500°C (see Sample Problem 6.3(c) for the initial history). The final microstructure will be:

70% fine pearlite + 30% tempered martensite

(b) Figure 6-13 indicates that complete transformation to fine pearlite occurs after 5 sec. at 500°C. Subsequent thermal history does not change that. The final microstructure is, then: 100% fine pearlite.

6.11 (a) A carbon steel with 1.13 wt % C is given the following heat treatment: (i) instantaneously quenched to 200°C, (ii) held for 1 day, and (iii) cooled slowly to room temperature. What is the resulting microstructure? (b) What microstructure would result if a carbon steel with 0.5 wt % C were given exactly the same heat treatment?

6.11

(a) By inspection of Figure 6–13, 100% bainite.

(b) By inspection of Figure 6–14, >90% martensite with balance retained austenite (in lieu of diffusional transformation information).

6.12 Three different eutectoid steels are given the following heat treatments: (a) instantaneously quenched to 600°C, held for 2 minutes, then cooled to room temperature; (b) instantaneously quenched to 400°C, held for 2 minutes, then cooled to room temperature; and (c) instantaneously quenched to 100°C, held for 2 minutes, then cooled to room temperature. List these heat treatments in order of decreasing hardness of the final product. Briefly explain your answer.

6.12

(3) → (2) → (1) → decreasing hardness

(3) will produce >90% martensite and high hardness

(2) will produce 100% fine pearlite and lower hardness

(1) will produce 100% coarse pearlite and even lower hardness

6.13 (a) A eutectoid steel is cooled at a steady rate from 727 to 200°C in exactly 1 day. Superimpose this cooling curve on the TTT diagram of Figure 6-10. (b) From the result of your plot for part (a), determine at what temperature a phase transformation would first be observed. (c) What would be the first phase to be observed? (Note the footnote on page 218 in regard to the approximate nature of an exercise such as this.)

6.13

(a) The cooling rate is:

$$\frac{\Delta T}{\Delta t} = \frac{727°C - 200°C}{24 h \times 3600 s/h} = 6.10 \times 10^{-3} \, °C/s$$

For various times along the cooling curve,

t (sec)	ΔT(°C) →	T(°C)
1	6.1×10^{-3}	~727
10	0.06	726.94
100	0.6	726.4
1,000	6.1	720.9
2,000	12.2	715
3,000	18.3	709
10,000	61	666

Giving the plot:

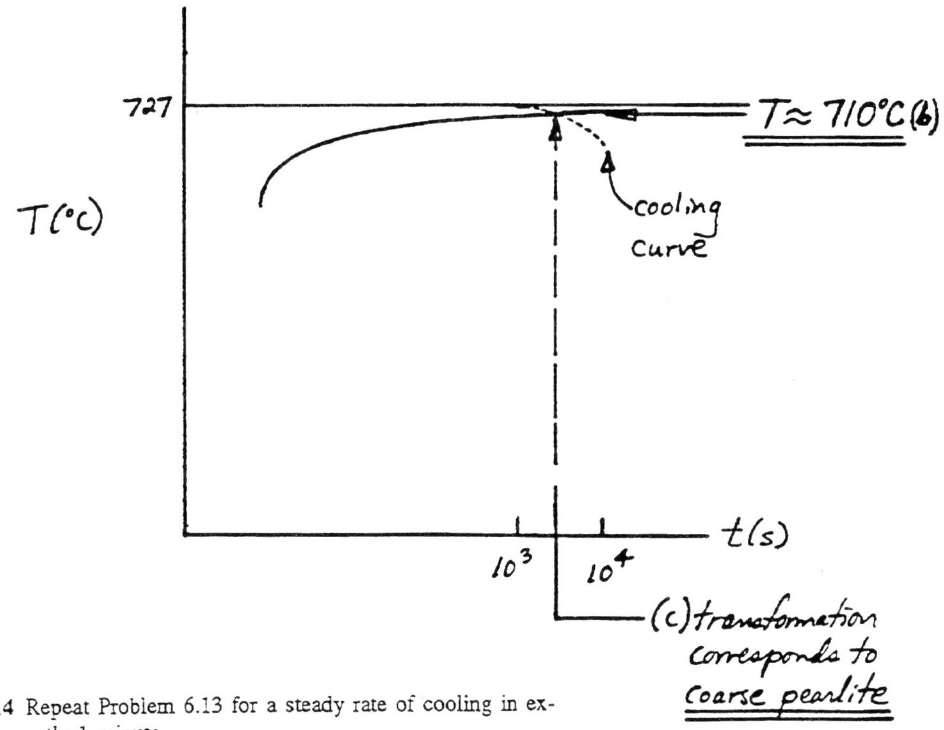

6.14 Repeat Problem 6.13 for a steady rate of cooling in exactly 1 minute.

6.14 (a) The cooling rate is:

$$\frac{\Delta T}{\Delta t} = \frac{727°C - 200°C}{1\ m \times 60\ s/m} = 8.78\ °C/s$$

→

t (s)	ΔT(°C)	T(°C)
1	8.78	718
5	43.9	683
10	87.8	639
20	176	551

→

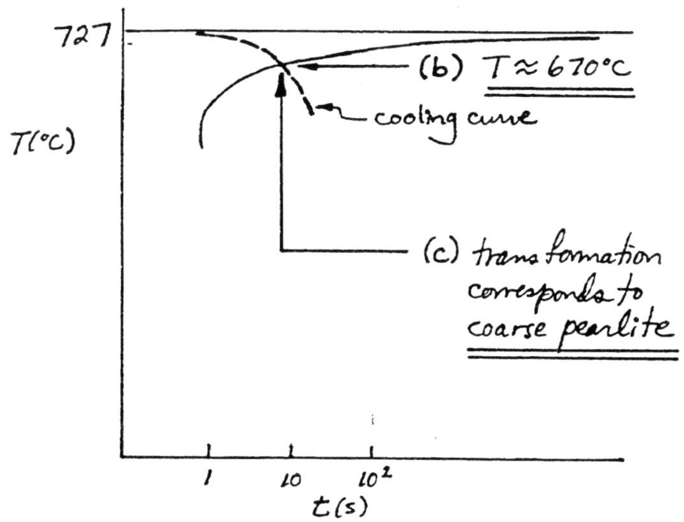

6.15 Repeat Problem 6.13 for a steady rate of cooling in exactly 1 second.

6.15

(a) The cooling rate is:

$$\frac{\Delta T}{\Delta t} = \frac{727°C - 200°C}{1 s} = 527°C/s$$

$t(s)$	$\Delta T(°C)$	$T(°C)$
0.1	52.7	674
0.5	264	463
1.0	527	200

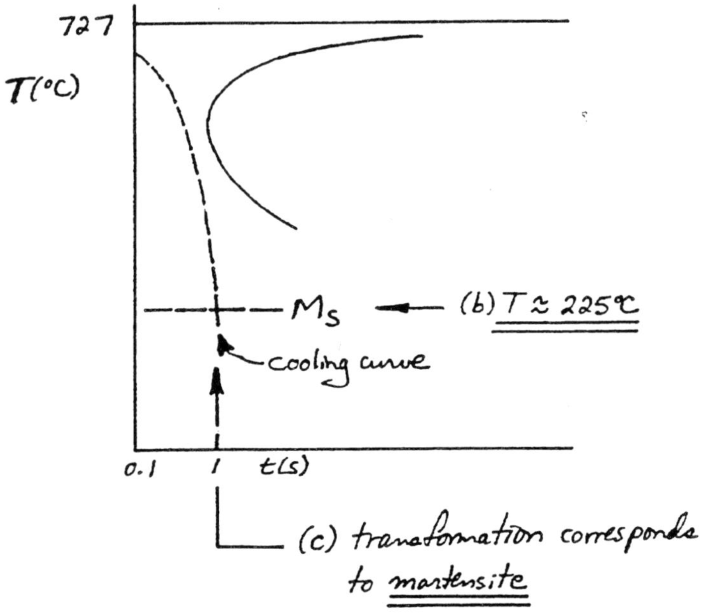

6.16 (a) Using Figures 6-10, 6-13, and 6-14 as data sources, plot M_s, the temperature at which the martensitic transformation begins, as a function of carbon content. (b) Repeat part (a) for M_{50}, the temperature at which the martensitic transformation is 50% complete. (c) Repeat part (a) for M_{90}, the temperature at which the martensitic transformation is 90% complete.

6.16

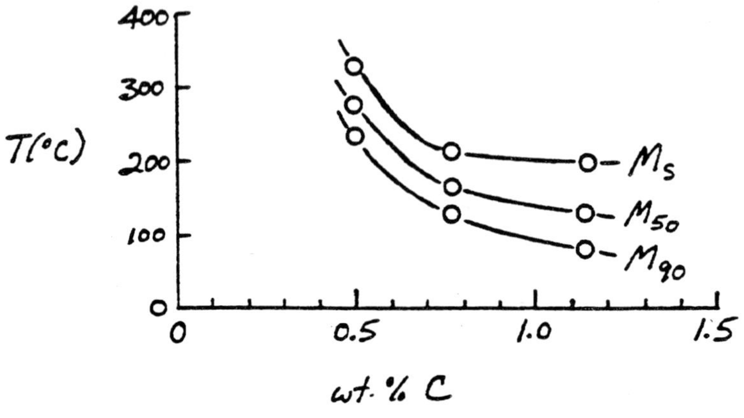

6.17 Using the trends of Figures 6-10, 6-13, and 6-14, sketch as specifically as you can a TTT diagram for a hypoeutectoid steel that has 0.6 wt % C. (Note Problem 6.16 for one specific compositional trend.)

6.17 M_s, M_{50}, and M_{90} can be interpolated from the results of Problem 6.16. We need similar trends for $t_{\gamma \to \alpha + Fe_3C}$ & $T_{\text{"knee"}}$:

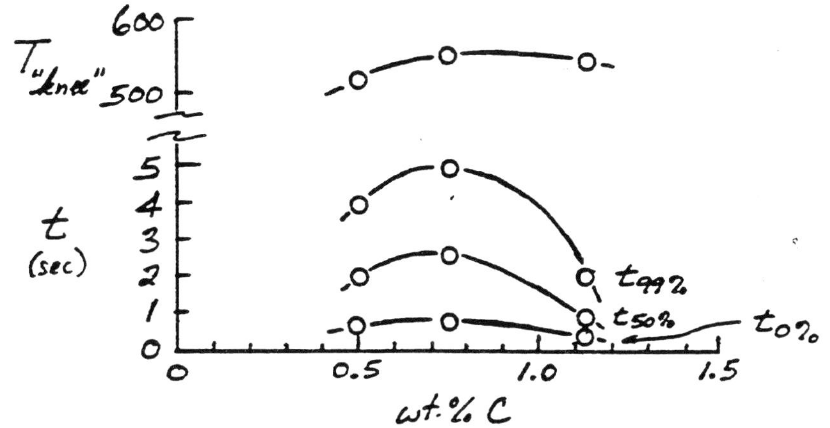

By interpolating on the above plot and that for Problem 6.16, we see that the key data for a 0.6 wt% C steel would be:

$M_s = 270°C$		$T_{\text{"knee"}}$	$= 540°C$
$M_{50} = 220°C$		$t_{99\%}$	$= 4.7\,s$
$M_{90} = 180°C$		$t_{50\%}$	$= 2.5\,s$
		$t_{1\%}$	$= 0.8\,s$

Giving a plot:

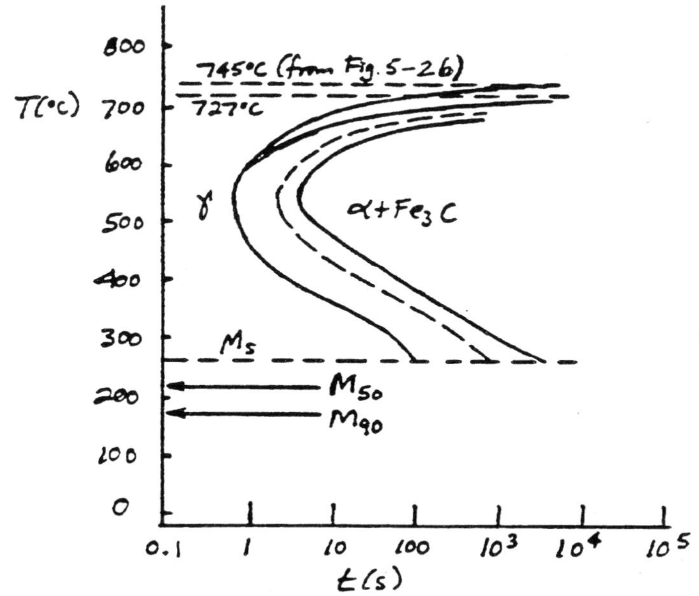

6.18 Repeat Problem 6.17 for a hypereutectoid steel with 0.9 wt % C.

6.18

By interpolating on the plots for Problems 6.16 & 6.17, we see that the key data for a 0.9 wt% C steel are:

$M_s = 200°C$ $\quad$ $T_{\text{"knee"}} = 550°C$

$M_{50} = 150°C$ $\quad$ $t_{99\%} = 4.7 s$

$M_{90} = 100°C$ $\quad$ $t_{50\%} = 2.2 s$

$t_{1\%} = 0.7 s$

Giving a plot

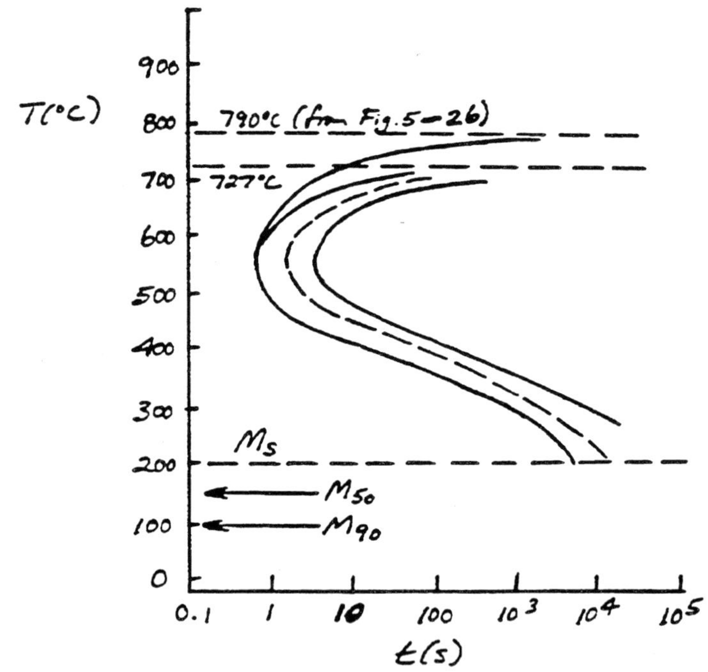

6.19 What is the final microstructure of a 0.6 wt % C hypoeutectoid steel given the following heat treatment: (i) instantaneously quenched to 500°C, (ii) held for 10 seconds, and (iii) instantaneously quenched to room temperature. (Note the TTT diagram developed in Problem 6.17.)

6.19

Inspection of the TTT diagram generated for Problem 6.17, we see that 10 s at 500°C would yield 100% $\alpha + Fe_3C$ which would be stable upon further quenching. Comparison to similar composition TTT diagrams in the text indicates that the $\alpha + Fe_3C$ at 500°C would be fine pearlite. Therefore, the resulting microstructure is 100% fine pearlite.

6.20 Repeat Problem 6.19 for a 0.9 wt % C hypereutectoid steel. (Note the TTT diagram developed in Problem 6.18.)

6.20

As for Problem 6.19, the microstructure will be 100 % fine pearlite.

6.21 It is worth noting that in TTT diagrams such as Figure 6-6, the 50% completion (dashed line) curve lies roughly midway between the onset (1%) curve and completion (99%) curve. It is also worth noting that the progress of transformation is not linear but sigmoidal (s-shaped) in nature. For example, careful observation of Figure 6-6 at 500°C shows that 1%, 50%, and 99% completion occur at 0.9 s, 3.0 s, and 9.0 s, respectively. Intermediate completion data, however, can be given as:

% completion	t(s)
20	2.3
40	2.9
60	3.2
80	3.8

Plot the % completion at 500°C versus log t to illustrate the sigmoidal nature of the transformation.

6.21

Complete data set:

% Completion	t(s)	$\log_{10} t(s)$
1	0.9	-0.0458
20	2.3	0.362
40	2.9	0.462
50	3.0	0.477
60	3.2	0.505
80	3.8	0.580
99	9.0	0.954

resulting plot:

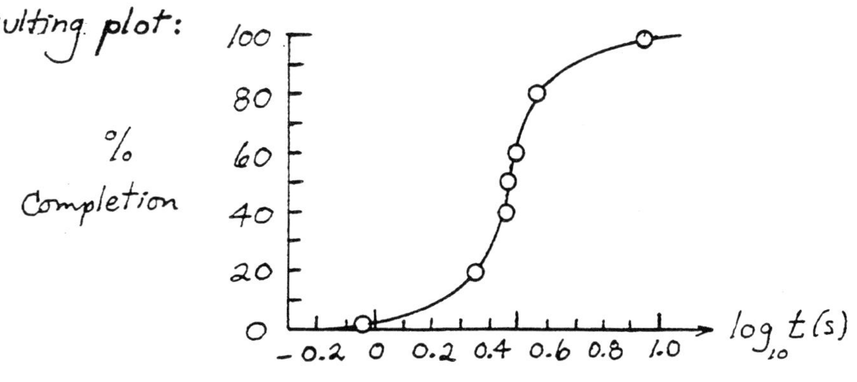

6.22 (a) Using the result of Problem 6.21, determine t for 25% and 75% completion. (b) Superimpose the result of (a) on a sketch of Figure 6-6 to illustrate that the 25% and 75% completion lines are much closer to the 50% line than to either the 1% or 99% completion lines.

6.22 (a) Superimposing on the plot of Problem 6.21:

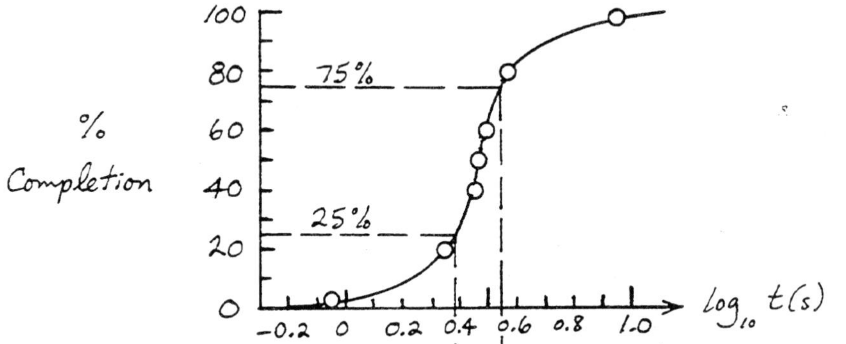

$\log_{10} t_{75\%} = 0.54 \rightarrow t_{75\%} = 3.5\,s$

$\log_{10} t_{25\%} = 0.38 \rightarrow t_{25\%} = 2.4\,s$

(b)

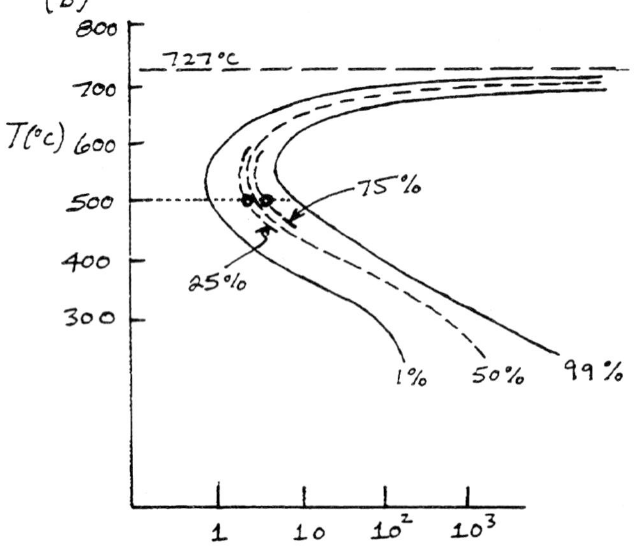

Section 6.3 – Hardenability

PP 6.6 In Sample Problem 6.6, we are able to estimate a quench rate that leads to a hardness of Rockwell C45 in a 4340 steel. What quench rate would be necessary to produce a hardness of (**a**) C50 and (**b**) C40?

PP 6.6

(a) From Figure 6–22, C50 occurs at $\approx \frac{13}{16}$ inch giving $D_{qe} = \frac{13}{16}$ in. x 25.4 mm/in. = 20.6 mm.

Figure 6–20 gives

quench rate = $\approx$ 7°C/s (at 700°C)

(b) Similarly, C40 occurs at $\approx \frac{32}{16}$ inch or

$D_{qe} = \frac{32}{16}$ x 25.4 mm = 50.8 mm giving

quench rate = $\approx$ 2.5°C/s (at 700°C)

PP 6.7 In Sample Problem 6.7, we find that the hardness of a 4140 steel is lower than that for a 4340 steel (given equal quench rates). Determine the corresponding hardness for (**a**) a 9840 steel, (**b**) an 8640 steel, and (**c**) a 5140 steel.

PP 6.7

(a) Again using a D_{qe} = 23/16 inch, Figure 6–20 gives:

hardness = Rockwell C38

(b) For 8640 steel:

hardness = Rockwell C25

(c) For 5140 steel:

hardness = Rockwell C21.5

6.23 (a) Specify a quench rate necessary to ensure a hardness of *at least* Rockwell C40 in a 4140 steel. (b) Specify a quench rate necessary to ensure a hardness of *no more than* Rockwell C40 in the same alloy.

6.23 (a) Figure 6-22 indicates that C40 occurs in a 4140 steel at $\approx$ 11/16 inch or $D_{qe} = \frac{11}{16} \times 25.4 =$ 17.5 mm. Figure 6-20 gives a corresponding quench rate of $\approx 10°C/s$ (at 700°C).

To ensure a hardness > C40 requires, then:

quench rate > $\approx 10°C/s$

(b) Similarly, to ensure a hardness < C40 requires:

quench rate < $\approx 10°C/s$.

(Note, again, the comments about "uncertainty bars" in Sample Problems 6.6 and 6.7.)

6.24 The surface of a forging made from a 4340 steel is unexpectedly subjected to a quench rate of 100°C/s (at 700°C). The forging is specified to have a hardness between Rockwell C46 and C48. Is the forging within specifications? Briefly explain your answer.

6.24 No

Figure 6-20 indicates that this quench rate corresponds to a $D_{qe} \approx 3$ mm (i.e., just over 1 inch). Figure 6-22 indicates that such a small D_{qe} corresponds to a Rockwell hardness of $\approx$ C53.

6.25 A flywheel shaft is made of 8640 steel. The surface hardness is found to be Rockwell C35. By what percentage would the cooling rate at the point in question have to change in order for the hardness to be increased to a more desirable value of Rockwell C45?

6.25

From Figure 6-22, we find that, for 8640, a hardness of C35 occurs at about 9.2 sixteenths distance. Figure 6-20 shows that this distance corresponds to a cooling rate of ≈ 13 °C/s (at 700 °C).

Similarly, a hardness of C45 occurs at a distance of about 5.8 sixteenths, corresponding to a cooling rate of ≈ 24 °C/s (at 700 °C).

The resulting increase in cooling rate would be:

$$\frac{24-13}{13} \times 100\% = \underline{\underline{84.6\% \text{ increase}}}$$

6.26 Repeat Problem 6.25 assuming that the shaft is made of 9840 steel.

6.26

From Figure 6-22, the hardness of C35 occurs at about 29.3 sixteenths for 9840 steel. Figure 6-20 gives a corresponding cooling rate of ≈ 2.6 °C/s (at 700 °C).

Similarly, a hardness of C45 occurs at a distance of about 13.8 sixteenths, corresponding to a cooling rate of ≈ 6.9 °C/s (at 700 °C).

The resulting increase in cooling rate would be:

$$\frac{6.9-2.6}{2.6} \times 100\% = \underline{\underline{165\% \text{ increase}}}$$

6.27 Quenching a bar of 4140 steel at 700°C into a stirred water bath produces an instantaneous quench rate at the surface of 100°C/s. Use Jominy test data to predict the surface hardness resulting from this quench.

Note. The quench described is not in the Jominy configuration. Nonetheless, Jominy data provide general information on hardness as a function of quench rate.

6.27

Using Figure 6-20, the given cooling rate corresponds to a distance of 3mm from end &

$$3\text{mm} \times \frac{1\text{in}}{25.4\text{mm}} \times \frac{16''}{16''} = 1.89 \text{ sixteenths of an inch}$$

for which Figure 6-22 indicates a hardness of Rockwell C53.

6.28 Repeat Problem 6.27 for a stirred oil bath that produces a surface quench rate of 20°C/s.

6.28

Figure 6-20 → cooling rate $\Rightarrow 11\text{mm} \times \frac{1\text{in}}{25.4\text{mm}} \times \frac{16''}{16''}$

= 6.9 sixteenths

for which Figure 6-22 → Rockwell C47.

6.29 A bar of steel quenched into a stirred liquid will cool much more slowly in the center than at the surface in contact with the liquid. The following limited data represent the initial quench rates at various points across the diameter of a bar of 4140 steel (initially at 700°C):

Position	Quench Rate (°C/s)
center	35
15 mm from center	55
30 mm from center (at surface)	200

(a) Plot the quench rate profile across the diameter of the bar. (Assume the profile to be symmetrical.) (b) Use Jominy test data to plot the resulting hardness profile across the diameter of the bar. (*Note:* The quench rates described are not in the Jominy configuration. Nonetheless, Jominy data provide general information on hardness as a function of quench rate.)

6.29

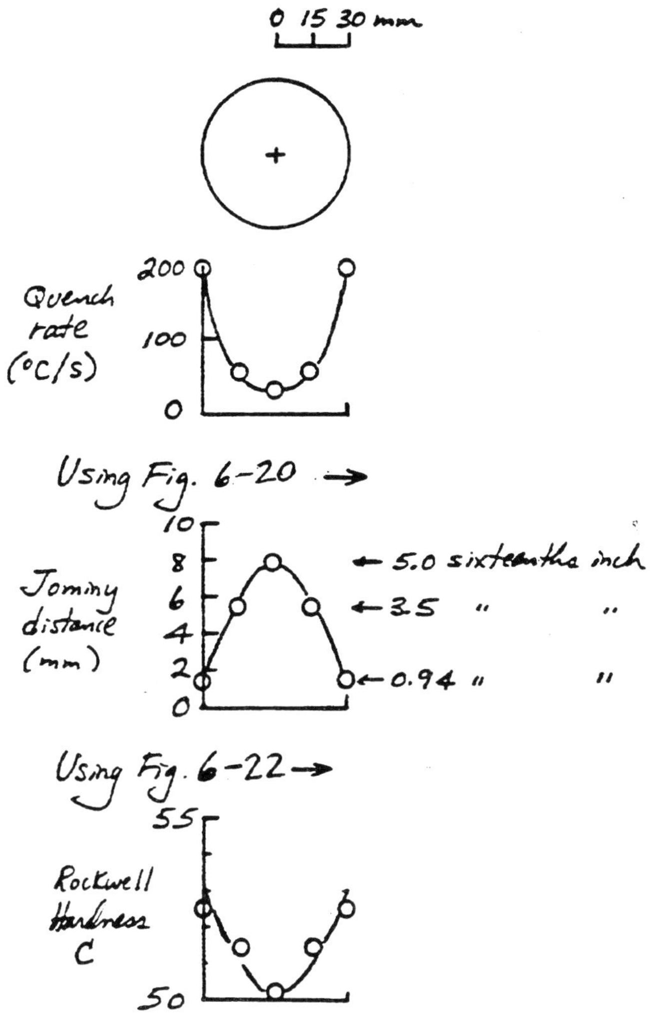

6.30 Repeat Problem 6.29b for a bar of 5140 steel that would have the same quench rate profile.

6.30

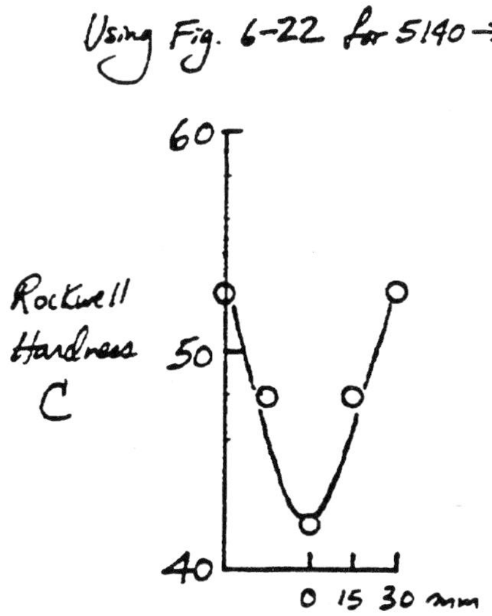

6.31 In heat-treating a complex-shaped part made from a 5140 steel, a final quench in stirred oil leads to a hardness of Rockwell C30 at 3 mm beneath the surface. This is unacceptable, as design specifications require a hardness of Rockwell C45 at that point. Select an alloy substitution to provide this hardness, assuming the heat treatment must remain the same.

6.31

By inspection of Figure 6-22 we see that the Jominy curve for 5140 steel has a hardness of C30 at ≈ 10/16 inch from the end of the water-quenched bar. Obviously, the part achieved the same cooling rate at only 3 mm beneath the surface. Using Figure 6-22 to monitor hardness achieved by other alloys at the same cooling rate, we observe two alloys with hardness > C45:

alloys 4340 and 9840

6.32 In general, hardenability decreases with decreasing alloy additions. Illustrate this by superimposing the following data for a plain carbon 1040 steel on the plot for other xx40 steels in Figure 6-22:

Distance from quenched end (in sixteenths of an inch)	Rockwell C hardness
2	44
4	27
6	22
8	18
10	16
12	13
14	12
16	11

6.32

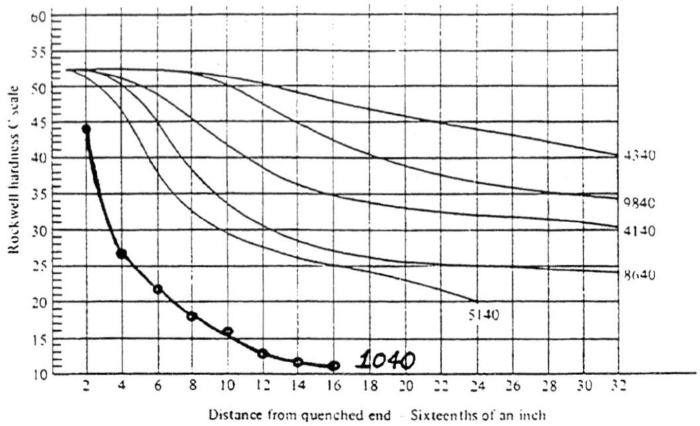

Section 6.4 – Precipitation Hardening

PP 6.8 The nature of precipitation in a 95.5 Al–4.5 Cu alloy is considered in Sample Problem 6.8. Repeat these calculations for a 96 Al–4 Cu alloy.

PP 6.8

(a) As before,

$$\text{wt.\% } \theta = \frac{4-0}{53-0} \times 100\% = \underline{\underline{7.55\%}}$$

(b) Again,

$$\text{wt.\% } \theta = \underline{\underline{7.55\%}}$$

6.33 (a) Calculate the maximum amount of second-phase precipitation in a 90 wt % Al–10 wt % Mg alloy at 100°C.
(b) What is the precipitate in this case?

6.33

(a) Using Figure 5–35, we obtain:

$$\text{wt.\% precipitate} = \frac{10-2}{36.1-2} \times 100\% = \underline{\underline{23.5\%}}$$

(b) $\underline{\underline{\beta \text{ phase}}}$

6.34 Repeat Problem 6.33 for stoichiometric γ phase, $Al_{12}Mg_{17}$.

6.34

(a) $$\text{at.\% Mg} = \frac{17}{12+17} \times 100\% = 58.6\%$$

$\therefore$ 100 g·atom $\gamma \longrightarrow$ 41.4 g·atom Al + 58.6 g·atom Mg

or (41.4 g·atom Al)(26.98 g Al/g·atom Al)

+ (58.6 g·atom Mg)(24.31 g Mg/g·atom Mg)

= 1117 g Al + 1425 g Mg

giving $$\frac{1425}{1117+1425} \times 100\% = 56.1 \text{ wt.\% Mg}$$

$$\therefore \text{wt.\% precipitate} = \frac{57.5-56.1}{57.5-38} \times 100\% = \underline{\underline{7.18\%}}$$

(b) $\underline{\underline{\beta \text{ phase}}}$

6.35 Specify an aging temperature for a 95 Al–5 Cu alloy that will produce a maximum of 5 wt % θ precipitate.

6.35

Taking x_θ = 53 wt.% Cu & using Fig. 5-34 at various temperatures:

At 100°C, $\text{wt.\% } \theta = \frac{5-0}{53-0} \times 100\% = 9.43\%$

" 300°C, " $= \frac{5-1}{53-1} \times " = 7.69\%$

" 400°C, " $= \frac{5-2.5}{53-2.5} \times " = 4.95\%$

" 500°C, " = 0% by inspection

→

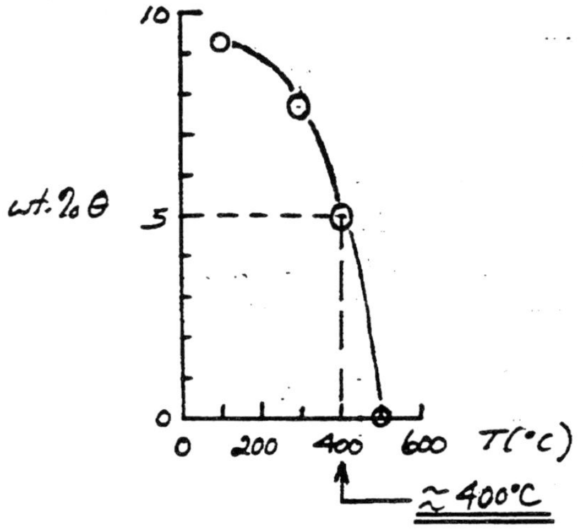

6.36 Specify an aging temperature for a 95 Al–5 Mg alloy that will produce a maximum of 5 wt % β precipitate.

6.36

Taking x_β = 36.1 wt.% Mg & using Fig. 5-35 at various temperatures:

At 100°C,

$$\text{wt.\% } \beta = \frac{5-2}{36.1-2} \times 100\% = 8.8\%$$

At 200°C,

$$\text{wt.\% } \beta = \frac{5-3}{36.1-3} \times " = 6.0\%$$

At 250°C,

$$\text{wt.\% } \beta = \frac{5-4.5}{36.1-4.5} \times " = 1.6\%$$

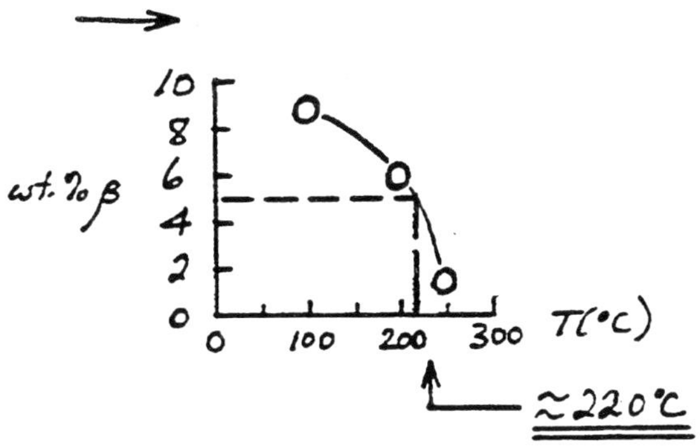

6.37 As second-phase precipitation is a thermally activated process, an Arrhenius expression can be used to estimate the time required to reach maximum hardness (see Figure 6-25b). To a first approximation, one can treat t_{max}^{-1} as a "rate," where t_{max} is the time to reach maximum hardness. For a given aluminum alloy, t_{max} is 40 hours at 150°C and only 4 hours at 190°C. Use Equation 4.1 to calculate the activation energy for this precipitation process.

6.37

$$t^{-1} = Ce^{-Q/RT}$$

$$\frac{(40^{-1})h^{-1}}{(4^{-1})h^{-1}} = \frac{Ce^{-Q/R(150+273)}}{Ce^{-Q/R(190+273)}} = 0.1$$

or

$$\ln(0.1) = -\frac{Q}{8.314\ J/mol \cdot K}(2.042 \times 10^{-4})K^{-1}$$

giving

$$Q = \underline{\underline{93.8\ kJ/mol}}$$

6.38 Estimate the time to reach maximum hardness, t_{max}, at 250°C for the aluminum alloy of Problem 6.37.

6.38 Substituting into the initial expression,

$$1/4\ h^{-1} = Ce^{-(93{,}800\ J/mol)/(8.314\ J/mol \cdot K)(463\ K)}$$

giving

$$C = 9.44 \times 10^{9}\ h^{-1}$$

$$\therefore 1/t_{250°C} = 9.44 \times 10^{9} h^{-1} e^{-(93{,}800)/(8.314)(250+273)}$$

$$= 4.085\ h^{-1}$$

or

$$t_{250°C} = 0.245\ h \times 60\ m/h = \underline{\underline{14.7\ m}}$$

Section 6.5 – Annealing

PP 6.9 Noting the result of Sample Problem 6.9, plot the estimated temperature range for recrystallization of Cu–Zn alloys as a function of composition over the entire range from pure Cu to pure Zn.

PP 6.9

wt.% Zn	T_{mp} (solidus) °C	T_{mp} (solidus) K	$\frac{1}{3}T_{mp}$ (K)	$\frac{1}{2}T_{mp}$ (K)
0	1084.5	1358	453	679
10	1040	1313	438	657
20	980	1253	418	627
30	920	1193	398	597
40	890	1163	388	582
50	870	1143	381	572
60	835	1108	369	554
70	700	973	324	487
80	598	871	290	436
90	424	697	232	349
100	420	693	231	347

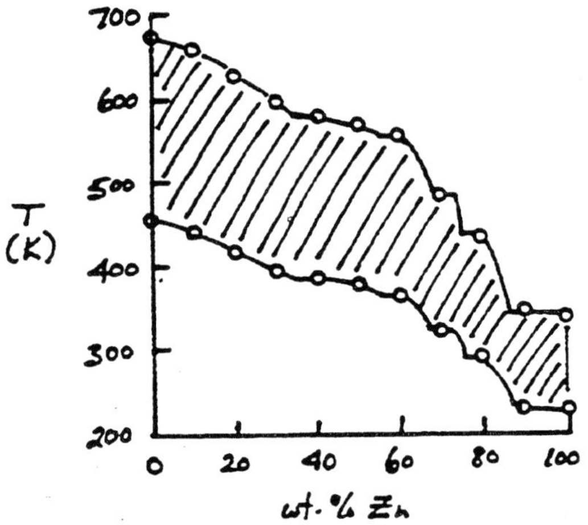

6.39 A 90:10 Ni–Cu alloy is heavily cold-worked. It will be used in a structural design that is occasionally subjected to 200°C temperatures for as much as 1 hour. Do you expect annealing effects to occur?

6.39 Inspection of Figure 5-36 indicates a solidus temperature for this alloy of ≈1410°C (= 1683 K). The resulting recrystallization range is 561 to 842 K ($\frac{1}{3}$ to $\frac{1}{2}$ T_{mp}) or 288 to 569°C. No, we do not expect annealing effects.

6.40 Repeat Problem 6.39 for a 90:10 Cu–Ni alloy.

6.40

Inspection of Figure 5-36 indicates a solidus temperature of ≈ 1130 °C (= 1403 K). The resulting recrystallization range is 468 to 702 K or 195 to 429 °C. <u>Yes</u>, we <u>may</u> expect annealing effects.

6.41 A 12.5-mm-diameter rod of steel is drawn through a 10-mm-diameter die. What is the resulting percentage cold work?

6.41

$$\% CW = \frac{A_0 - A_f}{A_0} \times 100\% = \frac{\pi (d_0/2)^2 - \pi (d_f/2)^2}{\pi (d_0/2)^2} \times 100\%$$

$$= \frac{d_0^2 - d_f^2}{d_0^2} \times 100\% = \frac{(12.5\,mm)^2 - (10\,mm)^2}{(12.5\,mm)^2} \times 100\%$$

$$= \underline{36\%}$$

• **6.42** An annealed copper alloy sheet is cold worked by rolling. The x-ray diffraction pattern of the original annealed sheet with an fcc crystal structure is represented schematically by Figure 3-40. Given that the rolling operation tends to produce the preferred orientation of the (220) planes parallel to the sheet surface, sketch the x-ray diffraction pattern you would expect for the rolled sheet. (To locate the 2θ positions for the alloy, assume pure copper and note the calculations for Problem 3.83.)

6.42

As the diffraction experiment is done relative to the surface of the specimen and the (111) planes are predominantly parallel to that surface, the (111) peak [and that for the parallel (222) planes] is more intense, with the other peaks being diminished (compared to Fig. 3-40):

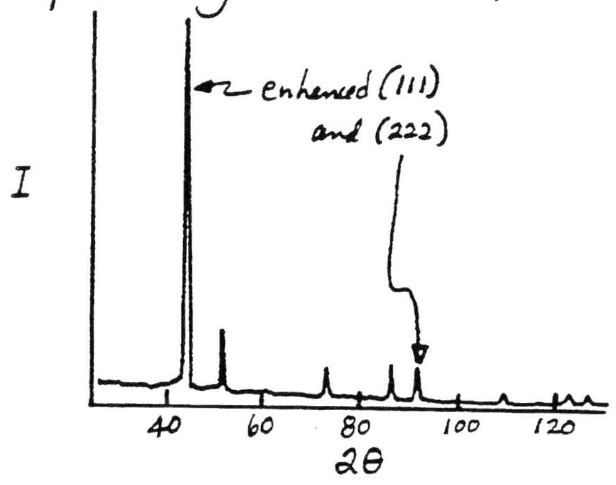

6.43 Recrystallization is a thermally activated process and, as such, can be characterized by the Arrhenius expression (Equation 4.1). To a first approximation, we can treat t_R^{-1} as a "rate," where t_R is the time necessary to fully recrystallize the microstructure. For a 75% cold-worked aluminum alloy, t_R is 100 hours at 256°C and only 10 hours at 283°C. Calculate the activation energy for this recrystallization process. (Note Problem 6.37 in which a similar method was applied to the case of precipitation hardening.)

6.43

$$t_R^{-1} = Ce^{-Q/RT}$$

$$\frac{10\,h}{100\,h} = \frac{Ce^{-Q/R(256+273)K}}{Ce^{-Q/R(283+273)K}} = 0.1$$

or

$$\ln 0.1 = -\frac{Q}{R}(9.17 \times 10^{-5}\,K^{-1})$$

giving

$$Q = \underline{\underline{209\,kJ/mol}}$$

6.44 Calculate the temperature at which complete recrystallization would occur for the aluminum alloy of Problem 6.43 within one hour.

6.44

Re-arranging the original expression and substituting gives:

$$C = (100\,h)^{-1} e^{+(209{,}000\,J/mol)/(8.314\,J/mol \cdot K)(529\,K)}$$

$$= 3.91 \times 10^{18}\,h^{-1}$$

$$\therefore 1/t_R = (1\,h)^{-1} = (3.91 \times 10^{18}\,h^{-1})e^{-(209{,}000)/(8.314\,T)}$$

or

$$T = 586\,K = \underline{\underline{313°C}}$$

Section 6.6 – The Kinetics of Phase Transformations for Nonmetals

PP 6.10 Convert the 62 mol % from Sample Problem 6.10 to weight percent.

PP 6.10

1 mol. ZrO_2 = $(91.22 + 2[16.00])$ amu = 123.22 amu

1 mol. CaO = $(40.08 + 16.00)$ amu = 56.08 amu

1 mol. of 15 mol. % alloy = 0.85 mol. ZrO_2 + 0.15 mol. CaO

$$\text{wt.\% CaO} = \frac{0.15(56.08)\text{ amu}}{[0.15(56.08) + 0.85(123.22)]\text{ amu}} \times 100\% = 7.43\%$$

1 mol. of 2 mol. % alloy = 0.98 mol. ZrO_2 + 0.02 mol. CaO

$$\text{wt.\% CaO} = \frac{0.02(56.08)\text{ amu}}{[0.02(56.08) + 0.98(123.22)]\text{ amu}} \times 100\% = 0.92\%$$

Therefore,

$$\text{wt.\% monoclinic} = \frac{7.4 - 3.4}{7.4 - 0.9} \times 100\% = \underline{\underline{62 \text{ wt.\%}}}$$

6.45 The sintering of ceramic powders is a thermally activated process and shares the "rule-of-thumb" about temperature with diffusion and recrystallization. Estimate a minimum sintering temperature for (a) pure Al_2O_3, (b) pure mullite, and (c) pure spinel. (See Figures 5-39 and 5-40.)

6.45

Rule of thumb is $\frac{1}{3}$ to $\frac{1}{2}$ T_m. Therefore, minimum sintering temperature is $\frac{1}{3}T_m$.

(a) T_{m, Al_2O_3} in Fig. 5-39 = 2054°C = 2327 K; $\frac{1}{3}T_m$ = 776 K = $\underline{\underline{503°C}}$

(b) $T_{m,\text{ mullite}}$ in Fig. 5-39 = 1890°C = 2163 K; $\frac{1}{3}T_m$ = 721 K = $\underline{\underline{448°C}}$

(c) $T_{m,\text{ spinel}}$ in Fig. 5-40 ≅ 2110°C = 2383 K; $\frac{1}{3}T_m$ = 794 K = $\underline{\underline{521°C}}$

6.46 Four ceramic phase diagrams were presented in Figures 5-39 to 5-42. In which systems would you expect precipitation hardening to be a possible heat treatment? Briefly explain your answer.

6.46

The essential feature of the phase diagram is a solid solution region with decreasing solubility limit with decreasing temperature. The 3 systems which display this feature are:

$\underline{\underline{Al_2O_3 - SiO_2 \text{ (Fig. 5-39)}}}$

$\underline{\underline{MgO - Al_2O_3 \text{ (Fig. 5-40)}}}$

$\underline{\underline{CaO - ZrO_2 \text{ (Fig. 5-42)}}}$

6.47 The total sintering rate for $BaTiO_3$ increases 10-fold between 750 and 794°C. Calculate the activation energy for sintering in $BaTiO_3$.

6.47

$$R_s = Ce^{-Q/RT}$$

$$\frac{R_{794}}{R_{750}} = \frac{e^{-Q/R(794+273)K}}{e^{-Q/R(750+273)K}} = 10$$

or

$$\ln 10 = +\frac{Q}{(8.314\ J/mol\cdot K)}(4.03\times 10^{-5}\ K^{-1})$$

giving

$$Q = 475\ kJ/mol$$

6.48 Given the data in Problem 6.47, predict the temperature at which the initial sintering rate for $BaTiO_3$ would have increased a hundredfold compared to 750°C.

6.48 Substituting into the original expression:

$$\frac{R_T}{R_{750}} = \frac{e^{-(475{,}000)/(8.314T)}}{e^{-(475{,}000)/(8.314\times 1023\ K)}} = 100$$

giving

$$T = 1115\ K = 842°C$$

6.49 At room temperature, moisture absorption will occur slowly in parts made of nylon, a common engineering polymer. Such absorption will increase dimensions and lower strength. To stabilize dimensions, nylon products are sometimes given a preliminary "moisture conditioning" by immersion in hot or boiling water. At 60°C, the time to condition a 5-mm-thick nylon part to a 2.5% moisture content is 20 hours. At 77°C, the time for the same size part is 7 hours. As the conditioning is diffusional in nature, the time required for other temperatures can be estimated from the Arrhenius expression (Equation 4.1). Calculate the activation energy for this moisture conditioning process. (Note the method used for similar kinetics examples in Problems 6.37 and 6.43.)

6.49

$$t_{condition}^{-1} = Ce^{-Q/RT}$$

$$\frac{t_{c}^{-1},_{60°C}}{t_{c}^{-1},_{77°C}} = \frac{Ce^{-Q/R(60+273)K}}{Ce^{-Q/R(77+273)K}} = \frac{7\ h}{20\ h}$$

or

$$\ln(7/20) = -\frac{Q}{8.314\ J/mol\cdot K}(1.458\times10^{-4})K^{-1}$$

giving

$$Q = 59.9\ kJ/mol$$

6.50 Estimate the conditioning time in boiling water (100°C) for the nylon part discussed in Problem 6.49.

6.50 Substituting into the initial expression:

$$(20h)^{-1} = Ce^{-(59,900)/(8.314)(333)}$$

giving

$$C = 1.22\times10^{8}\ h^{-1}$$

$$\therefore t_{c,100°C}^{-1} = (1.22\times10^{8}\ h^{-1})e^{-(59,900)/(8.314)(373)}$$

$$= 0.508\ h^{-1}$$

or $t_{c,100°C} = 1.97\ h$

Section 7.1 - Ferrous Alloys

PP 7.1 For every 100,000 atoms of an SAE J431 (F10009) gray cast iron, how many atoms of each main alloying element are present? (Use elemental compositions in the middle of the ranges given in Table 7.5.) (See Sample Problem 7.1.)

PP 7.1 From Table 7.5, wt.% C = 3.40, wt.% Mn = 0.75, wt.% Si = 1.85, wt.% P = 0.12, wt.% S = 0.12

For a 100 g alloy, there will be 3.40 g C, etc....

Assuming balance in Fe, there will be

$$100\,g - (3.40 + 0.75 + 1.85 + 0.12 + 0.12)\,g = 93.76\,g \text{ Fe}$$

Number of atoms in 100 g alloy are:

$$N_{Fe} = \frac{93.76}{55.85} \times 0.6023 \times 10^{24} \text{ atoms} = 1.0111 \times 10^{24} \text{ atoms}$$

$$N_C = \frac{3.40}{12.01} \times \text{ " } = 1.705 \times 10^{23} \text{ "}$$

$$N_{Mn} = \frac{0.75}{54.94} \times \text{ " } = 8.22 \times 10^{21} \text{ "}$$

$$N_{Si} = \frac{1.85}{28.09} \times \text{ " } = 3.97 \times 10^{22} \text{ "}$$

$$N_P = \frac{0.12}{30.97} \times \text{ " } = 2.33 \times 10^{21} \text{ "}$$

$$N_S = \frac{0.12}{32.06} \times \text{ " } = 2.25 \times 10^{21} \text{ "}$$

$$1.2341 \times 10^{24} \text{ "}$$

For 100,000 atom alloy,

$$N_{Fe} = (1.0111 \times 10^{24} / 1.231 \times 10^{24}) \times 10^5 \text{ atoms} = 82,136$$

$$N_C = (1.705 \times 10^{23} / \text{ " }) \times \text{ " } = 13,851$$

$$N_{Mn} = (8.22 \times 10^{21} / \text{ " }) \times \text{ " } = 668$$

$$N_{Si} = (3.97 \times 10^{22} / \text{ " }) \times \text{ " } = 3,225$$

$$N_P = (2.33 \times 10^{21} / \text{ " }) \times \text{ " } = 189$$

$$N_S = (2.25 \times 10^{21} / \text{ " }) \times \text{ " } = 183$$

7.1 (a) Estimate the density of 1040 carbon steel as the weighted average of the densities of constituent elements.

(b) The density of 1040 steel is what percentage of the density of pure Fe?

7.1 (a) $\rho \cong \sum_i w_i \rho_i$

$$= (0.004)(2.27 \text{ g/cm}^3) + (1-0.004)(7.87 \text{ g/cm}^3)$$

$$= \underline{\underline{7.85 \text{ g/cm}^3}}$$

(b) $$\frac{\rho_{1040}}{\rho_{Fe}} \times 100\% = \frac{7.85 \text{ g/cm}^3}{7.87 \text{ g/cm}^3} \times 100\% = \underline{\underline{99.7\%}}$$

7.2 Repeat Problem 7.1 for the type 304 stainless steel in Table 7.10.

7.2 (a) $\rho \cong \sum_i w_i \rho_i$

$$= (0.0008)(2.27 \text{ g/cm}^3) + (0.02)(7.47 \text{ g/cm}^3) + (0.01)(2.33 \text{ g/cm}^3) + (0.19)(7.19 \text{ g/cm}^3) + (0.09)(8.91 \text{ g/cm}^3) + (1-0.3108)(7.87 \text{ g/cm}^3)$$

$$= \underline{\underline{7.77 \text{ g/cm}^3}}$$

(b) $$\frac{\rho_{304}}{\rho_{Fe}} \times 100\% = \frac{7.77}{7.87} \times 100\% = \underline{\underline{98.7\%}}$$

7.3 Use the weighted average of the densities of constituent elements to estimate the density of T1 tool steel in Table 7.3.

7.3 $\rho \cong \sum_i w_i \rho_i$

$$= [(0.00725)(2.27) + (0.0025)(7.47) + (0.003)(2.33) + (0.03875)(7.19) + (0.003)(8.91) + (0.18)(19.25) + (0.011)(6.09) + (0.7545)(7.87)] \text{ g/cm}^3$$

$$= \underline{\underline{9.82 \text{ g/cm}^3}}$$

7.4 Use the weighted average of the densities of constituent elements to estimate the density of Incoloy 903 in Table 7.4.

7.4

$$\rho \cong \sum_i w_i \rho_i$$

$$= [(0.001)(7.19) + (0.38)(8.91) + (0.15)(8.8) + (0.001)(10.22) + (0.03)(8.58) + (0.014)(4.51) + (0.007)(2.70) + (0.41)(7.87) + (0.0004)(2.27)]\ g/cm^3$$

$$= \underline{\underline{8.29\ g/cm^3}}$$

Section 7.2 - Nonferrous Alloys

PP 7.2 A common basis for selecting nonferrous alloys is their low density as compared to structural steels. Alloy density can be approximated as a weighted average of the densities of the constituent elements. In this way, calculate the densities of the aluminum alloys given in Table 7.10.

PP 7.2

$$\rho_{3003} = 0.9775\rho_{Al} + 0.0125\rho_{Mn} + 0.01\rho_{Mg}$$

$$= [0.9775(2.70) + 0.0125(7.47) + 0.01(1.74)]\ Mg/m^3$$

$$= \underline{\underline{2.75\ Mg/m^3}}$$

$$\rho_{2048} = 0.948\rho_{Al} + 0.004\rho_{Mn} + 0.033\rho_{Cu} + 0.015\rho_{Mg}$$

$$= [0.948(2.70) + 0.004(7.47) + 0.033(8.93) + 0.015(1.74)]\ Mg/m^3 = \underline{\underline{2.91\ Mg/m^3}}$$

7.5 A prototype Al–Li alloy is being considered for replacement of a 7075 alloy in a commercial aircraft. The compositions are compared in the following table. (**a**) Assuming that the same volume of material is used, what percentage reduction in density would occur by this material substitution? (**b**) If a total mass of 75,000 kg of 7075 alloy is currently used on the aircraft, what net mass reduction would result from the substitution of the Al–Li alloy?

	Primary alloying elements (wt %)					
Alloy	**Li**	**Zn**	**Cu**	**Mg**	**Cr**	**Zr**
Al–Li	2.0		3.0			0.12
7075		5.6	1.6	2.5	0.23	

7.5

(a) $\rho_{Al-Li} = 0.9488\rho_{Al} + 0.02\rho_{Li} + 0.03\rho_{Cu} + 0.0012\rho_{Zr}$

$= [0.9488(2.70) + 0.02(0.533) + 0.03(8.93) + 0.0012(6.51)]\ Mg/m^3$

$= 2.85\ Mg/m^3$

$\rho_{7075} = 0.9007\rho_{Al} + 0.056\rho_{Zn} + 0.016\rho_{Cu} + 0.025\rho_{Mg} + 0.0023\rho_{Cr}$

$= [0.9007(2.70) + 0.056(7.13) + 0.016(8.93) + 0.025(1.74) + 0.0023(7.19)]\ Mg/m^3$

$= 3.03\ Mg/m^3$

$\therefore \%\ change = \dfrac{3.03 - 2.85}{3.03} \times 100\% = \underline{\underline{5.94\%}}$

(b) mass reduction $= 0.0594(75{,}000\ kg)$

$= \underline{\underline{4{,}455\ kg}}$

7.6 Estimate the alloy densities for (**a**) the magnesium alloys of Table 7.10 and (**b**) the titanium alloy of Table 7.10.

7.6

(a) $\rho_{AZ31B} = 0.958\rho_{Mg} + 0.002\rho_{Mn} + 0.03\rho_{Al} + 0.01\rho_{Zn}$

$= [0.958(1.74) + 0.002(7.47) + 0.03(2.70) + 0.01(7.13)]\ Mg/m^3 = \underline{\underline{1.83\ Mg/m^3}}$

$\rho_{AM100A} = 0.899\rho_{Mg} + 0.001\rho_{Mn} + 0.10\rho_{Al}$

$= [0.899(1.74) + 0.001(7.47) + 0.10(2.70)]\ Mg/m^3$

$= \underline{\underline{1.84\ Mg/m^3}}$

(b) $\rho_{Ti\text{-}5Al\text{-}2.5Sn} = 0.925\rho_{Ti} + 0.05\rho_{Al} + 0.025\rho_{Sn}$

$= [0.925(4.51) + 0.05(2.70) + 0.025(7.29)]$

$= \underline{\underline{4.49\ Mg/m^3}}$

7.7 Between 1975 and 1985, the volume of all iron and steel in a given automobile model decreased from 0.162 m^3 to 0.116 m^3. In the same time frame, the volume of all aluminum alloys increased from 0.012 m^3 to 0.023 m^3. Using the densities of pure Fe and Al, estimate the mass reduction resulting from this trend in materials substitution.

7.7

mass reduction due to iron removal $= (0.162 - 0.116)\,m^3 \times \rho_{Fe}$

$= (0.046\,m^3)(7.87\,Mg/m^3)$

$= 0.362\,Mg = 362\,kg$

mass increase due to Al addition $= (0.023 - 0.012)\,m^3 \times \rho_{Al}$

$= (0.011)\,m^3\,(2.70\,Mg/m^3)$

$= 0.030\,Mg = 30\,kg$

∴ net mass reduction $= 362\,kg - 30\,kg$

$= \underline{\underline{332\,kg}}$

7.8 It is estimated that for an automobile design equivalent to the model described in Problem 7.8, the volume of all iron and steel will be further reduced to 0.082 m^3 by the year 2000. In the same time frame, the total volume of aluminum alloys is expected to increase to 0.034 m^3. Estimate the mass reduction (compared to 1975) resulting from this projected materials substitution.

7.8

mass reduction due to iron removal $= (0.162 - 0.082)\,m^3 \times \rho_{Fe}$

$= (0.080)\,m^3\,(7.87\,Mg/m^3)$

$= 0.630\,Mg = 630\,kg$

mass increase due to Al addition $= (0.034 - 0.012)\,m^3 \times \rho_{Al}$

$= (0.022)\,m^3\,(2.70\,Mg/m^3)$

$= 0.059\,Mg = 59\,kg$

∴ net mass reduction $= 630\,kg - 59\,kg$

$= \underline{\underline{571\,kg}}$

Section 7.3 - Mechanical Properties of Metals

PP 7.3 In Sample Problem 7.3, the basic mechanical properties of a 2024-T81 aluminum are calculated based on its stress–strain curve (Figure 7-4). Given below is load–elongation data for a type 304 stainless steel similar to that presented in Figure 7-3. This steel is similar to alloy 3(a) in Table 7.11 except that it has a different thermomechanical history, giving it slightly higher strength with lower ductility. **(a)** Plot these data in a manner comparable to Figure 7-3. **(b)** Replot these data as a stress–strain curve similar to Figure 7-4. **(c)** Replot the initial strain data on an expanded scale, similar to that used for Figure 7-5. **(d)** Using the results of parts (a)–(c), calculate (i) E, (ii) Y.S., (iii) T.S., and (iv) percent elongation at failure for this 304 stainless steel. For parts (i)–(iii), express answers in both Pa and psi units.

Load (N)	Gage length (mm)	Load (N)	Gage length (mm)
0	50.8000	35,220	50.9778
4,890	50.8102	35,720	51.0032
9,779	50.8203	40,540	51.816
14,670	50.8305	48,390	53.340
19,560	50.8406	59,030	55.880
24,450	50.8508	65,870	58.420
27,620	50.8610	69,420	60.960
29,390	50.8711	69,670 (maximum)	61,468
32,680	50.9016	68,150	63.500
33,950	50.9270	60,810 (fracture)	66.040 (after fracture)
34,580	50.9524		

Original specimen diameter: 12.7 mm.

PP 7.3

LOAD (N) = P	L (mm)	ΔL (mm) = L − L_o	σ (MPa) = P/A_o *	ε = ΔL/L_o
0	50.8000	0	0	0
4,890	50.8102	0.0102	38.6	0.0002
9,779	50.8203	0.0203	77.2	0.0004
14,670	50.8305	0.0305	115.8	0.0006
19,560	50.8406	0.0406	154.4	0.0008
24,450	50.8508	0.0508	193	0.0010
27,620	50.8610	0.0610	218	0.0012
29,390	50.8711	0.0711	232	0.0014
32,680	50.9016	0.1016	258	0.0020
33,950	50.9270	0.1270	268	0.0025
34,580	50.9524	0.1524	273	0.0030
35,220	50.9778	0.1778	278	0.0035
35,720	51.0032	0.2032	282	0.0040
40,540	51.816	1.016	320	0.02
48,390	53.340	2.540	382	0.05
59,030	55.880	5.080	466	0.10
65,870	58.420	7.620	520	0.15
69,420	60.960	10.160	548	0.20
69,670	61.468	10.668	550	0.21
68,150	63.500	12.700	538	0.25
60,810	66.040	15.240	480	0.30

* $A_o = \pi/4\,(12.7\text{ mm})^2$

(a)

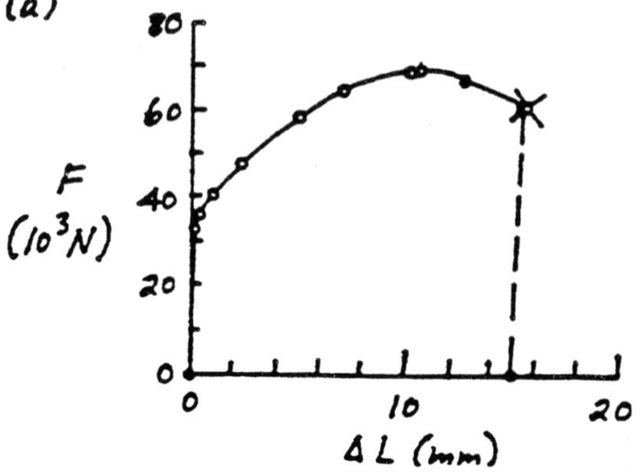

(b)

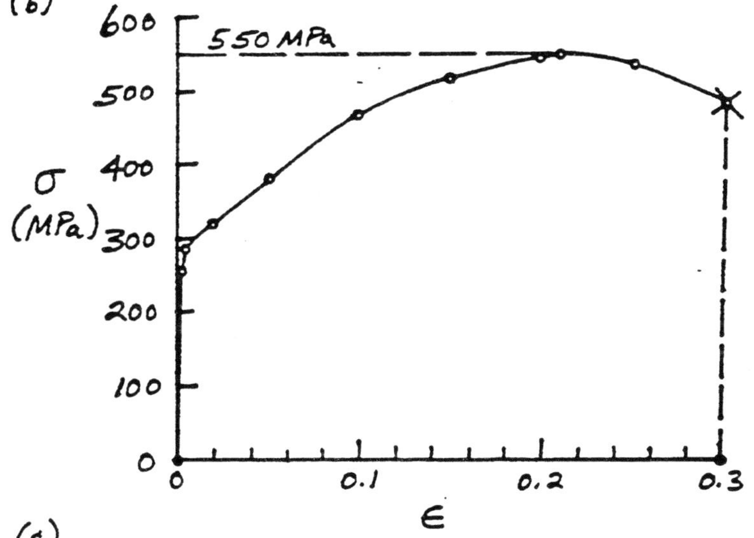

(c)

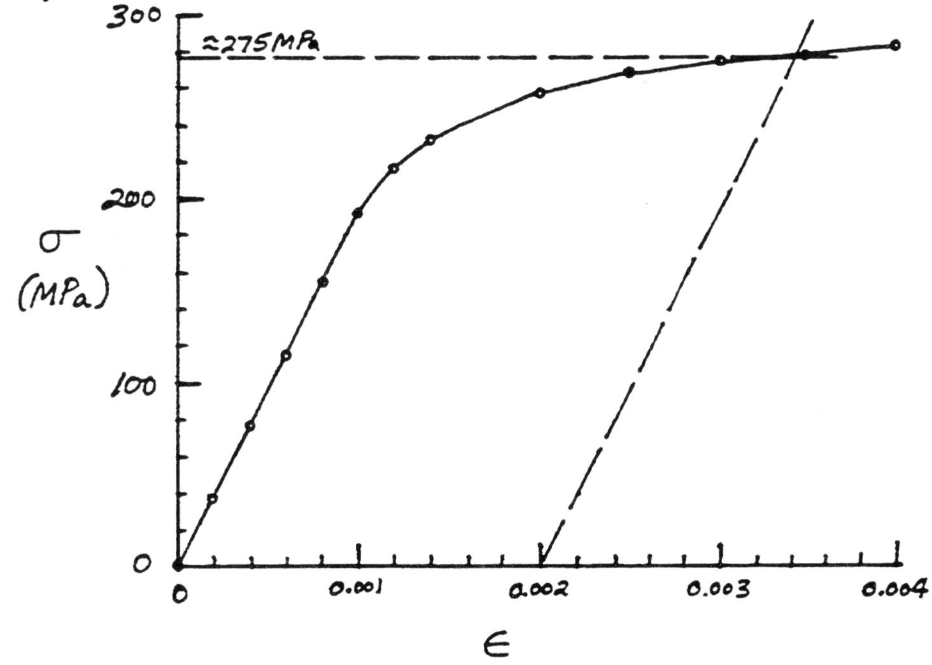

(d) (i) The construction in the graph of part (c) indicates that $E = \frac{193\ MPa}{0.001} = \underline{\underline{193 \times 10^3\ MPa}}$

$$\times \frac{10^6\ Pa}{1\ MPa} \times 0.145 \times 10^{-3}\ psi/Pa = \underline{\underline{28.0 \times 10^6\ psi}}$$

(ii) Again referring to the graph of part (c) we see that

$$Y.S. = \underline{\underline{275\,MPa}}$$

$$\times \frac{10^6\,Pa}{1\,MPa} \times 0.145 \times 10^{-3}\,psi/Pa = \underline{\underline{39,900\,psi}}$$

(iii) From part (b),

$$T.S. = \underline{\underline{550\,MPa}}$$

$$\times \frac{10^6\,Pa}{1\,MPa} \times 0.145 \times 10^{-3}\,psi/Pa = \underline{\underline{79,800\,psi}}$$

(iv) Using part (b) again,

$$\%\text{elongation at failure} = 100 \times \epsilon_{failure}$$

$$= 100 \times 0.3 = \underline{\underline{30\,\%}}$$

PP 7.4 For the 304 stainless steel introduced in Practice Problem 7.3, calculate the elastic recovery for the specimen upon removal of the load of **(a)** 35,720 N and **(b)** 69,420 N. (See Sample Problem 7.4.)

PP 7.4

(a) $\sigma = P/A_0$

For $P = 35{,}720\,N$, $\sigma = \dfrac{35{,}720\,N}{\pi/4\,(12.7 \times 10^{-3}\,m)^2} = 282\,MPa$

$\varepsilon = \sigma/E$

From PP 7.3, $E = 193 \times 10^3\,MPa$.

$\varepsilon = (282 \times 10^6\,Pa)/(193 \times 10^9\,MPa) = \underline{\underline{1.46 \times 10^{-3}}}$

(b) For $P = 69{,}420\,N$,

$$\sigma = \frac{69{,}420\,N}{\pi/4\,(12.7 \times 10^{-3}\,m)^2} = 548\,MPa$$

$\varepsilon = (548 \times 10^6\,Pa)/(193 \times 10^9\,MPa) = \underline{\underline{2.84 \times 10^{-3}}}$

PP 7.5

a. Calculate the center-to-center separation distance of two Fe atoms along the ⟨100⟩ direction in unstressed α-iron.

b. Calculate the separation distance along that direction under a tensile stress of 1000 MPa. (See Sample Problem 7.5.)

PP 7.5

(a) For a bcc structure, the separation distance = lattice parameter:

$$a = \frac{4R}{\sqrt{3}} = \frac{4(0.124\ \text{nm})}{\sqrt{3}} = \underline{\underline{0.2864\ \text{nm}}}$$

(b) $\epsilon = \sigma/E = (1000\ \text{MPa})/(125 \times 1000\ \text{MPa}) = 0.008$

or $a_{stretched} = 1.008 \times 0.2864\ \text{nm} = \underline{\underline{0.2887\ \text{nm}}}$

PP 7.6

For the alloy in Sample Problem 7.6, calculate the rod diameter at the (tensile) yield stress indicated in Table 7.11.

PP 7.6

The yield strength in Table 7.11 = 145 MPa.

$$\epsilon_z = \frac{\sigma}{E} = \frac{145\ \text{MPa}}{70 \times 10^3\ \text{MPa}} = 2.07 \times 10^{-3}$$

$$\epsilon_{dia} = -\nu\, \epsilon_z = -(0.33)(2.07 \times 10^{-3}) = -6.84 \times 10^{-4}$$

Then,

$$d_f = d_0(\epsilon_{dia} + 1) = 10\ \text{mm}\,(-6.84 \times 10^{-4} + 1)$$

$$= \underline{\underline{9.9932\ \text{mm}}}$$

PP 7.7

Suppose that a ductile iron (100-70-03, air-quenched) has a tensile strength of 700 MPa. What diameter impression would you expect the 3000-kg load to produce with the 10-mm-diameter ball? (See Sample Problem 7.7.)

PP 7.7

Figure 7-17(b) gives (for T.S. = 700 MPa) 220 BHN.

$$220 = \frac{2(3000)}{\pi(10)\left[10 - \sqrt{10^2 - d^2}\right]}$$

or $d = \underline{\underline{4.08\ \text{mm}}}$

PP 7.8 Find the necessary carbon level to ensure that a plain-carbon steel will be relatively ductile down to 0°C. (See Sample Problem 7.8.)

PP 7.8 As in Sample Problem 7.8, we see from Figure 7-20(a) that we need a carbon level $\leq 0.20\%$.

PP 7.9 What crack size is needed to produce catastrophic failure in the alloy in Sample Problem 7.9 at (a) $\frac{1}{3}$ Y.S. and (b) $\frac{3}{4}$ Y.S.?

PP 7.9 (a) $K_{IC} = \sigma_f \sqrt{\pi a}$ or $a = \dfrac{K_{IC}^2}{\pi \sigma_f^2}$

$$a = \frac{(98 \text{ MPa}\sqrt{\text{m}})^2}{\pi [\frac{1}{3}(1460 \text{ MPa})]^2} = 0.0129 \text{ m} = \underline{\underline{12.9 \text{ mm}}}$$

(b)

$$a = \frac{(98 \text{ MPa}\sqrt{\text{m}})^2}{\pi [\frac{3}{4}(1460 \text{ MPa})]^2} = 2.55 \times 10^{-3} \text{ m} = \underline{\underline{2.55 \text{ mm}}}$$

PP 7.10 In Sample Problem 7.10, a service stress is calculated with consideration for fatigue loading. Using the same considerations, estimate a maximum permissible service stress for an 80-55-06 as-cast ductile iron with a Brinell hardness number of 200 (see Figure 7-17).

PP 7.10 As in Sample Problem 7.10,

$$\text{Service stress} = \frac{\text{F.S.}}{2} = \frac{(\frac{1}{4}\text{T.S.})}{2} = \frac{\text{T.S.}}{8}$$

From Figure 7-17(b), a BHN=200 corresponds to a T.S. = 620 MPa.

Then,

$$\text{Service stress} = \frac{620 \text{ MPa}}{8} = \underline{\underline{77.5 \text{ MPa}}}$$

PP 7.11 Using an Arrhenius equation, we are able to predict the creep rate for a given alloy at 600°C in Sample Problem 7.11. For the same system, calculate the creep rate at (**a**) 700°C, (**b**) 800°C, and (**c**) 900°C. (**d**) Plot the results on an Arrhenius plot similar to Figure 7-36.

PP 7.11

(a) $\dot{\varepsilon}_{700°C} = (80.5\times10^{6}\ \%\ \text{per hour})\, e^{-(2\times10^{5})/(8.314)(973)}$

$= 1.47\times10^{-3}\ \%\ \text{per hour}$

(b) $\dot{\varepsilon}_{800°C} = (80.5\times10^{6}\ \%\ \text{per hour})\, e^{-(2\times10^{5})/(8.314)(1073)}$

$= 1.48\times10^{-2}\ \%\ \text{per hour}$

(c) $\dot{\varepsilon}_{900°C} = (80.5\times10^{6}\ \%\ \text{per hour})\, e^{-(2\times10^{5})/(8.314)(1173)}$

$= 9.98\times10^{-2}\ \%\ \text{per hour}$

(d)

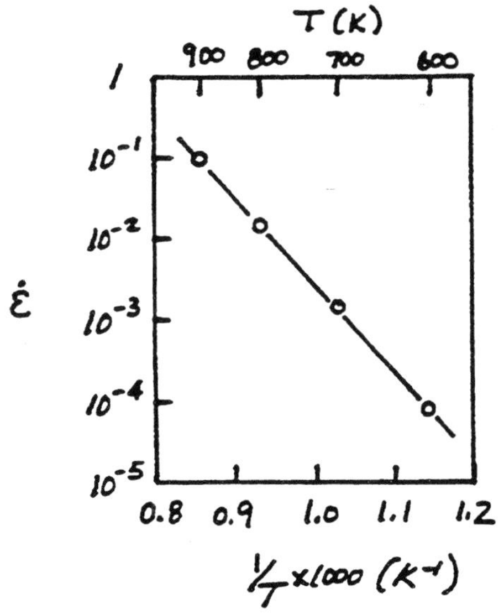

PP 7.12 In Sample Problem 7.12, we are able to estimate a maximum service temperature for Inconel 718 in order to survive a stress of 690 MPa (100,000 psi) for 10,000 h. What is the maximum service temperature that will allow this alloy to survive (**a**) 100,000 h and (**b**) 1000 h at the same stress?

PP 7.12

(a) For a rupture time of 10^5 h:

σ (ksi)	T (°C)
105	540
78	595
45	650

Plotting gives:

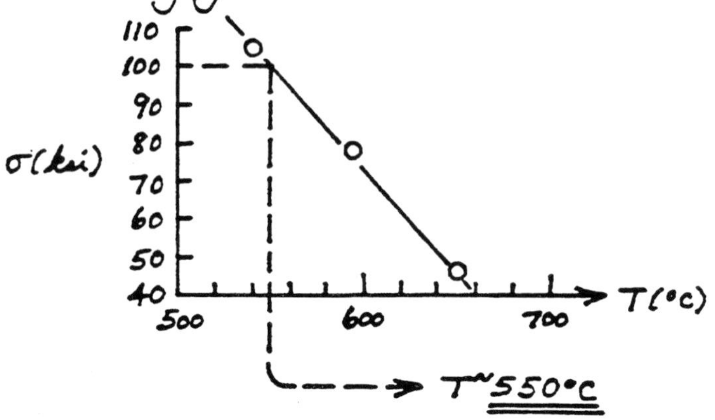

(b) For a rupture time of 10^3 h:

σ (ksi)	T(°C)
140	540
110	595
85	650
55	705

Plotting gives:

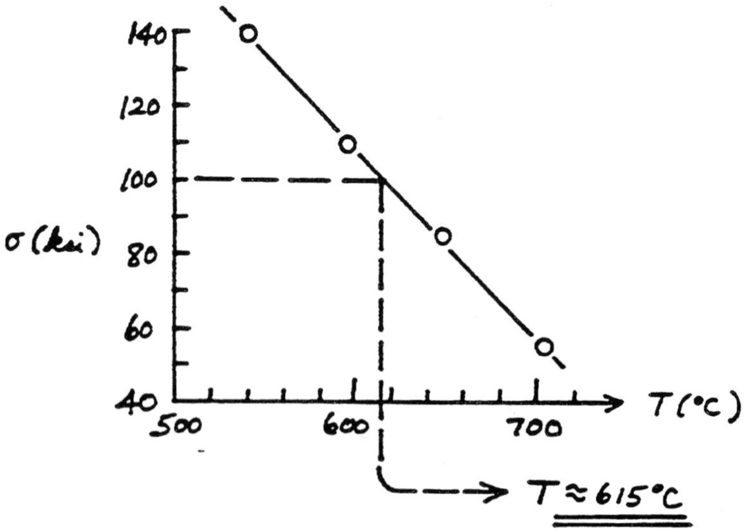

(In monitoring the results of Sample Problem 7.12 and Practice Problem 7.12, we note that, under the given stress, a change of service temperature of much less than 100°C leads to a two order-of-magnitude change in lifetime.)

7.9 The following three $\sigma - \epsilon$ data points are provided for a titanium alloy for aerospace applications: $\epsilon = 0.002778$ (at $\sigma = 300$ MPa), 0.005556 (600 MPa), 0.009897 (900 MPa). Calculate E for this alloy.

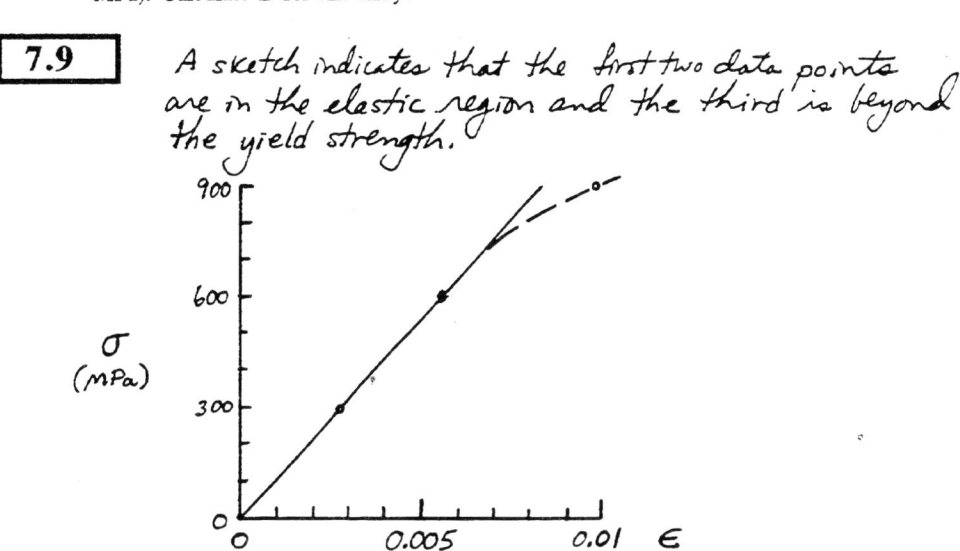

7.9 A sketch indicates that the first two data points are in the elastic region and the third is beyond the yield strength.

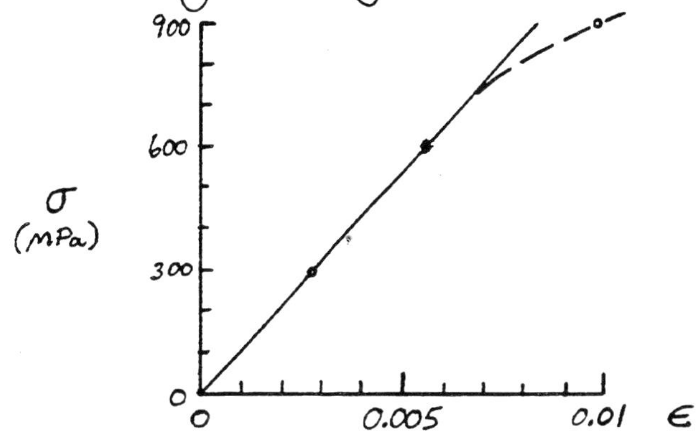

$$\therefore \quad E = \sigma/\epsilon = \frac{600\text{ MPa}}{0.005556} = 108 \times 10^3 \text{ MPa}$$

$$= \underline{\underline{108 \text{ GPa}}}$$

7.10 If the Poisson's ratio for the alloy in Problem 7.9 is 0.35, calculate (**a**) the shear modulus G, and (**b**) the shear stress τ necessary to produce an angular displacement, α, of 0.2865°.

7.10 (a) From Equation 7.8,

$$G = \frac{E}{2(1+\nu)}$$

Using the result of Problem 7.9,

$$G = \frac{108 \text{ GPa}}{2(1+0.35)} = \underline{\underline{40.0 \text{ GPa}}}$$

(b) From Equation 7.6,

$$\gamma = \tan\alpha = \tan(0.2865°) = 0.00500$$

allowing us to use Equation 7.7:

$$\tau = G\gamma = (40.0 \times 10^3 \text{ MPa})(0.00500)$$
$$= \underline{\underline{200 \text{ MPa}}}$$

7.11 In Section 4.4, the point was made that the theoretical strength (i.e., critical shear strength) of a material is roughly 0.1 G. **(a)** Use the result of Problem 7.10a to estimate the theoretical critical shear strength of the titanium alloy. **(b)** Comment on the relative value of the result in (a) compared to the apparent yield strength implied by the data given in Problem 7.9.

7.11 (a) $\tau_{theo} \cong 0.1 G$

$= 0.1 \times 40.0$

$= \underline{\underline{4.00 \text{ GPa}}}$

(b) Inspection of the sketch for Problem 7.9 indicates a Y.S. of roughly 700 to 800 MPa. Therefore the theoretical strength is roughly

$$\frac{4 \times 10^3 \text{ MPa}}{800 \text{ MPa}} \text{ to } \frac{4 \times 10^3 \text{ MPa}}{700 \text{ MPa}}$$

$\cong$ 5 to 6 times the yield strength indicating the substantial "weakening" of the alloy by dislocation motion.

7.12 Consider the 1040 carbon steel listed in Table 7.11. **(a)** A 20-mm-diameter bar of this alloy is used as a structural member in an engineering design. The unstressed length of the bar is precisely 1 m. The structural load on the bar is 9×10^4 N in tension. What will be the length of the bar under this structural load? **(b)** A design engineer is considering a structural change that will increase the tensile load on this member. What is the maximum tensile load that can be permitted without producing extensive plastic deformation of the bar? Give your answers in both newtons (N) and pounds force (lb_f).

7.12 (a) $\sigma = P/A_0 = (9 \times 10^4 \text{ N}) / \pi (10 \times 10^{-3} \text{ m})^2 = 286 \text{ MPa}$

$\varepsilon = \sigma/E = (286 \text{ MPa})/(200 \times 10^3 \text{ MPa}) = 1.43 \times 10^{-3}$

$L = L_0(1+\varepsilon) = \underline{\underline{1.00143 \text{ m}}}$

(b) Y.S. = 600 MPa

$P = \sigma A_0 = 600 \times 10^6 \frac{N}{m^2} \times \pi (10 \times 10^{-3} \text{ m})^2 = \underline{\underline{1.88 \times 10^5 \text{ N}}}$

$= 1.88 \times 10^5 \text{ N} \times 0.2248 \text{ lb}_f/\text{N} = \underline{\underline{4.24 \times 10^4 \text{ lb}_f}}$

7.13 Heat treatment of the alloy in Problem 7.12 does not significantly affect the modulus of elasticity but does change strength and ductility. For a quench-and-temper operation that produces a tempered martensite (see Figure 6-15), the corresponding mechanical property data are

Y.S. = 1100 MPa (159 ksi)

T.S. = 1380 MPa (200 ksi)

% elongation at failure = 12

Again considering a 20-mm-diameter by 1-m-long bar of this alloy, what is the maximum tensile load that can be permitted without producing extensive plastic deformation of the bar?

7.13

$$P = \sigma A_0 = (Y.S.)A_0 = 1100\times10^6\,N/m^2 \times \pi(10\times10^{-3}\,m)^2 =$$

$$= \underline{\underline{3.46\times10^5\,N}} \times 0.2248\,lb_f/N = \underline{\underline{7.77\times10^4\,lb_f}}$$

7.14 Repeat Problem 7.12 for the 2024-T81 aluminum illustrated in Figure 7-4 and Sample Problem 7.3.

7.14

(a) Again, $\sigma = 286\,MPa$.

$$\varepsilon = \sigma/E = 286\,MPa/(70\times10^3\,MPa) = 4.09\times10^{-3}$$

$$L = L_0(1+\varepsilon) = \underline{\underline{1.00409\,m}}$$

(b) $$P = \sigma A_0 = (Y.S.)A_0 = 410\times10^6\,N/m^2 \times \pi(10\times10^{-3}\,m)^2 =$$

$$= \underline{\underline{1.29\times10^5\,N}} \times 0.2248\,lb_f/N = \underline{\underline{2.90\times10^4\,lb_f}}$$

7.15 In normal motion, the load exerted on the hip joint is 2.5 times body weight. (a) Calculate the corresponding stress (in MPa) on an artificial hip implant with a cross-sectional area of 5.64 cm^2 in a patient weighing 150 lb_f. (b) Calculate the corresponding strain if the implant is made of Ti-6Al-4V which has an elastic modulus of 124 GPa.

7.15

(a) $$\sigma = \frac{F}{A} = \frac{2.5(150\,lb_f)(1N/0.2248\,lb_f)}{5.64\,cm^2(1m/100cm)^2} =$$

$$= 2.96\times10^6\,N/m^2 = \underline{\underline{2.96\,MPa}}$$

(b) $$\therefore \varepsilon = \sigma/E = \frac{2.96\,MPa}{124\,GPa} = \frac{2.96\,MPa}{124\times10^3\,MPa}$$

$$= \underline{\underline{2.39\times10^{-5}}}$$

7.16 Repeat Problem 7.15 for the case of an athlete who undergoes a hip implant. The same alloy is used but, because the athlete weighs 200 lb_f, a larger implant is required (with a cross-sectional area of 6.90 cm^2). Also, consider the situation in which the athlete slides into second base exerting a load of five times body weight.

7.16

(a) $$\sigma = \frac{F}{A} = \frac{5(200\ lb_f)(1N/0.2248\ lb_f)}{6.90\ cm^2(1m/100\ cm)^2}$$

$$= 6.45 \times 10^6\ N/m^2 = \underline{\underline{6.45\ MPa}}$$

(b) $$\therefore \epsilon = \sigma/E = \frac{6.45\ MPa}{124 \times 10^3\ MPa}$$

$$= \underline{\underline{5.20 \times 10^{-5}}}$$

7.17 Suppose that you were asked to select a material for a spherical pressure vessel to be used in an aerospace application. The stress in the vessel wall is

$$\sigma = \frac{pr}{2t}$$

where p is the internal pressure, r the outer radius of the sphere, and t the wall thickness. The mass of the vessel is

$$m = 4\pi r^2 t\rho$$

where ρ is the material density. The operating stress of the vessel will always be

$$\sigma \leq \frac{\text{Y.S.}}{S}$$

where S is a safety factor. (a) Show that the minimum mass of the pressure vessel will be

$$m = 2S\pi pr^3 \frac{\rho}{\text{Y.S.}}$$

(b) Given Table 7.11 and the following data, select the alloy that will produce the lightest vessel.

Alloy	ρ (Mg/m^3)	Cost[a] (\$/kg)
1040 carbon steel	7.8	0.63
304 stainless steel	7.8	3.70
3003-H14 aluminum	2.73	3.00
Ti–5 Al–2.5 Sn	4.46	15.00

[a]Approximate in U.S. dollars.

(c) Given Table 7.11 and the data in the preceding table, select the alloy that will produce the minimum cost vessel.

7.17 (a) $m = 4\pi r^2 t \rho$ or $t = \dfrac{m}{4\pi r^2 \rho}$

$$\sigma = \frac{pr}{2t} \le \frac{Y.S.}{S}$$

Therefore,

$$\frac{Y.S.}{S} \ge \frac{pr\,4\pi r^2 \rho}{2m} = \frac{2\pi r^3 p\rho}{m}$$

or,

$$m \ge 2S\pi r^3 p\left(\frac{\rho}{Y.S.}\right)$$

That is, the minimum mass is:

$$m = 2S\pi p r^3 \frac{\rho}{Y.S.}$$

(b) For a given design, S, p, and r are fixed. In effect, we are looking for the minimum ratio of $\rho/Y.S.$:

Alloy	ρ (Mg/m^3)	Y.S. (MPa)	ρ/Y.S. (Mg/[m^3·MPa])
1040	7.8	600	0.0130
304	7.8	205	0.0380
3003-H14	2.73	145	0.0188
Ti-5Al-2.5Sn	4.46	827	0.00539

Therefore, Ti-5Al-2.5Sn will produce the lightest vessel.

(c) Here, we are looking for the minimum product of ρ/Y.S. × cost:

Alloy	ρ/Y.S. (Mg/[m^3·MPa])	cost ($/kg)	ρ/Y.S. × cost
1040	0.0130	0.63	0.00819
304	0.0380	3.70	0.1406
3003-H14	0.0188	3.00	0.0564
Ti-5Al-2.5Sn	0.00539	15.00	0.0809

Therefore, 1040 carbon steel will produce the minimum cost vessel.

7.18 Many design engineers, especially in the aerospace field, are more interested in strength-per-unit density than strength or density individually. (If two alloys each have adequate strength, the lower density one is preferred for potential fuel savings.) Prepare a table comparing the tensile strength-per-unit density of the aluminum alloys of Table 7.10 with the 1040 steel in the same table. Note Problem 7.1 and Practice Problem 7.2 for density calculations. (The strength-per-unit density is generally termed *specific strength*, or *strength-to-weight ratio*, and is discussed relative to composite properties in Section 10.5.)

7.18

Alloy	T.S.(MPa)	ρ (Mg/m^3)	T.S./ρ (mm)*
1040	750	7.85	9.7×10^6
3003 Al	150	2.75	5.6×10^6
2048 Al	457	2.91	16.0×10^6

*See Sample Problem 10.10 for a discussion of the units for specific strength.

7.19 Expand on Problem 7.18 by including the magnesium alloys and the titanium alloy of Table 7.10 in the comparison of strength-per-unit density. (Note Problem 7.6 for additional density calculations.)

7.19

Alloy	T.S.(MPa)	ρ(Mg/m^3)	T.S./ρ (mm)
AZ31B Mg	290	1.83	16.2×10^6
AM100A Mg	150	1.84	8.3×10^6
Ti-5Al-2.5Sn	862	4.49	19.6×10^6

7.20 (a) Select the alloy in Problem 7.17 with the maximum tensile strength-per-unit density. (Note Problem 7.18 for a discussion of this quantity.) (b) Select the alloy in Problem 7.17 with the maximum (tensile strength-per-unit density)/unit cost.

7.20

Alloy	TS(MPa)	ρ(Mg/m^3)	TS/ρ (mm)	$\frac{TS/\rho}{cost}$ ($\frac{mm}{\$/kg}$)
1040	750	7.8	9.8×10^6	15.6×10^6 ← (b)
304	515	7.8	6.7×10^6	1.8×10^6
3003-H14	150	2.73	5.6×10^6	1.9×10^6
Ti-5Al-2.5Sn	862	4.46	19.7×10^6	1.3×10^6 ← (a)

• 7.21 The stress remaining within a structural material after all applied loads are removed is termed *residual stress*. This commonly occurs following various thermomechanical treatments such as welding and machining. In analyzing residual stress by x-ray diffraction, the following stress constant, K_1, is used:

$$K_1 = \frac{E\cot\theta}{2(1+\nu)\sin^2\psi}$$

where E and ν are the elastic constants defined in this chapter, θ is a Bragg angle (see Section 3.7), and ψ is an angle of rotation of the sample during the x-ray diffraction experiment (generally $\psi = 45°$). To maximize experimental accuracy, one prefers to use the largest possible Bragg angle, θ. However, hardware configuration (Figure 3-41) prevents θ from being greater than 80°. (a) Calculate the maximum θ for a 1040 carbon steel using CrK$_\alpha$ radiation (λ = 0.2291 nm). (Note that 1040 steel is nearly pure iron, which is a bcc metal, and that the reflection rules for a bcc metal are given in Table 3.3.) (b) Calculate the value of the stress constant for 1040 steel.

7.21 (a) The necessary calculations were made for Problem 3.80 which indicate that $(2\theta)_{211} = 157°$ or $\theta_{211} = \underline{78.5°}$. (The next higher diffraction line (220) does not satisfy the Bragg condition.)

(b)
$$K_1 = \frac{E\cot\theta}{2(1+\nu)\sin^2\psi}$$

$$= \frac{(200\times10^3\,\text{MPa})\cot 78.5°}{2(1+0.3)\sin^2(45°)}$$

$$= \underline{31.3\,\text{GPa}}$$

Note: In fact, the units for the stress constant are GPa/radian. This becomes apparent only by closer inspection of the derivation of the term.

● 7.22 Repeat Problem 7.21 for 2048 aluminum, which for purposes of the diffraction calculations can be approximated by pure aluminum. (Note that aluminum is an fcc metal and that the reflection rules for such materials are given in Table 3.3.)

7.22 (a) One can use the d spacing calculations from PP 3.23, leading to the calculation

$$\theta_{222} = \arcsin \frac{0.2291\ nm}{2 \times 0.117\ nm} = \underline{\underline{78.3°}}.$$

(The next higher diffraction line (400) does not satisfy the Bragg condition.)

(b) $$K_1 = \frac{E \cot \theta}{2(1+\nu)\sin^2 \psi}$$

As in 7.21 we can take E & ν data from Tables 7.11 and 7.12 for the most similar available aluminum alloys →

$$K_1 = \frac{(70.3 \times 10^3 MPa) \cot (78.3°)}{2(1+0.33)\sin^2(45°)}$$

$$= \underline{\underline{10.9\ GPa}}$$

7.23 You are provided an unknown alloy with a measured Brinell hardness value of 100. Having no other information than the data of Figure 7-17a, estimate the tensile strength of the alloy. (Express your answer in the form $x \pm y$.)

7.23 All data in Figure 7-17(a) fall within the band:

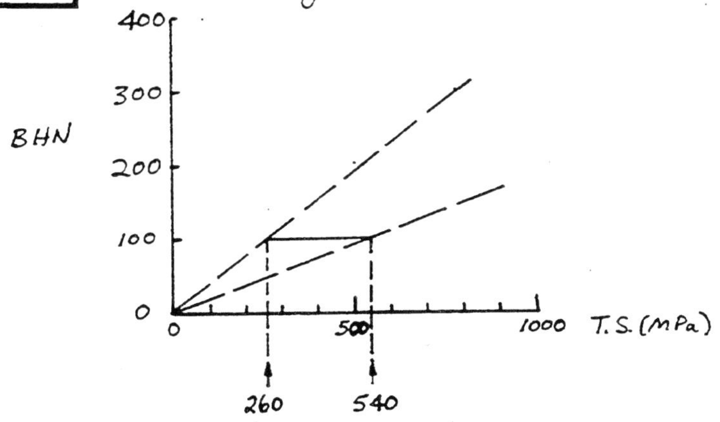

$\therefore$ avg. est. T.S. $= \frac{260+540}{2}$ MPa $= 400$ MPa

"error bar" $= 540$ MPa $- 400$ MPa $= 140$ MPa

or est. T.S. $=$ 400 MPa $\pm$ 140 MPa

7.24 Show that the data of Figure 7-17b are consistent with the plot of Figure 7-17a.

7.24 The T.S. data of Figure 7-17(b) follow a roughly linear band between (160 BHN, 400 MPa) and (380 BHN, 1100 MPa). Superimposing this band on the wider band of data shown in Figure 7-17(a) gives:

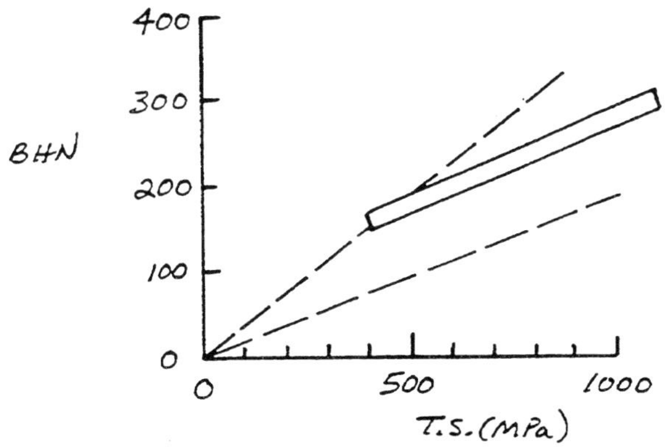

showing that the ductile iron data set largely falls within the band of the larger data set.

7.25 A ductile iron (65-45-12, annealed) is to be used in a spherical pressure vessel. The specific alloy obtained for the vessel has a Brinell hardness number of 200. The design specifications for the vessel include a spherical outer radius of 0.30 m, wall thickness of 20 mm, and a safety factor of 2. Using the information in Figure 7-17 and Problem 7.17, calculate the maximum operating pressure p for this vessel design.

7.25

Figure 7-17(b) indicates that BHN = 200 corresponds to Y.S. = 400 MPa.

From Problem 7.17,

$$\sigma_{max} = Y.S./S = 400\ MPa/2 = 200\ MPa.$$

Also,

$$p_{max} = \frac{2\sigma_{max} t}{r} = \frac{2(200\ MPa)(20 \times 10^{-3} m)}{0.30\ m}$$

$$= \underline{\underline{26.7\ MPa}}$$

7.26 Repeat Problem 7.25 for another ductile iron (grade 120-90-02, oil-quenched) with a Brinell hardness number of 280.

7.26

Figure 7-17(b) indicates that BHN = 280 corresponds to Y.S. = 850 MPa.

From Problem 7.17,

$$\sigma_{max} = Y.S./S = 850\ MPa/2 = 425\ MPa.$$

Also,

$$p_{max} = \frac{2\sigma_{max} t}{r} = \frac{2(425\ MPa)(20 \times 10^{-3} m)}{0.30\ m} = \underline{\underline{56.7\ MPa}}$$

7.27 The simple expressions for Rockwell hardness numbers in Table 7.13 involve indentation, t, expressed in millimeters. A given steel with a BHN of 235 is also measured by a Rockwell hardness tester. Using a $\frac{1}{16}$-in.-diameter steel sphere and a load of 100 kg, the indentation t is found to be 0.062 mm. What is the Rockwell hardness number?

7.27

$$\underline{\underline{R_B}} = 130 - 500t = 130 - 500(0.062) = \underline{\underline{99}}$$

7.28 An additional Rockwell hardness test is made on the steel considered in Problem 7.27. Using a diamond cone under a load of 150 kg, an indentation t of 0.157 mm is found. What is the resulting alternative Rockwell hardness value?

7.28

$$\underline{\underline{R_C}} = 100 - 500t = 100 - 500(0.157) = \underline{\underline{21.5}}$$

7.29 You are asked to nondestructively measure the yield strength and tensile strength of an annealed 65-45-12 cast iron structural member. Fortunately, a small hardness indentation in this structural design will not impair its future usefulness, which is a working definition of "non-destructive." A 10-mm-diameter tungsten carbide sphere creates a 4.26-mm-diameter impression under a 3000-kg load. What are the yield and tensile strengths?

7.29

$$BHN = \frac{2(3000)}{\pi(10)(10-\sqrt{10^2-4.26^2})} = 200$$

Giving (from Figure 7-17(b)):

$\underline{\underline{Y.S. = 400\,MPa}}$

and $\underline{\underline{T.S. \cong 550\,MPa}}$

7.30 As in Problem 7.29, calculate the yield and tensile strengths for the case of a 4.48-mm-diameter impression under identical conditions.

7.30

$$BHN = \frac{2(3000)}{\pi(10)(10-\sqrt{10^2-4.48^2})} = 180$$

Giving (from Figure 7-17(b)):

$Y.S. \cong \underline{\underline{360\,MPa}}$

and $T.S. \cong \underline{\underline{490\,MPa}}$

7.31 The Ti-6Al-4V orthopedic implant material introduced in Problem 7.15 gives a 3.27-mm-diameter impression when a 10-mm-diameter tungsten carbide sphere is applied to the surface with a 3000-kg load. What is the Brinell hardness number of this alloy?

7.31

$$BHN = \frac{2(3000)}{\pi(10)(10-\sqrt{10^2-3.27^2})}$$

$\underline{\underline{BHN}} = \underline{\underline{347}}$

7.32 Which of the alloys in Table 7.15 would you expect to exhibit ductile-to-brittle transition behavior? (State the basis of your selection.)

7.32 As stated in the text, ductile-to-brittle behavior is characteristic of bcc alloys. In Table 7.15, the alloys we expect to be bcc are those which are predominantly α-Fe or Nb:

1. 1040 carbon steel
2. 8630 low-alloy steel
4. L2 tool steel
6. (a) Ductile iron
14. Nb-1 Zr

7.33 (a) For the Fe–Mn–0.05 C alloys of Figure 7-20b, plot the ductile-to-brittle transition temperature (indicated by the sharp vertical rise in impact energy) against percentage Mn. (b) Using the plot from (a), estimate the percentage Mn level (to the nearest 0.1%) necessary to produce a ductile-to-brittle transition temperature of precisely 0°C.

7.33 (a) Taking the data from Fig. 7-20(b) →

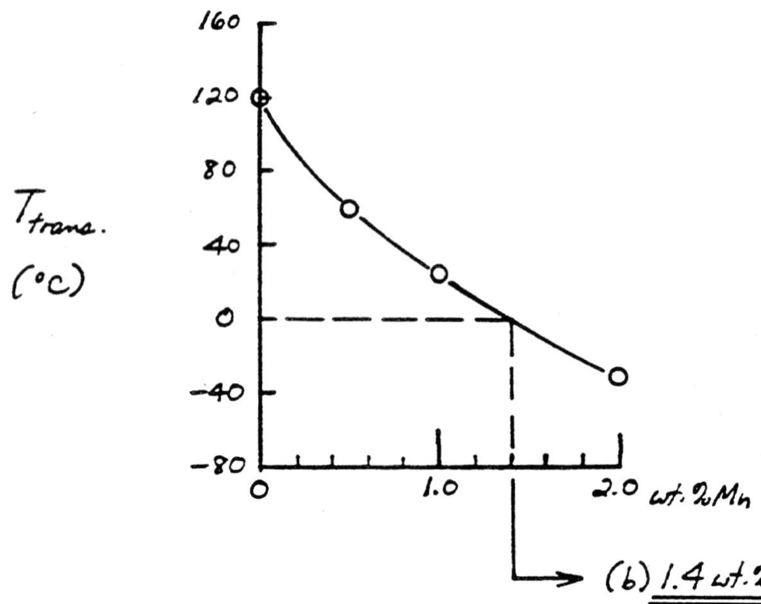

(b) 1.4 wt.% Mn

7.34 Estimate the percentage Mn level (to the nearest 0.1%) necessary to produce a ductile-to-brittle transition temperature of −25°C in the Fe–Mn–0.05 C alloy series of Figure 7-20b.

7.34

Using the plot generated for Problem 7.33(a), we find that a T_{trans} of −25°C corresponds to a Mn level of: approximately 1.8 wt.% Mn.

7.35 Using the footnote on page 292, calculate the specimen thickness necessary to make the plane strain assumption used in Sample Problem 7.9 valid.

7.35

$$t \geq 2.5\,(K_{IC}/Y.S.)^2$$
$$\geq 2.5\,(98\,MPa\sqrt{m}/1460\,MPa)^2$$
$$\geq 0.0113\ m$$
$$\geq 11.3\ mm$$

7.36 In the designing of a pressure vessel, it is convenient to plot operating stress (related to operating pressure) as a function of flaw size. (It is usually possible to ensure that flaws above a given size are not present by a careful inspection program.) General yielding (independent of a flaw) was given in Problem 7.17. Flaw-induced fracture is described by Equation 7.9. Taking Y in that equation as 1 gives the resulting schematic design plot:

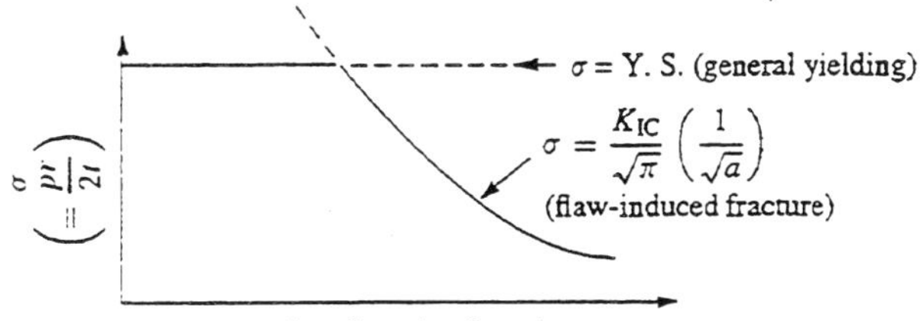

Generate such a design plot for a pressure vessel steel with Y.S. = 1000 MPa and K_{IC} = 170 MPa $\sqrt{m}$. For convenience, use the logarithmic scale for flaw size and cover a range of flaw sizes from 0.1 to 100 mm. (An additional practical point about the design plot is that failure by general yielding is preceded by observable deformation, whereas flaw-induced fracture occurs rapidly with no such warning. As a result, flaw-induced fracture is sometimes referred to as "fast fracture.")

7.36

In region of flaw-induced fracture,

$$\sigma = K_{IC}/\sqrt{\pi a} = 170\,MPa\sqrt{m}/\sqrt{\pi a}$$

a (mm)	σ (MPa)
0.1	9590
0.2	6780
0.5	4290
1.0	3030
2.0	2140
5.0	1360
10.0	959
20.0	678
50.0	429
100.0	303

Resulting plot:

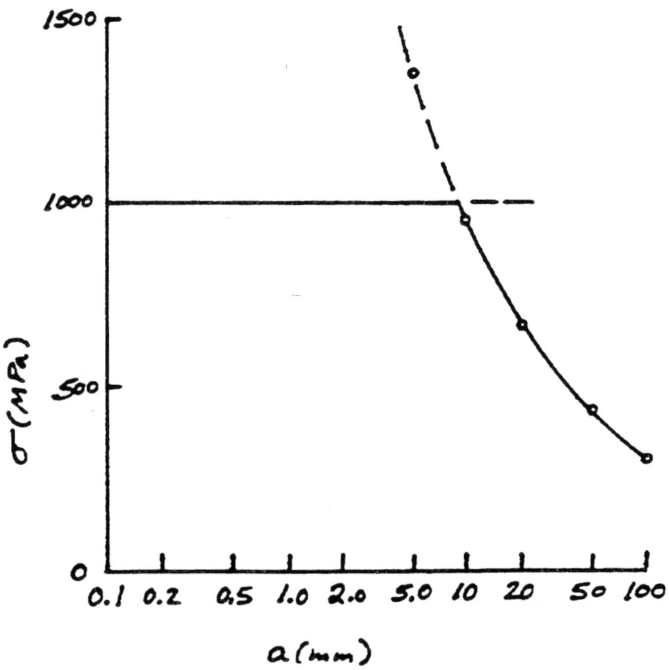

7.37 Repeat Problem 7.36 for an aluminum alloy with Y.S. = 400 MPa and K_{IC} = 25 MPa $\sqrt{m}$.

7.37

In region of flaw-induced fracture,
$\sigma = K_{IC}/\sqrt{\pi a} = 25\ MPa\sqrt{m}/\sqrt{\pi a}$

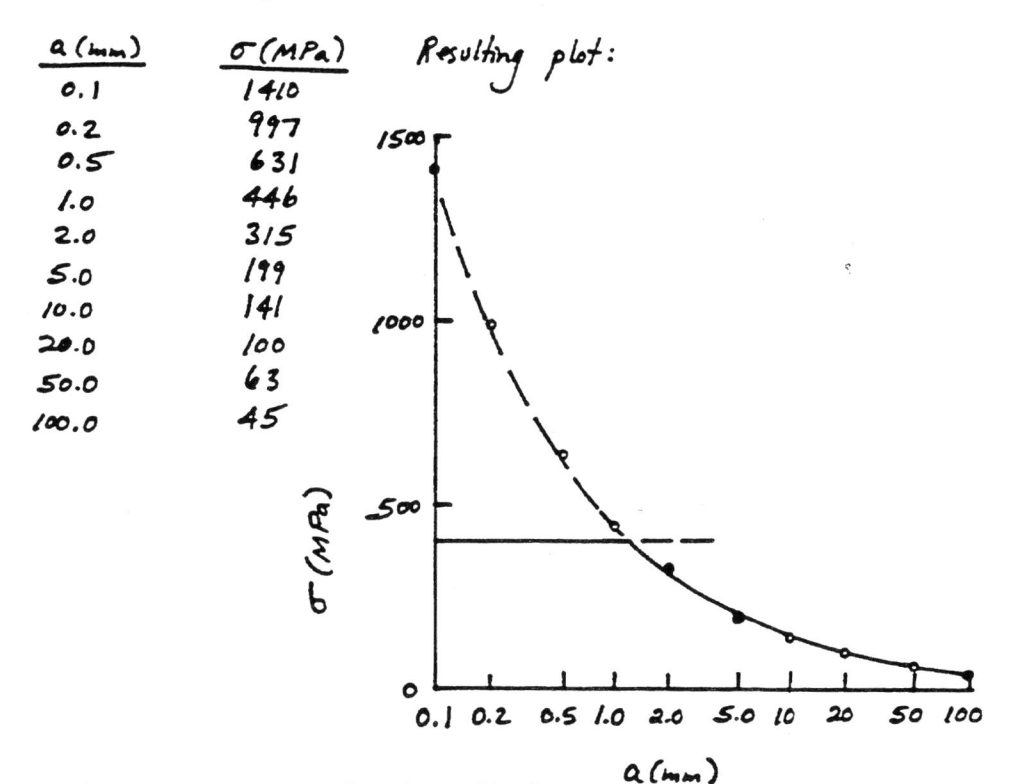

a (mm)	σ (MPa)
0.1	1410
0.2	997
0.5	631
1.0	446
2.0	315
5.0	199
10.0	141
20.0	100
50.0	63
100.0	45

Resulting plot:

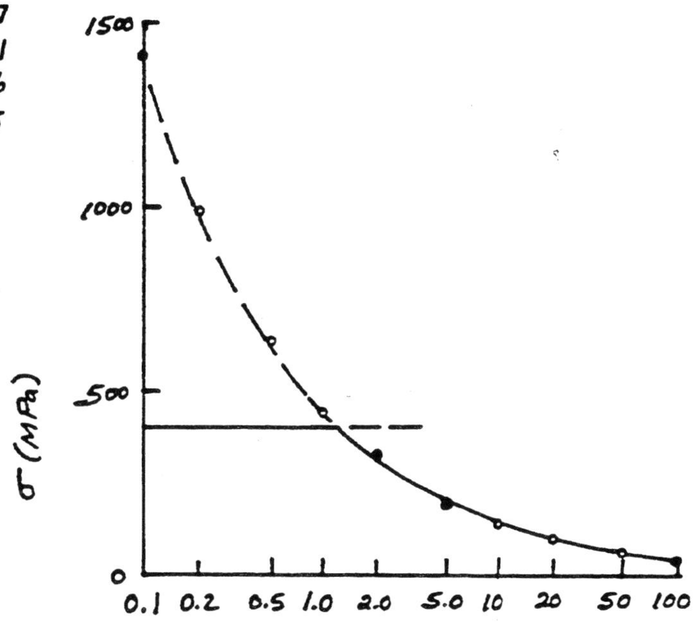

7.38 Critical flaw size corresponds to the transition between general yielding and fast fracture, as illustrated in Problem 7.36. If the fracture toughness of a high-strength steel can be increased by 50% (from 100 to 150 MPa $\sqrt{m}$) without changing its yield strength of 1250 MPa, by what percentage is its critical flaw size changed?

7.38

The initial a_{crit} will be, taking $Y = 1$:

$$a_{crit} = (1/\pi)\, K_{IC}^2/(Y.S.)^2$$

$$= (1/\pi)(100\ MPa\sqrt{m})^2/(1250\ MPa)^2$$

$$= 2.04 \times 10^{-3}\ m = 2.04\ mm$$

The improved a_{crit} will be

$$a_{crit} = (1/\pi)(150\ MPa\sqrt{m})^2/(1250\ MPa)^2$$

$$= 4.58 \times 10^{-3}\ m = 4.58\ mm$$

$$\% = \frac{4.58 - 2.04}{2.04} \times 100\% = \underline{\underline{125\%\ \text{increase}}}$$

7.39 A nondestructive testing program for a design component using the 1040 steel of Table 7.11 can ensure that no flaw greater than 1 mm will exist. If this steel has a fracture toughness of 120 MPa $\sqrt{m}$, can this inspection program prevent the occurrence of fast fracture? (See Problem 7.36.)

7.39

$$\sigma_{Y.S.} = \frac{K_{IC}}{\sqrt{\pi a}} \quad \text{or} \quad a = \frac{(K_{IC}/\sigma_{Y.S.})^2}{\pi}$$

$$= \frac{(120\,MPa\sqrt{m}/600\,MPa)^2}{\pi}$$

$$= 1.27 \times 10^{-2}\,m$$

$$= 12.7\,mm > 1\,mm \therefore \underline{\underline{Yes}}$$

7.40 Would the nondestructive testing program described in Problem 7.39 be adequate for the cast iron alloy labeled 6(b) in Table 7.11, given a fracture toughness of 15 MPa $\sqrt{m}$?

7.40

As in Problem 7.39,

$$a = \frac{(K_{IC}/\sigma_{Y.S.})^2}{\pi}$$

$$= \frac{(15\,MPa\sqrt{m}/329\,MPa)^2}{\pi}$$

$$= 6.6 \times 10^{-4}\,m = 0.66\,mm < 1\,mm \therefore \underline{\underline{No}}$$

7.41 In Problem 7.25, a ductile iron was evaluated for a pressure vessel application. For that alloy, determine the maximum pressure to which the vessel can be repeatedly pressurized without producing a fatigue failure.

7.41

Take $\sigma = F.S. = \frac{1}{4}\,T.S.$

From Figure 7-17(b), $T.S. \approx 550\,MPa$

or $\sigma = \frac{1}{4}(550\,MPa) = 138\,MPa$

From Problem 7.17,

$$\sigma = \frac{pr}{2t} \quad \text{or} \quad p = \frac{2\sigma t}{r} = \frac{2(138\,MPa)(0.020\,m)}{(0.30\,m)}$$

$$= \underline{18.3\,MPa}$$

(It is worthwhile to note that this value is lower than the design stress calculated in Problem 7.25, assuming static conditions.)

7.42 Repeat Problem 7.41 for the ductile iron of Problem 7.26.

7.42 From Figure 7-17(b), T.S. ≈ 1000 MPa

$$\sigma = \frac{1}{4}(1000\,MPa) = 250\,MPa$$

$$p = \frac{2\sigma t}{r} = \frac{2(250\,MPa)(0.020\,m)}{(0.30\,m)} = \underline{\underline{33.3\,MPa}}$$

7.43 A structural steel with a fracture toughness of 60 MPa $\sqrt{m}$ has no surface crack larger than 3 mm in length. By how much (in %) would this largest surface crack have to grow before the system would experience fast fracture under an applied stress of 500 MPa. (See Problem 7.36.)

7.43 As in Problem 7.39,

$$a = \frac{(K_{IC}/\sigma_{Y.S.})^2}{\pi}$$

$$= \frac{(60\,MPa\sqrt{m}/500\,MPa)^2}{\pi} = 4.58 \times 10^{-3}\,m = 4.58\,mm$$

or crack must grow by:

$$\frac{4.58\,mm - 3\,mm}{3\,mm} \times 100\,\% = \underline{\underline{52.8\,\%}}$$

7.44 For the conditions given in Problem 7.43, calculate the % increase in crack size if the applied stress in the structural design is 600 MPa.

7.44 Again,

$$a = \frac{(K_{IC}/\sigma_{Y.S.})^2}{\pi} = \frac{(60\,MPa\sqrt{m}/600\,MPa)^2}{\pi} = 3.18 \times 10^{-3}\,m = 3.18\,mm$$

or crack must grow by:

$$\frac{3.18\,mm - 3\,mm}{3\,mm} \times 100\,\% = \underline{\underline{6.10\,\%}}$$

7.45 The application of a C11000 copper wire in a control circuit design will involve cyclic loading for extended periods at the elevated temperatures of a production plant. Use the data of Figure 7-28b to specify an upper temperature limit to ensure a fatigue strength of at least 100 MPa for a stress life of 10^7 cycles.

7.45 From Figure 7-28 we note that:

F. S. (MPa)	T (°C)
132	21
116	65
94	100

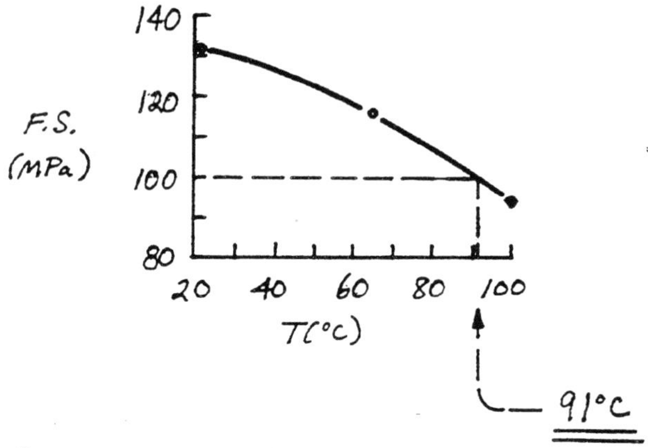

7.46 (a) The landing gear on a commercial aircraft experiences an impulse load upon landing. Assuming six such landings per day on average, how long would it take before the landing gear has been subjected to 10^8 load cycles? (b) The crankshaft in a given automobile rotates, on average, at 2000 revolutions per minute for a period of 2 h per day. How long would it take before the crankshaft has been subjected to 10^8 load cycles?

7.46 (a)

$$\frac{10^8 \text{ cycles}}{6 \text{ cycles/day}} = 1.67\times10^7 \text{ d} \times \frac{1 \text{ yr}}{365 \text{ d}} = \underline{\underline{45{,}700 \text{ yr}}}$$

(Note: One must be cautious about this apparently "safe" result. More frequent cyclic loads and high stress impulse loads can lead to failures after a much shorter period of time.)

(b) In one day, there will be:

$$2000 \text{ cycles/min} \times 120 \text{ min} = 240{,}000 \text{ cycles}$$

$$\therefore \frac{10^8 \text{ cycles}}{240{,}000 \text{ cycles/day}} = 416.7 \text{ days} = \underline{\underline{1 \text{ yr } \& \text{ } 52 \text{ days}}}$$

7.47 In analyzing a hip implant material for potential fatigue damage, it is important to note that an average person takes 4800 steps on an average day. How many stress cycles would this average person produce in (a) one year, (b) 10 years? (Note Problem 7.15.)

7.47 (a) $N = 4800 \text{ day}^{-1} \times 365 \text{ days/yr}$

$= \underline{\underline{1.75 \times 10^6}}$ per year

(b) $N = 1.75 \times 10^6/\text{yr} \times 10 \text{ yr}$

$= \underline{\underline{17.5 \times 10^6}}$ over 10 years

7.48 In analyzing a hip implant material for potential fatigue damage when used by an active athlete, we find that he takes a total of 10,000 steps and/or strides on an average day. How many stress cycles would this active person produce in (a) one year, (b) 10 years? (Note Problem 7.16.)

7.48 (a) $N = 10{,}000 \text{ day}^{-1} \times 365 \text{ days/yr}$

$= \underline{\underline{3.65 \times 10^6}}$ per year

(b) $N = 3.65 \times 10^6/\text{yr} \times 10 \text{ yr}$

$= \underline{\underline{36.5 \times 10^6}}$ over 10 years

7.49 An alloy is evaluated for potential creep deformation in a short-term laboratory experiment. The creep rate ($\dot{\epsilon}$) is found to be 1% per hour at 800°C and 5.5×10^{-2} % per hour at 700°C. (**a**) Calculate the activation energy for creep in this temperature range. (**b**) Estimate the creep rate to be expected at a service temperature of 500°C. (**c**) What important assumption underlies the validity of your answer in part (b)?

7.49

(a) $$\frac{\dot{\epsilon}_{800°C}}{\dot{\epsilon}_{700°C}} = \frac{Ce^{-Q/R(1073\,K)}}{Ce^{-Q/R(973\,K)}}$$

$$\frac{1\% \text{ per hr}}{5.5\times10^{-2}\% \text{ per hr}} = e^{-\frac{Q}{[8.314\,J/(mol\cdot K)]}\left(\frac{1}{1073}-\frac{1}{973}\right)K^{-1}}$$

or $Q = 2.52\times10^5\,J/mol = \underline{\underline{252\,kJ/mol}}$

(b) $C = \dot{\epsilon}e^{+Q/RT}$

$$= (1\% \text{ per hour})e^{+(2.52\times10^5)/(8.314)(1073)} = 1.80\times10^{12}\%\text{-hr}^{-1}$$

$$\dot{\epsilon}_{500°C} = (1.80\times10^{12}\% \text{ per hour})e^{-(2.52\times10^5)/(8.314)(773)}$$

$$= \underline{1.75\times10^{-5}\% \text{ per hour}}$$

(c) The creep mechanism (and Q) remains the same between 800 and 500°C.

7.50 Using the simple rule of thumb in the discussion of creep in Section 7.3 and the phase diagrams of Chapter 5, determine which of the following alloys might be a concern for creep deformation at 200°C: Cu–Ni (33:67), Cu–Zn (70:30), Al–Mg (40:60), or Al–Cu (95:5).

7.50

As the "high" temperature will be $\frac{1}{3}-\frac{1}{2}T_m$, a conservative estimate for the onset of creep will be $\frac{1}{3}T_m$.

Figure 5-36 indicates that T_m for Cu:Ni (33:67) is $\approx 1330°C = 1603\,K$ & $\frac{1}{3}T_m = 534\,K = 261°C$

Figure 5-37 indicates that T_m for Cu:Zn (70:30) is $\approx 920°C = 1193\,K$ & $\frac{1}{3}T_m = 398\,K = 125°C$

Figure 5-35 indicates that T_m for Al:Mg (40:60) is 437°C = 710 K & $\frac{1}{3} T_m$ = 237K = −36°C

Figure 5-34 indicates that T_m for Al:Cu (95:5) is ≈ 560°C = 833 K & $\frac{1}{3} T_m$ = 278K = 5°C

∴ <u>Cu:Zn (70:30), Al:Mg (40:60), and Al:Cu (95:5) will be a concern.</u>

7.51 As was shown in various problems for Chapter 6, the inverse of time to reaction (t_R^{-1}) can be used to approximate a rate and, consequently, can be estimated using the Arrhenius expression (Equation 7.10). The same is true for time-to-creep rupture, as defined in Figure 7-37. If the time to rupture for a given superalloy is 2000 h at 650°C and 50 h at 700°C, calculate the activation energy for the creep mechanism.

7.51

$$t_{CR}^{-1} = C e^{-Q/RT}$$

$$\frac{(t_{700°C})^{-1}}{(t_{650°C})^{-1}} = \frac{C e^{-Q/(8.314\ J/mol\cdot K)(973K)}}{C e^{-Q/(8.314 J/mol\cdot K)(923K)}} = \frac{2000h}{50h} = 40$$

or

$$\ln 40 = -\frac{Q}{(8.314\ J/mol)}\left(\frac{1}{973} - \frac{1}{923}\right)$$

giving

$$Q = \underline{\underline{551\ kJ/mol}}$$

7.52 Estimate the time to rupture at 750°C for the superalloy of Problem 7.51.

7.52 Substituting into the original expression gives

$$C = (50h)^{-1} e^{+(551{,}000)/(8.314)(973)}$$

$$= 7.62\times10^{27}\ h^{-1}$$

Then, we can solve for t_{CR} at 750°C as:

$$t_{CR,750°C}^{-1} = (7.62\times10^{27} h^{-1}) e^{-(551{,}000)/(8.314)(1023)}$$

$$= 0.558\ h^{-1}$$

or $t_{CR,750°C} = 1/(0.558 h^{-1}) = \underline{\underline{1.79\ hr}}$

• 7.53 Figure 7-35 indicates the dependence of creep on both stress (σ) and temperature (T). For many alloys, such dependence can be expressed in a modified form of the Arrhenius equation,

$$\dot{\epsilon} = C_1 \sigma^n e^{-Q/RT}$$

where $\dot{\epsilon}$ is the steady-state creep rate, C_1 is a constant, and n is a constant that usually lies within the range of 3 to 8. The exponential term ($e^{-Q/RT}$) is the same as in other Arrhenius expressions (see Equation 7.10). The product of $C_1\sigma^n$ is a temperature-independent term equal to the preexponential constant, C, in Equation 7.10. The presence of the σ^n term gives the name "power-law" creep to this expression. Given the power-law creep relationship with $Q = 250$ kJ/mol and $n = 4$, calculate what percentage increase in stress will be necessary to produce the same increase in $\dot{\epsilon}$ as a 10°C increase in temperature from 1000 to 1010°C.

7.53

$$\dot{\varepsilon} = C_1 \sigma^n e^{-Q/RT} \qquad Q = 250\times10^3 \text{ J/mol},\ n = 4$$

$$\frac{\dot{\varepsilon}_{1010°C}}{\dot{\varepsilon}_{1000°C}} = \frac{e^{-(250{,}000)/(8.314)(1283)}}{e^{-(250{,}000)/(8.314)(1273)}} = 1.202$$

$$\frac{\dot{\varepsilon}_{\sigma_2}}{\dot{\varepsilon}_{\sigma_1}} = \left(\frac{\sigma_2}{\sigma_1}\right)^4 = 1.202$$

$$\text{or } \sigma_2 = (1.202)^{1/4}\sigma_1 = 1.047\sigma_1$$

or **an increase of 4.7%**

• 7.54 Calculate the percentage increase in stress that would be necessary to produce the strain rate increase described in Problem 7.53 for a 100°C increase in temperature from 1000 to 1100°C.

7.54

$$\frac{\dot{\varepsilon}_{1100°C}}{\dot{\varepsilon}_{1000°C}} = \frac{e^{-(250{,}000)/(8.314)(1373)}}{e^{-(250{,}000)/(8.314)(1273)}} = 5.587$$

$$\text{or } \sigma_2 = (5.587)^{1/4}\sigma_1 = 1.537\sigma_1$$

or **an increase of 53.7%**

Section 8.1 - Ceramics - Crystalline Materials

PP 8.1 What is the weight fraction of Al_2O_3 in spinel ($MgAl_2O_4$)? (See Sample Problem 8.1.)

PP 8.1

mol. wt. $MgO = [24.31 + 16.00]$ amu $= 40.31$ amu

mol. wt. $Al_2O_3 = [2(26.98) + 3(16.00)]$ amu $= 101.96$ amu

$$\text{wt. fraction } Al_2O_3 = \frac{101.96}{101.96 + 40.31} = \underline{\underline{0.717}}$$

8.1 As pointed out in the discussion relative to the Al_2O_3–SiO_2 phase diagram (Figure 5-39), an alumina-rich mullite is desirable to ensure a more refractory (temperature-resistant) product. Calculate the composition of a refractory made by adding 2.5 kg of Al_2O_3 to 100 kg of stoichiometric mullite.

8.1

From Sample Problem 8.1,

wt frac. Al_2O_3 in stoichiometric mullite $= 0.718$

∴ " " SiO_2 " " " $= 1 - 0.718$
$= 0.282$

∴ 100 kg mullite → 71.8 kg Al_2O_3 + 28.2 kg SiO_2

addition of 2.5 kg Al_2O_3 → $71.8 + 2.5 = 74.3$ kg

$$\therefore \text{wt. \% } Al_2O_3 = \frac{74.3}{74.3 + 28.2} \times 100\% = \underline{\underline{72.5\%}}$$

$$\& \text{ wt. \% } SiO_2 = \frac{28.2}{74.3 + 28.2} \times 100\% = \underline{\underline{27.5\%}}$$

8.2 A fireclay refractory of simple composition can be produced by heating the raw material kaolinite, $Al_2(Si_2O_5)(OH)_4$, driving off the waters of hydration. Calculate the composition (weight percent basis) for the resulting refractory. (Note that this process was introduced in Sample Problem 5.11 relative to the Al_2O_3–SiO_2 phase diagram.)

8.2

$$Al_2(Si_2O_5)(OH)_4 = Al_2O_3 \cdot 2SiO_2 \cdot 2H_2O$$

$$Al_2O_3 \cdot 2SiO_2 \cdot 2H_2O \xrightarrow{\text{heat}} Al_2O_3 \cdot 2SiO_2 + 2H_2O\uparrow$$

1 mole $Al_2O_3 = [2(26.98) + 3(16.00)]$ amu $= 101.96$ amu

2 moles $SiO_2 = 2[28.09 + 2(16.00)]$ amu $= 120.18$ amu

$$\text{wt. \% } Al_2O_3 = \frac{101.96}{101.96 + 120.18} \times 100\% = \underline{\underline{45.9\%}}$$

$$\text{wt. \% } SiO_2 = \frac{120.18}{101.96 + 120.18} \times 100\% = \underline{\underline{54.1\%}}$$

8.3 Usingtheresultsof Problem 8.2 and Sample Problem 5.11, calculate the weight percent of SiO_2 and mullite present in the final microstructure of a fireclay refractory made by heating kaolinite.

8.3 From Problem 8.2, the overall composition is 45.9 wt. %.

From Sample Problem 5.11, the composition of SiO_2 is 0 mol.% Al_2O_3 = 0 wt. % Al_2O_3

and the composition of mullite is 60 mol.% Al_2O_3:

$$wt.\%Al_2O_3 = \frac{60[2(26.98)+3(16.00)]}{60[2(26.98)+3(16.00)]+40[28.09+2(16.00)]} \times 100\%$$
$$= 71.8 \text{ wt.}\%$$

From the lever rule,

$$wt.\%\,SiO_2 = \frac{71.8-45.9}{71.8-0} \times 100\% = 36.1\%$$

and

$$wt.\%\text{ mullite} = \frac{45.9-0}{71.8-0} \times 100\% = 63.9\%$$

(Note that this solution is an alternate route to the results obtained in PP 5.11.)

8.4 Estimate the density of (a) a partially stabilized zirconia (with 4 wt % CaO) as the weighted average of the densities of ZrO_2 (= 5.60 Mg/m^3) and CaO (= 3.35 Mg/m^3), and (b) a fully stabilized zirconia with 8 wt % CaO.

8.4 (a) Taking $\rho = \sum_i w_i \rho_i$,

$$\rho_{PSZ} = 0.04\rho_{CaO} + (1.00-0.04)\rho_{ZrO_2}$$
$$= 0.04(3.35 Mg/m^3) + 0.96(5.60 Mg/m^3)$$
$$= 5.51 Mg/m^3$$

(b) Similarly,

$$\rho_{SZ} = 0.08(3.35 Mg/m^3) + 0.92(5.60 Mg/m^3)$$
$$= 5.42 Mg/m^3$$

8.5 The primary reason for introducing ceramic components in automotive engine designs is the possibility of higher operating temperatures and, therefore, improved efficiencies. A by-product of this substitution, however, is mass reduction. For the case of 2 kg of cast iron (density = 7.15 Mg/m^3) being replaced by an equivalent volume of partially stabilized zirconia (density = 5.50 Mg/m^3), calculate the mass reduction.

8.5

$$V_{replaced} = \frac{m_{iron}}{\rho_{iron}} = \frac{2 \times 10^3 \text{ g}}{7.15 \times 10^6 \text{ g}} \text{ m}^3 = 2.80 \times 10^{-4} \text{ m}^3$$

$$m_{PSZ} = \rho_{PSZ} V = (5.50 \times 10^6 \text{ g/m}^3)(2.80 \times 10^{-4} \text{ m}^3)$$

$$= 1.54 \times 10^3 \text{ g}$$

$$\therefore \text{mass reduction} = 2.00 \text{ kg} - 1.54 \text{ kg}$$

$$= \underline{\underline{0.462 \text{ kg}}}$$

8.6 Calculate the mass reduction attained if silicon nitride (density = 3.18 Mg/m^3) is used in place of 2 kg of cast iron (density = 7.15 Mg/m^3).

8.6

$$m_{Si_3N_4} = \rho_{Si_3N_4} V = (3.18 \times 10^6 \text{ g/m}^3)(2.80 \times 10^{-4} \text{ m}^3)$$

$$= 0.890 \times 10^3 \text{ g}$$

$$\therefore \text{mass reduction} = 2.00 \text{ kg} - 0.89 \text{ kg}$$

$$= \underline{\underline{1.11 \text{ kg}}}$$

Section 8.2 - Glasses - Noncrystalline Materials

PP 8.2 In Sample Problem 8.2 we calculated a batch formula for a common soda–lime–silica glass. To improve chemical resistance and working properties, Al_2O_3 is often added to the glass. This can be done by adding soda feldspar (albite), $Na(AlSi_3)O_8$, to the batch formula. Calculate the formula of the glass produced when 2000 kg of the batch formula is supplemented with 100 kg of this feldspar.

PP 8.2 2000 kg of batch gives:

$$0.216 \times 2000\text{ kg} = 432\text{ kg } Na_2CO_3$$
$$0.150 \times \text{ "} = 300\text{ kg } CaCO_3$$
$$0.633 \times \text{ "} = 1266\text{ kg } SiO_2$$

Note that $Na(AlSi_3)O_8 = \frac{1}{2}Na_2O + \frac{1}{2}Al_2O_3 + 3SiO_2$

$$\text{mol. wt. } Na(AlSi_3)O_8 = [22.99 + 26.98 + 3(28.09) + 8(16.00)]\text{ amu}$$
$$= 262.24\text{ amu}$$

$$\text{mol. wt. } \tfrac{1}{2}Na_2O = \tfrac{1}{2}[2(22.99) + 16.00]\text{ amu} = 30.99\text{ amu}$$

$$\text{mol. wt. } \tfrac{1}{2}Al_2O_3 = \tfrac{1}{2}[2(26.98) + 3(16.00)]\text{ amu} = 50.98\text{ amu}$$

$$\text{mol. wt. } 3SiO_2 = 3[28.09 + 2(16.00)]\text{ amu} = 180.27\text{ amu}$$

Then, 100 kg $Na(AlSi_3)O_8$ yields:

$$\frac{30.99}{262.24} \times 100\text{ kg} = 11.8\text{ kg } Na_2O$$

$$\frac{50.98}{262.24} \times 100\text{ kg} = 19.4\text{ kg } Al_2O_3$$

$$\frac{180.27}{262.24} \times 100\text{ kg} = 68.7\text{ kg } SiO_2$$

Using calculations from Sample Problem 8.2,

$$432\text{ kg } Na_2CO_3 \rightarrow \frac{61.98}{105.98} \times 432\text{ kg} = 252.6\text{ kg } Na_2O$$

$$300\text{ kg } CaCO_3 \rightarrow \frac{56.08}{100.08} \times 300\text{ kg} = 168.1\text{ kg } CaO$$

For the final product:

$$m_{Na_2O} = (11.8 + 252.6)\text{ kg} = 264.4\text{ kg}$$

$$m_{CaO} = 168.1\text{ kg}$$

$m_{Al_2O_3} = 19.4\ kg$

$m_{SiO_2} = (68.7 + 1266)\ kg = 1334.7\ kg$

$m_{total} = (264.4 + 168.1 + 19.4 + 1334.7)\ kg = 1786.6\ kg$

The resulting glass formula is:

$$\text{wt.\% } Na_2O = \frac{264.4}{1786.6} \times 100\% = 14.8\%$$

$$\text{wt.\% } CaO = \frac{168.1}{1786.6} \times 100\% = 9.4\%$$

$$\text{wt.\% } Al_2O_3 = \frac{19.4}{1786.6} \times 100\% = 1.1\%$$

$$\text{wt.\% } SiO_2 = \frac{1334.7}{1786.6} \times 100\% = 74.7\%$$

8.7 A batch formula for a window glass contains 400 kg Na_2CO_3, 300 kg $CaCO_3$, and 1300 kg SiO_2. Calculate the resulting glass formuia.

8.7 As in PP 8.2,

$$400\ kg\ Na_2CO_3 \rightarrow \left(\frac{61.98}{105.98}\right) 400\ kg = 233.9\ kg\ Na_2O$$

$$300\ kg\ CaCO_3 \rightarrow \left(\frac{56.08}{100.08}\right) 300\ kg = 168.1\ kg\ CaO$$

$$1300\ kg\ SiO_2 \rightarrow 1300\ kg\ SiO_2$$

or $m_{total} = (233.9 + 168.1 + 1300.0)\ kg = 1702\ kg$

giving an overall glass formula:

$$\text{wt.\% } Na_2O = \frac{233.9}{1702} \times 100\% = 13.7\%$$

$$\text{wt.\% } CaO = \frac{168.1}{1702} \times 100\% = 9.9\%$$

$$\text{wt.\% } SiO_2 = \frac{1300}{1702} \times 100\% = 76.4\%$$

8.8 For the window glass in Problem 8.7, calculate the glass formula if the batch is supplemented by 100 kg of lime feldspar (anorthite), $Ca(Al_2Si_2)O_8$.

8.8

Note that $Ca(Al_2Si_2)O_8 = CaO + Al_2O_3 + 2\,SiO_2$

& mol. wt. $Ca(Al_2Si_2)O_8 = [40.08 + 2(26.98) + 2(28.09) + 8(16)]$ amu $= 278.22$ amu

& mol. wt. $CaO = [40.08 + 16]$ amu $= 56.08$ amu

mol. wt. $Al_2O_3 = [2(26.98) + 3(16)]$ amu $= 101.96$ amu

mol. wt. $2SiO_2 = 2[28.09 + 2(16)]$ amu $= 120.18$ amu

Then, 100 kg $Ca(Al_2Si_2)O_8 \rightarrow$

$$\frac{56.08}{(56.08 + 101.96 + 120.18)} \times 100\text{ kg} = 20.2\text{ kg } CaO$$

$$\frac{101.96}{278.22} \times 100\text{ kg} = 36.6\text{ kg } Al_2O_3$$

$$\frac{120.18}{278.22} \times 100\text{ kg} = 43.2\text{ kg } SiO_2$$

Combining with calculations for Problem 8.7:

$m_{Na_2O} = 233.9$ kg

$m_{CaO} = (168.1 + 20.2)$ kg $= 188.3$ kg

$m_{Al_2O_3} = 36.6$ kg

$m_{SiO_2} = (1300 + 43.2)$ kg $= 1343.2$ kg

or $m_{total} = 1802$ kg

giving a glass formula:

$$\text{wt.\% } Na_2O = \frac{233.9}{1802} \times 100\% = \underline{\underline{13.0\%}}$$

$$\text{wt.\% } CaO = \frac{188.3}{1802} \times 100\% = \underline{\underline{10.4\%}}$$

$$\text{wt.\% } Al_2O_3 = \frac{36.6}{1802} \times 100\% = \underline{\underline{2.0\%}}$$

$$\text{wt.\% } SiO_2 = \frac{1343.2}{1802} \times 100\% = \underline{\underline{74.5\%}}$$

8.9 An economical substitute for vitreous silica is a high-silica glass made by leaching the B_2O_3-rich phase from a two-phase borosilicate glass. (The resulting porous microstructure is densified by heating.) A typical starting composition is 81 wt % SiO_2, 4 wt % Na_2O, 2 wt % Al_2O_3, and 13 wt % B_2O_3. A typical final composition is 96 wt % SiO_2, 1 wt % Al_2O_3, and 3 wt % B_2O_3. How much product (in kilograms) would be produced from 100 kg of starting material, assuming no SiO_2 is lost by leaching?

8.9

Taking the amount of SiO_2 as fixed, we see that, for 100 kg starting material, we would still have $0.81 \times 100\,kg = 81\,kg$ in the final product. That is,

$$81\,kg = 0.96\, m_{product}$$

or

$$m_{product} = \underline{\underline{84.4\,kg}}$$

8.10 How much B_2O_3 (in kilograms) is removed by leaching in the glass-manufacturing process described in Problem 8.9?

8.10

The initial material contains $0.13 \times 100\,kg = 13\,kg\ B_2O_3$. Using the result of Problem 8.9, we see that the final product contains $0.03 \times 84.4\,kg = 2.53\,kg\ B_2O_3$.

$\therefore\ 13\,kg - 2.53\,kg = \underline{\underline{10.5\,kg\ B_2O_3}}$ are lost.

8.11 A novel electronic material involves the dispersion of small silicon particles in a glass matrix. These "quantum dots" are discussed in Section 12.5. If 4.85×10^{16} particles of Si are dispersed per mm^3 of glass corresponding to a total content of 5 wt %, calculate the average particle size of the quantum dots. (Assume spherical particles and note that the density of the silicate glass matrix is 2.60 Mg/m^3.)

8.11

$$1\ \text{mm}^3 \text{ of glass} \rightarrow 2.60 \times 10^6 \frac{\text{g}}{\text{m}^3} \times 1\ \text{mm}^3 \times \left(\frac{1\ \text{m}}{10^3\ \text{mm}}\right)^3$$

$$= 2.6 \times 10^{-3}\ \text{g}$$

For 5 wt.% Si, $m_{Si} = 0.05 \times 2.6 \times 10^{-3}\ \text{g} = 1.3 \times 10^{-4}\ \text{g}$

$V_{Si} = m_{Si}/\rho_{Si}$ & obtaining ρ from Appendix 1 →

$$V_{Si} = 1.3 \times 10^{-4}\ \text{g}/(2.33\ \text{g/cm}^3) = 5.58 \times 10^{-5}\ \text{cm}^3 \left(\frac{10^7\ \text{nm}}{\text{cm}}\right)^3$$

$$= 5.58 \times 10^{16}\ \text{nm}^3$$

$$V_{particle} = \frac{5.58 \times 10^{16}\ \text{nm}^3}{4.85 \times 10^{16}\ \text{particles}} = 1.15\ \text{nm}^3 = \frac{4}{3}\pi r^3$$

or $$r_{particle} = \left(\frac{3}{4\pi} \times 1.15\ \text{nm}^3\right)^{1/3} = 0.650\ \text{nm}$$

or $$d_{particle} = 2 r_{particle} = \underline{\underline{1.30\ \text{nm}}}$$

8.12 Calculate the average separation distance between the centers of adjacent Si particles in the quantum dot material of Problem 8.11. (For simplicity, assume a simple cubic array of dispersed particles.)

8.12

Assuming a simple cubic array of "dots":

$$d_{separation}^{-1} = (\text{density}_{dots})^{1/3}$$

$$= (4.85 \times 10^{16}\ \text{mm}^{-3})^{1/3}$$

$$= 3.65 \times 10^5\ \text{mm}^{-1}$$

or

$$d_{separation} = 0.274 \times 10^{-5}\ \text{mm} \times \frac{10^6\ \text{nm}}{\text{mm}}$$

$$= \underline{\underline{2.74\ \text{nm}}}$$

Section 8.3 - Glass-Ceramics

PP 8.3 What would be the mole percentage of Li_2O, Al_2O_3, SiO_2, and TiO_2 in the first commercial glass-ceramic composition of Table 8.7? (See Sample Problem 8.3.)

PP 8.3 For a 100 g of glass-ceramic, Table 8.7 gives:

$74 g\ SiO_2 + 4 g\ Li_2O + 16 g\ Al_2O_3 + 6 g\ TiO_2$

Using calculations from Sample Problem 8.3,

$74 g\ SiO_2 \times 1 mol/60.09 g = 1.231\ mol$

$4 g\ Li_2O \times 1 mol/29.88 g = 0.134\ mol$

$16 g\ Al_2O_3 \times 1 mol/101.96 g = 0.157\ mol$

Similarly,

mol. wt. $TiO_2 = [47.90 + 2(16.00)]\ amu = 79.90\ amu$

$6 g\ TiO_2 \times 1 mol/79.90 g = 0.075\ mol$

Total no. moles $= (1.231 + 0.134 + 0.157 + 0.075)\ mol$

$= 1.597\ mol$

As a result:

mol. % $SiO_2 = 1.231/1.597 \times 100\% = 77.1\%$

mol. % $Li_2O = 0.134/1.597 \times 100\% = 8.4\%$

mol. % $Al_2O_3 = 0.157/1.597 \times 100\% = 9.8\%$

mol. % $TiO_2 = 0.075/1.597 \times 100\% = 4.7\%$

8.13 Assuming the TiO_2 in a $Li_2O–Al_2O_3–SiO_2$ glass-ceramic is uniformly distributed with a dispersion of 10^{12} particles per cubic millimeter and a total amount of 6 wt %, what is the average particle size of the TiO_2 particles? (Assume spherical particles. The density of the glass-ceramic is 2.85 Mg/m^3 and of the TiO_2 is 4.26 Mg/m^3.)

8.13

1 m^3 of glass-ceramic → 2.85 Mg

For 6 wt.% TiO_2, $m_{TiO_2} = 0.06 \times 2.85\,Mg = 0.171\,Mg$

$$\text{Vol. } TiO_2 = 0.171\,Mg / 4.26\,Mg/m^3 = 0.0401\,m^3$$

$$V_{particle} = \frac{0.0401\,m^3 \qquad (1\,m/10^3\,mm)^3}{(10^{12}\,\text{particles}/mm^3\,\text{glass-cer.}) \times (1\,m^3\,\text{glass-cer.})} =$$

$$= 4.01 \times 10^{-23}\,m^3$$

$$= \tfrac{4}{3}\pi r^3$$

$$\text{or } r_{particle} = \left(\frac{3}{4\pi} \times 4.01 \times 10^{-23}\,m^3\right)^{1/3} = 2.12 \times 10^{-8}\,m$$

$$= 0.0212\,\mu m$$

or dia. $= 2r =$ **0.0425 μm**

8.14 Repeat Problem 8.13 for a 3 wt % dispersion of P_2O_5 with a concentration of 10^{12} particles per cubic millimeter. (The density of P_2O_5 is 2.39 Mg/m^3.)

8.14

1 m^3 of glass-ceramic → 2.85 Mg

For 3 wt.% P_2O_5, $m_{P_2O_5} = 0.03 \times 2.85\,Mg = 0.0855\,Mg$

$$\text{Vol. } P_2O_5 = 0.0855\,Mg / 2.39\,Mg/m^3 = 0.0358\,m^3$$

$$V_{particle} = \frac{0.0358\,m^3}{(10^{12}/mm^3)(10^9\,mm^3/m^3)(1\,m^3)}$$

$$= 3.58 \times 10^{-23}\,m^3$$

$$= \tfrac{4}{3}\pi r^3$$

$$r_{particle} = \left(\frac{3}{4\pi} \times 3.58 \times 10^{-23}\,m^3\right)^{1/3} = 2.04 \times 10^{-8}\,m$$

$$= 0.0204\,\mu m$$

dia. $= 2r =$ **0.0408 μm**

8.15 What is the overall volume percent of TiO_2 in the glass-ceramic described in Problem 8.13?

8.15

100 kg product would yield:

94 kg LAS + 6 kg TiO_2

The volume of each component would be:

$$V_{LAS} = \frac{94 \text{ kg}}{2.85 \text{ Mg/m}^3} \times \frac{1 \text{ Mg}}{1000 \text{ kg}} = 0.0333 \text{ m}^3$$

$$V_{TiO_2} = \frac{6 \text{ kg}}{4.26 \text{ Mg/m}^3} \times \frac{1 \text{ Mg}}{1000 \text{ kg}} = 0.0014 \text{ m}^3$$

$$0.0347 \text{ m}^3$$

$$\therefore \text{vol. \% } TiO_2 = \frac{0.0014}{0.0347} \times 100\% = \underline{\underline{4.0\%}}$$

8.16 What is the overall volume percent of P_2O_5 in the glass-ceramic described in Problem 8.14?

8.16

100 kg → 97 kg LAS + 3 kg P_2O_5

$$V_{LAS} = \frac{97 \text{ kg}}{2.85 \text{ Mg/m}^3} \times \frac{1 \text{ Mg}}{1000 \text{ kg}} = 0.0340 \text{ m}^3$$

$$V_{P_2O_5} = \frac{3 \text{ kg}}{2.39 \text{ Mg/m}^3} \times \frac{1 \text{ Mg}}{1000 \text{ kg}} = 0.0013 \text{ m}^3$$

$$0.0353 \text{ m}^3$$

$$\therefore \text{vol. \% } P_2O_5 = \frac{0.0013}{0.0353} \times 100\% = \underline{\underline{3.6\%}}$$

8.17 Calculate the average separation distance between the centers of adjacent TiO_2 particles in the glass-ceramic described in Problems 8.13 and 8.15. (Note Problem 8.12.)

8.17 Assuming a simple cubic array of particles:

$$d^{-1}_{separation} = (density_{ptcles})^{1/3}$$

$$= (10^{12}\ mm^{-3})^{1/3} = 10^{4}\ mm^{-1}$$

or

$$d_{separation} = 10^{-4}\ mm \times 10^{3}\ \frac{mm}{\mu m}$$

$$= \underline{\underline{0.100\ \mu m}}$$

8.18 Calculate the average separation distance between the centers of adjacent P_2O_5 particles in the glass-ceramic described in Problems 8.14 and 8.16. (Note Problem 8.12.)

8.18 As in Problem 8.17,

$$d^{-1}_{separation} = (density_{ptcles})^{1/3}$$

$$= (10^{12}\ mm^{-3})^{1/3} = 10^{4}\ mm^{-1}$$

or

$$d_{separation} = 10^{-4}\ mm \times 10^{3}\ mm/\mu m$$

$$= \underline{\underline{0.100\ \mu m}}$$

Section 8.4 - Mechanical Properties of Ceramics and Glasses

PP 8.4 Calculate the breaking strength of a given glass plate containing (**a**) a.5-μm-long surface crack and (**b**) a 5-μm-long surface crack. Except for the length of the crack, use the conditions described in Sample Problem 8.4.

PP 8.4 (a) $\sigma = \frac{1}{2}\sigma_m\left(\frac{\rho}{c}\right)^{1/2}$

$$= \frac{1}{2}(7.0\times10^{9}\ Pa)\left(\frac{0.264\times10^{-9}\ m}{0.5\times10^{-6}\ m}\right)^{1/2} = \underline{\underline{80\ MPa}}$$

(b) $\sigma = \frac{1}{2}(7.0\times10^{9}\ Pa)\left(\frac{0.264\times10^{-9}\ m}{5\times10^{-6}\ m}\right)^{1/2} = \underline{\underline{25\ MPa}}$

PP 8.5 In Sample Problem 8.5, maximum service stress for two structural ceramics is calculated based on the assurance of no flaws greater than 25 μm in size. Repeat these calculations given that a more economical inspection program can only guarantee detection of flaws greater than 100 μm in size.

PP 8.5 (a) As $\sigma_f \propto a^{-1/2}$, the maximum service strength will be reduced by a factor of $\sqrt{25/100} = 0.5$

$\therefore \sigma_{f,100\,\mu m} = 339\,MPa \times 0.5 = \underline{169\,MPa}$

(b) Similarly,

$\sigma_{f,100\,\mu m} = 1020\,MPa \times 0.5 = \underline{508\,MPa}$

PP 8.6 For the system discussed in Sample Problem 8.6, what would be the time to fracture **(a)** at 0°C and **(b)** at room temperature, 25°C?

PP 8.6 (a) $t^{-1} = C e^{-Q/RT}$

with $C = 5.15 \times 10^{12}\,s^{-1}$

and $Q = 78.6\,kJ/mol$

At 0°C,

$t^{-1} = (5.15 \times 10^{12}\,s^{-1}) e^{-(78{,}600\,J/mol)/[8.314\,J/(mol\cdot K)](273\,K)}$

$= 4.70 \times 10^{-3}\,s^{-1}$

or

$t = \underline{213\,s}$

(b) At 25°C,

$t^{-1} = (5.15 \times 10^{12}\,s^{-1}) e^{-(78{,}600)/(8.314)(298)}$

$= 8.59 \times 10^{-2}\,s^{-1}$

or

$t = \underline{11.6\,s}$

PP 8.7 In Sample Problem 8.7 the lifetime of an Al_2O_3 furnace tube is calculated. **(a)** Calculate the lifetime of a $MgAl_2O_4$ furnace tube with a 2- to 5-μm grain size (see Table 8.11). **(b)** Repeat the calculation for $MgAl_2O_4$ with a grain size of 1 to 3 mm. (c) Comment on the basis of the difference in creep behavior for the two different grain sizes.

PP 8.7 (a) From Table 8.11, $\dot{\varepsilon} = 263 \times 10^{-6}$ mm/(mm·h)

$$n_{life} = \frac{0.01}{263 \times 10^{-6}\,h^{-1}} = \underline{\underline{38.0\ hr}}$$

(b) In this case, $\dot{\varepsilon} = 1 \times 10^{-6}$ mm/(mm·h)

$$n_{life} = \frac{0.01}{1 \times 10^{-6}\,h^{-1}} = 1.00 \times 10^{4}\,h \times \frac{1\ day}{24\ h}$$

$$= \underline{\underline{417\ days}}$$

(c) Grain boundary sliding is a common creep mechanism for ceramics. A larger grain-size material [part (b) compared to part (a)] has less grain boundary area and less creep deformation (longer life).

PP 8.8 In Sample Problem 8.8 the stress in an Al_2O_3 tube is calculated as a result of constrained heating to 1000°C. To what temperature could the furnace tube be heated to be stressed to an acceptable (but not necessarily desirable) compressive stress of 2100 MPa?

PP 8.8 $\sigma = E\epsilon = E\alpha\Delta T$

$$\text{or } \Delta T = \frac{\sigma}{E\alpha} = \frac{2100\ MPa}{(370 \times 10^{3}\ MPa)(8.8 \times 10^{-6}\ ^{\circ}C^{-1})}$$

$$= 645\ ^{\circ}C$$

$$\text{or } T = (645 + 25)^{\circ}C = \underline{\underline{670\ ^{\circ}C}}$$

PP 8.9 In Sample Problem 8.9 a temperature drop of approximately 50°C caused by a water spray is seen to be sufficient to fracture an Al_2O_3 furnace tube originally at 1000°C. Approximately what temperature drop due to a 2.5-lb/(s · ft^2) airflow would cause a fracture?

PP 8.9 Again using Figure 8-15 and noting the broad range of $r_m h$ associated with this condition, the temperature drop would be:

$\underline{\underline{\approx 250\ ^{\circ}C \text{ to } 1000\ ^{\circ}C\ (\approx 700\ ^{\circ}C \text{ midrange})}}$

PP 8.10 In Sample Problem 8.10 various viscosity ranges are characterized for a soda–lime–silica glass. For this material, calculate the annealing range (see Figure 8-19).

PP 8.10 The text points out that the annealing range illustrated in Figure 8-19 is between $\eta = 10^{12.5}$ and $10^{13.5}$ poise. Using the data from Sample Problem 8.10,

$$\eta = \eta_0 e^{+Q/RT} \quad \text{with } \eta_0 = 3.31 \times 10^{-18} \text{ poise}$$

and $Q = 465 \text{ kJ/mol}$

Then,

$$10^{12.5} \text{ poise} = (3.31 \times 10^{-18} \text{ poise}) e^{+\left[\frac{465{,}000 \text{ J/mol}}{8.314 \text{ J/(mol}\cdot\text{K)}\, T_1}\right]}$$

or

$$T_1 = 810 \text{ K} = 537°\text{C}$$

$$10^{13.5} \text{ poise} = (3.31 \times 10^{-18} \text{ poise}) e^{+\left[\frac{465{,}000 \text{ J/mol}}{8.314 \text{ J/(mol}\cdot\text{K)}\, T_2}\right]}$$

or

$$T_2 = 784 \text{ K} = 511°\text{C}$$

giving an annealing range = 511 to 537°C

8.19 (a) The following data are collected for a modulus of rupture test on an MgO refractory brick (refer to Equation 8.1 and Figure 8-3):

$$F = 7.0 \times 10^4 \text{ N}$$

$$L = 178 \text{ mm}$$

$$b = 114 \text{ mm}$$

$$h = 76 \text{ mm}$$

Calculate the modulus of rupture. (b) Suppose that you are given a similar MgO refractory with the same strength and same dimensions except that its height, h, is only 64 mm. What would be the load (F) necessary to break this thinner refractory?

8.19 (a) $MOR = \frac{3FL}{2bh^2}$

$$= \frac{3(7.0 \times 10^4 N)(178 mm)}{2(114 mm)(76 mm)^2 (1 m^2/10^6 mm^2)}$$

$$= \underline{\underline{28.4 \ MPa}}$$

(b) $F = \frac{2}{3}\frac{(MOR)bh^2}{L} \propto h^2$

$$= 7 \times 10^4 N \left(\frac{64}{76}\right)^2$$

$$= \underline{\underline{4.96 \times 10^4 N}}$$

8.20 A single crystal Al_2O_3 rod (precisely 5 mm diameter × 50 mm long) is used to apply loads to small samples in a high-precision dilatometer (a length-measuring device). If the crystal is subjected to a 25-kN axial compression load, calculate the resulting rod dimensions.

8.20 $\epsilon_z = \frac{\sigma}{E} = \frac{P/\pi r^2}{E}$

Using Table 8.8 gives for the compressive (negative) load:

$$\epsilon_z = \frac{(-25 \times 10^3 N)/\pi(2.5 \times 10^{-3} m)^2}{380 \times 10^9 Pa \times (1 N/m^2)/Pa} = -3.35 \times 10^{-3}$$

$\therefore l_f = l_o(1 + \epsilon_z) = 50 mm (1 - 3.35 \times 10^{-3}) = \underline{\underline{49.832 mm}}$

& $\epsilon_{dia} = -\nu \epsilon_z$

Using ϵ_z from above & ν from Table 8.9:

$$\epsilon_{dia} = -(0.26)(-3.35 \times 10^{-3}) = +8.71 \times 10^{-4}$$

$\therefore d_f = d_o(1 + \epsilon_{dia}) = 5 mm (1 + 8.71 \times 10^{-4}) = \underline{\underline{5.0044 mm}}$

8.21 A freshly drawn glass fiber (100 μm diameter) breaks under a tensile load of 40 N. A similar fiber, after subsequent handling, breaks under a tensile load of 0.15 N. Assuming the first fiber was defect-free and that the second fiber broke due to an atomically sharp surface crack, calculate the length of that crack.

8.21

For the defect-free fiber, $\sigma_m = P_m/(\pi r^2)$.

For the defective fiber, $\sigma = P/(\pi r^2)$.

And, from the Griffith expression,

$$\sigma_m = 2\sigma (c/\rho)^{1/2}$$

or

$$(c/\rho)^{1/2} = \frac{\sigma_m}{2\sigma} = \frac{P_m}{2P}$$

or

$$c = \rho\left(\frac{P_m}{2P}\right)^2$$

Using the value of ρ from Sample Problem 8.4 gives:

$$c = (2.64\times10^{-10}\,m)\left(\frac{40\,N}{2\times0.15\,N}\right)^2$$

$$= 4.69\times10^{-6}\,m = \underline{\underline{4.69\,\mu m}}$$

8.22 A nondestructive testing program can ensure that a given 80-μm-diameter glass fiber will have no surface cracks longer than 5 μm. Given that the theoretical strength of the fiber is 5 GPa, what can you say about the expected breaking strength of this fiber?

8.22

With the crack tip radius always $\geq d_{O^{2-}}$ and the crack length always $\leq 5\,\mu m$, we can say

$$\sigma \geq \frac{1}{2}\sigma_m\left(\frac{\rho}{c}\right)^{1/2}$$

$$\geq \frac{1}{2}(5\times10^3\,MPa)\left(\frac{0.264\times10^{-9}\,m}{5.0\times10^{-6}\,m}\right)^{1/2}$$

$$\geq \underline{\underline{18.2\,MPa}}$$

8.23 A silicon nitride turbine rotor fractures at a stress level of 300 MPa. Estimate the flaw size responsible for this failure.

8.23

Re-arranging Equation 7.9 and taking Y = 1 gives:

$$a = \frac{1}{\pi}\left(\frac{K_{IC}}{\sigma_f}\right)^2$$

For the range of K_{IC} given in Table 8.10,

$$a \le \frac{1}{\pi}\left(\frac{5 \text{ MPa}\sqrt{m}}{300 \text{ MPa}}\right)^2$$

$$\le 88.4 \times 10^{-6} \text{ m}$$

$$\& \quad a \ge \frac{1}{\pi}\left(\frac{4 \text{ MPa}\sqrt{m}}{300 \text{ MPa}}\right)^2$$

$$\ge 56.6 \times 10^{-6} \text{ m}$$

or $\underline{\underline{57 \mu m \le a \le 88 \mu m}}$

8.24 Estimate the flaw size responsible for the failure of a turbine rotor made from alumina that fractures at a stress level of 300 MPa.

8.24

For the range of K_{IC} given in Table 8.10,

$$a \le \frac{1}{\pi}\left(\frac{5 \text{ MPa}\sqrt{m}}{300 \text{ MPa}}\right)^2 = 88.4 \times 10^{-6} \text{ m}$$

$$\& \quad a \ge \frac{1}{\pi}\left(\frac{3 \text{ MPa}\sqrt{m}}{300 \text{ MPa}}\right)^2 = 31.8 \times 10^{-6} \text{ m}$$

or $\underline{\underline{32 \mu m \le a \le 88 \mu m}}$

8.25 Estimate the flaw size responsible for the failure of a turbine rotor made from partially stabilized zirconia that fractures at a stress level of 300 MPa.

8.25

For the K_{IC} given in Table 8.10,

$$a = \frac{1}{\pi}\left(\frac{9 \text{ MPa}\sqrt{m}}{300 \text{ MPa}}\right)^2$$

$$= 286 \times 10^{-6} \text{ m} = \underline{\underline{286 \mu m}}$$

8.26 Plot the breaking stress for MgO as a function of flaw size, a, on a logarithmic scale using Equation 7.9 and taking $Y = 1$. Cover a range of a from 1 to 100 mm. (See Table 8.10 for fracture toughness data.) You may wish to review Problem 7.36.

8.26

$$\sigma = K_{IC}/\sqrt{\pi a} = 3\ \text{MPa}\sqrt{\text{m}}/\sqrt{\pi a}$$

a (mm)	σ (MPa)
1	53.5
2	37.8
5	23.9
10	16.9
20	12.0
50	7.6
100	5.4

Resulting plot:

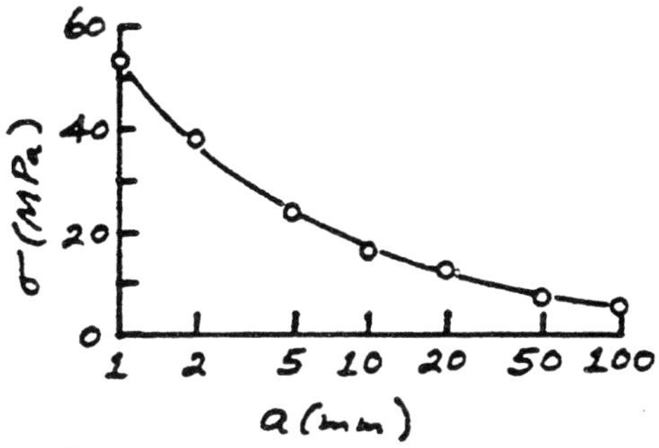

8.27 To appreciate the relatively low values of fracture toughness for traditional ceramics, plot, on a single graph, breaking stress versus flaw size, a, for an aluminum alloy with a K_{IC} of 30 MPa $\sqrt{\text{m}}$ and silicon carbide (see Table 8.10). Use Equation 7.9 and take $Y = 1$. Cover a range of a from 1 to 100 mm on a logarithmic scale. (Note also Problem 8.26.)

8.27

$$\sigma = K_{IC}/\sqrt{\pi a}$$

$K_{IC} = 3\ \text{MPa}\sqrt{\text{m}}$ for SiC

$= 30\ \text{MPa}\sqrt{\text{m}}$ for Al

a (mm)	σ_{SiC} (MPa)	σ_{Al} (MPa)
1	53.5	535
2	37.8	378
5	23.9	239
10	16.9	169
20	12.0	120
50	7.6	75.7
100	5.4	53.5

Resulting plot:

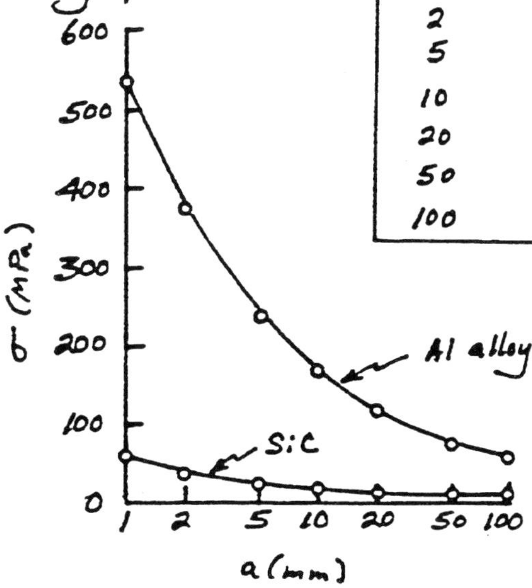

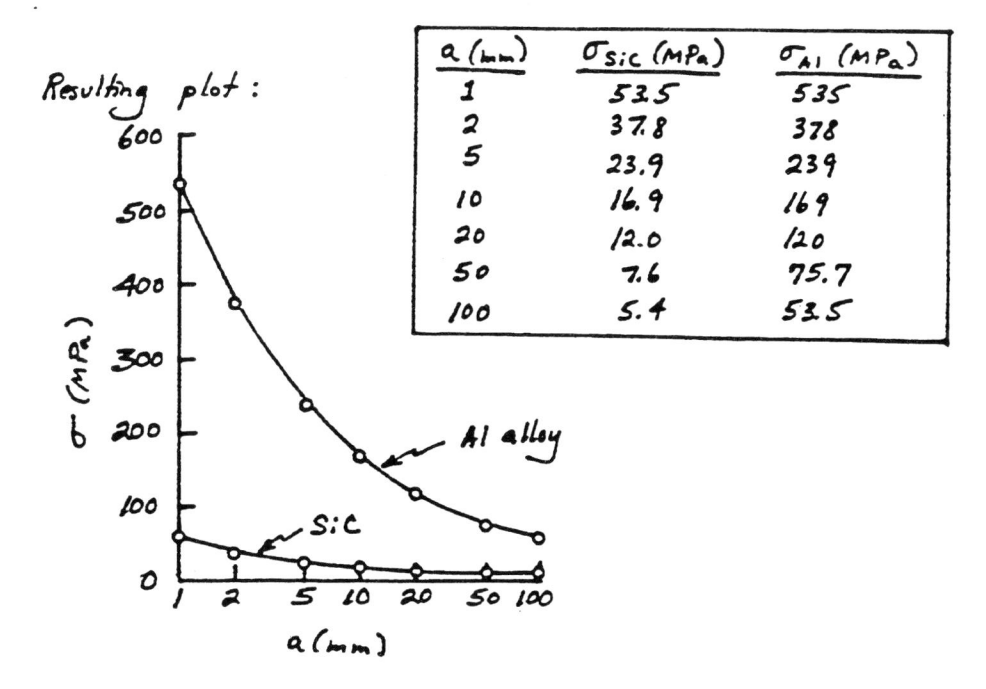

8.28 To appreciate the improved fracture toughness of the new generation of structural ceramics, superimpose a plot of breaking stress versus flaw size for partially stabilized zirconia on the result for Problem 8.27.

8.28

The additional data are:

$\sigma = K_{IC}/\sqrt{\pi a}$ with $K_{IC} = 9\ MPa\sqrt{m}$ for PSZ:

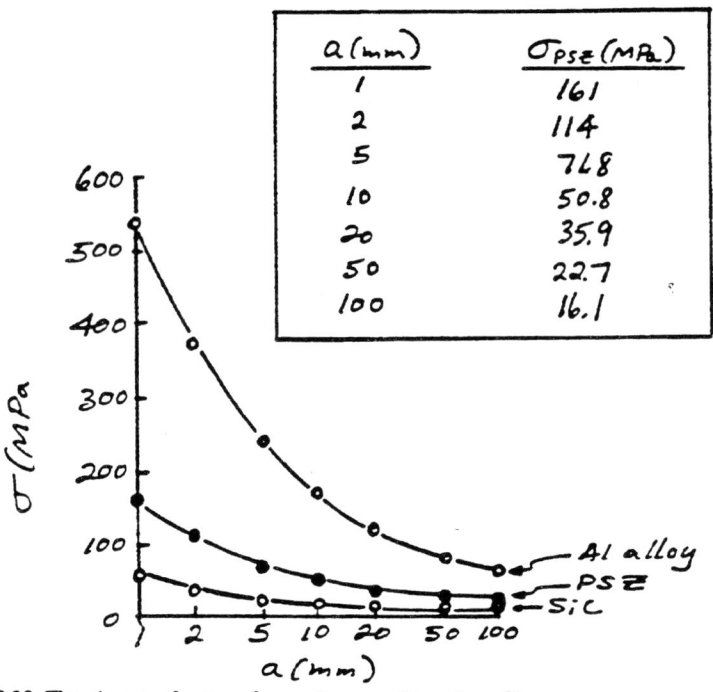

a (mm)	σ_{PSZ} (MPa)
1	161
2	114
5	72.8
10	50.8
20	35.9
50	22.7
100	16.1

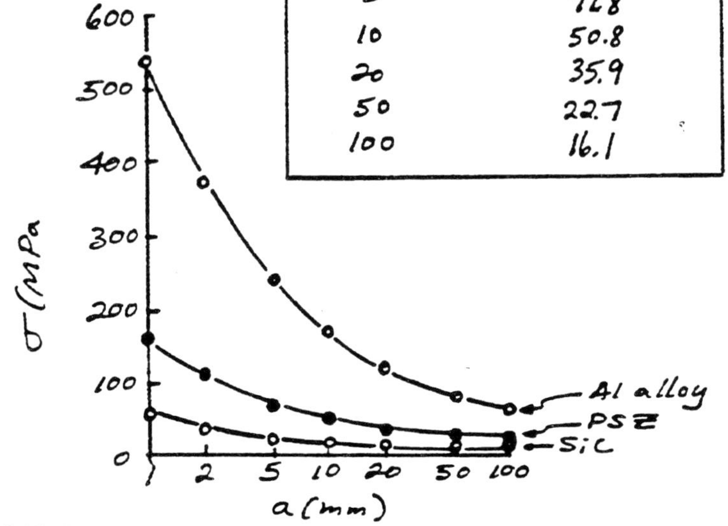

8.29 The time to fracture for a vitreous silica glass fiber at +50°C is 10^4 s. What will be the time to fracture at room temperature (25°C)? Assume the same activation energy as given in Sample Problem 8.6.

8.29

$$t^{-1} = Ce^{-Q/RT}$$

$$C = t^{-1} e^{+Q/RT}$$

$$= (10^{-4} s^{-1}) e^{+(78{,}600\ J/mol)/[8.314\ J/(mol \cdot K)](323\ K)}$$

$$= 5.15 \times 10^{8} s^{-1}$$

At 25°C,

$$t^{-1} = (5.15 \times 10^{8} s^{-1}) e^{-(78{,}600)/(8.314)(298)}$$

$$= 8.58 \times 10^{-6} s^{-1}$$

or

$$t = 1.17 \times 10^{5} s \times 1\ h/3600\ s$$

$$= \underline{\underline{32.4\ hr}}$$

8.30 To illustrate the very rapid nature of the water reaction with silicate glasses above 150°C, calculate the time to fracture for the vitreous silica fiber of Problem 8.29 at 200°C.

8.30 Following the calculations for Problem 8.29, at 200°C,

$$t^{-1} = (5.15 \times 10^8 s^{-1}) e^{-(78,600)/(8.314)(200+273)}$$

$$= 1.08 \ s^{-1}$$

or

$$t = 1/(1.08 \ s^{-1}) = \underline{\underline{0.930 \text{ second}}}$$

8.31 Using Table 8.11, calculate the lifetime of (a) a slip cast MgO refractory at 1300°C and 12.4 MPa if 1% total strain is permissible. (b) Repeat the calculation for a hydrostatically pressed MgO refractory. (c) Comment on the effect of processing on the relative performance of these two refractories.

8.31 (a) From Table 8.11, $\dot{\epsilon} = 330 \times 10^{-6}$ mm/(mm·h)

$$n_{life} = \frac{0.01}{330 \times 10^{-6} h^{-1}} = \underline{\underline{30.3 \text{ hr}}}$$

(b) In this case, $\dot{\epsilon} = 33 \times 10^{-6}$ mm/(mm·h)

$$n_{life} = \frac{0.01}{33 \times 10^{-6} h^{-1}} = \underline{\underline{303 \text{ hr}}}$$

(c) Hydrostatic pressing produces a more compact, less porous microstructure less susceptible to mechanisms of creep deformation.

8.32 Assume the activation energy for the creep of Al_2O_3 is 425 kJ/mol. (a) Predict the creep rate, $\dot{\epsilon}$, for Al_2O_3 at 1000°C and 1800 psi applied stress. (See Table 8.11 for data at 1300°C and 1800 psi.) (b) Calculate the lifetime of an Al_2O_3 furnace tube at 1000°C and 1800 psi if 1% total strain is permissible.

8.32 (a) $\dot{\epsilon} = Ce^{-Q/RT}$

$C = \dot{\epsilon}\, e^{+Q/RT}$

Using Table 8.11:

$C = (1.3 \times 10^{-6}\ h^{-1})e^{+(425,000)/(8.314)(1573)}$

$= 1.69 \times 10^{8}\ h^{-1}$

$\dot{\epsilon}_{1000°C} = (1.69 \times 10^{8}\ h^{-1})e^{-(425,000)/(8.314)(1273)}$

$= \underline{\underline{6.14 \times 10^{-10}\ mm/(mm \cdot h)}}$

(b) $n_{life} = \dfrac{0.01}{6.14 \times 10^{-10}\ h^{-1}} = 1.63 \times 10^{7}\ h \times \left(1\ year/8,760\ h\right)$

$= \underline{\underline{1,860\ years}}$

8.33 In Problem 7.53 "power law" creep was introduced in which

$$\dot{\epsilon} = c_1\sigma^n e^{-Q/RT}$$

(a) For a value of $n = 4$, calculate the creep rate, $\dot{\epsilon}$, for Al_2O_3 at 1300°C and 900 psi. (b) Calculate the lifetime of an Al_2O_3 furnace tube at 1300°C and 900 psi if 1% total strain is permissible.

8.33 (a) Using Table 8.11,

$\dot{\epsilon}_{1300°C,\ 1800\ psi} = 1.3 \times 10^{-6}\ mm/(mm \cdot h)$

$= c_1 \sigma^n e^{-Q/RT}$

Therefore,

$\dot{\epsilon}_{1300°C,\ 900\ psi} = \left[1.3 \times 10^{-6}\ mm/(mm \cdot h)\right]\left(\dfrac{900}{1800}\right)^4$

$= \underline{\underline{8.13 \times 10^{-8}\ mm/(mm \cdot h)}}$

(b) As in Sample Problem 8.7,

$n_{life} = \dfrac{0.01}{8.13 \times 10^{-8}}\ h = 1.23 \times 10^{5}\ h \times \left(1\ year/8,760\ h\right)$

$= \underline{\underline{14\ years}}$

• 8.34 (a) The creep plot in Figure 8-9 indicates a general "band" of data roughly falling between the two parallel lines. Calculate a general activation energy for the creep of oxide ceramics using the slope indicated by those parallel lines. (b) Estimate the uncertainty in the answer to part (a) by considering the maximum and minimum slopes within the band between temperatures of 1400 and 2200° C.

8.34 Note that the reference points for solving the problem are:

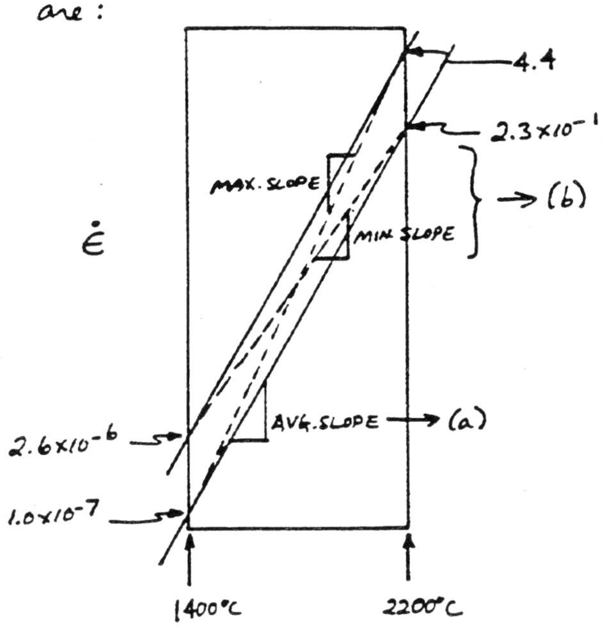

(a) For average slope → general activation energy

$$\frac{2.3\times10^{-1}}{1.0\times10^{-7}} = \frac{C e^{-Q/R(2200+273)K}}{C e^{-Q/R(1400+273)K}} = 2.3\times10^{+6}$$

or

$$\ln(2.3\times10^{+6}) = -\frac{Q}{R}\left(\frac{1}{2473} - \frac{1}{1673}\right)K^{-1}$$

giving

$$Q = (8.314\ J/mol\cdot K)\frac{\ln(2.3\times10^{6})}{(1/1673 - 1/2473)K^{-1}}$$

$$= \underline{\underline{630\ kJ/mol}}$$

(b) For maximum slope,

$$\frac{4.4}{1.0\times 10^{-7}} = \frac{Ce^{-Q/R(2473\,K)}}{Ce^{-Q/R(1673\,K)}} = 4.4\times 10^{7}$$

giving

$$Q = (8.314\ J/mol\cdot K)\frac{\ln(4.4\times 10^{7})}{(1/1673 - 1/2473)\ K^{-1}}$$

$$= 757\ kJ/mol$$

$\therefore$ upper uncertainty $= 757\ kJ/mol - 630\ kJ/mol$
$= +127\ kJ/mol$

For minimum slope,

$$\frac{2.3\times 10^{-1}}{2.6\times 10^{-6}} = \frac{Ce^{-Q/R(2473\,K)}}{Ce^{-Q/R(1673\,K)}} = 8.85\times 10^{4}$$

giving

$$Q = (8.314\ J/mol\cdot K)\frac{\ln(8.85\times 10^{4})}{(1/1673 - 1/2473)\ K^{-1}} = 490\ kJ/mol$$

$\therefore$ lower uncertainty $= 490\ kJ/mol - 630\ kJ/mol$
$= -140\ kJ/mol$

$\therefore$ overall uncertainty: $630\ kJ/mol\ ^{+127\ kJ/mol}_{-140\ kJ/mol}$

8.35 Calculate the rate of heat loss per square meter through the fireclay refractory wall of a furnace operated at 1000°C. The external face of the furnace wall is at 100°C, and the wall is 10 cm thick.

8.35

Taking the average thermal conductivity from Table 8.13,

$$\bar{k} = \frac{1.1 + 1.5}{2} \frac{J}{s \cdot m \cdot K} = 1.3 \frac{J}{s \cdot m \cdot K}$$

Using Equation 8.5 and re-arranging:

$$\frac{\Delta Q}{\Delta t} = -kA\left(\frac{\Delta T}{\Delta x}\right)$$

$$= -\left(1.3 \frac{J}{s \cdot m \cdot K}\right)(1\ m^2)\left(\frac{1000°C - 100°C}{10 \times 10^{-2}\ m}\right)$$

$$= -11.7 \times 10^3\ J/s$$

$$= -11.7\ kW$$

(Note that, for temperature differentials, 1 °C is equivalent to 1 K.)

8.36 Repeat Problem 8.35 for a 5-cm-thick refractory wall.

8.36

Again, $\bar{k} = 1.3 \frac{J}{s \cdot m \cdot K}$

Using Equation 8.5

$$\frac{\Delta Q}{\Delta t} = -\left(1.3 \frac{J}{s \cdot m \cdot K}\right)(1\ m^2)\left(\frac{1000°C - 100°C}{5 \times 10^{-2}\ m}\right)$$

$$= -23.4 \times 10^3\ J/s$$

$$= -23.4\ kW$$

8.37 Calculate the rate of heat loss per cm^2 through a stabilized zirconia lining of a high-temperature laboratory furnace operated at 1400°C. The external face of the lining is at 100°C and its thickness is 1 cm. (Assume the data for stabilized zirconia in Table 8.13 is linear with temperature and can be extrapolated to 1400°C.)

8.37 Extrapolating the data in Table 8.13 gives:

$$k_{1400°C} = \left(2.0 + [2.3-2.0]\frac{[1400°C - 100°C]}{[1000°C - 100°C]}\right)\frac{J}{s \cdot m \cdot K}$$

$$= 2.43\ J/s \cdot m \cdot K \longrightarrow \bar{k} = \frac{2.00 + 2.43}{2}\ \frac{J}{m \cdot s \cdot K} = 2.217\frac{J}{m \cdot s \cdot K}$$

Using Equation 8.5,

$$\frac{\Delta Q}{\Delta t} = -\left(2.217\frac{J}{s \cdot m \cdot K}\right)\left(10^{-2} m\right)^2\left(\frac{1400°C - 100°C}{1 \times 10^{-2}\ m}\right)$$

$$= -28.8\ J/s$$

$$= \underline{\underline{-28.8\ W}}$$

8.38 What would be the stress developed in a mullite furnace tube constrained in the way illustrated in Figure 8-13 if it were heated to 1000°C?

8.38 Table 8.12 gives $\alpha = 5.3 \times 10^{-6}\ mm/(mm \cdot °C)$
Table 8.8 gives $E = 69 \times 10^3 MPa$ (for mullite porcelain)

$$\sigma = E\epsilon = E\alpha\Delta T = (69 \times 10^3 MPa)(5.3 \times 10^{-6}\ °C^{-1})(1000-25)°C$$

$= \underline{\underline{357 MPa}}$ (compressive)

8.39 Repeat Problem 8.38 for silica glass.

8.39 Table 8.12 gives $\alpha = 0.5 \times 10^{-6}\ mm/(mm \cdot °C)$
Table 8.8 gives $E = 72.4 \times 10^3 MPa$

$$\sigma = E\epsilon = E\alpha\Delta T = (72.4 \times 10^3 MPa)(0.5 \times 10^{-6}\ °C^{-1})(1000-25)°C$$

$= \underline{\underline{35.3 MPa}}$ (compressive)

• **8.40** A textbook on the mechanics of materials gives the following expression for the stress due to thermal expansion mismatch in a coating of (thickness a) on a substrate (of thickness b) at a temperature T:

$$\sigma = \frac{E}{1-\nu}(T_0 - T)(\alpha_c - \alpha_s)\left[1 - 3\left(\frac{a}{b}\right) + 6\left(\frac{a}{b}\right)^2\right]$$

where E and ν are the elastic modulus and Poisson's ratio of the coating, respectively, T_0 the temperature at which the coating is applied (and the coating stress is initially zero), and α_c and α_s are the thermal expansion coefficients of the coating and the substrate, respectively. **(a)** Calculate the room-temperature (25°C) stress in a thin soda–lime–silica glaze applied at 1000°C on a porcelain ceramic. (Take $E = 65 \times 10^3$ MPa and $\nu = 0.24$ and see Table 8.12 for relevant thermal expansion data.) **(b)** Repeat part **(a)** for a special high-silica glaze with an average thermal expansion coefficient of 3×10^{-6}°C^{-1}. [Take $E = 72 \times 10^3$ MPa and $\nu = 0.24$.]

8.40 (a) $\sigma = \frac{E}{1-\nu}(T_0 - T)(\alpha_c - \alpha_s)\left[1 - 3\left(\frac{a}{b}\right) + 6\left(\frac{a}{b}\right)^2\right]$

For a "thin" glaze, take $\frac{a}{b} \to 0$

or

$$\sigma = \frac{E}{1-\nu}(T_0 - T)(\alpha_c - \alpha_s)$$

Using the given and suggested data,

$$\sigma = \frac{65 \times 10^3 \text{ MPa}}{1 - 0.24}(1000°\text{C} - 25°\text{C})(9.0 - 6.0) \times 10^{-6}\ °\text{C}^{-1}$$

$+250$ MPa

Note: The + sign indicates tensile stress (due to the higher thermal expansion coating being constrained upon cooling by the substrate).

(b) For this system,

$$\sigma = \frac{72 \times 10^3 \text{ MPa}}{1 - 0.24}(1000°\text{C} - 25°\text{C})(3.0 - 6.0) \times 10^{-6}\ °\text{C}^{-1}$$

$= -277$ MPa

Note: The negative sign indicates a compressive stress for this low thermal expansion coating.

8.41 (a) A processing engineer suggests that a fused SiO_2 crucible be used for a water quench from 500°C. Would you endorse this plan? Explain. (b) Another processing engineer suggests that a porcelain crucible be used for the water quench from 500°C. Would you endorse this plan? Again explain.

8.41 (a) Yes. Figure 8-15 indicates that fused SiO_2 can withstand a water quench of greater than 2000°C.

(b) No. Figure 8-15 indicates that porcelain can withstand a water quench of only ≈ 100°C.

8.42 An automobile engine seal made of stabilized zirconia is subjected to a sudden spray of cooling oil corresponding to a heat transfer parameter ($r_m h$) of 0.1 (see Figure 8-15). Will a temperature drop of 100°C fracture this seal?

8.42 Inspection of the behavior of ZrO_2 in Figure 8-15 indicates that, yes, the 100°C drop will fracture the seal.

Note: Thermal shock is a limitation for various zirconia ceramics, even partially-stabilized zirconia with its relatively high K_{IC}.

8.43 A borosilicate glass used for sealed-beam headlights has an annealing point of 544°C and a softening point of 780°C. Calculate (a) the activation energy for viscous deformation in this glass, (b) its working range, and (c) its melting range.

8.43 (a) $\eta = \eta_0 e^{+Q/RT}$

annealing pt. = 544°C = 817 K

softening pt. = 780°C = 1053 K

$$\frac{10^{13.4}\text{ poise}}{10^{7.6}\text{ poise}} = \frac{e^{+Q/[8.314\text{ J/(mol}\cdot\text{K)}](817\text{ K})}}{e^{+Q/[8.314\text{ J/(mol}\cdot\text{K)}](1053\text{ K})}}$$

or

$$Q = 405\text{ kJ/mol}$$

(b) $\eta_0 = \eta e^{-Q/RT}$

$= (10^{13.4} \text{ poise}) e^{-(405{,}000)/(8.314)(817)}$

$= 3.32 \times 10^{-13}$ poise

Then, $T = Q/[R \ln(\eta/\eta_0)]$

or

$$T_1 = \frac{(405{,}000 \text{ J/mol})}{[8.314 \text{ J/(mol}\cdot\text{K)}] \ln(10^4/3.32\times10^{-13})}$$

$= 1283 \text{ K} = 1010\,°\text{C}$

and

$$T_2 = \frac{(405{,}000 \text{ J/mol})}{[8.314 \text{ J/(mol}\cdot\text{K)}] \ln(10^8/3.32\times10^{-13})}$$

$= 1032 \text{ K} = 759\,°\text{C}$

Therefore,

working range = 759 to 1010 °C

(c) Furthermore,

$$T_3 = \frac{(405{,}000 \text{ J/mol})}{[8.314 \text{ J/(mol}\cdot\text{K)}] \ln(50/3.32\times10^{-13})}$$

$= 1491 \text{ K} = 1218\,°\text{C}$

and

$$T_4 = \frac{(405{,}000 \text{ J/mol})}{[8.314 \text{ J/(mol}\cdot\text{K)}] \ln(500/3.32\times10^{-13})}$$

$= 1393 \text{ K} = 1120\,°\text{C}$

Therefore,

melting range = 1120 to 1218 °C

8.44 The following viscosity data are available on a borosilicate glass used for vacuum-tight seals:

T (°C)	η (poise)
700	4.0×10^7
1080	1.0×10^4

Determine the temperatures at which this glass should be (a) melted and (b) annealed.

8.44 (a) $\eta = \eta_o e^{+Q/RT}$

$$\frac{4.0\times10^7 \text{ poise}}{1.0\times10^4 \text{ poise}} = \frac{e^{+Q/[8.314 \text{ J/(mol·K)}](973 \text{ K})}}{e^{+Q/[8.314 \text{ J/(mol·K)}](1353 \text{ K})}}$$

or $Q = 239 \text{ kJ/mol}$

$$\eta_o = \eta e^{-Q/RT} = (4.0\times10^7 \text{ poise}) e^{-(2.39\times10^5)/(8.314)(973)}$$

$$= 5.98\times10^{-6} \text{ poise}$$

$$T = Q/[R \ln(\eta/\eta_o)]$$

$$T_1 = \frac{(239{,}000 \text{ J/mol})}{[8.314 \text{ J/(mol·K)}] \ln(50/5.98\times10^{-6})}$$

$$= 1803 \text{ K} = 1530\ ^\circ\text{C}$$

$$T_2 = \frac{(239{,}000 \text{ J/mol})}{[8.314 \text{ J/(mol·K)}] \ln(500/5.98\times10^{-6})}$$

$$= 1575 \text{ K} = 1302\ ^\circ\text{C}$$

Therefore,

melting range = 1302 to 1530 °C

(b)
$$T_3 = \frac{(239{,}000 \text{ J/mol})}{[8.314 \text{ J/(mol·K)}] \ln(10^{12.5}/5.98\times10^{-6})}$$

$$= 704 \text{ K} = 431\ ^\circ\text{C}$$

$$T_4 = \frac{(239{,}000 \text{ J/mol})}{[8.314 \text{ J/(mol·K)}] \ln(10^{13.5}/5.98\times10^{-6})}$$

$$= 666 \text{ K} = 393\ ^\circ\text{C}$$

Therefore,

annealing range = 393 to 431 °C

8.45 For the vacuum sealing glass described in Problem 8.44, assume you have traditionally annealed the product at the viscosity of 10^{13} poise. After a cost–benefit analysis, you realize that it is more economical to anneal for a longer time at a lower temperature. If you decide to anneal at a viscosity of $10^{13.4}$ poise, how many degrees (°C) should your annealing furnace operator lower the furnace temperature?

8.45 Using the results of Problem 8.44, we can calculate the temperatures corresponding to viscosities of 10^{13} P and $10^{13.4}$ P:

$$\eta_1 = 10^{13}\,P = (5.98\times10^{-6}\,P)\,e^{\frac{+239{,}000\ J/mol}{8.314\ J/mol\cdot K\ T_1}}$$

or $$T_1 = \frac{239{,}000}{8.314\ \ln(10^{13}/5.98\times10^{-6})}\,K = 685\ K$$

and

$$\eta_2 = 10^{13.4}\,P = (5.98\times10^{-6}\,P)\,e^{\frac{+239{,}000\ J/mol}{8.314\ J/mol\cdot K\ T_2}}$$

or

$$T_2 = \frac{239{,}000}{8.314\ \ln(10^{13.4}/5.98\times10^{-6})}\,K = 670\ K$$

$$\therefore \Delta T = (685-670)\,K = 15\,K = \underline{\underline{15\,°C}}$$

8.46 You are asked to help design a manufacturing furnace for a new optical glass. Given that it has an annealing point of 460°C and a softening point of 647°C, calculate the temperature range in which the product shape would be formed (i.e., the working range).

8.46

$$\eta = \eta_0\, e^{+Q/RT}$$

annealing pt. = 460°C = 733 K

softening pt. = 647°C = 920 K

$$\frac{10^{13.4}\ \text{poise}}{10^{7.6}\ \text{poise}} = \frac{e^{+Q/[8.314\ J/(mol\cdot K)](733\,K)}}{e^{+Q/[8.314\ J/(mol\cdot K)](920\,K)}}$$

or $$Q = 400\ kJ/mol$$

$$\eta_0 = \eta\, e^{-Q/RT}$$

$$= (10^{13.4}\ \text{poise})\, e^{-400{,}000/(8.314)(733)} = 7.33\times10^{-16}\ \text{poise}$$

$$T = Q/[R\ln(\eta/\eta_0)]$$

working range corresponds to $10^4 \text{ poise} < \eta < 10^8 \text{ poise}$

or

$$T_1 = \frac{400{,}000 \text{ J/mol}}{[8.314 \text{ J/(mol}\cdot\text{K)}] \ln(10^4/7.33\times10^{-16})}$$

$$= 1092 \text{ K} = 819°\text{C}$$

and

$$T_2 = \frac{400{,}000 \text{ J/mol}}{[8.314 \text{ J/(mol}\cdot\text{K)}] \ln(10^8/7.33\times10^{-16})}$$

$$= 903 \text{ K} = 630°\text{C}$$

or working range = 630 to 819°C

Section 8.5 - Optical Properties of Ceramics and Glasses

PP 8.11 In Sample Problem 8.11 a critical angle of incidence is calculated for light refraction from silica glass to air. What would be the critical angle if the air were to be replaced by a water environment (with $n = 1.333$)?

PP 8.11 In this case,

$$\frac{\sin\theta_i}{\sin\theta_r} = \frac{n_{water}}{n_{glass}} = \frac{1.333}{1.458}$$

$$= \frac{\sin\theta_c}{\sin 90°} = \frac{\sin\theta_c}{1}$$

$$\text{or } \theta_c = \arcsin\left(\frac{1.333}{1.458}\right) = 66.1°$$

PP 8.12 What is the reflectance of single-crystal sapphire, which is widely used as an optical and electronic material? (Sapphire is nearly pure Al_2O_3.) (See Sample Problem 8.12.)

PP 8.12 Using Table 8.14,

$$R = \left(\frac{n-1}{n+1}\right)^2$$

$$= \left(\frac{1.76-1}{1.76+1}\right)^2 = 0.0758$$

PP 8.13 The relationship between photon energy and wavelength is outlined in Sample Problem 8.13. A useful rule of thumb is that E (in electron volts) $= K\lambda$, where λ is expressed in nanometers. What is the value of K?

PP 8.13

$$E = \frac{hc}{\lambda}$$

For a wavelength of x nm,

$$E = \frac{(0.6626\times10^{-33}\ J\cdot s)(0.2998\times10^{9}\ m/s)}{(x\ nm)(10^{-9}\ m/nm)} \times \frac{6.242\times10^{18}\ eV}{J}$$

$$= \frac{1240}{x}\ eV$$

or $\underline{\underline{K = 1240}}$

8.47 By what percentage is the critical angle of incidence different for a lead oxide-containing "crystal" glass (with $n = 1.7$) in comparison to plain silica glass?

8.47

The critical angle for silica glass was found to be 43.3° in Sample Problem 8.11. For the lead glass,

$$\theta_c = \arcsin\frac{1}{1.7} = 36.0°$$

or

$$\%\ \text{difference} = \frac{43.3 - 36.0}{43.3}\times100\%$$

$$= \underline{\underline{16.9\%\ (\text{smaller})}}$$

corresponding to the high degree of internal reflection and characteristic "sparkle" discussed in Section 8.5.

8.48 What would be the critical angle of incidence for light transmission from single-crystal Al_2O_3 into a coating of silica glass?

8.48

$$\frac{\sin\theta_i}{\sin\theta_r} = \frac{n_{silica}}{n_{Al_2O_3}} = \frac{\sin\theta_c}{\sin 90°} = \frac{\sin\theta_c}{1}$$

Using data from Table 8.14,

$$\theta_c = \arcsin\left(\frac{n_{silica}}{n_{Al_2O_3}}\right) = \arcsin\left(\frac{1.458}{1.76}\right) = \underline{\underline{55.9°}}$$

• 8.49 (a) Consider a translucent orthoclase ceramic with a thin orthoclase glass coating (glaze). What is the maximum angle of incidence at the ceramic–glaze interface to ensure that an observer can see any visible light transmitted through the product (into an air atmosphere)? (Consider only specular transmission through the ceramic.) (b) Would the answer to part (a) change if the glaze coating were eliminated? Briefly explain. (c) Would the answer to part (a) change if the product was an orthoclase glass with a thin translucent coating of crystallized orthoclase? Briefly explain.

8.49 (a)

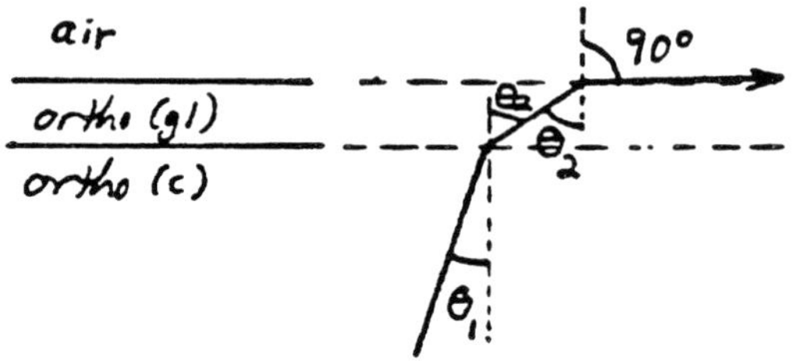

At air/glass interface,

$$\theta_2 = \arcsin\frac{1}{n_{gl}} = \arcsin\left(\frac{1}{1.51}\right) = 41.47°$$

At crystal/glass interface

$$\frac{\sin\theta_2}{\sin\theta_1} = \frac{\sin 41.47°}{\sin\theta_1} = \frac{n_1}{n_2} = \frac{1.525}{1.51}$$

or $\theta_1 = 40.98°$

(b) For a crystal/air interface,

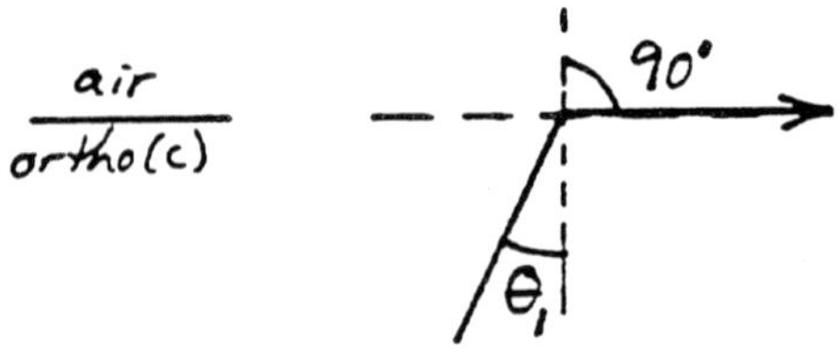

or $\theta_1 = \arcsin\left(\frac{1}{1.525}\right) = 40.98°$

∴ no, the answer is unchanged.

(c) For a glass/crystal/air system,

$$\theta_1 = \arcsin \frac{1.525}{1.51} \times \frac{1}{1.525} = \underline{\underline{41.47^\circ}}$$

∴ <u>yes</u>, the answer is changed.

8.50 By what percentage is the reflectance different for a lead oxide-containing "crystal" glass (with $n = 1.7$) in comparison to plain silica glass?

8.50 The reflectance of silica glass was found to be 0.035 in Sample Problem 8.12. For the lead glass,

$$R = \left(\frac{1.7 - 1}{1.7 + 1}\right)^2 = 0.0672$$

or

$$\% \text{ difference} = \frac{0.0672 - 0.035}{0.035} \times 100\,\%$$

$$= \underline{\underline{92.0\,\% \text{ (greater)}}}$$

again corresponding to a characteristic "sparkling" appearance.

8.51 Silica glass is frequently and incorrectly referred to as "quartz." This is the result of shortening the traditional term "fused quartz," which described the original technique of making silica glass by melting quartz powder. What is the percentage error in calculating the reflectance of silica glass by using the index of refraction of quartz?

8.51 Using data from Table 8.14,

$$R_{quartz} = \left(\frac{n-1}{n+1}\right)^2 = \left(\frac{1.55-1}{1.55+1}\right)^2 = 0.0465$$

$$R_{silica\ glass} = \left(\frac{1.458-1}{1.458+1}\right)^2 = 0.0347$$

$$\% \text{ error} = \frac{0.0465 - 0.0347}{0.0347} \times 100\,\% = \underline{\underline{34.0\,\%}}$$

8.52 For a single crystal Al_2O_3 disk used in a precision optical device, calculate (a) the angle of refraction for a light beam incident from a vacuum at $\theta_i = 30°$, and (b) the fraction of light transmitted into the Al_2O_3 at that angle.

8.52

(a) $$\frac{\sin\theta_i}{\sin\theta_r} = \frac{v_{vac}}{v_{Al_2O_3}} = n$$

or

$$\sin\theta_r = n^{-1}\sin\theta_i = 1.76^{-1}\sin 30°$$
$$= 0.284$$

or

$$\theta_r = \underline{\underline{16.5°}}$$

(b) $$\text{Fraction transmitted} = 1 - R = 1 - \left(\frac{n-1}{n+1}\right)^2$$
$$= 1 - \left(\frac{1.76-1}{1.76+1}\right)^2 = \underline{\underline{0.924}}$$

8.53 The highest-energy visible-light photon is the shortest wavelength light at the blue end of the spectrum. How many times greater is the energy of the CuK_α x-ray photon used in Section 3.7 for the analysis of crystal structure? (Note Sample Problem 8.13.)

8.53

As $E \propto 1/\lambda$,

$$\frac{E_{CuK\alpha}}{E_{blue}} = \frac{\lambda_{blue}}{\lambda_{CuK\alpha}} = \frac{400\text{ nm}}{0.1542\text{ nm}} = \underline{\underline{2{,}594}}$$

8.54 The lowest-energy visible-light photon is the longest wavelength light at the red end of the spectrum. How many times greater is this energy than that of a typical microwave photon with a wavelength of 10^7 nm?

8.54

As $E \propto 1/\lambda$

$$\frac{E_{red}}{E_{microwave}} = \frac{\lambda_{microwave}}{\lambda_{red}} = \frac{10^7\text{ nm}}{700\text{ nm}} = \underline{\underline{14{,}300}}$$

8.55 By what factor is the energy of a photon of red visible light (λ = 700 nm) greater than that of a photon of infrared light with $\lambda = 5\ \mu m$?

8.55

As $E \propto 1/\lambda$,

$$\frac{E_{red}}{E_{ir}} = \frac{\lambda_{ir}}{\lambda_{red}} = \frac{5 \times 10^{-6} m}{700 \times 10^{-9} m} = 7.14$$

8.56 What range of photon energies are absorbed in the blue glass of Figure 8-28?

8.56

Figure 8-26 shows that photons with wavelengths from roughly 600 nm to 700 nm are absorbed.

$$E_{700\,nm} = \frac{hc}{\lambda} = \frac{(0.6626 \times 10^{-33} J \cdot s)(0.2998 \times 10^{9} m/s)}{(700 \times 10^{-9} m)}$$

$$= 2.84 \times 10^{-19} J \times 6.242 \times 10^{18} eV/J = 1.77 eV$$

$$E_{600\,nm} = \frac{(0.6626 \times 10^{-33} J \cdot s)(0.2998 \times 10^{9} m/s)}{(600 \times 10^{-9} m)}$$

$$= 3.31 \times 10^{-19} J \times 6.242 \times 10^{18} eV/J = 2.07 eV$$

or

photon energies from roughly $2.84 - 3.31 \times 10^{-19}$ J are absorbed corresponding to roughly 1.77 - 2.07 eV.

Section 9.1 – Polymerization

PP 9.1 What would be the degree of polymerization of a polyvinyl chloride with an average molecular weight of 25,000 amu? (See Sample Problem 9.1.)

PP 9.1

$$n = \frac{\text{mol. wt. } (C_2H_3Cl)_n}{\text{mol. wt. } C_2H_3Cl}$$

$$= \frac{25{,}000 \text{ amu}}{[2(12.01) + 3(1.008) + 35.45] \text{ amu}} = 400$$

PP 9.2 How much H_2O_2 must be added to ethylene to yield an average degree of polymerization of (a) 500, and (b) 1000? (See Sample Problem 9.2.)

PP 9.2

(a) $$\text{wt.\% } H_2O_2 = \frac{\text{mol. wt. } H_2O_2}{500\ (\text{mol. wt. } C_2H_4)} \times 100\ \%$$

$$= \frac{2(1.008) + 2(16.00)}{500\ [2(12.01) + 4(1.008)]} \times 100\ \% = \underline{\underline{0.243\ \text{wt.\%}}}$$

(b) $$\text{wt.\% } H_2O_2 = \frac{2(1.008) + 2(16.00)}{1000\ [2(12.01) + 4(1.008)]} \times 100\ \% = \underline{\underline{0.121\ \text{wt.\%}}}$$

PP 9.3 What would be the mole percent of ethylene and vinyl chloride in an irregular copolymer that contains 50 wt % of each component? (See Sample Problem 9.3.)

PP 9.3

$$\text{mol. wt. } C_2H_4 = [2(12.01) + 4(1.008)]\ \text{amu} = 28.05\ \text{amu}$$

$$\text{mol. wt. } C_2H_3Cl = [2(12.01) + 3(1.008) + 35.45]\ \text{amu} = 62.49\ \text{amu}$$

For the copolymer in question,

$$1\ \text{kg copolymer} = 500\ \text{g } C_2H_4 + 500\ \text{g } C_2H_3Cl$$

or $$\frac{500\ \text{g}}{28.05\ \text{g/mol } C_2H_4} = 17.83\ \text{mol. } C_2H_4$$

and $$\frac{500\ \text{g}}{62.49\ \text{g/mol } C_2H_3Cl} = 8.00\ \text{mol. } C_2H_3Cl$$

giving

$$\text{mol. \% ethylene} = \frac{17.83}{17.83 + 8.00} \times 100\ \% = \underline{\underline{69.0\ \%}}$$

and

$$\text{mol. \% vinylchloride} = \frac{8.00}{17.83 + 8.00} \times 100\ \% = \underline{\underline{31.0\ \%}}$$

PP 9.4 Calculate the degree of polymerization for a polyacetal molecule with a molecular weight of 25,000 amu. (See Sample Problem 9.4.)

PP 9.4

$$n = \frac{\text{mol. wt. } (CH_2O)_n}{\text{mol. wt. } CH_2O}$$

$$= \frac{25{,}000\ \text{amu}}{[12.01 + 2(1.008) + 16.00]\ \text{amu}} = \underline{\underline{833}}$$

9.1 What is the average molecular weight of a polypropylene with a degree of polymerization of 500? (Note Table 9.1.)

9.1

$$\text{mol. wt. } (C_3H_6)_n = n\,(\text{mol. wt. } C_3H_6)$$

$$= 500\,[3(12.01) + 6(1.008)]\text{ amu}$$

$$= \underline{\underline{21{,}040 \text{ amu}}}$$

9.2 What is the average molecular weight of a polystyrene with a degree of polymerization of 500?

9.2

Table 9.1 indicates that styrene is C_8H_8. (Also, you should note that  $= C_6H_5$. One of the H in the benzene ring is lost in order to bond the ring to the "backbone" carbon.)

$$\text{mol. wt. } (C_8H_8)_n = n\,(\text{mol. wt. } C_8H_8)$$

$$= 500\,[8(12.01) + 8(1.008)]\text{ amu}$$

$$= \underline{\underline{52{,}070 \text{ amu}}}$$

9.3 How many grams of H_2O_2 would be needed to yield 1 kg of a polypropylene, $(C_3H_6)_n$, with an average degree of polymerization of 600? (Use the same assumptions used in Sample Problem 9.2.)

9.3

$$\frac{\text{mol. wt. } H_2O_2}{600 \times (\text{mol. wt. } C_3H_6)} = \frac{(x \text{ gm})(1 \text{ kg}/1000 \text{ gm})}{1 \text{ kg}}$$

$$\text{or } x = \frac{[2(1.008) + 2(16.00)](1000)}{600\,[3(12.01) + 6(1.008)]} = 1.35$$

or $\underline{\underline{1.35 \text{ g}}}$ H_2O_2 would be needed

9.4 A blend of polyethylene and polyvinyl chloride (see Figure 9-4) contains 10 wt % polyvinyl chloride. What is the molecular percentage of polyvinyl chloride?

9.4 mol. wt. C_2H_4 = [2(12.01) + 4(1.008)] amu = 28.05 amu

mol. wt. C_2H_3Cl = [2(12.01) + 3(1.008) + 35.45] amu

= 62.49 amu

For the copolymer in question,

1 kg copolymer = 900 g C_2H_4 + 100 g C_2H_3Cl

or $\frac{900\text{ g}}{28.05\text{ g/mol } C_2H_4} = 32.09 \text{ mol } C_2H_4$

and $\frac{100\text{ g}}{62.49\text{ g/mol } C_2H_3Cl} = 1.60 \text{ mol } C_2H_3Cl$

giving

mol. % polyvinylchloride $= \frac{1.60}{32.09 + 1.60} \times 100\% = 4.75\%$

9.5 A blend of polyethylene and polyvinyl chloride (see Figure 9-4) contains 10 mol % polyvinyl chloride. What is the weight percentage of polyvinyl chloride?

9.5 As calculated in Problem 9.4,

mol. wt. C_2H_4 = 28.05 amu

mol. wt. C_2H_3Cl = 62.49 amu

For the copolymer in question,

1 mol copolymer = 0.90 mol C_2H_4 + 0.10 mol C_2H_3Cl

or $0.90 \text{ mol } C_2H_4 \times \frac{28.05\text{ g}}{\text{mol}} = 25.25 \text{ g } C_2H_4$

and $0.10 \text{ mol } C_2H_3Cl \times \frac{62.49\text{ g}}{\text{mol}} = 6.25 \text{ g } C_2H_3Cl$

giving

wt % polyvinylchloride $= \frac{6.25}{25.25 + 6.25} \times 100\% = 19.8\%$

9.6 Calculate the degree of polymerization for (a) a low-density polyethylene with a molecular weight of 20,000 amu, (b) a high-density polyethylene with a molecular weight of 300,000 amu, and (c) an ultra-high molecular weight polyethylene with a molecular weight of 4,000,000 amu.

9.6

(a) $$n = \frac{\text{mol. wt. } (C_2H_4)_n}{\text{mol. wt. } C_2H_4} = \frac{20{,}000 \text{ amu}}{[2(12.01) + 4(1.008)] \text{ amu}} = \underline{\underline{713}}$$

(b) $$n = \frac{300{,}000 \text{ amu}}{28.05 \text{ amu}} = \underline{\underline{10{,}700}}$$

(c) $$n = \frac{4{,}000{,}000 \text{ amu}}{28.05 \text{ amu}} = \underline{\underline{142{,}600}}$$

9.7 A simplified mer formula for natural rubber (isoprene) is C_5H_8. (See Table 9.2 for a more detailed illustration.) Calculate the molecular weight for a molecule of isoprene with a degree of polymerization of 500.

9.7

$$\text{mol. wt. } (C_5H_8)_n = n(\text{mol. wt. } C_5H_8)$$

$$= (500)[5(12.01) + 8(1.008)] \text{ amu}$$

$$= \underline{\underline{34{,}060 \text{ amu}}}$$

9.8 Calculate the molecular weight for a molecule of chloroprene (a common synthetic rubber) with a degree of polymerization of 500. (See Table 9.2.)

9.8

$$\text{mol. wt. } (C_4H_5Cl)_n = n(\text{mol. wt. } C_4H_5Cl)$$

$$= 500[4(12.01) + 5(1.008) + 1(35.45)] \text{ amu}$$

$$= \underline{\underline{44{,}270 \text{ amu}}}$$

Section 9.2 – Structural Features of Polymers

PP 9.5 In Sample Problem 9.5, coiled and extended molecular lengths are calculated for a polyethylene with a degree of polymerization of 750. If the degree of polymerization of this material is increased by one-third (to $n = 1000$), by what percentage is **(a)** the coiled length and **(b)** the extended length increased?

PP 9.5 (a) As $\bar{L} \propto \sqrt{n}$,

$$\frac{\bar{L}_2}{\bar{L}_1} = \sqrt{\frac{n_2}{n_1}} = \sqrt{\frac{1000}{750}} = 1.155 \text{ or an increase of } \underline{15.5\%}$$

(b) As $L_{ext} \propto n$,

it will experience (along with n) an increase of $\underline{33.3\%}$

PP 9.6 A fraction of cross-link sites is calculated in Sample Problem 9.6. What actual number of sites does this represent in the 100 g of isoprene?

PP 9.6 Inspection of Figure 9-13 indicates one cross-link per mer.

Using the result of Sample Problem 9.6,

$$N_{sites} = 0.425\, N_{mers} = (0.425) \frac{100\,g\ (0.6023 \times 10^{24}\ mol^{-1})}{[5(12.01) + 8(1.008)]\,g/mol} =$$

$$= \underline{3.76 \times 10^{23}}$$

9.9 The data given in Figure 9-8 can be represented in tabular form as follows:

n Range	n_i (Midvalue)	Population Fraction
1–100	50	—
101–200	150	—
201–300	250	0.01
301–400	350	0.10
401–500	450	0.21
501–600	550	0.22
601–700	650	0.18
701–800	750	0.12
801–900	850	0.07
901–1000	950	0.05
1001–1100	1050	0.02
1101–1200	1150	0.01
1201–1300	1250	0.01
		$\sum = 1.00$

Calculate the average degree of polymerization for this system.

9.9 $\bar{n} = \sum_i p_i n_i$ where p_i is the population fraction with mid-value n_i.

$$\bar{n} = (0.01)(250) + (0.10)(350) + (0.21)(450) + (0.22)(550) + (0.18)(650) + (0.12)(750) + (0.07)(850) + (0.05)(950) + (0.02)(1050) + (0.01)(1150) + (0.01)(1250)$$

$$= \underline{\underline{612}}$$

9.10 If the polymer evaluated in Problem 9.9 is polypropylene, what would be the (a) coiled length and (b) extended length of the average molecule?

9.10 (a) As in Sample Problem 9.5(a),

$$\bar{L} = l\sqrt{2n} = (0.154\text{ nm})\sqrt{2(612)}$$
$$= \underline{\underline{5.39\text{ nm}}}$$

(b) As in Sample Problem 9.5(b),

$$L_{ext} = 2nl \sin\frac{109.5°}{2}$$
$$= 2(612)(0.154\text{ nm}) \sin\frac{109.5°}{2}$$
$$= \underline{\underline{154\text{ nm}}}$$

9.11 What would be the maximum fraction of cross-link sites that would be connected in 1 kg of chloroprene with the addition of 250 g of sulfur?

9.11 The amount of sulfur needed for full cross-linking of 1 kg of chloroprene would be:

$$m_S = \frac{\text{mol. wt. S}}{\text{mol. wt. chloroprene}} \times 1000\text{ g}$$
$$= \frac{32.06}{4(12.01) + 5(1.008) + 35.45} \times 1000\text{ g}$$
$$= 362\text{ g}$$

$$\text{Fraction} = \frac{\text{amount S added}}{\text{amount S in fully-cross-linked system}}$$
$$= \frac{250\text{ g}}{362\text{ g}} = \underline{\underline{0.690}}$$

9.12 Calculate the average molecular length (extended) for a polyethylene with a molecular weight of 20,000 amu.

9.12 As in Sample Problem 9.1,

$$n = \frac{\text{mol. wt. } (C_2H_4)_n}{\text{mol. wt. } C_2H_4} = \frac{20{,}000 \text{ amu}}{[2(12.01) + 4(1.008)] \text{ amu}} = 713$$

As in Sample Problem 9.5(b),

$$L_{ext} = 2nl \sin \frac{109.5°}{2}$$

$$= 2(713)(0.154 \text{ nm}) \sin \frac{109.5°}{2}$$

$$= \underline{\underline{179 \text{ nm}}}$$

9.13 Calculate the average molecular length (extended) for polyvinyl chloride with a molecular weight of 20,000 amu.

9.13 As in Sample Problem 9.1,

$$n = \frac{\text{mol. wt. } (C_2H_3Cl)}{\text{mol. wt. } C_2H_3Cl} = \frac{20{,}000 \text{ amu}}{[2(12.01) + 3(1.008) + 35.45] \text{ amu}}$$

$$= 320$$

As in Sample Problem 9.5(b),

$$L_{ext} = 2nl \sin \frac{109.5°}{2}$$

$$= 2(320)(0.154 \text{ nm}) \sin \frac{109.5°}{2} = \underline{\underline{80.5 \text{ nm}}}$$

9.14 If 0.2 g of H_2O_2 is added to 100 g of ethylene to establish the degree of polymerization, what would be the resulting average molecular length (coiled)? (Use the assumptions of Sample Problem 9.2.)

9.14 As in Sample Problem 9.2,

$$\frac{0.2 \text{ g}}{100 \text{ g}} = \frac{\text{mol. wt. } H_2O_2}{n(\text{mol. wt. } C_2H_4)} = \frac{2(1.008) + 2(16.00)}{n[2(12.01) + 4(1.008)]}$$

or $n = 606$

As in Sample Problem 9.5(a),

$$\bar{L} = l\sqrt{2n} = (0.154 \text{ nm})\sqrt{(2)(606)} = \underline{\underline{5.36 \text{ nm}}}$$

9.15 What would be the extended length of the average molecule described in Problem 9.14?

9.15 As in Sample Problem 9.5(b),

$$L_{ext} = 2nl \sin \frac{109.5^\circ}{2}$$

$$= 2(606)(0.154\ nm) \sin \frac{109.5^\circ}{2} = \underline{\underline{152\ nm}}$$

• **9.16** The acetal polymer in Figure 9-5 contains, of course, C—O bonds rather than C—C bonds along its molecular chain backbone. As a result, there are two types of bond angles to consider. The O—C—O bond angle is approximately the same as the C—C—C bond angle (109.5°) because of the tetrahedral bonding configuration in carbon (see Figure 2-19). However, the C—O—C bond is a flexible one with a possible bond angle ranging up to 180°. (a) Make a sketch similar to Figure 9-10 for a fully extended polyacetal molecule. (b) Calculate the extended length of a molecule with a degree of polymerization of 500. (Refer to Table 2.2 for bond length data.) (c) Calculate the coiled length of the molecule in part (b).

9.16 (a)

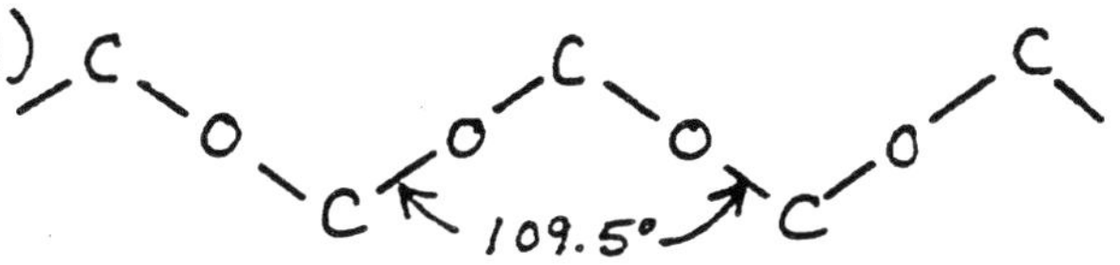

(b)

or $\sin\left(\frac{109.5^\circ}{2}\right) = \frac{x}{2(0.14\ nm)}$

giving

$$x = 0.229\ nm$$

$$\therefore L_{ext} = 500 \times 0.229\ nm = \underline{\underline{114\ nm}}$$

(c) As in Equation 9.4, $\bar{L} = l\sqrt{m}$

and for $(CH_2O)_n$, $m = 2n$

$$\therefore \bar{L} = 0.14\ nm\sqrt{2 \times 500} = \underline{\underline{4.43\ nm}}$$

Section 9.3 – Thermoplastic Polymers

PP 9.7 Calculate the weight fractions for an ABS copolymer that has equal mole fractions of each component. (See Sample Problem 9.7.)

PP 9.7 Following the calculations of Sample Problem 9.7,

$$1 \text{ mole } A = [3(12.01) + 3(1.008) + 14.01] \text{ amu} = 53.06 \text{ amu}$$

$$1 \text{ mole } B = [4(12.01) + 6(1.008)] \text{ amu} = 54.09 \text{ amu}$$

$$1 \text{ mole } S = [8(12.01) + 8(1.008)] \text{ amu} = 104.14 \text{ amu}$$

$$\text{wt.\% } A = \frac{53.06}{53.06 + 54.09 + 104.14} \times 100\% = 25.1\%$$

$$\text{wt.\% } B = \frac{54.09}{53.06 + 54.09 + 104.14} \times 100\% = 25.6\%$$

$$\text{wt.\% } S = \frac{104.14}{53.06 + 54.09 + 104.14} \times 100\% = 49.3\%$$

PP 9.8 What would be the molecular weight of a PPO polymer with a degree of polymerization of 700? (See Sample Problem 9.8.)

PP 9.8

$$\text{mol. wt.} = n(\text{mol. wt. PPO mer})$$

$$= 700[8(12.01) + 8(1.008) + 16.00] \text{ amu}$$

$$= 84{,}100 \text{ amu}$$

9.17 Calculate (**a**) the molecular weight, (**b**) coiled molecular length, and (**c**) extended molecular length for a polytetrafluoroethylene polymer with a degree of polymerization of 500.

9.17 (a) mol. wt. = $n \times$ mer wt.

Table 9.1 indicates that tetrafluorethylene is C_2F_4. Then,

$$\text{mol. wt.} = 500[2(12.01) + 4(19.00)] \text{ amu}$$

$$= 50{,}010 \text{ amu}$$

(b) $$\bar{L} = l\sqrt{2n} = (0.154 \text{ nm})\sqrt{2(500)} = 4.87 \text{ nm}$$

(c) $$L_{ext} = 2nl \sin \frac{109.5°}{2}$$

$$= 2(500)(0.154 \text{ nm}) \sin \frac{109.5°}{2} = 126 \text{ nm}$$

9.18 Repeat Problem 9.17 for a polypropylene polymer with a degree of polymerization of 700.

9.18

(a) $\text{mol. wt.} = 700\,[3(\text{m.wt})_C + 6(\text{m.wt.})_H]$

$= 700\,[3(12.01) + 6(1.008)]\,\text{amu}$

$= \underline{\underline{29{,}450\ \text{amu}}}$

(b) $\bar{L} = l\sqrt{2n} = (0.154\ \text{nm})\sqrt{2(700)} = \underline{\underline{5.76\ \text{nm}}}$

(c) $L_{ext} = 2nl \sin\dfrac{109.5°}{2} = 2(700)(0.154\,\text{nm})\sin\dfrac{109.5°}{2}$

$= \underline{\underline{176\ \text{nm}}}$

9.19 Calculate the degree of polymerization of a polycarbonate polymer with a molecular weight of 100.000 amu.

9.19

Taking the benzene rings for this molecule in Table 9.1 to provide 6 carbon atoms and 4 hydrogen atoms each, the mer formula for polycarbonate is:

$C_{16}H_{14}O_3$

or

$100{,}000\ \text{amu} = n\,[16(12.01) + 14(1.008) + 3(16.00)]\,\text{amu}$

giving $n = \underline{\underline{393}}$

9.20 Calculate the molecular weight for a poly(methyl methacrylate) polymer with a degree of polymerization of 500.

9.20

Table 9.1 indicates that methyl methacrylate is $C_5H_8O_2$

$\text{mol. wt.} = 500\,[5(12.01) + 8(1.008) + 2(16.00)]\,\text{amu}$

$= \underline{\underline{50{,}060\ \text{amu}}}$

• 9.21 The reaction of two molecules to form a nylon monomer is shown in Table 9.1. (The H and OH units enclosed by a dashed line become an H_2O reaction by-product and are replaced by a C—N bond in the middle of the monomer.) (a) Sketch the reaction of nylon monomers to form a nylon polymer. (This occurs by one H and one OH at each end of the monomer being removed to become a reaction by-product.) (b) Calculate the molecular weight of the nylon mer.

9.21

(a)

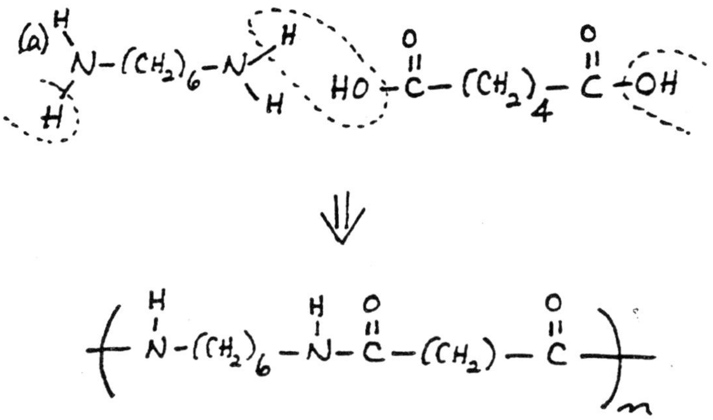

(b)

$$\text{mol. wt. mer} = [12(\text{mol. wt. C}) + 22(\text{mol. wt. H}) + 2(\text{mol. wt. N}) + 2(\text{mol. wt. O})]$$

$$= [12(12.01) + 22(1.008) + 2(14.01) + 2(16.00)]\ \text{amu}$$

$$= \underline{\underline{226.3\ \text{amu}}}$$

• 9.22 A high toughness alloy of nylon and PPO contains 10 wt % PPO. Calculate the mole fraction of PPO in this alloy. (Note Sample Problem 9.8 and Problem 9.21.)

9.22

Using the results of Sample Problem 9.8 and Problem 9.21:

100 g alloy → 90 g nylon + 10 g PPO

$$\text{no. moles nylon} = \frac{90\ \text{g}}{226.3\ \text{g}}\ \text{mol} = 0.398\ \text{mol}$$

$$\text{no. moles PPO} = \frac{10\ \text{g}}{120.1\ \text{g}}\ \text{mol} = 0.083\ \text{mol}$$

$$\therefore \text{mol. frac. PPO} = \frac{0.083\ \text{mol}}{(0.083 + 0.398)\ \text{mol}} = \underline{\underline{0.172}}$$

9.23 In Problem 7.7, the mass reduction in an automobile design was calculated based on trends in metal alloy selection. A more complete picture is obtained by noting that in the same 1975 to 1985 period, the volume of polymers used increased from 0.064 m^3 to 0.100 m^3. Estimate the mass reduction (compared to 1975) including this additional polymer data. (Approximate the polymer density as 1 Mg/m^3.)

9.23

As calculated in Problem 7.7,

mass reduction due to iron removal = 362 kg

mass increase due to Al addition = 30 kg

mass increase due to polymer addition =

$$(0.100 - 0.064)\,m^3 \times \rho_{polymer}$$

$$= (0.036\,m^3)(1\,Mg/m^3)$$

$$= 0.036\,Mg = 36\,kg$$

$\therefore$ net mass reduction = $(362 - 30 - 36)\,kg = \underline{\underline{296\,kg}}$

9.24 Repeat Problem 9.23 for the projected mass reduction in the year 2000 given the data in Problem 7.8 and the fact that the total volume of polymer used at that time is projected to be 0.122 m^3.

9.24

As calculated in Problem 7.8,

mass reduction due to iron removal = 630 kg

mass increase due to Al addition = 59 kg

mass increase due to polymer addition =

$$(0.122 - 0.064)\,m^3 \times \rho_{polymer}$$

$$= (0.058\,m^3)(1\,Mg/m^3)$$

$$= 0.058\,Mg = 58\,kg$$

$\therefore$ net mass reduction = $(630 - 59 - 58)\,kg = \underline{\underline{513\,kg}}$

Section 9.4 – Thermosetting Polymers

PP 9.9 The molecular weight of a product of phenol-formaldehyde is calculated in Sample Problem 9.9. How much water by-product is produced in the polymerization of this product?

PP 9.9

As shown in Sample Problem 9.9, 1.5 molecules of H_2O are produced along with one mer of phenol-formaldehyde:

112.12 amu phenol-formaldehyde along with

$1.5[2(1.008) + 16.00]$ amu $= 27.02$ amu H_2O

Therefore,

$$m_{H_2O\,produced} = m_{product} \times \frac{27.02\text{ amu}}{112.12\text{ amu}}$$

$$= 14\text{g}\left(\frac{27.02}{112.12}\right)$$

$$= \underline{\underline{3.37\text{g}}}$$

PP 9.10 For an elastomer similar to the one in Sample Problem 9.10, calculate the molecular fraction of each component if there are equal weight fractions of vinylidene fluoride and hexafluoropropylene.

PP 9.10

100g → 50g vinylidene fluoride
+ 50g hexafluoropropylene

Using the results of Sample Problem 9.10,

$$\text{mol. vinylidene fluoride} = \frac{50\text{g}}{64.04\text{g}}\text{mol} = 0.781\text{ mol}$$

$$\text{mol. hexafluoropropylene} = \frac{50\text{g}}{150.0\text{g}}\text{mol} = 0.333\text{ mol}$$

$$\overline{1.114\text{ mol}}$$

$$\therefore \text{mol. frac. vinylidene fluoride} = \frac{0.781\text{ mol}}{1.114\text{ mol}} = \underline{\underline{0.701}}$$

and

$$\text{mol. frac. hexafluoropropylene} = \frac{0.333\text{ mol}}{1.114\text{ mol}} = \underline{\underline{0.299}}$$

9.25 What would be the molecular weight of a 50-cm^3 plate made from urea-formaldehyde? (The density of the urea-formaldehyde is 1.50 Mg/m^3.)

9.25 Again (as in Sample Problem 9.9), assume urea to be trifunctional. Then, with the aid of Table 9.2,

$$(m.w.)_{mer} = (m.w.)_{urea} + 1.5(m.w.)_{formaldehyde} - 1.5(m.w.)_{H_2O}$$

$$= [12.01 + 4(1.008) + 2(14.01) + 16.00]\text{ amu}$$
$$+ 1.5[12.01 + 2(1.008) + 16.00]\text{ amu}$$
$$- 1.5[2(1.008) + 16.00]\text{ amu} = 78.08\text{ amu}$$

$$m = \rho V = 1.50\text{ g/cm}^3 \times 50\text{ cm}^3 = 75\text{ g}$$

$$n = \frac{75\text{ g}}{78.08\text{ g}/0.6023 \times 10^{24}\text{ mers}} = 5.79 \times 10^{23}\text{ mers}$$

This gives:

$$\text{mol. wt.} = 5.79 \times 10^{23}\text{ mers} \times 78.08\text{ amu/mer} = \underline{\underline{4.52 \times 10^{25}\text{ amu}}}$$

9.26 How much water by-product would be produced in the polymerization of the urea-formaldehyde product in Problem 9.25?

9.26 There will be 1.5 molecules of H_2O produced along with one mer of urea-formaldehyde. As calculated in Problem 9.25,

$$(m.w.)_{mer} = 78.08\text{ amu}$$

Also, $1.5 \times \text{mol. wt. } H_2O = 1.5[2(1.008) + 16.00]\text{ amu} = 27.02\text{ amu}$

Therefore,

$$m_{H_2O\text{ produced}} = 1.50\text{ g/cm}^3 \times 50\text{ cm}^3 \times \frac{27.02\text{ amu}}{78.08\text{ amu}}$$

$$= \underline{\underline{26.0\text{ g}}}$$

9.27 Polyisoprene loses its elastic properties with 1 wt % O_2 addition. If we assume that this is due to a cross-linking mechanism similar to that for sulfur, what fraction of the cross-link sites are occupied in this case?

9.27

Following a path similar to that in Sample Problem 9.6, we find that the amount of O needed for complete cross-linking of 100g of isoprene is

$$m_O = \frac{\text{mol. wt. O}}{\text{mol. wt. isoprene}} \times 100g$$

$$= \frac{16.00}{5(12.01) + 8(1.008)} \times 100g = 23.5g$$

So,

$$\text{fraction} = \frac{\text{amount O added (as } O_2)}{\text{amount O in fully cross-linked system}}$$

$$= \frac{1g}{23.5g} = \underline{\underline{0.0426}}$$

9.28 Repeat the calculation of Problem 9.27 for the case of the oxidation of polychloroprene by 1 wt % O_2.

9.28

Complete cross-linking comes from:

$$m_O = \frac{\text{mol. wt. O}}{\text{mol. wt. chloroprene}} \times m_{chloroprene}$$

$$= \frac{16.0}{4(12.01) + 5(1.008) + 35.45} \times \text{''}$$

$$= 0.181\, m_{chloroprene}$$

Then,

$$\text{fraction} = \frac{0.01\, m_{chloroprene}}{0.181\, \text{''}} = \underline{\underline{0.0553}}$$

Section 9.5 – Additives

PP 9.11 Sample Problem 9.11 describes a high strength and stiffness engineering polymer. Strength and stiffness can be further increased by a greater "loading" of glass fibers. Calculate the density of a nylon 66 with 43 wt % glass fibers.

PP 9.11

1 kg product → 0.43 kg glass + (1−0.43) kg nylon

Also,

$$V_{product} = V_{nylon} + V_{glass}$$

$$= \frac{m_{nylon}}{\rho_{nylon}} + \frac{m_{glass}}{\rho_{glass}}$$

$$= \left(\frac{0.57\,kg}{1.14\,Mg/m^3} + \frac{0.43\,kg}{2.54\,Mg/m^3}\right) \times \frac{1\,Mg}{1000\,kg}$$

$$= 6.69 \times 10^{-4}\,m^3$$

Giving a density of:

$$\rho = \frac{1\,kg}{6.69\times10^{-4}\,m^3} \times \frac{1\,Mg}{1000\,kg} = \underline{\underline{1.49\,Mg/m^3}}$$

9.29 An epoxy (density = 1.1 Mg/m^3) is reinforced with 25 vol % E-glass fibers (density = 2.54 Mg/m^3). Calculate (a) the weight percent E-glass fibers and (b) the density of the reinforced polymer.

9.29

(a) For 1 m^3 of reinforced polymer, we will have 0.25 m^3 E-glass and (1.00 − 0.25) m^3 = 0.75 m^3 epoxy.

The mass of each component will be:

$$m_{E\text{-}glass} = \frac{2.54\,Mg}{m^3} \times 0.25\,m^3 = 0.635\,Mg$$

$$m_{epoxy} = \frac{1.1\,Mg}{m^3} \times 0.75\,m^3 = 0.825\,Mg$$

or

$$\underline{\underline{\text{wt. \% glass}}} = \frac{0.635\,Mg}{(0.635+0.825)\,Mg} \times 100\% = \underline{\underline{43.5\%}}$$

(b)

$$\rho = \frac{m}{V} = \frac{(0.635+0.825)\,Mg}{1\,m^3} = \underline{\underline{1.46\,Mg/m^3}}$$

9.30 Calculate the % mass savings that would occur if the reinforced polymer described in Problem 9.29 was used to replace a steel gear. (Assume the gear volume is the same for both materials and approximate the density of steel by pure iron.)

9.30

$$\% \text{ mass savings} = \frac{(\text{mass})_{Fe} - (\text{mass})_{poly}}{(\text{mass})_{Fe}} \times 100\%$$

$$= \frac{\rho_{Fe} - \rho_{poly}}{\rho_{Fe}} \times 100\%$$

Using Appendix 1 and the result of Problem 9.29,

$$\% \text{ mass savings} = \frac{7.87 - 1.46}{7.87} \times 100\% = \underline{\underline{81.4\%}}$$

9.31 Repeat Problem 9.30 if the polymer replaces an aluminum alloy. (Again, assume the gear volume is the same and approximate the density of the alloy by pure aluminum.)

9.31

Again using Appendix 1 and the result of Problem 9.29,

$$\% \text{ mass savings} = \frac{m_{Al} - m_{poly}}{m_{Al}} \times 100\% = \frac{\rho_{Al} - \rho_{poly}}{\rho_{Al}} \times 100\%$$

$$= \frac{2.70 - 1.46}{2.70} \times 100\% = \underline{\underline{45.9\%}}$$

9.32 Some injection moldable nylon 66 contains 40 wt % glass spheres as a filler. Improved mechanical properties are the result. If the average glass sphere diameter is 100 μm, estimate the density of such particles per cubic millimeter.

9.32

Consider 1 kg product = 0.40 kg glass + 0.60 kg nylon

$$V_{glass} = \frac{0.40\times10^3\,g}{2.54\times10^6\,g}\,m^3 = 1.57\times10^{-4}\,m^3$$

$$V_{nylon} = \frac{0.60\times10^3\,g}{1.14\times10^6\,g}\,m^3 = 5.26\times10^{-4}\,m^3$$

$$V_{total} = (1.57+5.26)\times10^{-4}\,m^3 = 6.83\times10^{-4}\,m^3$$

$$V_{single\ glass\ sphere} = \frac{4}{3}\pi r^3 = \frac{4}{3}\pi\left[100\,\mu m/2 \times 10^{-6}\,m/\mu m\right]^3$$

$$= 5.24\times10^{-13}\,m^3$$

$$n_{spheres} = \frac{V_{glass}}{V_{sphere}} = \frac{1.57\times10^{-4}\,m^3}{5.24\times10^{-13}\,m^3} = 3.00\times10^8$$

$$\text{particle density} = \frac{n_{spheres}}{V_{total}} = \frac{3.00\times10^8}{6.83\times10^{-4}\,m^3}\times\frac{1\,m^3}{10^9\,mm^3}$$

$$= \underline{\underline{448\ \text{particles}/mm^3}}$$

9.33 Calculate the average separation distance between the centers of adjacent glass spheres in the nylon described in Problem 9.32. (Note the assumptions in Problem 8.12.)

9.33

Assuming a simple cubic array of spheres:

$$d^{-1}_{separation} = (\text{density}_{spheres})^{1/3}$$

$$= (448\,mm^{-3})^{1/3} = 7.65\,mm^{-1}$$

or

$$d_{separation} = \frac{1}{7.65\,mm^{-1}} = \underline{\underline{0.131\,mm}}$$

9.34 Repeat Problem 9.33 for the case of the same wt % spheres but an average glass sphere diameter of 50 μm.

9.34

As in Problem 9.32,

$1\ kg\ product \rightarrow 0.40\ kg\ glass\ with\ V_{glass} = 1.57 \times 10^{-4}\ m^3$

$+ 0.60\ kg\ nylon\ with\ V_{nylon} = 5.26 \times 10^{-4}\ m^3$

In the current case,

$$V_{single\ glass\ sphere} = \frac{4}{3}\pi r^3 = \frac{4}{3}\pi\left[\frac{50\ \mu m}{2} \times 10^{-6}\frac{m}{\mu m}\right]^3$$

$$\frac{4}{3}\pi\left[1.56 \times 10^{-14}\ m^3\right] = 6.54 \times 10^{-14}\ m^3$$

$$\therefore n_{spheres} = \frac{V_{glass}}{V_{sphere}} = \frac{1.57 \times 10^{-4}\ m^3}{6.54 \times 10^{-14}\ m^3} = 2.40 \times 10^9$$

giving

$$\text{particle density} = \frac{n_{spheres}}{V_{total}} = \frac{2.40 \times 10^9}{6.83 \times 10^{-4}\ m^3} \times \frac{1\ m^3}{10^9\ mm^3}$$

$$= 3512\ \text{particles}/mm^3$$

Then,

$$d^{-1}_{separation} = (3512\ mm^{-3})^{1/3} = 15.2\ mm^{-1}$$

or

$$d_{separation} = \frac{1}{15.2\ mm^{-1}} = 0.0658\ mm \times 10^3\frac{\mu m}{mm}$$

$$= \underline{\underline{65.8\ \mu m}}$$

9.35 Bearings and other parts requiring exceptionally low friction and wear can be fabricated from an acetal polymer with an addition of polytetrafluoroethylene (PTFE) fibers. The densities of acetal and PTFE are 1.42 Mg/m^3 and 2.15 Mg/m^3, respectively. If the density of the polymer with additive is 1.54 Mg/m^3, calculate the weight percent of the PTFE addition.

9.35

$1\ kg\ product = x\ kg\ PTFE + (1-x)\ kg\ acetal$

$$V_{product} = \left(\frac{x\ kg}{2.15\ Mg/m^3} + \frac{(1-x)\ kg}{1.42\ Mg/m^3}\right) \times \frac{1\ Mg}{1000\ kg}$$

$$= \frac{m_{product}}{\rho_{product}} = \frac{1\ kg}{1.54\ Mg/m^3} \times \frac{1\ Mg}{1000\ kg}$$

$$= 6.493 \times 10^{-4}\ m^3$$

$$\therefore 6.493 \times 10^{-4} \times 1000 = \frac{x}{2.15} + \frac{1-x}{1.42}$$

or

$x = 0.230$ or PTFE is $\underline{\underline{23.0\ wt.\%}}$

9.36 Iron oxide is a common pigment for polymers. Use information from Section 8.5 to suggest the resulting color you would expect from this additive.

9.36 The answer is suggested by Table 8.15. In iron oxide, the iron can exist in both Fe^{2+} and Fe^{3+} valences. Depending on network or modifier position and coordination number, deep brown, blue-green, and weak yellow are possible. (The deep brown coloration is perhaps the most widely used.)

Section 9.6 – Mechanical Properties of Polymers

PP 9.12 The data in Sample Problem 9.12 permit the flexural modulus to be calculated. For the configuration described, an applied force of 680 N causes fracture of the nylon sample. Calculate the corresponding flexural strength.

PP 9.12 Using Equation 9.7,

$$F.S. = \frac{3FL}{2bh^2} = \frac{3(680\,N)(50\times10^{-3}\,m)}{2(13\times10^{-3}\,m)(7\times10^{-3}\,m)^2} =$$

$$= 80.1\times10^6\,N/m^2 = \underline{80.1\,MPa}$$

PP 9.13 In Sample Problem 9.13 strain is calculated for various materials under a stress of 1 MPa. While the strain is relatively large for polymers, there are some high-modulus polymers with substantially lower results. Calculate the strain in a cellulosic fiber with a modulus of elasticity of 28,000 MPa (under a uniaxial stress of 1 MPa).

PP 9.13 $$\epsilon = \frac{\sigma}{E} = \frac{1\,MPa}{28{,}000\,MPa} = \underline{3.57\times10^{-5}}$$

PP 9.14 In Sample Problem 9.14a the time for relaxation of stress to 1 MPa at 25°C is calculated. **(a)** Calculate the time for stress to relax to 0.5 MPa at 25°C. **(b)** Repeat part (a) for 35°C using the result of Sample Problem 9.14b.

PP 9.14 (a) $\sigma = \sigma_0 e^{-t/\tau}$

$$0.5\,MPa = 2\,MPa\; e^{-t/(60\,d)}$$

$$\text{or } t = -(60\,d)\ln\left(\frac{0.5}{2}\right) = \underline{83.2\text{ days}}$$

(b) $\sigma = \sigma_0 e^{-t/\tau}$

$$0.5\,MPa = 2\,MPa\; e^{-t/(40.5\,d)}$$

$$\text{or } t = -(40.5\,d)\ln\left(\frac{0.5}{2}\right) = \underline{56.1\text{ days}}$$

9.37 The following data are collected in a flexural test of a polyester to be used in the exterior trim of an automobile:

Test piece geometry: 5 mm × 15 mm × 50 mm
Distance between supports = L = 50 mm
Initial slope of load-deflection curve = 538 × 10^3 N/M

Calculate the flexural modulus of this engineering polymer.

9.37

Using Equation 9.8,

$$E_{flex} = \frac{L^3 m}{4bh^3}$$

$$= \frac{(50\times10^{-3}m)^3(538\times10^3 N/m)}{4(15\times10^{-3}m)(5\times10^{-3}m)^3}$$

$$= 8970\times10^6 N/m^2 = \underline{\underline{8970 MPa}}$$

9.38 The following data are collected in a flexural test of a polyester to be used in the fabrication of molded office furniture:

Test piece geometry: 10 mm × 15 mm × 100 mm
Distance between supports = L = 50 mm
Load at fracture = 3000 N

Calculate the flexural strength of this engineering polymer.

9.38

$$F.S. = \frac{3FL}{2bh^2}$$

$$= \frac{3(3\times10^3 N)(50\times10^{-3}m)}{2(15\times10^{-3}m)(10\times10^{-3}m)^2}$$

$$= 150\times10^6 N/m^2 = \underline{\underline{150 MPa}}$$

9.39 Figure 9-21 illustrates the effect of humidity on stress-versus-strain behavior for a nylon 66. In addition, the distinction between tensile and compressive behavior is shown. Approximating the data between 0 and 20 MPa as a straight line, calculate (a) the initial elastic modulus in tension and (b) the initial elastic modulus in compression for the nylon at 60% relative humidity.

9.39

For the 60% relative humidity curve,

(a) $\epsilon_{20MPa} \approx 1.8\% \rightarrow E_{tension} = \frac{20 MPa}{0.018}$

$= 1{,}110 MPa = \underline{\underline{1.11 GPa}}$

(b) $\epsilon_{-20MPa} \approx -1.3\% \rightarrow E_{compression} = \frac{-20 MPa}{-0.013}$

$= 1{,}540 MPa = \underline{\underline{1.54 GPa}}$

9.40 Using Figure 9-20 and approximating the data up to 1% strain as a straight line, plot the elastic modulus of the polyester as a function of temperature.

9.40

For 1% strain,

σ(MPa)	T(°C)	E = σ/0.01 (MPa)
100	−40	10.0×10^3
81	23	8.1×10^3
38	93	3.8×10^3
25	149	2.5×10^3

Giving the plot:

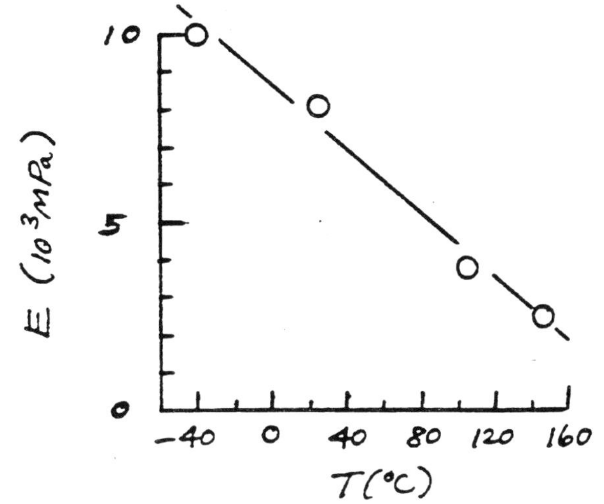

9.41 An acetal disk precisely 5 mm thick by 25 mm diameter is used as a cover plate in a mechanical loading device. If a 20-kN load is applied to the disk, calculate the resulting dimensions.

9.41

To obtain the compressive stress under this load:

$$\sigma = \frac{F}{A} = \frac{-20 \times 10^3 N}{\pi \left(\frac{25}{2} \times 10^{-3} m\right)^2} = -40.7 MPa$$

Using Table 9.3, we can estimate the modulus giving:

$$\epsilon_z = \frac{\sigma}{E} = \frac{-40.7 MPa}{3100 MPa} = -0.0131$$

$$\therefore t_f = t_o (1 + \epsilon_z) = 5 mm (1 - 0.0131) = \underline{\underline{4.934 mm}}$$

Again using Table 9.3,

$$\epsilon_{dia} = -\nu \epsilon_z = -(0.35)(-0.0131) = +0.00460$$

$$\therefore (dia)_f = (dia)_o (1 + \epsilon_{dia}) = 25 mm (1 + 0.0046) = \underline{\underline{25.12 mm}}$$

9.42 In Section 7.3 a useful correlation between hardness and tensile strength was demonstrated for metallic alloys. Plot hardness versus tensile strength for the data given in Table 9.3 and comment on whether a similar trend is shown for these common thermoplastic polymers. (You can compare this plot with Figure 7-17a.)

9.42

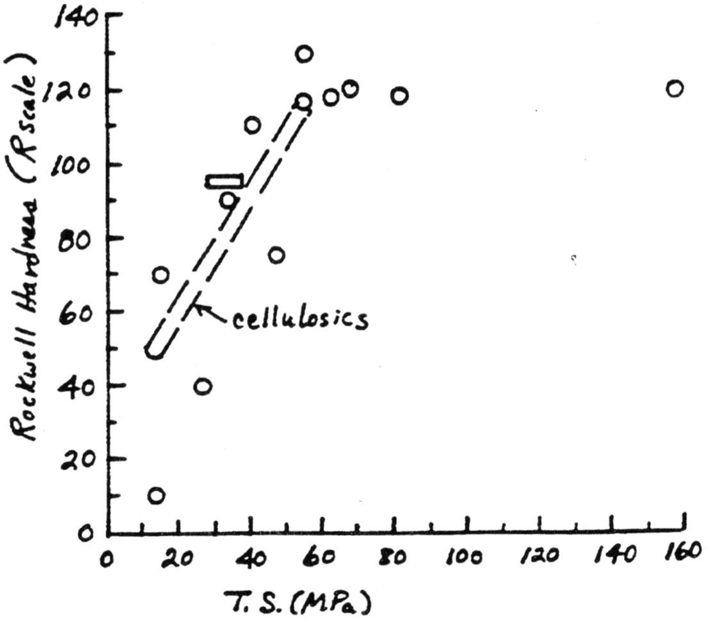

The trend is similar (higher hardness with higher strength) although somewhat more scattered.

9.43 Some fracture mechanics data are given in Table 9.3. Plot the breaking stress for low-density polyethylene as a function of flaw size, a (on a logarithmic scale), using Equation 7.9 and taking $Y = 1$. Cover a range of a from 1 to 100 mm. (See Problems 7.36 and 8.26.)

9.43

$$\sigma = K_{IC}/\sqrt{\pi a} = 1\,MPa\sqrt{m}/\sqrt{\pi a}$$

a (mm)	σ (MPa)
1	17.8
2	12.6
5	8.0
10	5.6
20	4.0
50	2.5
100	1.8

Resulting plot:

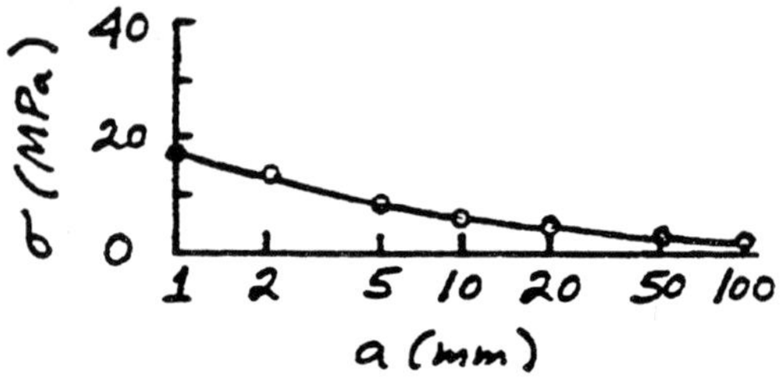

9.44 Superimpose a breaking stress plot for high-density polyethylene and ABS polymer on the result for Problem 9.43.

9.44

As in Problem 9.43,

$\sigma_{HDPE} = 2\,MPa\sqrt{m}/\sqrt{\pi a}$ & $\sigma_{ABS} = 4\,MPa\sqrt{m}/\sqrt{\pi a}$

a (mm)	σ_{HDPE} (MPa)	σ_{ABS} (MPa)
1	35.7	71.4
2	25.2	50.5
5	16.0	31.9
10	11.3	22.6
20	8.0	16.0
50	5.0	10.0
100	3.6	7.1

Resulting plot:

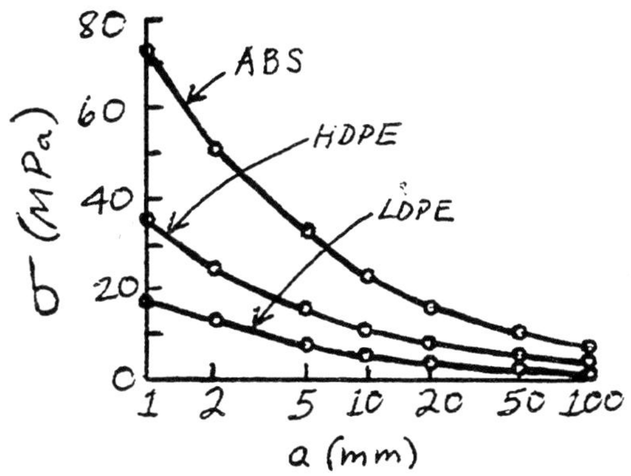

9.45 Calculate the breaking stress for a rod of ABS with a surface flaw size of 100 μm.

9.45

$\sigma = K_{IC}/\sqrt{\pi a}$

$= 4\,MPa\sqrt{m}/\sqrt{\pi\, 100\times10^{-6}\,m}$

$= \underline{226\,MPa}$

9.46 A nondestructive testing program can ensure that a thermoplastic polyester part will have no flaws greater than 0.1 mm in size. Calculate the maximum service stress available with this engineering polymer.

9.46

Using Equation 7.9 and taking $Y = 1$ and K_{IC} from Table 9.3:

$$\sigma_f = \frac{K_{IC}}{\sqrt{\pi a}} = \frac{0.5\,\text{MPa}\sqrt{\text{m}}}{\sqrt{\pi \times 0.1 \times 10^{-3}\,\text{m}}} = \underline{\underline{28.2\,\text{MPa}}}$$

9.47 Repeat Problem 9.46 with nylon 66 substituted for the polyester.

9.47

As in Problem 9.46,

$$\sigma_f = \frac{K_{IC}}{\sqrt{\pi a}} = \frac{3.0\,\text{MPa}\sqrt{\text{m}}}{\sqrt{\pi \times 0.1 \times 10^{-3}\,\text{m}}} = \underline{\underline{169\,\text{MPa}}}$$

9.48 A small pressure vessel is fabricated from an acetal polymer. The stress in the vessel wall is

$$\sigma = \frac{pr}{2t}$$

where p is the internal pressure, r the outer radius of the sphere, and t the wall thickness. For the vessel in question, $r = 30$ mm and $t = 2$ mm. What is the maximum permissible internal pressure for this design if the application is at room temperature and the wall stress is only tensile (due to internal pressurizations that will occur no more than 10^6 times)? (See Figure 9-22 for relevant data.)

9.48

Figure 9-22 indicates that for room temperature ($\approx 23°C$), fatigue fracture occurs (under tension only) after 10^6 cycles at $\sigma \approx 50\,\text{MPa}$.

$$\therefore\ 50\,\text{MPa} = \frac{p_{max}\, r}{2t}$$

$$\text{or } p_{max} = 100\,\text{MPa}\,\frac{t}{r}$$

$$= 100\,\text{MPa}\,\frac{2\,\text{mm}}{30\,\text{mm}} = \underline{\underline{6.67\,\text{MPa}}}$$

9.49 The stress on a rubber disk is seen to relax from 0.75 to 0.5 MPa in 100 days. (a) What is the relaxation time, τ, for this material? (b) What will be the stress on the disk after (i) 50 days, (ii) 200 days, or (iii) 365 days? (Consider time = 0 to be at the stress level of 0.75 MPa.)

9.49

(a) $\sigma = \sigma_0 e^{-t/\tau}$

$0.50\,MPa = 0.75\,MPa\; e^{-(100d)/\tau}$

$-\frac{100d}{\tau} = \ln\left(\frac{0.50}{0.75}\right)$

or $\tau = \underline{\underline{247\ days}}$

(b): (i) $\sigma_{50d} = 0.75\,MPa\; e^{-(50d)/(247d)} = \underline{\underline{0.612\,MPa}}$

(ii) $\sigma_{200d} = 0.75\,MPa\; e^{-(200d)/(247d)} = \underline{\underline{0.333\,MPa}}$

(iii) $\sigma_{365d} = 0.75\,MPa\; e^{-(365d)/(247d)} = \underline{\underline{0.171\,MPa}}$

9.50 Increasing temperature from 20 to 30°C decreases the relaxation time for a polymeric fiber from 3 to 2 days. Determine the activation energy for relaxation.

9.50

$1/\tau = Ce^{-Q/RT}$

$\frac{(1/2)d^{-1}}{(1/3)d^{-1}} = \frac{e^{-Q/[8.314\,J/(mol\cdot K)](303\,K)}}{e^{-Q/[8.314\,J/(mol\cdot K)](293\,K)}}$

or $Q = \underline{\underline{29.9\ kJ/mol}}$

9.51 Given the data in Problem 9.50, calculate the expected relaxation time at 40°C.

9.51

From Problem 9.50,

$C = \left(\frac{1}{\tau}\right) e^{+Q/RT}$

$= (1/2) d^{-1} e^{+(29{,}900\,J/mol)/[8.314\,J/(mol\cdot K)](303K)}$

$= 7.14 \times 10^4\, d^{-1}$

$\therefore \left(\frac{1}{\tau}\right)_{40°C} = (7.14 \times 10^4\, d^{-1})\, e^{-(29{,}900\,J/mol)/[8.314\,J/(mol\cdot K)](313K)}$

$= 0.731\ d^{-1}$

$\therefore \tau_{40°C} = \underline{\underline{1.37\ days}}$

9.52 A spherical pressure vessel is fabricated from nylon 66 and will be used at 60°C and 50% relative humidity. The vessel dimensions are 50 mm outer radius and 2 mm wall thickness. (**a**) What internal pressure is required to produce a stress in the vessel wall of 6.9 MPa (1000 psi)? (Note Problem 9.48.) (**b**) Calculate the circumference of the sphere after 10,000 h at this pressure. (Note Figure 9-19.)

9.52 (a) $\sigma = \frac{pr}{2t}$ or $p = 2\sigma \frac{t}{r} = 2(6.9\,MPa)\frac{2\,mm}{50\,mm}$

$= \underline{\underline{0.552\,MPa}}$

(b) Figure 9-19 indicates that, after 10,000h, the strain will be $\simeq 2\%$

or $(circ)_f = (circ)_o (1 + 0.02)$

$= \pi d\,(1.02) = \pi\,(2 \times 50\,mm)(1.02)$

$= \underline{\underline{320\,mm}}$

Section 9.7 - Optical Properties of Polymers

PP 9.15 Using Fresnel's formula, calculate the reflectance, R, of (**a**) a sheet of polypropylene and (**b**) a sheet of polytetrafluoroethylene (with an average refractive index of 1.35). (See Sample Problem 9.15.)

PP 9.15 (a) Using Table 9.5,

$R = \left(\frac{n-1}{n+1}\right)^2 = \left(\frac{1.47-1}{1.47+1}\right)^2 = \underline{\underline{0.0362}}$

(b) $R = \left(\frac{1.35-1}{1.35+1}\right)^2 = \underline{\underline{0.0222}}$

9.53 Calculate the critical angle of incidence for the air–nylon 66 interface. (The critical angle of incidence was defined in Sample Problem 8.11, along with a discussion of its relevance to the function of optical fibers.)

9.53 $\theta_c = \arcsin \frac{1}{n}$

$= \arcsin \frac{1}{1.53} = \underline{\underline{40.8°}}$

9.54 Calculate the critical angle if water ($n = 1.333$) replaced the air in Problem 9.53.

9.54

$$\frac{\sin\theta_i}{\sin\theta_r} = \frac{n_{water}}{n_{polymer}} = \frac{1.333}{1.53} = \frac{\sin\theta_c}{\sin 90^\circ} = \frac{\sin\theta_c}{1}$$

or

$$\theta_c = \arcsin\left(\frac{1.333}{1.53}\right) = \underline{\underline{60.6^\circ}}$$

9.55 We shall find in Chapter 14 that polymers are susceptible to ultraviolet-light damage. Calculate the wavelength of ultraviolet light necessary to break the C—C single bond. (Note that bond energies are given in Table 2.2 and the energy of a photon of electromagnetic radiation is given in Sample Problem 8.13. For comparison, the range of ultraviolet radiation is illustrated in Figure 3-35.)

9.55

Table 2.2 indicates that the C–C bond energy is 370 kJ/mol.

$$E_{bond} = 370\times10^{3}\,J/mol \times \frac{1\,mol}{0.6023\times10^{24}\,bonds}$$

$$= 6.14\times10^{-19}\,J/bond$$

$$E_{photon} = \frac{hc}{\lambda} \quad \text{or} \quad \lambda = \frac{hc}{E_{photon}}$$

$$\lambda = \frac{(0.6626\times10^{-33}\,J\cdot s)(0.2998\times10^{9}\,m/s)}{6.14\times10^{-19}\,J}$$

$$= 323\times10^{-9}\,m = \underline{\underline{323\,nm}}$$

9.56 Repeat Problem 9.55 for the C=C double bond.

9.56

Table 2.2 indicates that the C=C bond energy is 680 kJ/mol. Then,

$$\lambda = \frac{(0.6626\times10^{-33}\,J.s)(0.2998\times10^{9}\,m/s)}{(680\times10^{3}\,J/mol)/(0.6023\times10^{24}\,mol^{-1})}$$

$$= 176\times10^{-9}\,m = \underline{\underline{176\,nm}}$$

Section 10.1 - Synthetic Fiber-Reinforced Composites

PP 10.1 In Sample Problem 10.1 we found the density of a typical fiberglass composite. Repeat the calculations for **(a)** 50 vol % and **(b)** 75 vol % E-glass fibers in an epoxy matrix.

PP 10.1

(a) 1 m^3 composite gives us 0.50 m^3 E-glass and 0.5 m^3 epoxy.

$$m_{E\text{-}glass} = \frac{2.54\ Mg}{m^3} \times 0.50\ m^3 = 1.27\ Mg$$

$$m_{epoxy} = \frac{1.1\ Mg}{m^3} \times 0.50\ m^3 = 0.55\ Mg$$

$$\rho = \frac{m}{V} = \frac{(1.27 + 0.55)}{m^3} Mg = \underline{\underline{1.82\ Mg/m^3}}$$

(b) 1 m^3 composite gives us 0.75 m^3 E-glass and 0.25 m^3 epoxy.

$$m_{E\text{-}glass} = \frac{2.54\ Mg}{m^3} \times 0.75\ m^3 = 1.905\ Mg$$

$$m_{epoxy} = \frac{1.1\ Mg}{m^3} \times 0.25\ m^3 = 0.275\ Mg$$

$$\rho = \frac{m}{V} = \frac{(1.905 + 0.275)}{m^3} Mg = \underline{\underline{2.18\ Mg/m^3}}$$

10.1 Calculate the density of a fiber-reinforced composite composed of 14 vol % Al_2O_3 whiskers in a matrix of epoxy. The density of Al_2O_3 is 3.97 Mg/m^3 and of epoxy is 1.1 Mg/m^3.

10.1

1 m^3 composite gives us 0.14 m^3 Al_2O_3 and 0.86 m^3 epoxy.

$$m_{Al_2O_3} = \frac{3.97\ Mg}{m^3} \times 0.14\ m^3 = 0.556\ Mg$$

$$m_{epoxy} = \frac{1.1\ Mg}{m^3} \times 0.86\ m^3 = 0.946\ Mg$$

$$\rho = \frac{m}{V} = \frac{(0.556 + 0.946)}{m^3} Mg$$

$$= \underline{\underline{1.50\ Mg/m^3}}$$

10.2 Calculate the density of a boron filament reinforced epoxy composite containing 70 vol % filaments. The density of epoxy is 1.1 Mg/m^3.

10.2

1 m^3 composite gives us 0.70 m^3 B and 0.30 m^3 epoxy. Using the density of epoxy given in Problem 10.1,

$$m_{epoxy} = \frac{1.1\ Mg}{m^3} \times 0.30\ m^3 = 0.33\ Mg$$

Using the density of boron from Appendix 1,

$$m_B = \frac{2.47\ Mg}{m^3} \times 0.70\ m^3 = 1.73\ Mg$$

$$\rho = \frac{m}{V} = \frac{(0.33 + 1.73)\ Mg}{m^3} = \underline{\underline{2.06\ Mg/m^3}}$$

10.3 Using the information in the footnote on page 396, calculate the molecular weight of an aramid polymer with an average degree of polymerization of 500.

10.3

$$\left[HN-C_6H_4-\overset{H}{\overset{|}{N}}-\underset{\underset{O}{\|}}{C}-C_6H_4-CO \right]_{500}$$

$$\therefore \text{mol. wt.} = 500[10(1.008) + 2(14.01) + 14(12.01) + 2(16.00)]\ amu$$

$$= \underline{\underline{119,100\ amu}}$$

10.4 Calculate the density of the Kevlar fiber-reinforced epoxy composite in Table 10.12. The density of Kevlar is 1.44 Mg/m^3 and of epoxy is 1.1 Mg/m^3.

10.4

From Table 10.12, the vol.% Kevlar = 82%

$\therefore$ $1\,m^3$ composite = $0.82\,m^3$ Kevlar + $0.18\,m^3$ epoxy

and

$$m_{composite} = m_{Kevlar} + m_{epoxy}$$

$$= \rho_K V_K + \rho_e V_e$$

$$= (1.44\,Mg/m^3)(0.82\,m^3) + (1.1\,Mg/m^3)(0.18\,m^3)$$

$$= 1.38\,Mg$$

or $$\rho_{composite} = \frac{m_{composite}}{V_{composite}} = \frac{1.38\,Mg}{1\,m^3}$$

$$= \underline{\underline{1.38\,Mg/m^3}}$$

10.5 In a contemporary commercial aircraft, a total of 0.25 m^3 of its exterior surface is constructed of a Kevlar/epoxy composite, rather than a conventional aluminum alloy. Calculate the mass savings using the density calculated in Problem 10.4 and approximating the alloy density by that of pure aluminum.

10.5

mass savings = mass aluminum − mass composite

$$= \rho_{Al} \times V_{Al} - \rho_{comp} \times V_{comp}$$

$$= \left(2.70\,\frac{Mg}{m^3}\right)(0.25\,m^3)$$

$$- \left(1.38\,\frac{Mg}{m^3}\right)(0.25\,m^3) = 0.330\,Mg = \underline{\underline{330\,kg}}$$

10.6 What would be the mass savings (relative to the aluminum alloy) if a carbon/epoxy composite with a density of 1.5 Mg/m^3 is used rather than the Kevlar/epoxy composite of Problem 10.5?

10.6

$$\text{mass savings} = \left(2.70\,\frac{Mg}{m^3}\right)(0.25\,m^3) - \left(1.50\,\frac{Mg}{m^3}\right)(0.25\,m^3)$$

$$= 0.300\,Mg = \underline{\underline{300\,kg}}$$

Section 10.2 - Wood - A Natural Fiber-Reinforced Composite

PP 10.2 Calculate the molecular weight of a hemicellulose molecule with a degree of polymerization (a) $n = 150$ and (b) $n = 250$. (See Sample Problem 10.2.)

PP 10.2

$$\text{(a) mol. wt.} = (150)[6(12.01) + 10(1.008) + 5(16.00)]\,g/mol$$

$$= \underline{\underline{24{,}320\,g/mol}}$$

$$\text{(b) mol. wt.} = (250)[6(12.01) + 10(1.008) + 5(16.00)]\,g/mol$$

$$= \underline{\underline{40{,}540\,g/mol}}$$

10.7 Calculate the degree of polymerization of a cellulose molecule in the cell wall of a wood. The average molecular weight is 95,000 amu.

10.7

$$(\text{mol. wt.})_{polymer} = n\,(\text{mol. wt.})_{mer}$$

$$\text{or}\quad n = \frac{(\text{mol. wt.})_{polymer}}{(\text{mol. wt.})_{mer}} = \frac{95{,}000}{[6(12.01) + 10(1.008) + 5(16.00)]}$$

$$= \underline{\underline{586}}$$

10.8 A small industrial building can be constructed of steel or wood framing. A decision is made to use steel for fire resistance. If 0.60 m^3 of steel is required for the same structural support as 1.25 m^3 of wood, how much more mass will the building have as a result of this decision? (Approximate the density of steel by that of pure iron. The density of wood can be taken to be 0.42 Mg/m^3.)

10.8

$$\text{extra mass} = \text{mass steel} - \text{mass wood}$$

$$= \rho_{st} V_{st} - \rho_w V_w$$

$$= \left(7.87\,\frac{Mg}{m^3}\right)(0.60\,m^3) - \left(0.42\,\frac{Mg}{m^3}\right)(1.25\,m^3)$$

$$= \underline{\underline{4.20\,Mg}}$$

Section 10.3 - Aggregate Composites

PP 10.3 Calculate the weight percent of $CaO + Al_2O_3 + SiO_2$ in type III portland cement. (See Sample Problem 10.3.)

PP 10.3 Following the procedure of Sample Problem 10.3, 100 kg Type III cement yields 53 kg C_3S, 19 kg C_2S, 11 kg C_3A, and 9 kg C_4AF.

Total mass $CaO = (0.737)(53\,kg) + (0.651)(19\,kg) + (0.623)(11\,kg) + (0.462)(9\,kg) = 62.4\,kg$

Total mass $Al_2O_3 = (0.377)(11\,kg) + (0.210)(9\,kg) = 6.0\,kg$

Total mass $SiO_2 = (0.263)(53\,kg) + (0.349)(19\,kg) = 20.6\,kg$

Then,

total wt.% $(CaO + Al_2O_3 + SiO_2) = (62.4 + 6.0 + 20.6)\% = 89.0\%$

PP 10.4 Calculate the density of a particulate composite containing 50 vol % W particles in a copper matrix. (See Sample Problem 10.4.)

PP 10.4 From Appendix 1,

$$\rho_W = 19.25\,Mg/m^3 \text{ and } \rho_{Cu} = 8.93\,Mg/m^3$$

$1\,m^3$ composite yields $0.50\,m^3$ W + $0.50\,m^3$ Cu

or,

$$\rho = [0.5(19.25) + 0.5(8.93)]\,Mg/m^3 = 14.1\,Mg/m^3$$

10.9 Calculate the combined weight percent of CaO, Al_2O_3, and SiO_2 in type II portland cement.

10.9

As in Sample Problem 10.3,

100 kg Type II cement yields 44 kg C_3S, 31 kg C_2S, 5 kg C_3A, and 13 kg C_4AF.

Total mass CaO = (0.737)(44 kg) + (0.651)(31 kg) + (0.623)(5 kg) + (0.462)(13 kg) = 61.7 kg

Total mass Al_2O_3 = (0.377)(5 kg) + (0.210)(13 kg) = 4.6 kg

Total mass SiO_2 = (0.263)(44 kg) + (0.349)(31 kg) = 22.4 kg

Then,

total wt.% ($CaO + Al_2O_3 + SiO_2$) = (61.7 + 4.6 + 22.4) %

= **88.7 %**

10.10 A reinforced concrete beam has a cross-sectional area of 0.0323 m^2, including four steel reinforcing bars 19 mm in diameter. What is the overall density of this support structure, assuming a uniform cross-section? (The density of the nonreinforced concrete is 2.30 Mg/m^3. The density of the steel can be approximated by that of pure iron.)

10.10

Assuming a 1 m length of beam, first consider the cross-sectional area of reinforcing bars:

$$A_{bars} = 4\pi\left(\frac{d}{2}\right)^2 = 4\pi\left(\frac{19.0\times10^{-3}\,m}{2}\right)^2 = 1.13\times10^{-3}\,m^2$$

∴ net area of concrete is:

$$A_{concrete} = 3.23\times10^{-2}\,m^2 - 1.13\times10^{-3}\,m^2 = 3.12\times10^{-2}\,m^2$$

$$\therefore m_{bars} = \rho_{Fe}V_{Fe} = 7.87\,\frac{Mg}{m^3}\times1.13\times10^{-3}\,m^3 = 8.89\times10^{-3}\,Mg$$

and

$$m_{concrete} = \rho_{Con}V_{Con} = 2.30\,\frac{Mg}{m^3}\times3.12\times10^{-2}\,m^3 = 7.18\times10^{-2}\,Mg$$

giving

$$m_{beam} = (8.89\times10^{-3} + 7.18\times10^{-2})\,Mg = 8.07\times10^{-2}\,Mg$$

and a resulting density of

$$\rho_{beam} = 8.07\times10^{-2}\,Mg / 3.23\times10^{-2}\,m^3 = \underline{\underline{2.50\,Mg/m^3}}$$

10.11 Suppose larger reinforcing bars are specified for the concrete structure in Problem 10.10. If 31.8-mm diameter reinforcing bars are used, what would be the overall density of the structure?

10.11 In this case,

$$A_{bars} = 4\pi\left(\frac{31.8\times10^{-3}\,m}{2}\right)^2 = 3.18\times10^{-3}\,m^2$$

$$A_{concrete} = 3.23\times10^{-2}\,m^2 - 3.18\times10^{-3}\,m^2 = 2.91\times10^{-2}\,m^2$$

$$\therefore\ m_{bars} = 7.87\,Mg/m^3 \times 3.18\times10^{-3}\,m^3 = 2.50\times10^{-2}\,Mg$$

$$m_{concrete} = 2.30\,Mg/m^3 \times 2.91\times10^{-2}\,m^3 = 6.69\times10^{-2}\,Mg$$

and

$$m_{beam} = (2.50\times10^{-2} + 6.69\times10^{-2})\,Mg = 9.19\times10^{-2}\,Mg$$

giving

$$\rho_{beam} = 9.19\times10^{-2}\,Mg / 3.23\times10^{-2}\,m^3 = \underline{\underline{2.85\,Mg/m^3}}$$

10.12 Calculate the density of a particulate composite composed of 50 vol % Mo in a copper matrix.

10.12 From Appendix 1, $\rho_{Mo} = 10.22\,\frac{Mg}{m^3}$ and $\rho_{Cu} = 8.93\,\frac{Mg}{m^3}$

Then,

$$\rho = [0.5(10.22) + 0.5(8.93)]\,\frac{Mg}{m^3} = \underline{\underline{9.58\,Mg/m^3}}$$

10.13 Calculate the density of a dispersion-strengthened copper with 10 vol % Al_2O_3.

10.13 From Appendix 1, $\rho_{Cu} = 8.93\,Mg/m^3$. Then,

$$\rho = [0.1(3.97) + 0.9(8.93)]\,Mg/m^3 = \underline{\underline{8.43\,Mg/m^3}}$$

10.14 Calculate the density of a WC/Co cutting tool material with 60 vol % WC in a Co matrix. (The density of WC is 15.7 Mg/m^3.)

10.14

$1\,m^3$ composite $\rightarrow$ $0.60\,m^3$ WC and $0.40\,m^3$ Co.

$$m_{WC} = \frac{15.7\,Mg}{m^3} \times 0.60\,m^3 = 9.42\,Mg$$

Using the density of Co from Appendix 1,

$$m_{Co} = \frac{8.8\,Mg}{m^3} \times 0.40\,m^3 = 3.52\,Mg$$

$$\rho = \frac{m}{V} = \frac{(9.42 + 3.52)\,Mg}{1\,m^3} = \underline{\underline{12.94\,Mg/m^3}}$$

Section 10.4 - Property Averaging

PP 10.5 Calculate the composite modulus for a composite with 50 vol % E-glass in a polyester matrix. (See Sample Problem 10.5.)

PP 10.5

$$E_c = v_m E_m + v_f E_f$$
$$= (0.5)(6.9 \times 10^3\,MPa) + (0.5)(72.4 \times 10^3\,MPa)$$
$$= \underline{\underline{39.7 \times 10^3\,MPa}}$$

PP 10.6 The thermal conductivity of a particular fiberglass composite is calculated in Sample Problem 10.6. Repeat this calculation for a composite with 50 vol % E-glass in a polyester matrix.

PP 10.6

$$k_c = v_m k_m + v_f k_f$$
$$= (0.5)[0.17\,W/(m \cdot K)] + (0.5)[0.97\,W/(m \cdot K)]$$
$$= \underline{\underline{0.57\,W/(m \cdot K)}}$$

PP 10.7 Calculate the elastic modulus and thermal conductivity perpendicular to continuous reinforcing fibers for a composite with 50 vol % E-glass in a polyester matrix. (See Sample Problem 10.7.)

PP 10.7

$$E_c = \frac{E_m E_f}{v_m E_f + v_f E_m}$$

$$= \frac{(6.9\times10^3 MPa)(72.4\times10^3 MPa)}{(0.5)(72.4\times10^3 MPa) + (0.5)(6.9\times10^3 MPa)}$$

$$= 12.6\times10^3 MPa$$

$$k_c = \frac{k_m k_f}{v_m k_f + v_f k_m}$$

$$= \frac{[0.17\ W/(m\cdot K)][0.97\ W/(m\cdot K)]}{(0.5)[0.97\ W/(m\cdot K)] + (0.5)[0.17\ W/(m\cdot K)]}$$

$$= 0.29\ W/(m\cdot K)$$

PP 10.8 In Sample Problem 10.8 the case of a modulus equation with $n = 0$ is treated. Estimate the composite modulus for a reciprocal case in which 50 vol % Co aggregate is dispersed in a WC matrix. For this case, the value of n can be taken as $\frac{1}{2}$.

PP 10.8

In this case,

$$E_c^{1/2} = v_l E_l^{1/2} + v_h E_h^{1/2}$$

$$= (0.5)(207\times10^3 MPa)^{1/2} + (0.5)(704\times10^3 MPa)^{1/2}$$

$$= 20.46\ (10^3 MPa)^{1/2}$$

or $E_c = 419\times10^3 MPa$

10.15 Calculate the composite modulus for epoxy reinforced with 70 vol % boron filaments under isostrain conditions.

10.15

Using the data in Tables 10.10 and 10.11, we have:

$$E_c = v_m E_m + v_f E_f$$

$$= (0.30)(6.9\times10^3 MPa) + (0.70)(410\times10^3 MPa)$$

$$= 289\times10^3 MPa$$

10.16 Calculate the composite modulus for aluminum reinforced with 50 vol % boron filaments under isostrain conditions.

10.16

Using the data in Tables 10.10 and 10.11, we have:

$$E_c = (0.50)(69 \times 10^3 \text{ MPa}) + (0.50)(410 \times 10^3 \text{ MPa})$$
$$= \underline{\underline{240 \times 10^3 \text{ MPa}}}$$

10.17 Calculate the modulus of elasticity of a metal–matrix composite under isostrain conditions. Assume an aluminum matrix is reinforced by 60 vol % SiC fibers.

10.17

Using the data in Tables 10.10 & 10.11, we have:

$$E_c = (0.40)(69 \times 10^3 \text{ MPa}) + (0.60)(430 \times 10^3 \text{ MPa})$$
$$= \underline{\underline{286 \times 10^3 \text{ MPa}}}$$

10.18 Calculate the modulus of elasticity of a ceramic–matrix composite under isostrain conditions. Assume an Al_2O_3 matrix is reinforced by 60 vol % SiC fibers.

10.18

Using the data in Table 10.11 for both matrix and fiber:

$$E_c = (0.40)(430 \times 10^3 \text{ MPa}) + (0.60)(430 \times 10^3 \text{ MPa})$$
$$= \underline{\underline{430 \times 10^3 \text{ MPa}}}$$

10.19 On a plot similar to Figure 10-11, show the composite modulus for **(a)** 60 vol % fibers (the result of Sample Problem 10.5) and **(b)** 50 vol % fibers (the result of Practice Problem 10.5). Include the individual glass and polymer plots.

10.19

For 60 vol.% fibers (from Sample Problem 10.5),

$\sigma = (46.2 \times 10^3 \text{ MPa})(0.02) = 924 \text{ MPa}$

For 50 vol.% fibers (from PP 10.5),

$\sigma = (39.7 \times 10^3 \text{ MPa})(0.02) = 794 \text{ MPa}$

For E-glass (from Sample Problem 10.5),

$\sigma = (72.4 \times 10^3 \text{ MPa})(0.02) = 1{,}448 \text{ MPa}$

For polyester (from Sample Problem 10.5),

$\sigma = (6.9 \times 10^3 \text{ MPa})(0.02) = 138 \text{ MPa}$

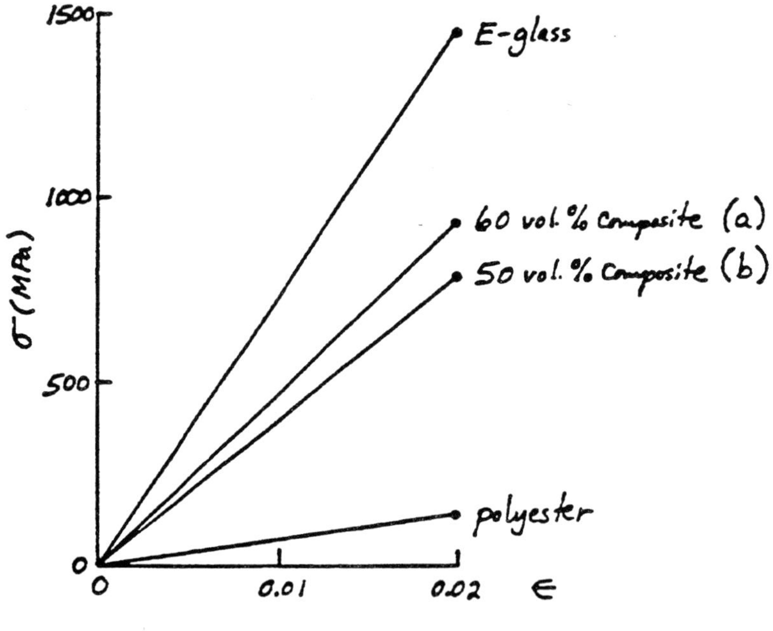

10.20 On a plot similar to Figure 10-11, show the composite modulus for an epoxy reinforced with 70 vol % carbon fibers under isostrain conditions. (Use the midrange value for carbon modulus in Table 10.11. Epoxy data are given in Table 10.10. For effectiveness of comparison with the case in Figure 10-11, use the same stress and strain scales. Include the individual matrix and fiber plots.)

10.20

For C: $E = 360 \times 10^3 MPa$

epoxy: $E = 6.9 \times 10^3 MPa$

$$E_{composite} = (0.3)(6.9 \times 10^3 MPa) + (0.7)(360 \times 10^3 MPa)$$
$$= 254 \times 10^3 MPa$$

At $\epsilon = 0.02$,

$$\sigma_C = (360 \times 10^3 MPa)(0.02) = 7200 \text{ MPa}$$
$$\sigma_{composite} = (254 \times 10^3 MPa)(0.02) = 5080 \text{ MPa}$$
$$\sigma_{epoxy} = (6.9 \times 10^3 MPa)(0.02) = 138 \text{ MPa}$$

(Because of the high modulus, plots should indicate a limit corresponding to the T.S. of fiber and composite which is ≈ 2300 MPa and 1200 MPa, as seen in Tables 10.11 and 10.12.)

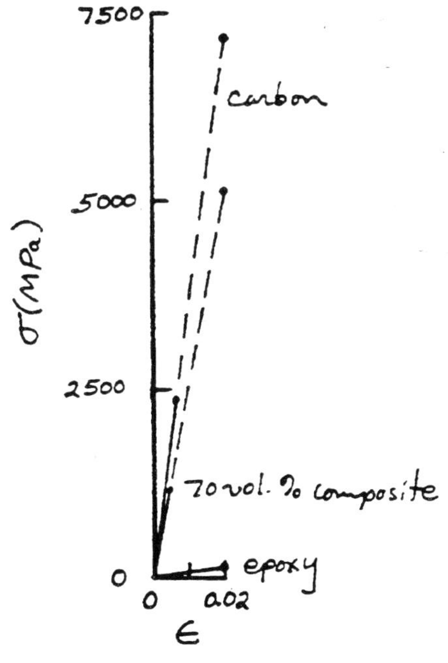

10.21 Calculate the composite modulus for polyester reinforced with 10 vol % Al_2O_3 whiskers under isostrain conditions. (See Tables 10.10 and 10.11 for appropriate moduli.)

10.21

$$E_c = v_m E_m + v_f E_f$$
$$= (0.90)(6.90 \times 10^3 MPa) + (0.10)(430 \times 10^3 MPa)$$
$$= \underline{\underline{49.2 \times 10^3 MPa}}$$

10.22 Calculate the composite modulus for epoxy reinforced with 70 vol % boron filaments under isostress conditions.

10.22 Using the data in Tables 10.10 and 10.11:

$$E_c = \frac{E_m E_f}{v_m E_f + v_f E_m}$$

$$= \frac{(6.9 \times 10^3 \text{ MPa})(410 \times 10^3 \text{ MPa})}{(0.30)(410 \times 10^3 \text{ MPa}) + (0.70)(6.9 \times 10^3 \text{ MPa})}$$

$$= 22.1 \times 10^3 \text{ MPa}$$

(One might note that this is more than an order of magnitude lower than the isostrain result of Problem 10.15.)

10.23 Calculate the modulus of elasticity of a metal–matrix composite under isostress conditions. Assume an aluminum matrix is reinforced by 60 vol % SiC fibers.

10.23

$$E_c = \frac{E_m E_f}{v_m E_f + v_f E_m}$$

$$= \frac{(69 \times 10^3 \text{ MPa})(430 \times 10^3 \text{ MPa})}{(0.40)(430 \times 10^3 \text{ MPa}) + (0.60)(69 \times 10^3 \text{ MPa})}$$

$$= 139 \times 10^3 \text{ MPa}$$

10.24 Calculate the modulus of elasticity of a ceramic–matrix composite under isostress conditions. Assume an Al_2O_3 matrix is reinforced by 60 vol % SiC fibers.

10.24

$$E_c = \frac{E_m E_f}{v_m E_f + v_f E_m}$$

$$= \frac{(430 \times 10^3 \text{ MPa})(430 \times 10^3 \text{ MPa})}{(0.40)(430 \times 10^3 \text{ MPa}) + (0.60)(430 \times 10^3 \text{ MPa})}$$

$$= 430 \times 10^3 \text{ MPa}$$

(Obviously when $E_m = E_f$, either isostrain or isostress averaging will lead to the same result: $E_m = E_f = E_c$.)

10.25 On a plot similar to Figure 10-11, show the isostress composite modulus for 50 vol % E-glass fibers in a polyester matrix (the result of Problem 10.19). It is interesting to compare the appearance of the resulting plot with that from Problem 10.19b.

10.25

At $\epsilon = 0.02$, $\sigma_{50\%\,fibers} = (12.6 \times 10^3\,\text{MPa})(0.02) = 252\,\text{MPa}$

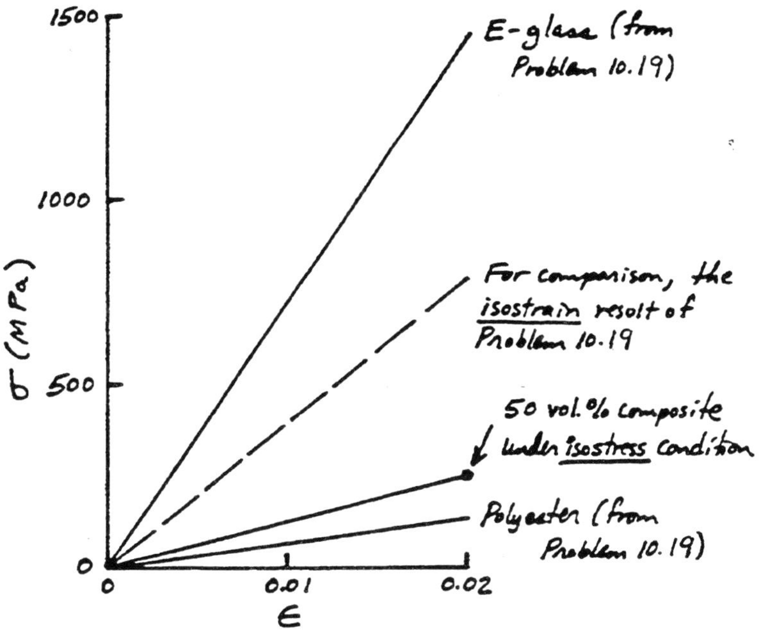

10.26 Repeat Problem 10.25 for an epoxy reinforced with 70 vol % carbon fibers and compare with the isostrain results from Problem 10.20.

10.26

Taking data from Problem 10.20,

$$E_{composite} = \frac{E_m E_f}{v_m E_f + v_f E_m}$$

$$= \frac{(6.9 \times 10^3 \text{ MPa})(360 \times 10^3 \text{ MPa})}{(0.3)(360 \times 10^3 \text{ MPa}) + (0.7)(6.9 \times 10^3 \text{ MPa})}$$

$$= 22.0 \times 10^3 \text{ MPa}$$

At $\epsilon = 0.02$, $\sigma_{composite} = (22.0 \times 10^3 \text{ MPa})(0.02)$
$= 440 \text{ MPa}$

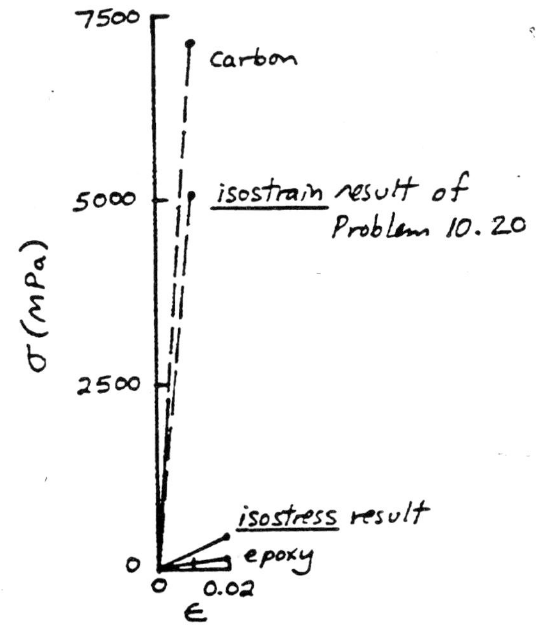

10.27 Plot Poisson's ratio as a function of reinforcing fiber content for an SiC fiber-reinforced Si_3N_4 composite system loaded parallel to the fiber direction and SiC contents between 50 and 75 vol %. (Note the discussion relative to Equation 10.8 and data in Table 8.9.)

10.27

$$\nu = v_m \nu_m + v_f \nu_f$$

$$\nu_{50\%\,SiC} = (0.5)(0.24) + (0.5)(0.19) = 0.215$$
$$\nu_{75\%\,SiC} = (0.25)(0.24) + (0.75)(0.19) = 0.2025$$

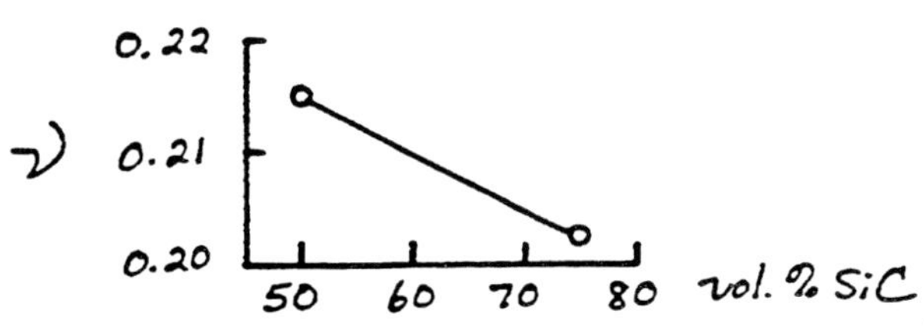

10.28 Calculate the composite modulus of polyester reinforced with 10 vol % Al_2O_3 whiskers under isostress conditions. (Refer to Problem 10.21.)

10.28

$$E_c = \frac{E_m E_f}{v_m E_f + v_f E_m}$$

$$= \frac{(6.9\times10^3\,\text{MPa})(430\times10^3\,\text{MPa})}{(0.9)(430\times10^3\,\text{MPa}) + (0.1)(6.9\times10^3\,\text{MPa})}$$

$$= \underline{\underline{7.65\times10^3\,\text{MPa}}}$$

10.29 Generate a plot similar to Figure 10-13 for the case of epoxy reinforced with Al_2O_3 whiskers. (Refer to Problems 10.21 and 10.28.)

10.29

As epoxy and polyester have the same elastic moduli (Table 10.10), the calculations of Problems 10.21 & 10.28 apply to this system. To monitor the overall isostress curve, we need a few more points. Consider $v_f = 0.5, 0.8,$ and 0.9:

$$E_{c,0.5} = \frac{(6.9\times10^3)(430\times10^3)}{[(0.5)(430\times10^3) + (0.5)(6.9\times10^3)]}\,\text{MPa}$$

$$= 13.6\times10^3\,\text{MPa}$$

$$E_{c,0.8} = \frac{(6.9\times10^3)(430\times10^3)}{[(0.2)(430\times10^3) + (0.8)(6.9\times10^3)]}\,\text{MPa}$$

$$= 32.4\times10^3\,\text{MPa}$$

$$E_{c,0.9} = \frac{(6.9\times10^3)(430\times10^3)}{[(0.1)(430\times10^3) + (0.9)(6.9\times10^3)]}\,\text{MPa}$$

$$= 60.3\times10^3\,\text{MPa}$$

The resulting plot:

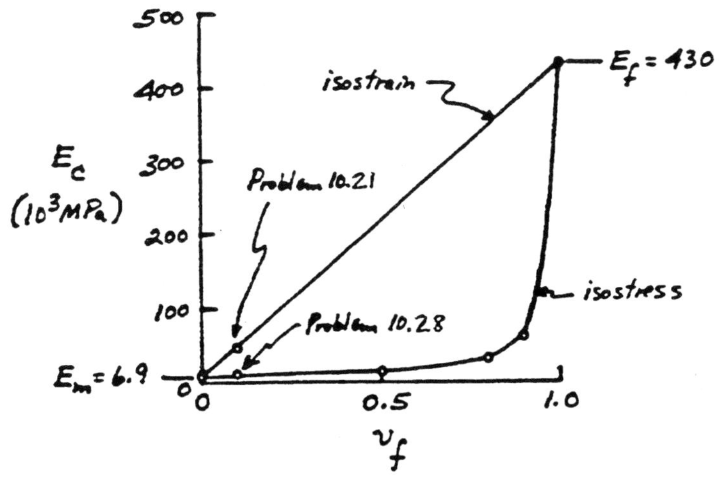

10.30 Calculate the modulus of elasticity of a metal–matrix composite composed of 50 vol % SiC whiskers in a matrix of aluminum. Assume that the modulus lies exactly midway between the isostress and isostrain values.

10.30 Using data in Tables 10.10 and 10.11,

$$E_{c,\,isostrain} = (0.50)(69 \times 10^3\ MPa) + (0.50)(430 \times 10^3\ MPa)$$

$$= 250 \times 10^3\ MPa$$

$$E_{c,\,isostress} = \frac{(69 \times 10^3\ MPa)(430 \times 10^3\ MPa)}{(0.50)(430 \times 10^3\ MPa) + (0.50)(69 \times 10^3\ MPa)}$$

$$= 119 \times 10^3\ MPa$$

$$\therefore\ E_c = \left(\frac{250 + 119}{2}\right) \times 10^3\ MPa = \underline{\underline{184 \times 10^3\ MPa}}$$

10.31 Calculate the composite modulus for 20 vol % SiC whiskers in an Al_2O_3 matrix.

10.31 We do not know a priori if the whisker configuration is closer to isostrain or isostress conditions. However, inspection of Table 10.11, we see that both materials, SiC and Al_2O_3, have a modulus of 430×10^3 MPa. Therefore, by inspection, the average modulus will be:

$$\underline{\underline{430 \times 10^3\ MPa}}.$$

(Clearly, this and many other ceramic-matrix composites are developed for improved strength and toughness, rather than improved modulus.)

10.32 Generate a plot similar to Figure 10-15 for the case of Co–WC composites. For Co being the matrix, the value of n in Equation 10.20 is zero (see Sample Problem 10.8). For WC being the matrix, the value of n is $\frac{1}{2}$ (see Practice Problem 10.8). (The extreme cases of $n = 1$ and $n = -1$ should not be plotted for this system with components having relatively similar modulus values.)

10.32

We need only deal with $n = 0$ and $1/2$.
To monitor $n = 0$, we can use an approximation such as $n = +0.01$ with Equation 10.20.

$$E_{c,(v_h=0.25)}^{0.01} = 0.75(207)^{0.01} + 0.25(704)^{0.01} = 1.058$$

or $E_{c,(v_h=0.25)} = 282$ [in units of 10^3 MPa]

$$E_{c,(v_h=0.5)}^{0.01} = 0.50(207)^{0.01} + 0.50(704)^{0.01} = 1.061$$

or $E_{c,(v_h=0.5)} = 382$

$$E_{c,(v_h=0.75)}^{0.01} = 0.25(207)^{0.01} + 0.75(704)^{0.01} = 1.065$$

or $E_{c,(v_h=0.75)} = 519$

$$E_{c,(v_h=0.25)}^{1/2} = 0.75(207)^{1/2} + 0.25(704)^{1/2} = 17.42 \text{ or } E_{c,(v_h=0.25)} = 304$$

$$E_{c,(v_h=0.5)}^{1/2} = 0.5(207)^{1/2} + 0.5(704)^{1/2} = 20.46 \text{ or } E_{c,(v_h=0.5)} = 419$$

$$E_{c,(v_h=0.75)}^{1/2} = 0.25(207)^{1/2} + 0.75(704)^{1/2} = 23.50 \text{ or } E_{c,(v_h=0.75)} = 552$$

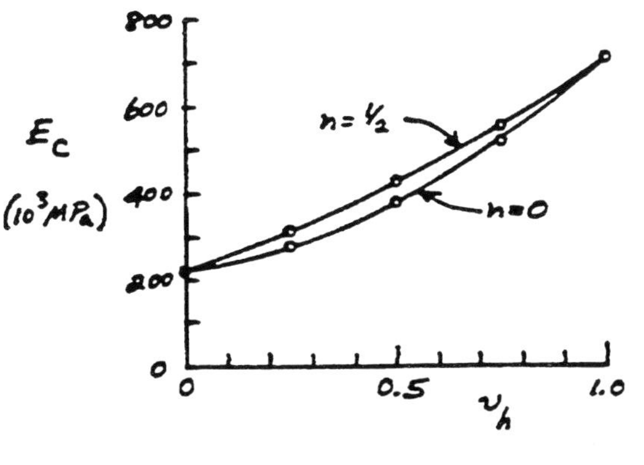

• 10.33 Consider further the discussion of interfacial strength in Section 10.4. The axial loading of a reinforcing fiber under ideal (isostrain) conditions leads to a shear stress at the fiber surface which, in turn, leads to the build-up of tensile stress in the fiber. (a) Taking the tensile stress in the fiber (with radius r) at a distance x from either end of the fiber to be σ_x, use a force balance between tensile and shear components to derive an expression for σ_x in terms of the fiber geometry and the interfacial shear stress, τ (which is uniform along the entire interface). (b) Make a schematic plot of the tensile stress in a short fiber (in which σ_x is always less than $\sigma_{critical}$, the failure stress of the fiber).

10.33 (a) At x,

$$F_{tensile} = F_{shear}$$

or

$$\sigma_x A_{cross\text{-}sec} = \tau A_{interface}$$

or

$$\sigma_x \pi r^2 = \tau(2\pi r x)$$

or

$$\sigma_x = \frac{2\pi \tau r x}{\pi r^2} = \frac{2\tau x}{r}$$

(b)

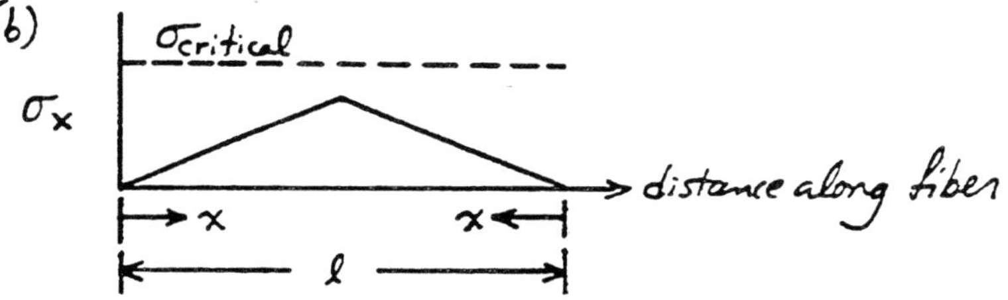

• 10.34 (a) Referring to Problem 10.33, sketch the tensile stress distribution in a long fiber (in which the stress in the middle portion of the fiber reaches a maximum, constant value, corresponding to fiber failure). (b) Using the result of Problem 10.33a, derive an expression for the critical stress transfer length, l_c, the minimum fiber length that must be exceeded if fiber failure is to occur; that is, if σ_x is to reach $\sigma_{critical}$. (For maximum efficiency in reinforcement, the fiber length should be much greater than l_c to ensure that the average tensile stress in the fiber is near $\sigma_{critical}$. For fiber length $= l_c$, the average tensile stress is, of course, only $\sigma_{critical}/2$.)

10.34 (a)

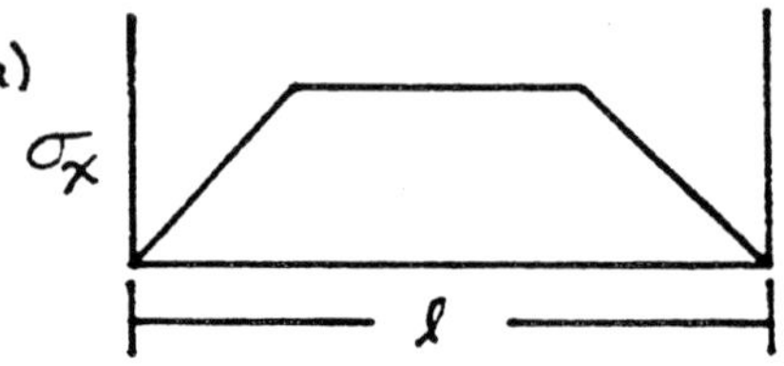

(b) $\sigma_x = \dfrac{2\tau x}{r}$

At l_c,

$$\sigma_{crit} = \frac{2\tau(l_c/2)}{r} \quad \text{or} \quad l_c = \frac{\sigma_{crit}\, r}{\tau}$$

Section 10.5 - Mechanical Properties of Composites

PP 10.9 In Sample Problem 10.9 the isostrain modulus for a fiberglass composite is shown to be close to a calculated value. Repeat this comparison for the isostrain modulus of B (70 vol %)/epoxy composite given in Table 10.12.

PP 10.9 In Table 10.12, we find $E_c = 210 - 280 \times 10^3$ MPa.

Using data for B and epoxy from Tables 10.10 and 10.11:

$$E_c = v_m E_m + v_f E_f$$

$$= (0.30)(6.9 \times 10^3 \text{ MPa}) + (0.70)(410 \times 10^3 \text{ MPa})$$

$$= 289 \times 10^3 \text{ MPa}$$

$$\% \text{ error} = \frac{289 - 210}{210} \times 100\% = 38\%$$

to

$$\% \text{ error} = \frac{289 - 280}{280} \times 100\% = 3.2\%$$

or <u>3.2 to 38 % error</u>

PP 10.10 In Sample Problem 10.10 we find that dispersion-strengthened aluminum has a substantially higher specific strength than pure aluminum. In a similar way, calculate the specific strength of the E-glass/epoxy composite of Table 10.12 compared to the pure epoxy of Table 10.10. For density information, refer to Sample Problem 10.1. (You may wish to compare your calculations with the values in Table 10.13.)

PP 10.10 $\text{sp. str.} = \frac{T.S.}{\rho}$

Using Table 10.10 and Sample Problem 10.1 data:

$$\text{sp. str., epoxy} = \frac{(69 \text{ MPa})(1.02 \times 10^{-1} \text{ kg/mm}^2)/\text{MPa}}{(1.1 \text{ Mg/m}^3)(10^3 \text{ kg/Mg})(1 \text{ m}^3/10^9 \text{ mm}^3)}$$

$$= \underline{6.40 \times 10^6 \text{ mm}}$$

(equal to the value in Table 10.13)

For 73.3 vol.% E-glass/epoxy composite (after Sample Problem 10.1):

$$\rho = \frac{(1.1\,Mg/m^3)(0.267\,m^3) + (2.54\,Mg/m^3)(0.733\,m^3)}{m^3}$$

$$= 2.15\,Mg/m^3$$

Table 10.12 gives T.S. = 1640 MPa. Then,

$$\text{sp. str., E-glass/epoxy} = \frac{(1640)(1.02 \times 10^{-1})}{(2.15)(10^{-6})}\,mm$$

$$= \underline{\underline{77.8 \times 10^6\,mm}}$$

($\cong$ Table 10.13 value of 77.2×10^6 mm)

10.35 Compare the calculated value of the isostrain modulus of a W fiber (50 vol %)/copper composite with that given in Table 10.12. The modulus of tungsten is 407×10^3 MPa.

10.35 In Table 10.12, we find $E_c = 260 \times 10^3$ MPa.

Noting E_{Cu} given in Table 10.10,

$$E_c = v_m E_m + v_f E_f$$

$$= (0.50)(115 \times 10^3\,MPa) + (0.50)(407 \times 10^3\,MPa)$$

$$= 261 \times 10^3\,MPa$$

$$\underline{\underline{\%\,error}} = \frac{261 - 260}{260} \times 100\% = \underline{\underline{0.38\%}}$$

10.36 Determine the error made in Problem 10.15 in calculating the isostrain modulus of the B/epoxy composite of Table 10.12.

10.36 Table 10.12 gives $E_c = 210 - 280 \times 10^3$ MPa

Problem 10.15 gave $E_c = 289 \times 10^3$ MPa

$$\therefore \%\,error = \frac{289 - 280}{289} \times 100\% \quad \text{to} \quad \frac{289 - 210}{289} \times 100\%$$

$$= \underline{\underline{3.1\%}} \text{ to } \underline{\underline{27\%}}$$

10.37 Calculate the error in assuming the isostrain modulus of an epoxy reinforced with 67 vol % C fibers is given by Equation 10.6. (Note Table 10.12 for experimental data.)

10.37

Table 10.12 indicates that $E_{comp} = 221\,GPa$

To calculate using data from Tables 10.10 and 10.11:

$$E_{comp} = v_m E_m + v_f E_f$$

$$= (0.33)(6.9\,GPa) + (0.67)(340 - 380\,GPa)$$

$$= 230\,GPa \rightarrow 257\,GPa$$

$$\therefore \%error = \frac{230-221}{221} \times 100\% \longrightarrow \frac{257-221}{221} \times 100\%$$

$$= \underline{\underline{4.1\%}} \longrightarrow \underline{\underline{16\%}}$$

10.38 (a) Calculate the error in assuming that the modulus for the Al_2O_3 whiskers (14 vol %)/epoxy composite in Table 10.12 is represented by isostrain conditions. (b) Calculate the error in assuming that the composite represents isostress conditions. (c) Comment on the nature of the agreement or disagreement indicated by your answers in parts (a) and (b).

10.38

(a) In Table 10.12, we find $E_c = 41 \times 10^3\,MPa$.

Using data from Tables 10.10 and 11 for isostrain,

$$E_c = v_m E_m + v_f E_f$$

$$= (0.86)(6.9 \times 10^3\,MPa) + (0.14)(430 \times 10^3\,MPa)$$

$$= 66 \times 10^3\,MPa$$

$$\underline{\underline{\%error}} = \frac{66-41}{41} \times 100\% = \underline{\underline{61\%}}$$

(b) For isostress,

$$E_c = \frac{E_m E_f}{v_m E_f + v_f E_m}$$

$$= \frac{(6.9 \times 10^3\,MPa)(430 \times 10^3\,MPa)}{(0.86)(430 \times 10^3\,MPa) + (0.14)(6.9 \times 10^3\,MPa)}$$

$$= 8.0 \times 10^3\,MPa$$

$$\underline{\underline{\%\,error}} = \frac{41-8}{41} \times 100\% = \underline{\underline{80\%}}$$

(c) The random orientation of whiskers gives a result which falls roughly midway between the extremes for isostrain and isostress orientations (refer to Figure 10-13).

• **10.39** Determine the appropriate value of n in Equation 10.20 to describe the modulus of the W particles (50 vol %)/copper composite given in Table 10.12. (The modulus of tungsten is 407×10^3 MPa.)

10.39

As in Sample Problem 10.8, consider:

$$E_c^n = v_\ell E_\ell^n + v_h E_h^n$$

Using data from Tables 10.10 and 12 and Problem 10.35 (in units of 10^3 MPa):

$$(190)^n = 0.5(115)^n + 0.5(407)^n$$

n	$(190)^n = A$	$0.5(115)^n + 0.5(407)^n = B$	B/A
+1	190	261	1.37
+1/2	13.78	15.45	1.12
+0.01	1.054	1.055	1.00
−0.01	0.949	0.948	1.00
−1	5.26×10^{-3}	5.41×10^{-3}	1.03

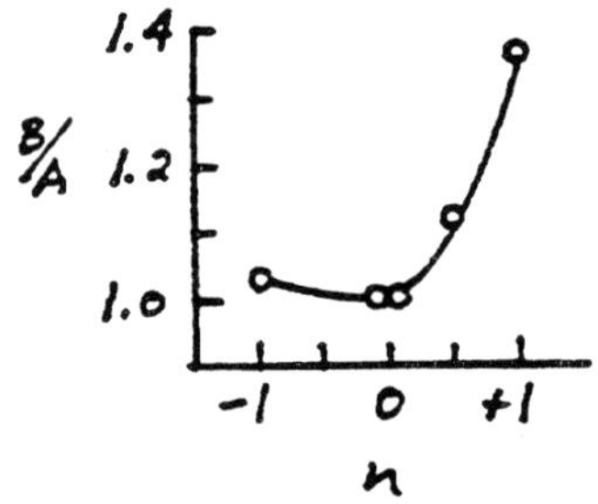

Therefore, $n \simeq 0$

10.40 Calculate the percent increase in composite modulus that would occur due to changing from a dispersed aggregate to an isostrain fiber loading of 50 vol % W in a copper matrix.

10.40

The measured data for these two cases are given in Table 10.12:

$$E_{agg.\ comp.} = 190 \times 10^3 \text{ MPa}$$

$$E_{isostrain} = 260 \times 10^3 \text{ MPa}$$

$$\therefore \%\ \text{increase} = \frac{260 - 190}{190} \times 100\ \% = 36.8\ \%$$

10.41 Calculate the fraction of the composite load carried by the W fibers in the isostrain case of Problem 10.40.

10.41 Using Equation 10.7 and the data from Problem 10.35 and Table 10.12:

$$\frac{P_f}{P_c} = \frac{E_f}{E_c} v_f = \frac{407 \times 10^3 \, MPa}{260 \times 10^3 \, MPa}(0.5) = 0.783$$

10.42 Calculate the specific strength of the Kevlar/epoxy composite in Table 10.12. (Note Problem 10.4.)

10.42 Using the result of Problem 10.4 and the data of Table 10.12,

$$sp. str. = \frac{(1517 \, MPa) \times (1.02 \times 10^{-1} \, kg/mm^2)/MPa}{(1.38 \, Mg/m^3)(10^3 \, kg/Mg)(1 \, m^3/10^9 \, mm^3)} = 112 \times 10^6 \, mm$$

10.43 Calculate the specific strength of the B/epoxy composite in Table 10.12. (Note Problem 10.2.)

10.43 From Problem 10.2, $\rho = 2.06 \, Mg/m^3$, giving

$$sp. str. = \frac{(1400 - 2100) \, MPa \, (1.02 \times 10^{-1} \, kg/mm^2)/MPa}{(2.06 \, Mg/m^3)(10^3 \, kg/Mg)(1 \, m^3/10^9 \, mm^3)} = (69.3 \text{ to } 104) \times 10^6 \, mm$$

10.44 Calculate the specific strength of the W particles (50 vol %)/copper composite listed in Table 10.12 (see Practice Problem 10.4).

10.44 From PP10.4, $\rho = 14.1 \, Mg/m^3$. Then,

$$sp. str. = \frac{(380 \, MPa)(1.02 \times 10^{-1} \, kg/mm^2)/MPa}{(14.1 \, Mg/m^3)(10^3 \, kg/Mg)(1 \, m^3/10^9 \, mm^3)} = 2.75 \times 10^6 \, mm$$

10.45 Calculate the specific strength for the W fibers (50 vol %)/copper composite listed in Table 10.12.

10.45 The ρ (= 14.1 Mg/m³) calculated in PP10.4 applies here also. (Geometry is not considered.) Then,

$$\text{sp. str.} = \frac{(1100\text{ MPa})(1.02\times10^{-1}\text{ kg/mm}^2)/\text{MPa}}{(14.1\text{ Mg/m}^3)(10^3\text{ kg/Mg})(1\text{ m}^3/10^9\text{ mm}^3)} = \underline{\underline{7.96\times10^6\text{ mm}}}$$

(Note that this is higher than the value for the much lower density aluminum of Sample Problem 10.10.)

10.46 Calculate the (flexural) specific strength of the SiC/Al_2O_3 ceramic–matrix composite in Table 10.12, assuming 50 vol % whiskers. (The density of SiC is 3.21 Mg/m³. The density of Al_2O_3 is 3.97 Mg/m³.)

10.46 First, calculate the density of the composite:

1 m³ composite → 0.5 m³ SiC and 0.5 m³ Al_2O_3

$$\rho_c = \frac{m_{SiC} + m_{Al_2O_3}}{V}$$

$$= \frac{(3.21\text{ Mg/m}^3)(0.5\text{ m}^3) + (3.97\text{ Mg/m}^3)(0.5\text{ m}^3)}{1\text{ m}^3}$$

$$= 3.59\text{ Mg/m}^3$$

Using the strength value from Table 10.12 gives

$$\text{sp. str.} = \frac{800\text{ MPa}\,(1.02\times10^{-1}\text{ kg/mm}^2)/\text{MPa}}{(3.59\text{ Mg/m}^3)(10^3\text{ kg/Mg})(1\text{ m}^3/10^9\text{ mm}^3)}$$

$$= \underline{\underline{22.7\times10^6\text{ mm}}}$$

10.47 Calculate the (tensile) specific strength of the Douglas fir (loaded parallel to the grain) in Table 10.12. (The density of this wood is 0.42 Mg/m³.)

10.47

$$\text{sp. str.} = \frac{85.5\text{ MPa}\,(1.02\times10^{-1}\text{ kg/mm}^2)/\text{MPa}}{(0.42\text{ Mg/m}^3)(10^3\text{ kg/Mg})(1\text{ m}^3/10^9\text{ mm}^3)}$$

$$= \underline{\underline{20.8\times10^6\text{ mm}}}$$

10.48 Calculate the (compressive) specific strength of the standard concrete (without air entrainer) in Table 10.12. (The density of this concrete is 2.30 Mg/m^3.)

10.48

$$\text{sp. str.} = \frac{41\ MPa\ (1.02 \times 10^{-1}\ kg/mm^2)/MPa}{(2.30\ Mg/m^3)(10^3\ kg/Mg)(1\ m^3/10^9\ mm^3)}$$

$$= 1.8 \times 10^6\ mm$$

10.49 To appreciate the relative toughness of (i) traditional ceramics, (ii) high-toughness, unreinforced ceramics, and (iii) ceramic–matrix composites, plot the breaking stress versus flaw size, a, for **(a)** silicon carbide, **(b)** partially stabilized zirconia, and **(c)** silicon carbide reinforced with SiC fibers. Use Equation 7.9 and take $Y = 1$. Cover a range of a from 1 to 100 mm on a logarithmic scale. (See Tables 8.10 and 10.12 for data.)

10.49

$$\sigma = K_{IC}/\sqrt{\pi a}$$

For SiC, $K_{IC} = 3\ MPa\sqrt{m}$
PSZ, " = 9 "
SiC/SiC, " = 25 "

a (mm)	σ_{SiC} (MPa)	σ_{PSZ} (MPa)	$\sigma_{SiC/SiC}$ (MPa)
1	53.5	161	446
2	37.8	114	315
5	23.9	71.8	199
10	16.9	50.8	141
20	12.0	35.9	100
50	7.6	22.7	63
100	5.4	16.1	45

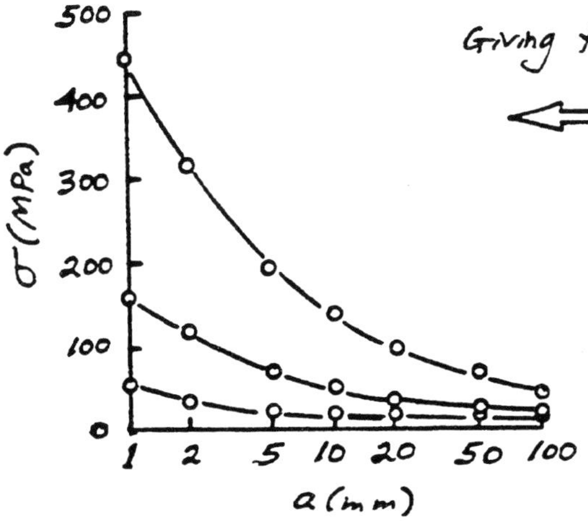

Giving the plot

10.50 In Problem 7.17 a competition among various metallic pressure vessel materials was illustrated. We can expand the selection process by including some composites, as listed in the following table:

Material	ρ(Mg/m^3)	Cost ($/kg)	Y.S. (MPa)
1040 carbon steel	7.8	0.63	
304 stainless steel	7.8	3.70	
3003-H14 aluminum	2.73	3.00	
Ti–5Al–2.5Sn	4.46	15.00	
Reinforced concrete	2.5	0.40	200
Fiberglass	1.8	3.30	200
Carbon fiber- reinforced polymer	1.5	270.00	600

(a) From this expanded list, select the material that will produce the lightest vessel. **(b)** Select the material that will produce the minimum-cost vessel.

10.50 (a) As in Problem 7.17, we are looking for the minimum ratio of ρ/Y.S. Following Problem 7.17:

Material	ρ (Mg/m^3)	Y.S. (MPa)	ρ/Y.S. (Mg/[m^3·MPa])
1040	7.8	600	0.0130
304	7.8	205	0.0380
3003-H14	2.73	145	0.0188
Ti-5Al-2.5Sn	4.46	827	0.00539
Concrete	2.5	200	0.0125
fiberglass	1.8	200	0.009
CFRP	1.5	600	0.0025

Therefore, CFRP will produce the lightest vessel.

(b) Here, we are looking for the minimum [ρ/Y.S. × cost] product.

Material	ρ/Y.S.	cost	[ρ/Y.S. × cost]
1040	0.0130	0.63	0.00819
304	0.0380	3.70	0.1406
3003-H14	0.0188	3.00	0.0564
Ti-5Al-2.5Sn	0.00539	15.00	0.0809
Concrete	0.0125	0.40	0.0050
fiberglass	0.009	3.30	0.0297
CFRP	0.0025	270.00	0.675

Therefore, reinforced concrete will produce the minimum cost vessel.

Section 11.1 - Charge Carriers and Conduction

PP 11.1 (**a**) The wire described in Sample Problem 11.1 shows a voltage drop of 432 mV. Calculate the voltage drop to be expected in a 0.5-mm-diameter (× 1-m-long) wire of the same alloy, also carrying a current of 10 A. (**b**) Repeat part (a) for a 2-mm-diameter wire.

PP 11.1

(a) $V = IR$

$$R = \frac{\rho l}{A} = \frac{(33.9\times10^{-9}\ \Omega\cdot m)(1\ m)}{\pi(0.25\times10^{-3}\ m)^2} = 0.173\ \Omega$$

$$V = (10\ A)(0.173\ \Omega) = \underline{\underline{1.73\ V}}$$

(b)

$$R = \frac{(33.9\times10^{-9}\ \Omega\cdot m)(1\ m)}{\pi(1\times10^{-3}\ m)^2} = 0.0108\ \Omega$$

$$V = (10\ A)(0.0108\ \Omega) = 0.108\ V = \underline{\underline{108\ mV}}$$

PP 11.2 How many free electrons would there be in a spool of high-purity copper wire (1 mm diameter ×10 m long)? (See Sample Problem 11.2.)

PP 11.2

$$\underline{\underline{\text{no. } e^-}} = n \times V_{wire}$$

$$= (104\times10^{27}\ m^{-3}) \times [\pi(0.5\times10^{-3}\ m)^2](10\ m)$$

$$= \underline{\underline{8.17\times10^{23}}}$$

PP 11.3 In Sample Problem 11.3, we compare the density of free electrons in copper with the density of atoms. How many copper atoms would be in the spool of wire described in Practice Problem 11.2?

PP 11.3

$$\underline{\underline{\text{no. atoms}}} = (\text{atom density}) \times V_{wire}$$

$$= (84.6\times10^{27}\ m^{-3}) \times [\pi(0.5\times10^{-3}\ m)^2](10\ m)$$

$$= \underline{\underline{6.65\times10^{23}}}$$

PP 11.4 The drift velocity of the free electrons in copper is calculated in Sample Problem 11.4. How long would a typical free electron take to move along the entire length of the spool of wire described in Practice Problem 11.2, under the voltage gradient of 0.5 V/m?

PP 11.4

$$t = l/\bar{v}$$

$$= (10\ m)/(1.75\times10^{-3}\ m/s)$$

$$= 5.71\times10^3\ s \times 1\ h/3600\ s$$

$$= \underline{\underline{1.59\ hr.}}$$

11.1 **(a)** Assume that the circuit in Figure 11-1 contains, as a sample, a cylindrical steel bar 1 cm diameter × 10 cm long with a conductivity of $7.00 \times 10^6\ \Omega^{-1}\cdot m^{-1}$. What would be the current in this bar due to a voltage of 10 mV? **(b)** Repeat part (a) for a bar of high-purity silicon of the same dimensions. (See Table 11.1.) **(c)** Repeat part (a) for a bar of borosilicate glass of the same dimensions. (Again, see Table 11.1.)

11.1

(a) $R = \frac{\rho l}{A} = \frac{l}{\sigma A} = \frac{10 \times 10^{-2} m}{(7.00 \times 10^6 \Omega^{-1} m^{-1})[\pi(0.5 \times 10^{-2} m)^2]}$

$= 1.82 \times 10^{-4} \Omega$

$I = \frac{V}{R} = \frac{0.010 V}{1.82 \times 10^{-4} \Omega} = \underline{\underline{55.0 A}}$

(b) From Table 11.1, $\sigma = 0.40 \times 10^{-3} \Omega^{-1} m^{-1}$

$R = \frac{10 \times 10^{-2} m}{(0.40 \times 10^{-3} \Omega^{-1} m^{-1})[\pi(0.5 \times 10^{-2} m)^2]} = 3.18 \times 10^6 \Omega$

$I = \frac{0.010 V}{3.18 \times 10^6 \Omega} = \underline{\underline{3.14 \times 10^{-9} A}}$

(c) From Table 11.1, $\sigma = 10^{-13} \Omega^{-1} m^{-1}$

$R = \frac{10 \times 10^{-2} m}{(10^{-13} \Omega^{-1} m^{-1})[\pi(0.5 \times 10^{-2} m)^2]} = 1.27 \times 10^{16} \Omega$

$I = \frac{0.010 V}{1.27 \times 10^{16} \Omega} = \underline{\underline{7.85 \times 10^{-19} A}}$

11.2 A light bulb operates with a line voltage of 110 V. If the filament resistance is 200 Ω, calculate the number of electrons per second traveling through the filament.

11.2

$I = \frac{V}{R} = \frac{110 V}{200 \Omega} = 0.55 A = 0.55 \frac{C}{s} \times \frac{1 e^-}{0.16 \times 10^{-18} C}$

$= \underline{\underline{3.44 \times 10^{18} e^-/s}}$

11.3 A semiconductor wafer is 0.5 mm thick. A potential of 100 mV is applied across this thickness. **(a)** What is the electron drift velocity if their mobility is $0.2\ m^2/(V \cdot s)$? **(b)** How much time is required for an electron to move across this thickness?

11.3

(a) $\mu = \bar{v}/E$ or $\bar{v} = \mu E = \mu \frac{V}{l} = (0.2 \frac{m^2}{V \cdot s}) \frac{(0.100 V)}{(0.5 \times 10^{-3} m)}$

$= \underline{\underline{40 m/s}}$

(b) $t = l/\bar{v} = (0.5 \times 10^{-3} m)/(40 m/s)$

$= 1.25 \times 10^{-5} s = \underline{\underline{12.5 \mu s}}$

11.4 A 1-mm-diameter wire is required to carry a current of 10 A, but the wire must not have a power dissipation (I^2R) greater than 10 W/m of wire. Of the materials listed in Table 11.1, which are suitable for this wire application?

11.4

For 1 m of wire,

$$\text{pow. dis.} = 10\ W = I^2R = (10\ A)^2 \frac{\rho l}{A}$$

$$= 100\ A^2 \frac{\rho\ (1\ m)}{\pi\ (0.5 \times 10^{-3}\ m)^2}$$

or

$$\rho = \frac{(10\ W/m)[\pi(0.5 \times 10^{-3}\ m)^2]}{100\ A^2} = 7.85 \times 10^{-8}\ \Omega \cdot m$$

or

$$\sigma = \frac{1}{\rho} = 1.27 \times 10^7\ \Omega^{-1} \cdot m^{-1} = 12.7 \times 10^6\ \Omega^{-1} \cdot m^{-1}$$

Inspection of Table 11.1 indicates that only aluminum and copper have conductivities $> 12.7 \times 10^6\ \Omega^{-1} \cdot m^{-1}$ and would, therefore, have power dissipations < 10 W/m.

11.5 A strip of aluminum metallization on a solid-state device is 1 mm long with a thickness of 1 μm and a width of 5 μm. What is the resistance of this strip?

11.5

Combining Equations 11.2 and 11.3,

$$R = \frac{\rho l}{A} = \frac{l}{\sigma A}$$

Using Table 11.1,

$$R = \frac{1 \times 10^{-3}\ m}{(35.36 \times 10^6\ \Omega^{-1} \cdot m^{-1})(1 \times 10^{-6}\ m)(5 \times 10^{-6}\ m)}$$

$$= 5.66\ \Omega$$

11.6 For a current of 10 mA along the aluminum strip in Problem 11.5, calculate (a) the voltage along the length of the strip and (b) the power dissipated (I^2R).

11.6

(a) $V = IR$

$$= (10 \times 10^{-3}\ A)(5.66\ \Omega)$$

$$= 0.0566\ V = 56.6\ mV$$

(b) $P = I^2R = I(IR)$

$$= (10 \times 10^{-3}\ A)(5.66 \times 10^{-2}\ V)$$

$$= 5.66 \times 10^{-4}\ W$$

11.7 A structural design involves a steel wire 2 mm in diameter that will carry an electrical current. If the resistance of the wire must be less than 25 Ω, calculate the maximum length of the wire, given the data in Table 11.1.

11.7

$$R = \frac{\rho l}{A}$$

or

$$l = \frac{RA}{\rho} = \frac{RA}{(1/\sigma)} = RA\sigma$$

Using the maximum value of σ in Table 11.1 to give the maximum value of l:

$$l = (25\,\Omega)[\pi(1 \times 10^{-3}\,m)^2](9.35 \times 10^6\,\Omega^{-1} \cdot m^{-1})$$

$$= \underline{\underline{734\,m}}$$

11.8 For the design discussed in Problem 11.7, calculate the allowable wire length if a 3-mm diameter is permitted.

11.8

In this case,

$$l = (25\,\Omega)[\pi\,(1.5 \times 10^{-3}\,m)^2](9.35 \times 10^6\,\Omega^{-1} \cdot m^{-1})$$

$$= 1.65 \times 10^3\,m = \underline{\underline{1.65\,km}}$$

Section 11.2 - Energy Levels and Energy Bands

PP 11.5 What is the probability of an electron's being promoted to the conduction band in diamond at 50°C? (See Sample Problem 11.5.)

PP 11.5

$$f(E) = \frac{1}{e^{(E-E_F)/kT} + 1}$$

$$= \frac{1}{e^{(2.8\,eV)/(86.2\times10^{-6}\,eV/K)(323\,K)} + 1}$$

$$= \underline{\underline{2.11\times10^{-44}}}$$

PP 11.6 What is the probability of an electron's being promoted to the conduction band in silicon at 50°C? (See Sample Problem 11.6.)

PP 11.6

$$f(E) = \frac{1}{e^{(0.5535\,eV)/(86.2\times10^{-6}\,eV/K)(323\,K)} + 1}$$

$$= \underline{\underline{2.32\times10^{-9}}}$$

11.9 At what temperature will the 5.60-eV energy level for electrons in silver be 25% filled? (The Fermi level for silver is 5.48 eV.)

11.9

$$f(E) = \frac{1}{e^{(E-E_F)/kT} + 1}$$

$$0.25 = \frac{1}{e^{(5.60-5.48)eV/(86.20\times10^{-6}\,eV/K)T} + 1}$$

or

$$e^{0.12\,eV/(86.20\times10^{-6}\,eV/K)T} = \frac{1}{0.25} - 1 = 3$$

$$e^{1392\,K/T} = 3$$

or

$$\frac{1392\,K}{T} = \ln 3 = 1.0986$$

or

$$T = 1267\,K = \underline{\underline{994\,°C}}$$

11.10 Generate a plot comparable to Figure 11-7 at a temperature of 1000 K for copper, which has a Fermi level of 7.04 eV.

11.10

$$f(E) = \frac{1}{e^{(E-E_F)/kT} + 1}$$

with $kT = 86.2 \times 10^{-6}$ eV/K $\times 1000$K $= 0.0862$ & $E_F = 7.04$eV

E(eV)	f(E)	E(eV)	f(E)
6.5	0.998	7.1	0.333
6.6	0.994	7.3	0.047
6.7	0.981	7.4	0.015
7.0	0.614	7.5	0.005
7.04	0.500		

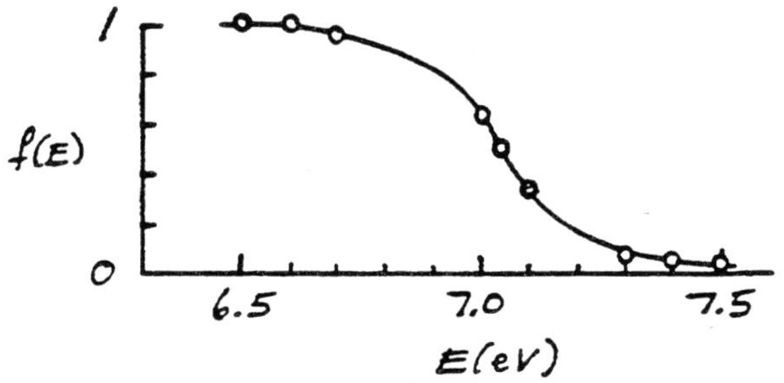

11.11 What is the probability of an electron's being promoted to the conduction band in indium antimonide, InSb, at **(a)** 25°C and **(b)** 50°C? (The band gap of InSb is 0.17 eV.)

11.11 (a) $f(E) = \frac{1}{e^{(0.17eV/2)/(86.2\times10^{-6}eV/K)(298K)} + 1} = \underline{\underline{3.53\times10^{-2}}}$

(b) $f(E) = \frac{1}{e^{(0.17eV/2)/(86.2\times10^{-6}eV/K)(323K)} + 1} = \underline{\underline{4.51\times10^{-2}}}$

11.12 At what temperature will diamond have the same probability for an electron's being promoted to the conduction band as silicon has at 25°C? (The answer to this question indicates the temperature range in which diamond can be properly thought of as a semiconductor rather than as an insulator.)

11.12

The probability for silicon at 25°C was calculated in Sample Problem 11.6. The band gap for diamond was given in Sample Problem 11.5.

$$f(E) = \frac{1}{e^{(E-E_F)/kT} + 1}$$

$$4.39\times10^{-10} = \frac{1}{e^{(2.8eV)/(86.2\times10^{-6}eV/K)T} + 1}$$

or

$$e^{(2.8eV)/(86.2\times10^{-6}eV/K)T} + 1 = \frac{1}{4.39\times10^{-10}} = 2.28\times10^{9}$$

or

$$\frac{2.8eV}{(86.2\times10^{-6}eV/K)T} = \ln(2.28\times10^{9}) = 21.55$$

or

$$T = 1508\,K = \underline{\underline{1235°C}}$$

11.13 Gallium forms semiconducting compounds with various group VA elements. The band gap systematically drops with increasing atomic number of the VA elements. For example, the band gaps for the III–V semiconductors GaP, GaAs, and GaSb are 2.25 eV, 1.47 eV, and 0.68 eV, respectively. Calculate the probability of an electron's being promoted to the conduction band in each of these semiconductors at 25°C.

11.13

$$f(E) = \frac{1}{e^{(E-E_F)/kT} + 1}$$

For GaP ($E_g = 2.25eV$)

$$f(E) = \frac{1}{e^{(2.25/2\ eV)/(86.2\times10^{-6}eV/K)(298K)} + 1}$$

$$= \underline{\underline{9.55\times10^{-20}}}$$

For GaAs ($E_g = 1.47eV$)

$$f(E) = \frac{1}{e^{(1.47/2\ eV)/(86.2\times10^{-6}eV/K)(298)} + 1}$$

$$= \underline{\underline{3.75\times10^{-13}}}$$

For GaSb ($E_g = 0.68$ eV)

$$f(E) = \frac{1}{e^{(0.68/2\ \mathrm{eV})/(86.2\times10^{-6}\ \mathrm{eV/K})(298)} + 1}$$

$$= \underline{\underline{1.79 \times 10^{-6}}}$$

11.14 The trend discussed in Problem 11.13 is a general one. Calculate the probability of an electron's being promoted to the conduction band at 25°C in the II–VI semiconductors CdS and CdTe, which have band gaps of 2.59 eV and 1.50 eV, respectively.

11.14

$$f(E) = \frac{1}{e^{(E-E_F)/kT} + 1}$$

For CdS ($E_g = 2.59$ eV)

$$f(E) = \frac{1}{e^{(2.59/2\ \mathrm{eV})/(86.2\times10^{-6}\ \mathrm{eV/K})(298\ \mathrm{K})} + 1}$$

$$= \underline{\underline{1.28 \times 10^{-22}}}$$

For CdTe ($E_g = 1.50$ eV)

$$f(E) = \frac{1}{e^{(1.50/2\ \mathrm{eV})/(86.2\times10^{-6}\ \mathrm{eV/K})(298\ \mathrm{K})} + 1}$$

$$= \underline{\underline{2.09 \times 10^{-13}}}$$

Section 11.3 - Conductors

PP 11.7 Calculate the conductivity at 200°C of **(a)** copper (annealed standard) and **(b)** tungsten. (See Sample Problem 11.7.)

PP 11.7 (a) $\rho = \rho_{rt}[1 + \alpha(T - T_{rt})]$

$$= (17.24\times10^{-9}\ \Omega\cdot\mathrm{m})[1 + 0.00393\ ^\circ\mathrm{C}^{-1}(200-20)^\circ\mathrm{C}]$$

$$= 2.94\times10^{-8}\ \Omega\cdot\mathrm{m}$$

$$\sigma = 1/\rho = \underline{\underline{34.0\times10^{6}\ \Omega^{-1}\cdot\mathrm{m}^{-1}}}$$

(b) $\rho = (55.1\times10^{-9}\ \Omega\cdot\mathrm{m})[1 + 0.0045\ ^\circ\mathrm{C}^{-1}(200-20)^\circ\mathrm{C}]$

$$= 9.97\times10^{-8}\ \Omega\cdot\mathrm{m}$$

$$\sigma = 1/\rho = \underline{\underline{10.0\times10^{6}\ \Omega^{-1}\cdot\mathrm{m}^{-1}}}$$

PP 11.8 Estimate the resistivity of a copper–0.06 wt % phosphorus alloy at 200°C. (See Sample Problem 11.8.)

PP 11.8 From Figure 11-12,

$$\rho_{20°C,\ Cu-0.06P} \cong 25.3 \times 10^{-9}\ \Omega\cdot m$$

Following Sample Problem 11.8,

$$\rho_{200°C,\ Cu-0.06P} = (25.3\times10^{-9}\ \Omega\cdot m)[1 + 0.00393\ °C^{-1}(200-20)°C]$$
$$= 43.2\times10^{-9}\ \Omega\cdot m$$

PP 11.9 In Sample Problem 11.9, we find the output from a type K thermocouple at 800°C. What would be the output from a Pt/90 Pt–10 Rh thermocouple?

PP 11.9 Table 11.3 shows that the Pt/90Pt-10Rh thermocouple is "type S." Figure 11-15 shows that the type S thermocouple has an output of 7 mV at 800°C.

PP 11.10 When the 1-2-3 superconductor in Sample Problem 11.10 is fabricated in a bulk specimen with dimensions 5 mm × 5 mm × 20 mm, the current in the long dimension at which superconductivity is lost is found to be 3.25×10^3 A. What is the critical current density for this configuration?

PP 11.10 As in Sample Problem 11.10,

$$\text{critical current density} = \frac{3.25\times10^3 A}{(5\times10^{-3} m)^2}$$
$$= 1.30\times10^{-8} A/m^2$$

which is roughly two orders of magnitude lower than for the value for the thin film configuration.

11.15 A strip of copper metallization on a solid-state device is 1 mm long with a thickness of 1 μm and a width of 5 μm. If a voltage of 0.1 V is applied along the long dimension, what is the resulting current?

11.15 Similar to Problem 11.5, the resistance will be (using Table 11.2):

$$R = \frac{\rho l}{A} = \frac{(17.24\times10^{-9}\ \Omega\cdot m)(1\times10^{-3} m)}{(1\times10^{-6} m)(5\times10^{-6} m)} = 3.45\ \Omega$$

$$I = V/R = (1\times10^{-1} V)/(3.45\ \Omega) = 29.0\times10^{-3} V$$
$$= 29.0\ mV$$

11.16 A metal wire 1 mm in diameter × 10 m long carries a current of 0.1 A. If the metal is pure copper at 30°C, what is the voltage drop along this wire?

11.16

$$\rho = \rho_{rt}\,[1+\alpha(T-T_{rt})]$$

$$= (17.24\times10^{-9}\,\Omega\cdot m)[1+(0.00393\,°C^{-1})(30-20)°C]$$

$$= 17.92\times10^{-9}\,\Omega\cdot m$$

$$R = \frac{\rho l}{A} = \frac{(17.92\times10^{-9}\,\Omega\cdot m)(10\,m)}{\pi(0.5\times10^{-3}\,m)^2} = 0.2281\,\Omega$$

$$V = IR = (0.1\,A)(0.2281\,\Omega) = 0.02281\,V = \underline{\underline{22.8\,mV}}$$

11.17 Repeat Problem 11.16 assuming that the wire is a Cu–0.1 wt % Al alloy at 30°C.

11.17

From Figure 11-12, $\rho_{20°C,\,Cu-0.1Al} \cong 18.6\times10^{-9}\,\Omega\cdot m$

$$\rho = (18.6\times10^{-9}\,\Omega\cdot m)[1+(0.00393\,°C^{-1})(30-20)°C]$$

$$= 19.33\times10^{-9}\,\Omega\cdot m$$

$$R = \frac{\rho l}{A} = \frac{(19.33\times10^{-9}\,\Omega\cdot m)(10\,m)}{\pi(0.5\times10^{-3}\,m)^2} = 0.2461\,\Omega$$

$$V = IR = (0.1\,A)(0.2461\,\Omega) = 0.02461\,V = \underline{\underline{24.6\,mV}}$$

11.18 A type K thermocouple is operated with a reference temperature of 100°C (established by the boiling of distilled water). What is the temperature in a crucible for which a thermocouple voltage of 30 mV is obtained?

11.18

Relative to Figure 11-15, we see that the emf (relative to a 0°C reference) of a type K thermocouple at 100°C is 4 mV. The temperature corresponding to 4 mV + 30 mV (= 34 mV) is $\underline{\underline{\approx 825\,°C}}$.

11.19 Repeat Problem 11.18 for the case of a chromel/constantan thermocouple.

11.19

Table 11.3 indicates chromel/constantan is "type E." Figure 11-15 shows the emf (relative to a 0°C reference) is 8 mV at 100°C.

The temperature corresponding to 8 mV + 30 mV (= 38 mV) is $\underline{\underline{\approx 500\,°C}}$.

11.20 A furnace for oxidizing silicon is operated at 1000°C. What would be the output (relative to an ice–water bath) for **(a)** a type S, **(b)** a type K, and **(c)** a type J thermocouple?

11.20

(a) Figure 11-15 shows the emf would be 9 mV for the Type S thermocouple.

(b) Similarly, the emf would be 41 mV for Type K.

(c) Table 11.3 indicates that there would be no output for the Type J thermocouple as 1000°C is well above its maximum service temperature.

11.21 An important application of metal conductors in the field of materials processing is in the form of metal wire for resistance-heated furnace elements. Some of the alloys used as thermocouples also serve as furnace elements. For example, consider the use of a 1-mm-diameter chromel wire to produce a 1-kW furnace coil in a laboratory furnace operated at 110 V. What length of wire is required for this furnace design? (*Note:* The power of the resistance-heated wire is equal to I^2R, and the resistivity of the chromel wire is $1.08 \times 10^6\ \Omega \cdot m$.)

11.21

$$\text{Power} = I^2R = (IR)I = VI$$

$$\text{or } I = \text{Power}/V = 1000\ W/110\ V = 9.09\ A$$

$$R = V/I = 110\ V/9.09\ A = 12.1\ \Omega = \frac{\rho l}{A} = \frac{(1.08\times10^{-6}\ \Omega\cdot m)\, l}{\pi(0.5\times10^{-3}\ m)^2}$$

$$\text{or } l = 8.80\ m$$

11.22 Given the information in Problem 11.21, calculate the power requirement of a furnace constructed of a chromel wire 5 m long and 1 mm diameter operating at 110 V.

11.22

$$\text{Power} = I^2R = \left(\frac{V}{R}\right)^2 R = \frac{V^2}{R} = \frac{V^2}{(\rho l/A)}$$

$$= \frac{V^2 A}{\rho l}$$

Using data from Problem 11.21,

$$\text{Power} = \frac{(110\ V)^2\left[\pi(0.5\times10^{-3}\ m)^2\right]}{(1.08\times10^{-6}\ \Omega\cdot m)(5\ m)}$$

$$= 1.76\times 10^3 \ \frac{V^2}{\Omega} = 1.76\times 10^3 \ V\cdot C/s$$

$$= 1.76\times 10^3 \ J/s = \underline{\underline{1.76 \ kW}}$$

11.23 What would be the power requirement for the furnace in Problem 11.22 if it is operated at 208 V?

11.23

$$\text{Power} = \frac{(208\ V)^2 [\pi (0.5\times 10^{-3}\ m)^2]}{(1.08\times 10^{-6}\ \Omega\cdot m)(5\ m)}$$

$$= 6.29\times 10^3 \ \frac{V^2}{\Omega} = \underline{\underline{6.29 \ kW}}$$

11.24 A tungsten light bulb filament is 10 mm long and 100 μm diameter. What is the current in the filament when operating at 1000°C with a line voltage of 110 V?

11.24

$$I = \frac{V}{R} = \frac{V}{(\rho l/A)} = \frac{VA}{\rho l}$$

Taking the 20°C data from Table 11.2,

$$I = \frac{(110\ V)\pi(50\times 10^{-6}\ m)^2}{(55.1\times 10^{-9}\ \Omega\ m)(10\times 10^{-3}\ m)}$$

$$= 1.57\times 10^{-3}\ A = \underline{\underline{1.57\ mA}}$$

11.25 What is the power dissipation (I^2R) in the filament of Problem 11.24?

11.25

$$\text{Power Dissipation} = I^2R = I^2\rho l/A$$

$$= \frac{(1.57\times 10^{-3}\ A)^2(55.1\times 10^{-9}\ \Omega\cdot m)(10\times 10^{-3}\ m)}{\pi(50\times 10^{-6}\ m)^2}$$

$$= 1.73\times 10^{-7}\ A^2\Omega = 1.73\times 10^{-7}\ VA$$

$$= \underline{\underline{1.73\times 10^{-7}\ W}}$$

11.26 For a bulk 1-2-3 superconductor with a critical current density of 1×10^8 A/m^2, what is the maximum supercurrent that could be carried in a 1-mm-diameter wire of this material?

11.26

I = Current density × area

$= (10^8 \, A/m^2)(\pi[1 \times 10^{-3} m/2]^2)$

$= \underline{\underline{78.5 \, A}}$

11.27 If progress in increasing T_c for superconductors had continued at the linear rate followed through 1975, by what year would a T_c of 95 K be achieved?

11.27

A reasonable "best-fit" straight line to the open circle data of Figure 11-17 can be obtained by drawing a straight line between the first and last data points, viz. (1911, 4.12K) and (1975, 23.3 K). The slope is:

$$\frac{23.3 K - 4.12 K}{1975 - 1911} = 0.300 \, K/year$$

Then, in x years after 1975:

$$95 K = 23.3 K + (0.300 K/yr) x$$

or

$$x = 239 \text{ years}$$

Therefore, $T_c = 95K$ would not have been reached until $1975 + 239 \cong \underline{\underline{2210}}$

11.28 Verify the footnote on page 446 regarding the presence of one Cu^{3+} valence in the $YBa_2Cu_3O_7$ unit cell.

11.28

First, consider the following "normal" valences from Appendix 2:

Y: 3+ ; Ba: 2+ ; O : 2−

The non-copper cations give a total charge of:

$$(1 Y)(3+/Y) + (2 Ba)(2+/Ba) = 7+$$

The oxygen anions give:

$$(7 O)(2-/O) = 14-$$

Charge neutrality will, then, require

$$(2\,Cu)(2+/Cu) + (1\,Cu)(3+/Cu) = 7+$$

i.e., one Cu^{3+} to provide charge neutrality (7+ and 7+ balancing 14−).

11.29 Verify the chemical formula for $YBa_2Cu_3O_7$ using the unit cell geometry of Figure 11-19.

11.29 Y: There is 1 Y at the center of the unit cell.

Ba: There are 2 Ba, one centered in the upper one-third of the unit cell and one centered in the lower one-third.

Cu: There are $8 \times \frac{1}{8} + 8 \times \frac{1}{4} = 3$ Cu.

O: There are $12 \times \frac{1}{4} + 8 \times \frac{1}{2} = 7$ O.

• **11.30** Describe the similarities and differences between the perovskite unit cell of Figure 3-22 and **(a)** the upper and lower thirds and **(b)** the middle third of the $YBa_2Cu_3O_7$ unit cell of Figure 11-19.

11.30 (a) similarities: a metal ion in the center and another metal ion at the corners.

differences: a non-planar metal-oxygen plane (at the bottom of the upper third and the top of the lower third); oxygen ions in the center of edges rather than the face-centered positions of simple perovskite. (Also, not all edge-center positions are occupied.)

(b) similarities: a metal ion in the center and another metal ion at the corners.

differences: The top and bottom of the middle third are non-planar metal-oxygen planes; oxygen ion at edge-centers rather than face-centers of perovskite. (Again, not all edge-center positions are occupied.)

Section 11.4 - Insulators

PP 11.11 Using the result of Sample Problem 11.11, calculate the total dipole moment for a 2-mm-thick × 2-cm-diameter disk of $BaTiO_3$, to be used as an ultrasonic transducer.

PP 11.11

From Sample Problem 11.11, the total dipole moment for a unit cell is 10.56×10^{-30} C·m.

From Figure 11-22, $V_{unit\,cell} = (0.399)^2(0.403)(10^{-9}\,m)^3$

$= 6.416 \times 10^{-29}\,m^3$

$V_{transducer} = (2 \times 10^{-3}\,m)[\pi(1 \times 10^{-2}\,m)^2] = 6.28 \times 10^{-7}\,m^3$

Therefore,

$$\text{Total dipole moment, transducer} = (10.56 \times 10^{-30}\,C \cdot m)\left(\frac{6.28 \times 10^{-7}\,m^3}{6.42 \times 10^{-29}\,m^3}\right)$$

$$= 1.034 \times 10^{-7}\,C \cdot m$$

PP 11.12 The inherent polarization of the unit cell of $BaTiO_3$ is calculated in Sample Problem 11.12. Under an applied electrical field, the polarization of the unit cell is increased to 0.180 C/m^2. Calculate the unit cell geometry under this condition and use sketches similar to Figures 11-21b and 11-22 to illustrate your results.

PP 11.12

First, assume $a \approx$ unchanged.

Second, assume all vertical dimensions $\propto$ polarization.

$$c = 0.403\,nm\left(\frac{0.180\,C/m^2}{0.165\,C/m^2}\right) = 0.440\,nm$$

$\Delta Ba^{2+} = 0.009\,nm$ (") $= 0.0098\,nm$

$\Delta Ti^{4+} = 0.006\,nm$ (") $= 0.0065\,nm$

$\Delta O^{2-} = 0.006\,nm$ (") $= 0.0065\,nm$

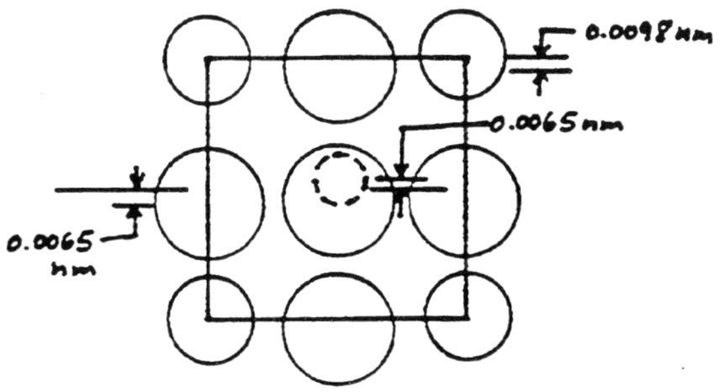

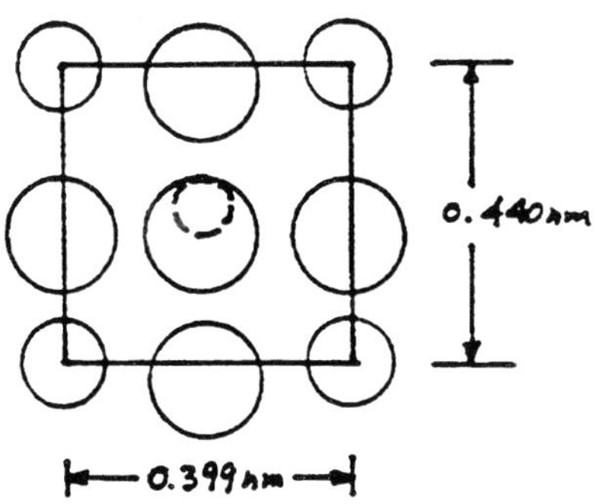

11.31 Calculate the charge density on a 2 mm-thick capacitor made of 99.5% Al_2O_3 under an applied voltage of 1 kV.

11.31 Using Equation 11.13 and Table 11.4,

$$D = \epsilon_0 K E$$
$$= (8.854 \times 10^{-12}\ C/V{\cdot}m)(9.8)(1 \times 10^{3}\ V/[2 \times 10^{-3}\ m])$$
$$= 4.34 \times 10^{-5}\ C/m^2$$

11.32 Repeat Problem 11.31 for the same material at its breakdown voltage gradient (= dielectric strength).

11.32 In this case,

$$D = \epsilon_0 K E$$
$$= (8.854 \times 10^{-12}\ C/V{\cdot}m)(9.8)(9.5 \times 10^{3}\ V/[1 \times 10^{-3}\ m])$$
$$= 8.24 \times 10^{-4}\ C/m^2$$

11.33 Calculate the charge density on a capacitor made of cordierite at its breakdown dielectric strength of 3 kV/mm. The dielectric constant is 4.5.

11.33

$$D = \epsilon_0 K E$$
$$= (8.854 \times 10^{-12}\ C/V{\cdot}m)(4.5)(3 \times 10^{3}\ V/[1 \times 10^{-3}\ m])$$
$$= 1.20 \times 10^{-4}\ C/m^2$$

11.34 By improved processing, a new cordierite capacitor can be made with properties superior to the one described in Problem 11.33. If the dielectric constant is increased to 5.0, calculate the charge density on a capacitor operated at a voltage gradient of 3 kV/mm (which is now below the breakdown strength).

11.34

$$D = \epsilon_0 K E$$
$$= (8.854 \times 10^{-12}\ C/V{\cdot}m)(5.0)(3 \times 10^{3}\ V/[1 \times 10^{-3}\ m])$$
$$= 1.33 \times 10^{-4}\ C/m^2$$

11.35 An alternate definition of polarization (introduced in Sample Problems 11.11 and 11.12) is

$$P = (\kappa - 1)\epsilon_0 E$$

where κ, ϵ_0, and E were defined relative to Equations 11.11 and 11.13. Calculate the polarization for 99.9% Al_2O_3 under a field strength of 5 kV/mm. (You might note the magnitude of your answer in comparison to the inherent polarization of tetragonal $BaTiO_3$ in Sample Problem 11.12.)

11.35 Using the data from Table 11.4 gives:

$$P = (\kappa - 1)\epsilon_0 E$$

$$= (10.1-1)(8.854\times10^{-12}\,C/V\cdot m)(5\times10^{3}\,V/10^{-3}\,m)$$

$$= \underline{\underline{4.03\times10^{-4}\,C/m^2}}$$

Note: This is quite small in comparison to the value of $0.165\,C/m^2$ for $BaTiO_3$ in Sample Problem 11.12.

11.36 Calculate the polarization of the acetal engineering polymer in Table 11.4 at its breakdown voltage gradient (= dielectric strength). (See Problem 11.35.)

11.36 Using the equation from Problem 11.35 and data from Table 11.4,

$$P = (\kappa - 1)\epsilon_0 E$$

$$= (3.7-1)(8.854\times10^{-12}\,C/V\cdot m)(19.7\times10^{3}\,V/10^{-3}\,m)$$

$$= \underline{\underline{4.71\times10^{-4}\,C/m^2}}$$

11.37 As in Problem 11.36, consider the polarization at the breakdown voltage gradient. By how much does this value increase for the nylon polymer in Table 11.4 in the humid environment, as compared to the dry condition?

11.37 As in Problem 11.36, for nylon 66 (dry):

$$P = (\kappa - 1)\epsilon_0 E$$

$$= (3.7-1)(8.854\times10^{-12}\,C/V\cdot m)(20.5\times10^{6}\,V/m)$$

$$= 4.90\times10^{-4}\,C/m^2$$

For nylon 66 (50% rel. hum.):

$$P = (7.8-1)(8.854\times10^{-12}\,C/V\cdot m)(17.3\times10^{6}\,V/m)$$

$$= 10.4\times10^{-4}\,C/m^2$$

$$\therefore \ \% \text{ increase} = \frac{10.4 - 4.90}{4.90} \times 100\% = 112\%$$

11.38 By heating $BaTiO_3$ to 100°C, the unit cell dimensions change to $a = 0.400$ nm and $c = 0.402$ nm (compared to the values in Figure 11-21). In addition, the ion shifts shown in Figure 11-20b are reduced by half. Calculate **(a)** the dipole moment and **(b)** the polarization of the $BaTiO_3$ unit cell at 100°C.

11.38 (a) To calculate the dipole moment, we can note that it must, by definition, be one-half of that calculated in Sample Problem 11.11 as all shifts are reduced by one-half.

Therefore,

$$\sum Qd = \frac{1}{2}(10.56 \times 10^{-30}\ C \cdot m) = 5.28 \times 10^{-30}\ C \cdot m$$

(b) As shown in Sample Problem 11.12,

$$P = \frac{\sum Qd}{V}$$

$$= \frac{5.28 \times 10^{-30}\ C \cdot m}{(0.402 \times 10^{-9}\ m)(0.400 \times 10^{-9}\ m)^2}$$

$$= 0.0821\ C/m^2$$

11.39 If the elastic modulus of $BaTiO_3$ in the c-direction is 109×10^3 MPa, what stress is necessary to reduce its polarization by 0.1%?

11.39 The percentage reduction in polarization corresponds to the elastic strain. Using Hooke's law:

$$\sigma = E\epsilon$$

$$= (109 \times 10^3\ MPa)(0.001)$$

$$= 109\ MPa$$

• **11.40** A central part of the appreciation of the mechanisms of ferroelectricity and piezoelectricity is the visualization of the material's crystal structure. For the case of the tetragonal modification of the perovskite structure (Figure 11-21b), sketch the atomic arrangements in the (**a**) (100), (**b**) (001), (**c**) (110), (**d**) (101), (**e**) (200), and (**f**) (002) planes.

11.40

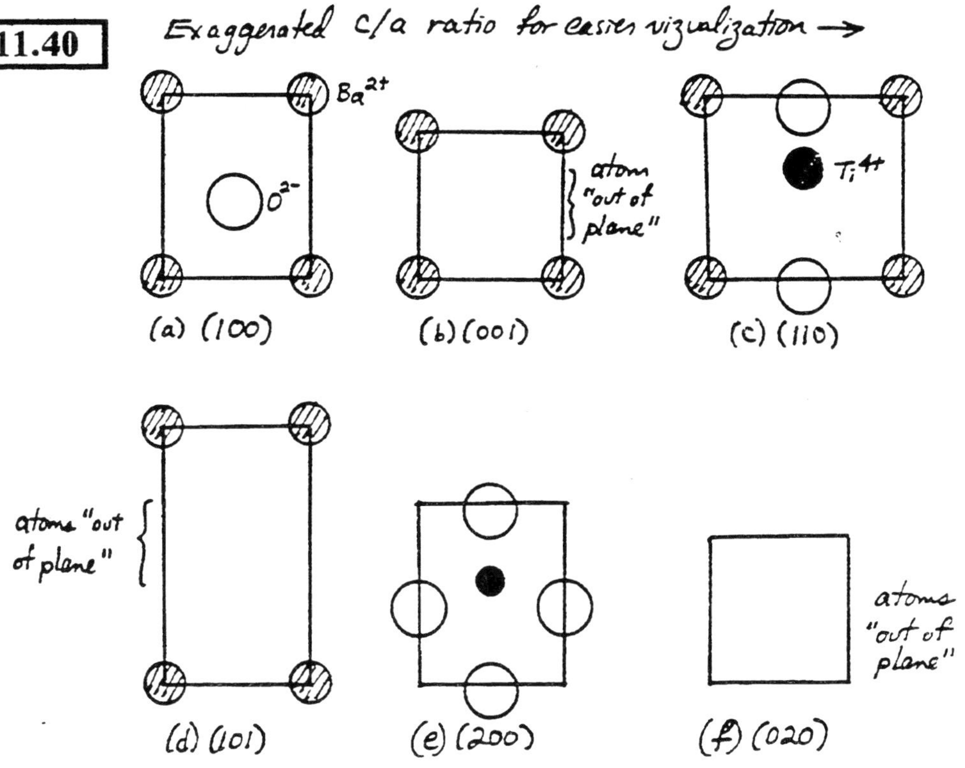

• **11.41** As in Problem 11.40, sketch the atomic arrangements in the **(a)** (100), **(b)** (001), **(c)** (110), and **(d)** (101) planes in the *cubic* perovskite structure (Figure 11-21a).

11.41

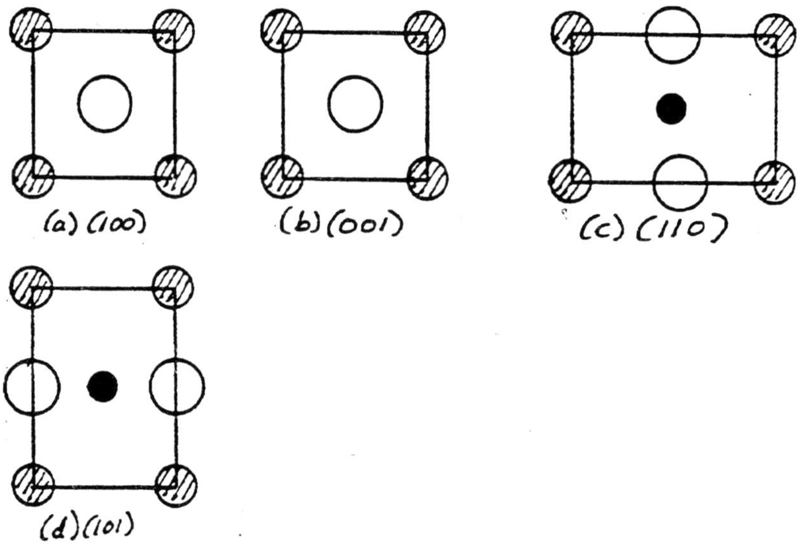

• **11.42** As in Problem 11.40, sketch the atomic arrangements in the **(a)** (200), **(b)** (002), and **(c)** (111) planes in the *cubic* perovskite structure (Figure 11-21a).

11.42

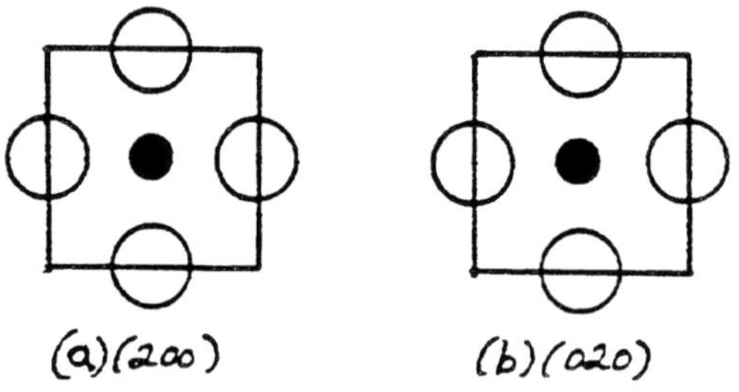

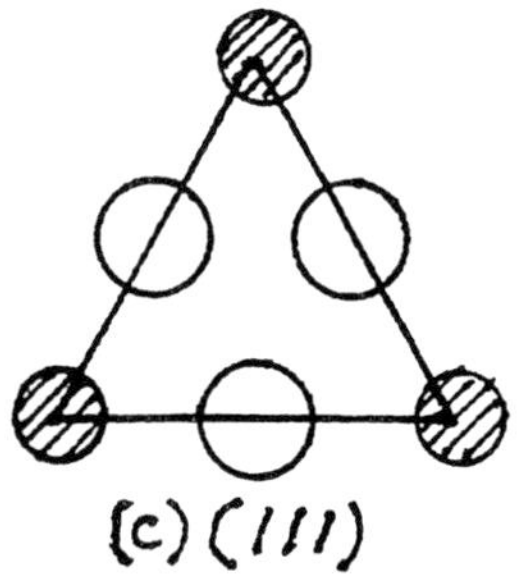

Section 11.5 - Semiconductors

PP 11.13 Using the data in Table 11.5, calculate **(a)** the total conductivity and **(b)** the resistivity of Si at room temperature. (See Sample Problem 11.13.)

PP 11.13

(a) $\sigma = nq(\mu_e + \mu_h)$

$= (14 \times 10^{15}\ m^{-3})(0.1602 \times 10^{-18}\ C)(0.140 + 0.038)\ \frac{m^2}{V \cdot s}$

$= \underline{\underline{3.99 \times 10^{-4}\ \Omega^{-1} \cdot m^{-1}}}$

(Consistent with the value given in Table 11.1)

(b) $\rho = 1/\sigma = \underline{\underline{2.50 \times 10^{3}\ \Omega \cdot m}}$

11.43 Calculate the fraction of Ge atoms that provides a conduction electron at room-temperature.

11.43

Using Appendix 1,

$\rho_{atomic} = 5.32\ \frac{g}{cm^3} \times 10^6\ \frac{cm^3}{m^3} \times \frac{1\ g\cdot atom}{72.59\ g} \times 0.6023 \times 10^{24}\ \frac{atoms}{g\cdot atom}$

$= 44.1 \times 10^{27}\ atoms/m^3$

Table 11.13 gives $n_e = 23 \times 10^{18}\ m^{-3}$

$\therefore$ fraction atoms providing conduction electrons:

$\text{fraction} = \frac{23 \times 10^{18}\ m^{-3}}{44.1 \times 10^{27}\ m^{-3}} = \underline{\underline{5.2 \times 10^{-10}}}$

11.44 What fraction of the conductivity of intrinsic silicon at room-temperature is due to (a) electrons and (b) electron holes?

11.44 (a) $\sigma = \sigma_e + \sigma_h = nq(\mu_e + \mu_h) = nq\mu_e + nq\mu_h$

or $\dfrac{\sigma_e}{\sigma} = \dfrac{nq\mu_e}{nq(\mu_e+\mu_h)} = \dfrac{\mu_e}{\mu_e+\mu_h}$

Using Table 11.13,

$$\frac{\sigma_e}{\sigma} = \frac{0.140}{0.140+0.038} = \underline{\underline{0.787}}$$

(b) Similarly,

$$\frac{\sigma_h}{\sigma} = \frac{nq\mu_h}{nq(\mu_e+\mu_h)} = \frac{\mu_h}{\mu_e+\mu_h}$$

$$= \frac{0.038}{0.140+0.038} = \underline{\underline{0.213}}$$

11.45 What fraction of the conductivity at room temperature for (a) germanium and (b) CdS is contributed by (i) electrons and (ii) electron holes?

11.45 (a) $\dfrac{\sigma_e}{\sigma} = \dfrac{\mu_e}{\mu_e+\mu_h} = \dfrac{0.364}{0.364+0.190} = \underline{\underline{0.657}}$

$\dfrac{\sigma_h}{\sigma} = \dfrac{\mu_h}{\mu_e+\mu_h} = \dfrac{0.190}{0.364+0.190} = \underline{\underline{0.343}}$

(b) $\dfrac{\sigma_e}{\sigma} = \dfrac{\mu_e}{\mu_e+\mu_h} = \dfrac{0.034}{0.034+0.0018} = \underline{\underline{0.950}}$

$\dfrac{\sigma_h}{\sigma} = \dfrac{\mu_h}{\mu_e+\mu_h} = \dfrac{0.0018}{0.034+0.0018} = \underline{\underline{0.050}}$

11.46 Using the data in Table 11.5, calculate the room temperature conductivity of intrinsic gallium arsenide.

11.46 $\sigma = nq(\mu_e+\mu_h)$

$= (1.4\times10^{12}\ m^{-3})(0.1602\times10^{-18}\ C)(0.720+0.020)\ \dfrac{m^2}{V\cdot s}$

$= \underline{\underline{1.66\times10^{-7}\ \Omega^{-1}\cdot m^{-1}}}$

11.47 Using the data in Table 11.5, calculate the room temperature conductivity of intrinsic InSb.

11.47

$$\sigma = nq(\mu_e + \mu_h)$$

$$= (13.5 \times 10^{21}\,m^{-3})(0.1602 \times 10^{-18}\,C)(8.00 + 0.045)\frac{m^2}{V \cdot s}$$

$$= \underline{\underline{1.74 \times 10^4\,\Omega^{-1} \cdot m^{-1}}}$$

11.48 What fraction of the conductivity calculated in Problem 11.47 is contributed by (**a**) electrons and (**b**) electron holes?

11.48

(a) $$\frac{\sigma_e}{\sigma} = \frac{\mu_e}{\mu_e + \mu_h} = \frac{8.00}{8.00 + 0.045} = \underline{\underline{0.994}}$$

(b) $$\frac{\sigma_h}{\sigma} = \frac{\sigma_h}{\sigma_e + \sigma_h} = \frac{0.045}{8.00 + 0.045} = \underline{\underline{0.006}}$$

Section 11.6 - Composites

PP 11.14 In Sample Problem 11.14, we calculate the electrical conductivity of an Al/Al_2O_3 composite parallel to the reinforcing fibers. Calculate the conductivity of this composite perpendicular to the reinforcing fibers.

PP 11.14 Using Equation 10.19:

$$\sigma_c = \frac{\sigma_m \sigma_f}{v_m \sigma_f + v_f \sigma_m}$$

$$= \frac{(35.36 \times 10^6\ \Omega^{-1} \cdot m^{-1})(10^{-11}\ \Omega^{-1} \cdot m^{-1})}{0.5(35.36 \times 10^6 + 10^{-11})\Omega^{-1} \cdot m^{-1}}$$

$$= \underline{\underline{2.0 \times 10^{-11}\ \Omega^{-1} \cdot m^{-1}}}$$

11.49 Calculate the conductivity at 20°C (a) parallel and (b) perpendicular to the W filaments in the Cu–matrix composite in Table 10.12.

11.49 (a) Taking resistivities for the 50 vol. % composite from Table 11.2 and using Equation 10.8:

$$\sigma_c = v_m \sigma_m + v_f \sigma_f$$

$$= (0.5)\left(\frac{1}{\rho_{Cu}}\right) + (0.5)\left(\frac{1}{\rho_W}\right)$$

$$= (0.5)\left(\frac{1}{17.24 \times 10^{-9}\ \Omega \cdot m}\right) + (0.5)\left(\frac{1}{55.1 \times 10^{-9}\ \Omega \cdot m}\right)$$

$$= \underline{\underline{38.1 \times 10^6\ \Omega^{-1} \cdot m^{-1}}}$$

(b) Using Equation 10.19:

$$\sigma_c = \frac{\sigma_m \sigma_f}{v_m \sigma_f + v_f \sigma_m}$$

$$= \frac{(1/\rho_{Cu})(1/\rho_W)}{(0.5)(1/\rho_W)+(0.5)(1/\rho_{Cu})}$$

$$= \frac{(17.24\times10^{-9}\,\Omega\cdot m)^{-1}(55.1\times10^{-9}\,\Omega\cdot m)^{-1}}{(0.5)(55.1\times10^{-9}\,\Omega\cdot m)^{-1}+(0.5)(17.24\times10^{-9}\,\Omega\cdot m)^{-1}}$$

$$= \underline{\underline{27.6\times10^{6}\,\Omega^{-1}\cdot m^{-1}}}$$

11.50 Using the form of Equation 10.20 as a guide, estimate the electrical conductivity of the dispersion-strengthened aluminum in Table 10.12. (Assume the exponent, n, in Equation 10.20 to be $\frac{1}{2}$.)

11.50

Using Equation 10.20:

$$\sigma_c^{1/2} = v_\ell \sigma_\ell^{1/2} + v_h \sigma_h^{1/2}$$

$$= (0.1)(10^{-11}\,\Omega^{-1}\cdot m^{-1})^{1/2} + (0.9)(35.36\times10^{6}\,\Omega^{-1}\cdot m^{-1})^{1/2}$$

$$= 5.36\times10^{3}\,\Omega^{-1/2}\cdot m^{-1/2}$$

or

$$\sigma_c = \underline{\underline{28.7\times10^{6}\,\Omega^{-1}\cdot m^{-1}}}$$

11.51 Calculate the conductivity at 20°C for the composite in Table 10.12 in which W particles are dispersed in a copper matrix. (Use the same assumptions as in Problem 11.50.)

11.51

Using Equation 10.20:

$$\sigma_c^{1/2} = v_\ell \sigma_\ell^{1/2} + v_h \sigma_h^{1/2}$$

$$= (0.5)(1/[55.1\times10^{-9}\,\Omega\cdot m])^{1/2} + (0.5)(1/[17.24\times10^{-9}\,\Omega\cdot m])^{1/2}$$

$$= (0.5)(4.26\times10^{3}\,\Omega^{-1/2}\cdot m^{-1/2}) + (0.5)(7.62\times10^{3}\,\Omega^{-1/2}\cdot m^{-1/2})$$

$$= 5.94\times10^{3}\,\Omega^{-1/2}\cdot m^{-1/2}$$

or

$$\sigma_c = \underline{\underline{3.53\times10^{7}\,\Omega^{-1}\cdot m^{-1}}}$$

11.52 Plot the conductivity at 20°C of a series of composites composed of W filaments in a Cu matrix. Show the extreme cases of conductivity **(a)** parallel and **(b)** perpendicular to the filaments. As in Figure 10-13, allow the volume fraction of filaments to vary from 0 to 1.0.

11.52

Using Equations 10.8 and 10.19 along with the data from Table 11.2, we have:

$$\sigma_W = \frac{1}{\rho_W} = \frac{1}{55.1\times10^{-9}\,\Omega\cdot m} = 1.81\times10^{7}\,\Omega^{-1}\cdot m^{-1} \quad \text{(filaments)}$$

$$\sigma_{Cu} = \frac{1}{\rho_{Cu}} = \frac{1}{17.24\times10^{-9}\,\Omega\cdot m} = 5.80\times10^{7}\,\Omega^{-1}\cdot m^{-1} \quad \text{(matrix)}$$

v_f	$\sigma_c(\parallel) = v_m\sigma_m + v_f\sigma_f$	$\sigma_c(\perp) = \dfrac{\sigma_m\sigma_f}{v_m\sigma_f + v_f\sigma_m}$
0	$5.80\times10^{7}\,\Omega^{-1}\cdot m^{-1}$	$5.80\times10^{7}\,\Omega^{-1}\cdot m^{-1}$
0.1	5.40×10^{7} "	4.75×10^{7} "
0.3	4.60×10^{7} "	3.49×10^{7} "
0.5	3.81×10^{7} "	2.76×10^{7} "
0.7	3.01×10^{7} "	2.28×10^{7} "
0.9	2.21×10^{7} "	1.94×10^{7} "
1.0	1.81×10^{7} "	1.81×10^{7} "

Giving the plot:

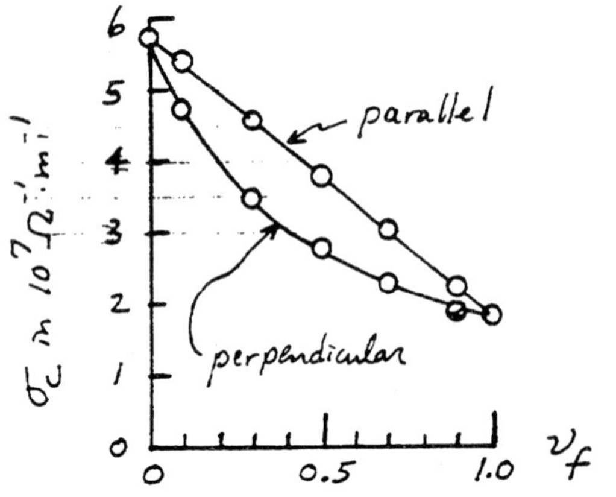

Note: This is an unusual case compared to plots of modulus in that the fiber has a lower conductivity than the matrix.

Section 12.1 - Intrinsic, Elemental Semiconductors

PP 12.1 In Sample Problem 12.1, we calculate the density of conduction electrons in silicon. The result is in agreement with data in Tables 11.5 and 12.5. In Problem 11.43, the fraction of germanium atoms contributing conduction electrons at room temperature was calculated. Make a similar calculation for germanium at 150°C. (Ignore the effect of thermal expansion of germanium.)

PP 12.1 The atomic density was calculated in Problem 11.43 as

$$\rho_{atomic} = 44.1 \times 10^{27} \text{ atoms}/m^3$$

As in Sample Problem 12.2, we can calculate the number of conduction electrons at an elevated temperature using Equation 12.1 and the band gap, 0.66 eV.

$$n \propto e^{-E_g/2kT}$$

or
$$\frac{n_{150°C}}{n_{27°C}} = \frac{e^{-(0.66\,eV/2)/k(150+273)K}}{e^{-(0.66\,eV/2)/k(27+273)K}}$$

or
$$n_{150°C} = n_{300K}\, e^{-(0.66\,eV/2k)\left(\frac{1}{423K} - \frac{1}{300K}\right)}$$

$$= (23 \times 10^{18} m^{-3})\, e^{-\left(\frac{0.66\,eV}{2 \times 86.2 \times 10^{-6} eV/K}\right)\left(\frac{1}{423K} - \frac{1}{300K}\right)}$$

$$= 9.4 \times 10^{20} m^{-3}$$

By ignoring thermal expansion, we can use the 300K atomic density giving:

$$\text{fraction} = \frac{9.4 \times 10^{20} m^{-3}}{44.1 \times 10^{27} m^{-3}} = \underline{\underline{2.1 \times 10^{-8}}}$$

(Note: This compares to only 5.2×10^{-10} at 300K, as calculated in Problem 11.43.)

PP 12.2 **(a)** Calculate the conductivity of germanium at 100°C, and **(b)** plot the conductivity over the range of 27 to 200°C as an Arrhenius-type plot similar to Figure 12-3. (See Sample Problem 12.2.)

PP 12.2 (a) $\sigma = \sigma_0 e^{-E_g/kT}$

$$= (7.11 \times 10^5 \Omega^{-1} \cdot m^{-1})\, e^{-(0.66\,eV)/2(86.2 \times 10^{-6} eV/K)(373K)}$$

$$= \underline{\underline{24.8\ \Omega^{-1} \cdot m^{-1}}}$$

(b)

$T(°C)$	$T(K)$	$1/T\ (K^{-1})$	$\sigma\ (\Omega^{-1}\cdot m^{-1})$	$\ln\sigma\ (\Omega^{-1}\cdot m^{-1})$
27	300	3.33×10^{-3}	2.04	0.7129
100	373	2.68×10^{-3}	24.8	3.211
200	473	2.11×10^{-3}	217	5.380

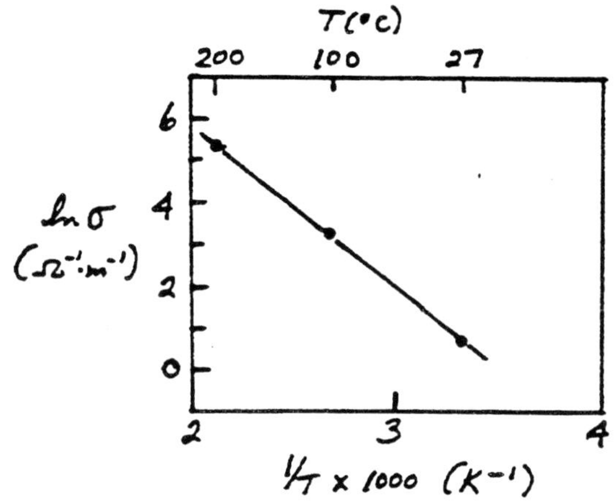

PP 12.3 In characterizing a semiconductor in Sample Problem 12.3, we calculate its band gap. Using that result, calculate its conductivity at 50°C.

PP 12.3

$$\sigma_0 = \sigma e^{+E_g/2kT}$$

$$= (250\ \Omega^{-1}\cdot m^{-1})\, e^{+(0.349\,eV)/2(86.2\times10^{-6}\,eV/K)(293\,K)}$$

$$= 2.50\times10^{5}\ \Omega^{-1}\cdot m^{-1}$$

$$\sigma_{50°C} = \sigma_0\, e^{-E_g/2k(323\,K)}$$

$$= (2.50\times10^{5}\ \Omega^{-1}\cdot m^{-1})\, e^{-(0.349\,eV)/2(86.2\times10^{-6}\,eV/K)(323\,K)}$$

$$= \underline{\underline{474\ \Omega^{-1}\cdot m^{-1}}}$$

12.1 In a 5-cm-diameter × 0.5-mm-thick wafer of pure silicon at room temperature, **(a)** how many conduction electrons would be present, and **(b)** how many electron holes would be present?

12.1 (a) From Table 11.5 (or 12.1),

$$n_e = 14\times10^{15}\ m^{-3}$$

$$\text{no. electrons} = n_e \times V = (14\times10^{15}\ m^{-3})(5\times10^{-4}\ m)\left[\pi\left(\frac{5\times10^{-2}\ m}{2}\right)^2\right]$$

$$= \underline{\underline{1.37\times10^{10}}}$$

(b) As $n_h = n_e$,

$$\text{no. electron holes} = \underline{\underline{1.37\times10^{10}}}$$

12.2 In a 5-cm-diameter × 0.5-mm-thick wafer of pure germanium at room temperature, (a) how many conduction electrons would be present, and (b) how many electron holes would be present?

12.2 Table 11.5 (or 12.1) indicates that

$$n_e = n_h = 23 \times 10^{18}\ m^{-3}$$

Therefore, no. electrons = no. electron holes =

$$(23 \times 10^{18}\ m^{-3})(5 \times 10^{-4}\ m)\left[\pi\left(\frac{5 \times 10^{-2}\ m}{2}\right)^2\right]$$

$$= 2.26 \times 10^{13}$$

12.3 Using data from Table 11.5, make a plot similar to Figure 12-3 showing both intrinsic silicon and intrinsic germanium over the temperature range of 27 to 200°C.

12.3 The appropriate calculations for Ge were made in Practice Problem 12.2. To summarize:

$T(°C)$	$T(K)$	$1/T\ (K^{-1})$	$\sigma_{Ge}\ (\Omega^{-1} \cdot m^{-1})$	$\ln \sigma_{Ge}\ (\Omega^{-1} \cdot m^{-1})$
27	300	3.33×10^{-3}	2.04	0.7129
100	373	2.68×10^{-3}	24.8	3.211
200	473	2.11×10^{-3}	217	5.380

Similarly, for Si at 300 K:

$$\sigma_{300K} = nq(\mu_e + \mu_h)$$

$$= (14 \times 10^{15}\ m^{-3})(0.16 \times 10^{-18}\ C)(0.140 + 0.038)\ \frac{m^2}{V \cdot s}$$

$$= 3.99 \times 10^{-4}\ \Omega^{-1} \cdot m^{-1}$$

$$\sigma_0 = \sigma e^{+E_g/2kT}$$

$$= (3.99 \times 10^{-4}\ \Omega^{-1} \cdot m^{-1})e^{+(1.107\ eV)/2(86.2 \times 10^{-6}\ eV/K)(300K)}$$

$$= 7.87 \times 10^5\ \Omega^{-1} \cdot m^{-1}$$

$$\sigma_{200°C} = \sigma_0 e^{-E_g/2k(473\ K)}$$

$$= (7.87 \times 10^5\ \Omega^{-1} \cdot m^{-1})e^{-(1.107\ eV)/2(86.2 \times 10^{-6}\ eV/K)(473\ K)}$$

$$= 1.00\ \Omega^{-1} \cdot m^{-1}$$

For Si, then:

$T(°C)$	$T(K)$	$1/T\ (K^{-1})$	$\sigma_{Si}\ (\Omega^{-1} \cdot m^{-1})$	$\ln \sigma_{Si}\ (\Omega^{-1} \cdot m^{-1})$
27	300	3.33×10^{-3}	3.99×10^{-4}	−7.827
200	473	2.11×10^{-3}	1.00	0

The resulting plot:

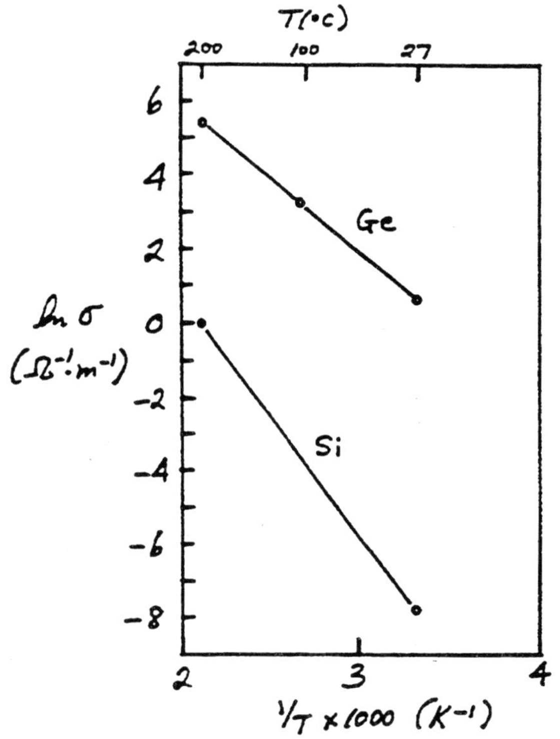

12.4 Superimpose a plot of the intrinsic conductivity of GaAs on the result of Problem 12.3.

12.4 For GaAs at 300K:

$$\sigma_{300K} = nq(\mu_e + \mu_h)$$

$$= (1.4\times10^{12}\,m^{-3})(0.16\times10^{-18}\,C)(0.720+0.020)\frac{m^2}{V\cdot s}$$

$$= 1.66\times10^{-7}\,\Omega^{-1}\cdot m^{-1}$$

$$\sigma_0 = \sigma e^{+E_g/2kT}$$

$$= (1.66\times10^{-7}\,\Omega^{-1}\cdot m^{-1})e^{+(1.47\,eV)/2(86.2\times10^{-6}\,eV/K)(300K)}$$

$$= 3.66\times10^{5}\,\Omega^{-1}\cdot m^{-1}$$

$$\sigma_{200^\circ C} = \sigma_0 e^{-E_g/2k(473K)}$$

$$= (3.66\times10^{5}\,\Omega^{-1}\cdot m^{-1})e^{-(1.47\,eV)/2(86.2\times10^{-6}\,eV/K)(473K)}$$

$$= 5.42\times10^{-3}\,\Omega^{-1}\cdot m^{-1}$$

In summary:

$T(°C)$	$T(K)$	$1/T\ (K^{-1})$	$\sigma_{GaAs}\ (\Omega^{-1}\cdot m^{-1})$	$\ln \sigma_{GaAs}\ (\Omega^{-1}\cdot m^{-1})$
27	300	3.33×10^{-3}	1.66×10^{-7}	-15.611
200	473	2.11×10^{-3}	5.42×10^{-3}	-5.218

Giving the resulting plot:

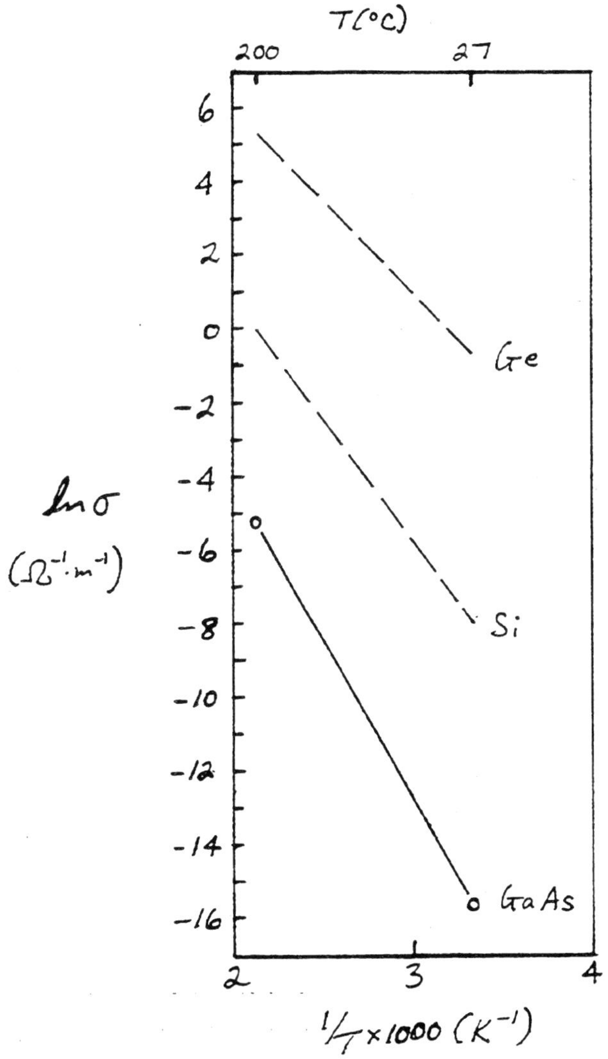

12.5 Starting from an ambient temperature of 300 K, what temperature increase is necessary to double the conductivity of pure silicon?

12.5

$$\sigma = \sigma_0 e^{-E_g/2kT}$$

$$\frac{\sigma_T}{\sigma_{300}} = 2 = \frac{\sigma_0 e^{-E_g/2kT}}{\sigma_0 e^{-E_g/2k[300\,K]}} = e^{-\frac{E_g}{2k}\left(\frac{1}{T} - \frac{1}{300\,K}\right)}$$

or

$$\ln 2 = -\frac{E_g}{2k}\left(\frac{1}{T} - \frac{1}{300\,K}\right)$$

or

$$\frac{1}{T} - \frac{1}{300\,K} = -\frac{\ln 2}{(E_g/2k)}$$

or

$$1/T = 1/300\,K - \frac{2k\ln 2}{E_g}$$

giving

$$T = \left[\frac{1}{300\,K} - \frac{2k\ln 2}{E_g}\right]^{-1}$$

$$= \left[\frac{1}{300\,K} - \frac{2(86.2\times10^{-6}\ eV/K)\ln 2}{1.107\ eV}\right]^{-1}$$

$$= 310\,K$$

$$\therefore \Delta T = 310\,K - 300\,K = \underline{\underline{10\,K}}$$

12.6 Starting from an ambient temperature of 300 K, what temperature increase is necessary to double the conductivity of pure germanium?

12.6

As in Problem 12.5,

$$T = \left[\frac{1}{300\,K} - \frac{2k\ln 2}{E_g}\right]^{-1}$$

$$= \left[\frac{1}{300\,K} - \frac{2(86.2\times10^{-6}\ eV/K)\ln 2}{0.66\ eV}\right]^{-1}$$

$$= 317\,K$$

$$\therefore \Delta T = 317\,K - 300\,K = \underline{\underline{17\,K}}$$

12.7 There is a slight temperature dependence for the band gap of a semiconductor. For silicon, this dependence can be expressed as

$$E_g(T) = 1.152 \text{ eV} - \frac{AT^2}{T + B}$$

where $A = 4.73 \times 10^{-4}$ eV/K, $B = 636$ K, and T is in Kelvin. What is the percentage error in taking the band gap at 200°C to be the same as that at room temperature?

12.7

$$E_g(473\text{ K}) = 1.152\text{ eV} - \frac{(4.73\times10^{-4}\text{ eV/K})(473\text{ K})^2}{473\text{ K} + 636\text{ K}}$$

$$= 1.057\text{ eV}$$

% error comes from comparison with Table 12.1 value:

$$\%\text{ error} = \frac{1.107\text{ eV} - 1.057\text{ eV}}{1.057\text{ eV}} \times 100\% = \underline{\underline{4.77\%}}$$

12.8 Repeat Problem 12.7 for GaAs, in which

$$E_g(T) = 1.567 \text{ eV} - \frac{AT^2}{T + B}$$

where $A = 5.405 \times 10^{-4}$ eV/K and $B = 204$ K.

12.8

$$E_g(473\text{ K}) = 1.567\text{ eV} - \frac{(5.405\times10^{-4}\text{ eV/K})(473\text{ K})^2}{473\text{ K} + 204\text{ K}}$$

$$= 1.388\text{ eV}$$

$$\%\text{ error} = \frac{1.47\text{ eV} - 1.388\text{ eV}}{1.388\text{ eV}} \times 100\% = \underline{\underline{5.88\%}}$$

Section 12.2 - Extrinsic, Elemental Semiconductors

PP 12.4 A 100-ppb doping of Al, in Sample Problem 12.4, is found to represent a 10.4×10^{-6} mol % addition. What is the atomic density of Al atoms in this extrinsic semiconductor? (Compare your answer with the maximum solid solubility level given in Table 12.2.)

PP 12.4 $\rho_{Si} = 2.33 \frac{g}{cm^3} \times 10^6 \frac{cm^3}{m^3} \times \frac{0.6023 \times 10^{24}\text{ atoms}}{28.09\text{ g}}$

$= 50.0 \times 10^{27}$ atoms/m³

Using the result of Sample Problem 12.4,

$\rho_{Al} = 10.4 \times 10^{-8} \times 50.0 \times 10^{27}$ atoms/m³

$= \underline{\underline{5.20 \times 10^{21}\text{ atoms/m}^3}}$

($\ll 20 \times 10^{24}$ atoms/m³ in Table 12.2)

PP 12.5 In Sample Problem 12.5, we calculate the probability of an electron being thermally promoted to the conduction band in a P-doped silicon at 25°C. What is the probability at 50°C?

PP 12.5 $f(E) = \frac{1}{e^{(E-E_F)/kT} + 1}$

$= \frac{1}{e^{(0.4535\,eV)/(86.2 \times 10^{-6}\,eV/K)(323\,K)} + 1}$

$= \underline{\underline{8.44 \times 10^{-8}}}$

PP 12.6 The conductivity of an n-type semiconductor at 25°C and 30°C can be found in Sample Problem 12.6. **(a)** Make a similar calculation at 50°C and **(b)** plot the conductivity over the range of 25 to 50°C as an Arrhenius-type plot similar to Figure 12-8. **(c)** What important assumption underlies the validity of your results in parts (a) and (b)?

PP 12.6 (a) $\sigma = \sigma_0 e^{-(E_g - E_d)/kT}$

$\sigma_{50^\circ C} = (4.91 \times 10^3\,\Omega^{-1} \cdot m^{-1}) e^{-(0.1\,eV)/(86.2 \times 10^{-6}\,eV/K)(323\,K)}$

$= \underline{\underline{135\,\Omega^{-1} \cdot m^{-1}}}$

(b)

T(°C)	T(K)	1/T (K⁻¹)	σ (Ω⁻¹·m⁻¹)	ln σ (Ω⁻¹·m⁻¹)
25	298	3.36×10^{-3}	100	4.605
30	303	3.30×10^{-3}	107	4.673
50	323	3.10×10^{-3}	135	4.905

Resulting plot:

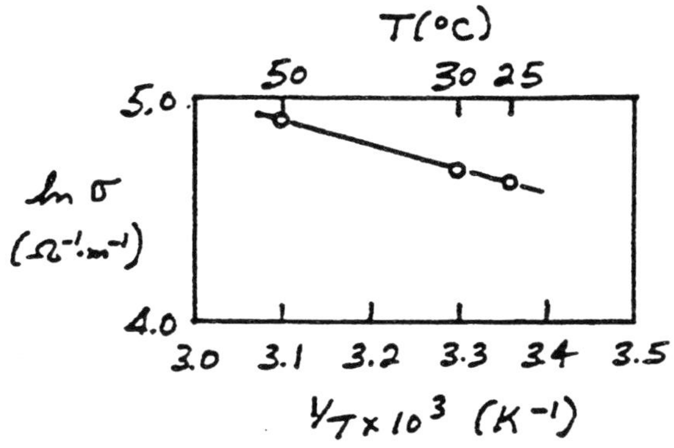

(c) The extrinsic behavior extends to 50 °C. Otherwise the slope of the plot in (b) would not be constant.

PP 12.7 In Sample Problems 12.7–12.9, detailed calculations about a P-doped Ge semiconductor are made. Assume now that the upper temperature limit of extrinsic behavior for an aluminum-doped germanium is also 100°C with an extrinsic conductivity at that point again being 60 $\Omega^{-1} \cdot m^{-1}$. Calculate **(a)** the level of aluminum doping in parts per billion (ppb) by weight, **(b)** the upper temperature for the saturation range, **(c)** the extrinsic conductivity at 300 K; **(d)** make a plot of the results similar to that in Sample Problem 12.9 and Figure 12-13.

PP 12.7

(a) $$n_h = \frac{60\ \Omega^{-1}\cdot m^{-1}}{(0.16\times10^{-18}\ C)(0.190\ m^2 V^{-1} s^{-1})} = 1.97\times10^{21}\ m^{-3}$$

$$[Al] = 1.97\times10^{21}\ m^{-3} \times \frac{26.98\ g\ Al}{0.6023\times10^{24}\ atoms\ Al} \times \frac{1\ cm^3\ Ge}{5.32\ g\ Ge} \times \frac{1\ m^3}{10^6\ cm^3}$$

$$= 16.6\times10^{-9}\ g\ Al/g\ Ge = \underline{\underline{16.6\ ppb}}$$

(b) This calculation is identical to that for Sample Problem 12.8(a). Again, $\underline{\underline{T = 135\ °C}}$.

(c) Table 12.3 gives $E_a = 0.01\ eV$ for Al in Ge, or

$$\sigma_0 = \sigma e^{+E_a/kT}$$

$$= (60\ \Omega^{-1} m^{-1})\, e^{+(0.01\ eV)/(86.2\times10^{-6}\ eV/K)(373\ K)}$$

$$= 81.89\ \Omega^{-1} m^{-1}$$

At 300 K,

$$\sigma = \sigma_0 e^{-E_a/kT}$$

$$= (81.89\ \Omega^{-1}\cdot m^{-1})\, e^{-(0.01\ eV)/(86.2\times10^{-6}\ eV/K)(300\ K)}$$

$$= 55.6\ \Omega^{-1}\cdot m^{-1}$$

(d) The key data are:

Conduction type	T(°C)	T(K)	1/T(K⁻¹)	$\sigma(\Omega^{-1}\cdot m^{-1})$	$\ln\sigma(\Omega^{-1}\cdot m^{-1})$
extrinsic	100	373	2.68×10^{-3}	60	4.09
extrinsic	27	300	3.33×10^{-3}	55.6	4.02
intrinsic	135	408	2.45×10^{-3}	60	4.09
intrinsic	27	300	3.33×10^{-3}	2.04	0.713

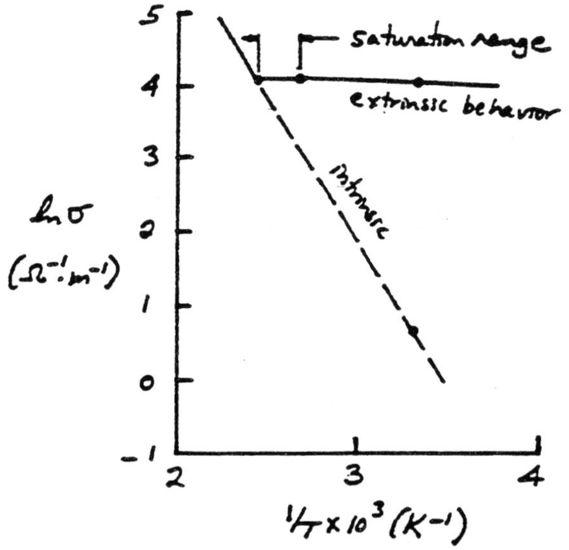

PP 12.8 As in Sample Problem 12.10, calculate **(a)** the photon wavelength (in nm) necessary to promote an electron to the conduction band in intrinsic germanium and **(b)** the wavelength necessary to promote a donor electron to the conduction band in arsenic-doped germanium.

PP 12.8 (a) $\lambda = \frac{hc}{E_g}$

Using Table 12.1,

$$\lambda = \frac{(0.663\times10^{-33}\ J\cdot s)(3.00\times10^{8}\ m/s)}{(0.66\ eV)(0.16\times10^{-18}\ J/eV)} \times 10^{9}\ \frac{nm}{m}$$

$$= 1880\ nm$$

(b) $\lambda = \frac{hc}{E_g - E_d}$

Using Table 12.3,

$$\lambda = \frac{(0.663\times10^{-33}\ \text{J}\cdot\text{s})(3.00\times10^{8}\ \text{m/s})}{(0.013\ \text{eV})(0.16\times10^{-18}\ \text{J/eV})} \times 10^{9}\ \frac{\text{nm}}{\text{m}}$$

$$= \underline{\underline{95{,}600\ \text{nm}}}$$

12.9 An n-type semiconductor consists of 100 ppb of P doping, by weight, in silicon. What is **(a)** the mole percentage P and **(b)** the atomic density of P atoms? Compare your answer in part (b) with the maximum solid solubility level given in Table 12.2.

12.9

(a) For 100 g of doped silicon, there will be $\frac{100}{10^9} \times 100\ \text{g P} = 1\times10^{-5}\ \text{g}$

$$\text{no. moles P} = \frac{1\times10^{-5}\ \text{g P}}{30.97\ \text{g/mol}} = 3.23\times10^{-7}\ \text{mol}$$

$$\text{no. moles Si} = \frac{(100 - 1\times10^{-5})\ \text{g Si}}{28.09\ \text{g}} = 3.56\ \text{mol}$$

Then,

$$\text{mol.\% P} = \frac{3.23\times10^{-7}\ \text{mol}}{(3.56 + 3.23\times10^{-7})\ \text{mol}} \times 100 = \underline{\underline{9.07\times10^{-6}\ \text{mol.\%}}}$$

(b) $$\rho_{Si} = 2.33\ \frac{\text{g}}{\text{cm}^3} \times 10^{6}\ \frac{\text{cm}^3}{\text{m}^3} \times \frac{0.6023\times10^{24}\ \text{atoms}}{28.09\ \text{g}}$$

$$= 50.0\times10^{27}\ \text{atoms/m}^3$$

Using the result of part (a),

$$\rho_P = 9.07\times10^{-8} \times 50.0\times10^{27}\ \text{atoms/m}^3$$

$$= \underline{\underline{4.54\times10^{21}\ \text{atoms/m}^3}}$$

($\ll 1000\times10^{24}\ \text{atoms/m}^3$ in Table 12.2)

12.10 An As-doped silicon has a conductivity of 2.00×10^{-2} $\Omega^{-1} \cdot m^{-1}$ at room temperature. (a) What is the predominant charge carrier in this material? (b) What is the density of these charge carriers? (c) What is the drift velocity of these carriers under an electrical field strength of 200 V/m? (The μ_e and μ_h values given in Table 11.5 also apply for an extrinsic material with low impurity levels.)

12.10

(a) Arsenic (As) is from Group VA of the periodic table. Therefore, it yields an n-type material with the electron as the predominant charge carrier.

(b) Conductivity follows Equation 12.4:

$$\sigma = nq\mu_e \quad \text{or} \quad n = \frac{\sigma}{q\mu_e}$$

Table 11.5 (or 12.1) gives $\mu_e = 0.140\ m^2/(V\cdot s)$ or

$$n = \frac{2.00\times 10^{-2}\ \Omega^{-1}\cdot m^{-1}}{(0.1602\times 10^{-18}\ C)(0.140\ m^2/[V\cdot s])} = 8.92\times 10^{17}\ m^{-3}$$

(c) From Equation 11.5,

$$\bar{v} = \mu E = 0.140\ \frac{m^2}{V\cdot s} \times 200\ \frac{V}{m} = 28\ m/s$$

12.11 Repeat Problem 12.10 for the case of a Ga-doped silicon with a conductivity of 2.00×10^{-2} $\Omega^{-1} \cdot m^{-1}$ at room temperature.

12.11

(a) Gallium (Ga) is from Group IIIA giving a p-type material with the electron hole as the predominant charge carrier.

(b) Conductivity follows Equation 12.6:

$$\sigma = nq\mu_h \quad \text{or} \quad n = \frac{\sigma}{q\mu_h}$$

Table 11.5 (or 12.1) gives $\mu_h = 0.038\ m^2/(V\cdot s)$ or

$$n = \frac{2.00\times 10^{-2}\ \Omega^{-1}\cdot m^{-1}}{(0.1602\times 10^{-18}\ C)(0.038\ m^2/[V\cdot s])} = 3.29\times 10^{18}\ m^{-3}$$

(c) From Equation 11.5,

$$\bar{v} = \mu E = 0.038\ \frac{m^2}{V\cdot s} \times 200\ \frac{V}{m} = 7.6\ m/s$$

12.12 Calculate the conductivity for the saturation range of silicon doped with 10-ppb boron.

12.12

10 ppb B doping corresponds to $\frac{10 g\ B}{10^9 g\ Si}$

which must equal to: (using data from Appendix 1)

$$\frac{n \text{ atoms B}}{m^3} \times \frac{10.81\ g\ B}{0.6023 \times 10^{24} \text{ atoms B}} \times \frac{1\ cm^3\ Si}{2.33 g\ Si} \times \frac{1\ m^3}{10^6\ cm^3}$$

or $n = 10^{-8} \times \frac{0.6023 \times 10^{24}}{10.81} \times 2.33 \times 10^6 = 1.30 \times 10^{21}$

This density of B atoms/m^3 is also the density of electron holes in this p-type semiconductor, at saturation.

Equation 12.6 gives (using data from Table 12.1)

$$\sigma = n q \mu_h$$

$$= (1.30 \times 10^{21} m^{-3})(0.16 \times 10^{-18} C)(0.038 \frac{m^2}{V \cdot s})$$

$$= \underline{\underline{7.89\ \Omega^{-1} \cdot m^{-1}}}$$

12.13 Calculate the conductivity for the saturation range of silicon doped with 20-ppb boron. (Note Problem 12.12)

12.13

By inspection of the calculations for Problem 12.12, doubling the doping concentration will double the carrier concentration, thereby doubling the conductivity:

$$\sigma = 2 \times (7.89\ \Omega^{-1} \cdot m^{-1}) = \underline{\underline{15.8\ \Omega^{-1} \cdot m^{-1}}}$$

12.14 Calculate the conductivity for the exhaustion range of silicon doped with 10-ppb antimony.

12.14 As calculated in Problem 12.12;

$$\frac{10\,g\,Sb}{10^9\,g\,Si} = n\,\frac{\text{atoms Sb}}{m^3} \times \frac{121.75\,g\,Sb}{0.6023\times10^{24}\,\text{atoms Sb}} \times \frac{1\,cm^3\,Si}{2.33\,g\,Si} \times \frac{1\,m^3}{10^6\,cm^3}$$

or,

$$n = 1.15\times10^{20}$$

The density of Sb atoms/m^3 is also the density of electrons in this n-type semiconductor, at exhaustion.

Equation 12.4 gives (using data from Table 12.1)

$$\sigma = nq\mu_e$$

$$= (1.15\times10^{20}\,m^{-3})(0.16\times10^{-18}\,C)(0.140\,\frac{m^2}{V\cdot s})$$

$$= \underline{\underline{2.59\,\Omega^{-1}\cdot m^{-1}}}$$

12.15 Calculate the upper temperature limit of the saturation range for silicon doped with 10-ppb boron. (Note Problem 12.12)

12.15 As indicated by Figure 12-13, the upper limit for the saturation range will be the temperature at which the conductivity of intrinsic Si will be $7.89\,\Omega^{-1}\cdot m^{-1}$ (the value calculated in Problem 12.12).

The data needed for calculating the conductivity of intrinsic Si were generated for Problem 12.3.

$$\sigma = \sigma_0 e^{-E_g/2kT}$$

$$7.89\,\Omega^{-1}\cdot m^{-1} = (7.87\times10^5\,\Omega^{-1}\cdot m^{-1}) \times e^{-(1.107\,eV)/2(86.2\times10^{-6}\,eV/K)T}$$

or

$$T = -\frac{1.107\,eV}{2(86.2\times10^{-6}\,eV/K)\ln(7.89/7.87\times10^5)}$$

$$= 558\,K = \underline{\underline{285\,^\circ C}}$$

12.16 Calculate the upper temperature limit of the exhaustion range for silicon doped with 10-ppb antimony. (Note Problem 12.14)

12.16

As indicated by Figure 12-9, the upper limit for the exhaustion range will be the temperature at which the conductivity of intrinsic Si will be $2.59\ \Omega^{-1}\cdot m^{-1}$ (as calculated in Problem 12.14).

Again, using data from Problem 12.3,

$$\sigma = \sigma_0 e^{-E_g/2kT}$$

$$2.59\ \Omega^{-1}\cdot m^{-1} = (7.87\times10^5\ \Omega^{-1}\cdot m^{-1}) \times e^{-(1.107\ eV)/2(86.2\times10^{-6}\ eV/K)T}$$

or

$$T = -\frac{1.107\ eV}{2(86.2\times10^{-6}\ eV/K)\ln(2.59/7.87\times10^5)}$$

$$= 509\ K = \underline{\underline{236\ °C}}$$

12.17 If the lower temperature limit of the saturation range for silicon doped with 10-ppb boron is 110°C, calculate the extrinsic conductivity at 300 K. (Note Problems 12.12 and 12.15)

12.17

From Problem 12.12, we see that the conductivity for the saturation range is $7.89\ \Omega^{-1}\cdot m^{-1}$

In the extrinsic region,

$$\sigma = \sigma_0 e^{-E_a/kT}$$

Using given data and Table 12.3:

$$\frac{\sigma_{300\ K}}{\sigma_{110°C}} = \frac{\sigma_0 e^{-E_a/k(300\ K)}}{\sigma_0 e^{-E_a/k[(273+110)K]}}$$

or

$$\sigma_{300K} = \sigma_{110°C}\ e^{-(E_a/k)\left(\frac{1}{300K} - \frac{1}{383\ K}\right)}$$

$$= 7.89\ \Omega^{-1}\cdot m^{-1}\ e^{-\left[\frac{0.045\ eV}{86.2\times10^{-6}\ eV/K}\right]\left(\frac{1}{300K} - \frac{1}{383K}\right)}$$

$$= \underline{\underline{5.41\ \Omega^{-1}\cdot m^{-1}}}$$

12.18 Plot the conductivity of the B-doped Si of Problem 12.17 in a manner similar to Figure 12-13.

12.18

Using the results of Problems 12.3, 12.12, 12.15 along with Problem 12.17 gives:

T(°C)	T(K)	$1/T$(K^{-1})	σ ($\Omega^{-1}\cdot m^{-1}$)	ln σ	
200	473	2.11×10^{-3}	1.00	0	intrinsic
27	300	3.33×10^{-3}	3.99×10^{-4}	−7.827	↑
285	558	1.79×10^{-3}	7.89	2.066	saturation
110	383	2.61×10^{-3}	7.89	2.066	↓
	300	3.33×10^{-3}	5.41	1.688	extrinsic

And a plot:

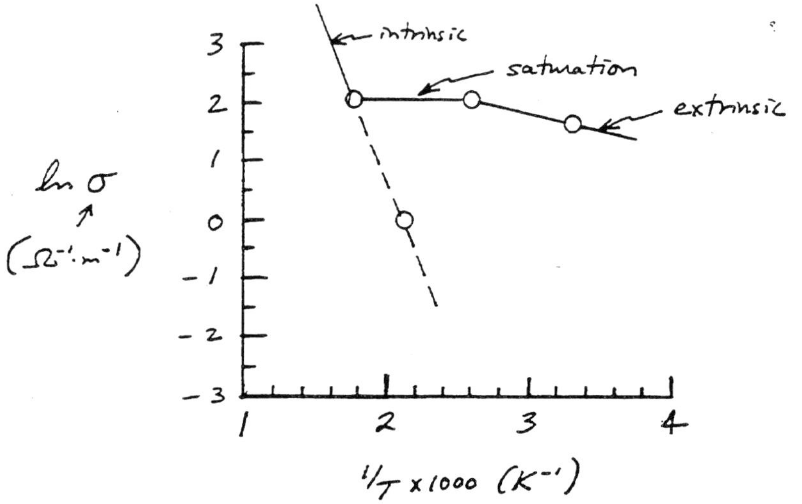

12.19 If the lower temperature limit of the exhaustion range for silicon doped with 10-ppb antimony is 80°C, calculate the extrinsic conductivity at 300 K. (Note Problems 12.14 and 12.16)

12.19

From Problem 12.14, we see that the conductivity for the exhaustion range is $2.59\ \Omega^{-1}\cdot m^{-1}$

In the extrinsic region,

$$\sigma = \sigma_0 e^{-(E_g - E_d)/kT}$$

Using given data and Table 12.3:

$$\frac{\sigma_{300K}}{\sigma_{80°C}} = \frac{\sigma_0 e^{-E_a/k(300K)}}{\sigma_0 e^{-E_a/k[(273+80)K]}}$$

or

$$\sigma_{300K} = \sigma_{80°C}\, e^{-[(E_g - E_d)/k]\left(\frac{1}{300K} - \frac{1}{353K}\right)}$$

$$= 2.59\ \Omega^{-1}\cdot m^{-1}\, e^{-\left[\frac{0.039\ eV}{86.2\times10^{-6}\ eV/K}\right]\left(\frac{1}{300K} - \frac{1}{353K}\right)}$$

$$= \underline{\underline{2.07\ \Omega^{-1}\cdot m^{-1}}}$$

12.20 Plot the conductivity of the Sb-doped Si of Problem 12.19 in a manner similar to Figure 12-9.

12.20

Using the results of Problems 12.3, 12.14, 12.16 along with Problem 12.19 gives:

$T(°C)$	$T(K)$	$1/T\ (K^{-1})$	$\sigma\ (\Omega^{-1} \cdot m^{-1})$	$\ln \sigma$	
200	473	2.11×10^{-3}	1.00	0	intrinsic ↑
236	509	1.96×10^{-3}	2.59	0.951	exhaustion
80	353	2.83×10^{-3}	2.59	0.951	↓
	300	3.33×10^{-3}	2.07	0.728	extrinsic

And a plot:

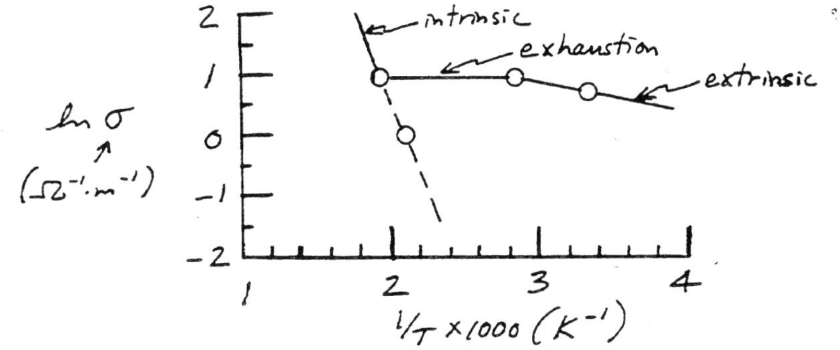

12.21 In designing a solid-state device using B-doped Si, it is important that the conductivity not increase more than 10% (relative to the value at room temperature) during the operating lifetime. For this factor alone, what is the maximum operating temperature to be specified for this design?

12.21

Using Equation 12.7 and Table 12.3, we note that:

$$\sigma = \sigma_0 e^{-E_a/kT}$$

or

$$1.1 = \frac{\sigma_{T_{max}}}{\sigma_{RT=300K}} = \frac{\sigma_0 e^{-E_a/kT_m}}{\sigma_0 e^{-E_a/k(300K)}}$$

or

$$\ln 1.1 = -\frac{E_a}{k}\left(\frac{1}{T_m} - \frac{1}{300K}\right)$$

$$= -\frac{0.045\,eV}{86.2\times10^{-6}\,eV/K}\left(\frac{1}{T_m} - \frac{1}{300K}\right)$$

giving

$$T_m = 317K\ (=44°C)$$

12.22 In designing a solid-state device using As-doped Si, it is important that the conductivity not increase more than 10% (relative to the value at room temperature) during the operating lifetime. For this factor alone, what is the maximum operating temperature to be specified for this design?

12.22 Using Equation 12.5 and Table 12.3:

$$\sigma = \sigma_0 e^{-(E_g - E_d)/kT}$$

or

$$1.1 = \frac{\sigma_{T_{max}}}{\sigma_{RT=300K}} = \frac{\sigma_0 e^{-(E_g-E_d)/kT_m}}{\sigma_0 e^{-(E_g-E_d)/k(300K)}}$$

or

$$\ln 1.1 = -\frac{0.049 eV}{86.2\times10^{-6} eV/K}\left(\frac{1}{T_m} - \frac{1}{300K}\right)$$

giving

$$T_m = 316 K (= 43°C)$$

• 12.23 (a) It was pointed out in Section 11.3 that the temperature sensitivity of conductivity in semiconductors makes them superior to traditional thermocouples for certain high-precision temperature measurements. Such devices are referred to as thermistors. As a simple example, consider a wire 0.5 mm in diameter × 10 mm long made of intrinsic silicon. If the resistance of the wire can be measured to within 10^{-3} Ω, calculate the temperature sensitivity of this device at 300 K. (**HINT:** The very small differences here may make you want to develop an expression for $d\sigma/dT$.) (**b**) Repeat the calculation for an intrinsic germanium wire of the same dimensions. (**c**) For comparison with the temperature sensitivity of a metallic conductor, repeat the calculation for a copper (annealed standard) wire of the same dimensions. (The necessary data for this case can be found in Table 11.2.)

12.23 (a) $\sigma = \sigma_0 e^{-E_g/2kT}$

$$\frac{d\sigma}{dT} = \sigma_0 e^{-E_g/2kT}\left(+\frac{E_g}{2kT^2}\right) \cong \frac{\Delta\sigma}{\Delta T}$$

or

$$\Delta T = \frac{\Delta\sigma\,(2kT^2/E_g)}{\sigma_0 e^{-E_g/2kT}} = \frac{\Delta\sigma}{\sigma}\left(\frac{2kT^2}{E_g}\right)$$

$\Delta R = 1\times10^{-3}\,\Omega$, or

$$\Delta\rho = \frac{\Delta R\,A}{l} = \frac{(1\times10^{-3}\,\Omega)[\pi\,(0.25\times10^{-3}\,m)^2]}{1\times10^{-2}\,m}$$

$$= 1.96\times10^{-8}\,\Omega\cdot m$$

As $\sigma_{Si, 300K} = 4 \times 10^{-4}\ \Omega^{-1}\cdot m^{-1}$ (from Table 11.1):

$$\rho_{Si, 300K} = 1/\sigma = 2.50 \times 10^{3}\ \Omega\cdot m$$

Therefore,

$$\frac{\Delta\rho}{\rho} = \frac{1.96 \times 10^{-8}\ \Omega\cdot m}{2.5 \times 10^{3}\ \Omega\cdot m} = 7.84 \times 10^{-12}$$

This fractional decrease in ρ with temperature is equal to the fractional increase in σ, i.e.

$$\frac{\Delta\sigma}{\sigma} = 7.84 \times 10^{-12}$$

Then,

$$\Delta T = (7.84 \times 10^{-12}) \times \left(\frac{2 \times 86.2 \times 10^{-6}\ eV/K\ [300\ K]^2}{1.107\ eV}\right)$$

$$= \underline{\underline{1.11 \times 10^{-10}\ K}}$$

(b) For Ge, we again have $\Delta\rho = 1.98 \times 10^{-8}\ \Omega\cdot m$

From Table 11.1, $\sigma_{Ge, 300K} = 2.0\ \Omega^{-1}\cdot m^{-1}$

and $\rho = 1/\sigma = 0.5\ \Omega\cdot m$. Then,

$$\frac{\Delta\rho}{\rho} = \frac{\Delta\sigma}{\sigma} = \frac{1.98 \times 10^{-8}\ \Omega\cdot m}{0.5\ \Omega\cdot m} = 3.96 \times 10^{-8}$$

Finally,

$$\Delta T = (3.96 \times 10^{-8})\left(\frac{2 \times 86.2 \times 10^{-6}\ eV/K\ [300\ K]^2}{0.66\ eV}\right)$$

$$= \underline{\underline{9.31 \times 10^{-7}\ K}}$$

(c) For a metal, Equation 11.9 applies:

$$\rho = \rho_{rt}[1 + \alpha(T - T_{rt})]$$

or $$\frac{\Delta\rho}{\Delta T} = \rho_{rt}\,\alpha$$

Using Table 11.2 for copper data:

$$\frac{\Delta\rho}{\Delta T} = (17.24\times10^{-9}\,\Omega\cdot m)(0.00393\,°C)$$

$$= 6.78\times10^{-11}\,\frac{\Omega\cdot m}{°C}$$

Then,

$$\Delta T = \left(\frac{\Delta T}{\Delta\rho}\right)\Delta\rho$$

$$= \frac{1.96\times10^{-8}\,\Omega\cdot m}{6.78\times10^{-11}\,\Omega\cdot m/°C} = 289\,°C = \underline{\underline{289\,K}}$$

Note: Although metals are made to look like poor choices for temperature measurement by this example, do not forget the important thermocouples in Chapter 11.3.

12.24 An application of semiconductors of great use to materials engineers is the "lithium-drifted silicon," Si(Li), solid-state photon detector. This is the basis for the detection of microstructural-scale elemental distributions as illustrated in Figure 4-65. A characteristic x-ray photon striking the Si(Li) promotes a number of electrons (N) to the conduction band creating a current pulse, where

$$N = \frac{\text{photon energy}}{\text{band gap}}$$

For an Si(Li) detector operating at liquid nitrogen temperature (77 K), the band gap is 3.8 eV. What would be the size of a current pulse (N) created by **(a)** a copper K_α characteristic x-ray photon ($\lambda = 0.1542$ nm) and **(b)** an iron K_α characteristic x-ray photon ($\lambda = 0.1938$ nm)? (By the way, "lithium-drifted" refers to the Li dopant being diffused into the Si under an electrical potential. The result is a highly uniform distribution of the dopant.)

12.24 (a)

$$N = \frac{\text{photon energy}}{\text{band gap}} = \frac{(hc/\lambda)}{E_g} = \frac{(0.663\times10^{-33}\,J\cdot s)(3.00\times10^{8}\,m/s)}{(0.1542\times10^{-9}\,m)(3.8\,eV)(0.16\times10^{-18}\,J/eV)}$$

$$= \underline{\underline{2120}}$$

(b)

$$N = \frac{(0.663\times10^{-33}\,J\cdot s)(3.00\times10^{8}\,m/s)}{(0.1938\times10^{-9}\,m)(3.8\,eV)(0.16\times10^{-18}\,J/eV)}$$

$$= \underline{\underline{1690}}$$

• 12.25 Using the information from Problem 12.24, sketch a "spectrum" produced by chemically analyzing a stainless steel with the use of characteristic x-rays. Assume the yield of x-rays is proportional to the atomic fraction of elements in the sample being bombarded by an electron beam. The spectrum itself consists of sharp spikes of height proportional to the x-ray yield (number of photons). The spikes are located along a "current pulse (N)" axis. For an "18–8" stainless steel (18 wt % Cr, 8 wt % Ni, bal. Fe), the following spikes are observed:

FeK_α ($\lambda = 0.1938$ nm)

FeK_β ($\lambda = 0.1757$ nm)

CrK_α ($\lambda = 0.2291$ nm)

CrK_β ($\lambda = 0.2085$ nm)

NiK_α ($\lambda = 0.1659$ nm)

NiK_β ($\lambda = 0.1500$ nm)

(K_β photon production is less probable than K_α production. Take the height of a K_β spike to be only 10% of that of the K_α spike for the same element.)

12.25 First, we need the atomic fraction of elements:

100 g steel → 18 g Cr + 8 g Ni + (100 − 18 − 8) g Fe

or $\text{g·atoms Cr} = \frac{18\text{ g}}{52.00\text{ g}}\text{ g·atom} = 0.346\text{ g·atom}$

$\text{g·atoms Ni} = \frac{8\text{ g}}{58.71\text{ g}}\text{ g·atom} = 0.136\text{ g·atom}$

$\text{g·atoms Fe} = \frac{74\text{ g}}{55.85\text{ g}}\text{ g·atom} = \frac{1.325\text{ g·atom}}{1.807\text{ g·atom}}$

or at. fraction Cr $= 0.346/1.807 = 0.191$

" " Ni $= 0.136/1.807 = 0.075$

" " Fe $= 1.325/1.807 = 0.733$

Then:

Photon	$E\,(= hc/\lambda)$	$I\,(= \text{at. frac.} \times 1(\alpha),\ \times 0.1(\beta))$	$100\ I/I_{FeK_\alpha}$
FeK_α	6410 eV	0.733	100
FeK_β	7080 eV	0.073	10
CrK_α	5430 eV	0.191	26
CrK_β	5960 eV	0.019	3
NiK_α	7490 eV	0.075	10
NiK_β	8290 eV	0.008	1

Giving the "spectrum" plot:

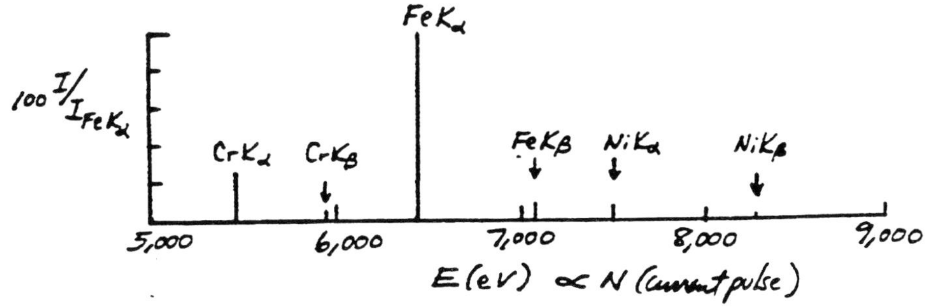

• **12.26** Using the information from Problems 12.24 and 12.25, sketch a spectrum produced by chemically analyzing a specialty alloy (75 wt % Ni, 25 wt % Cr) using characteristic x-rays.

12.26

To obtain atomic fractions of elements:

$$100\,g \text{ alloy} \rightarrow 75\,g\text{ Ni} + 25\,g\text{ Cr}$$

or $g\cdot\text{atoms Ni} = \frac{75\,g}{58.71\,g}\,g\cdot\text{atom} = 1.277\,g\cdot\text{atom}$

$g\cdot\text{atoms Cr} = \frac{25\,g}{52.00\,g}\,g\cdot\text{atom} = \frac{0.481\,g\cdot\text{atom}}{1.758\,g\cdot\text{atom}}$

or atomic fraction Ni $= 1.277/1.758 = 0.726$

" " Cr $= 0.481/1.758 = 0.274$

Then:

Photon	$E(=hc/\lambda)$	I (=at. frac $\times 1(\alpha)$, $\times 0.1(\beta)$)	$100\, I/I_{NiK_\alpha}$
Ni K_α	7490 eV	0.726	100
Ni K_β	8290 eV	0.0726	10
Cr K_α	5430 eV	0.274	38
Cr K_β	5960 eV	0.0274	4

Giving the "spectrum" plot:

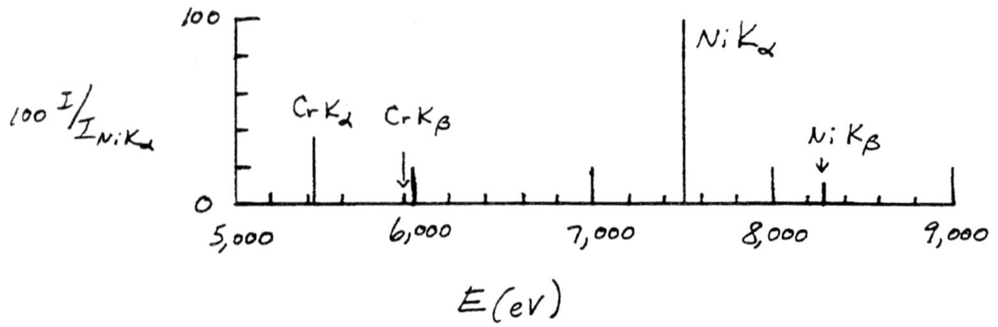

Section 12.3 - Compound Semiconductors

PP 12.9 Sample Problem 12.11 describes a GaAs semiconductor with 100-ppb Se doping. What is the atomic density of Se atoms in this extrinsic semiconductor? (The density of GaAs is 5.32 Mg/m^3.)

PP 12.9 From Sample Problem 12.11, there are 1.27×10^{-7} g-atom Se / 100 g GaAs.

$$V_{GaAs} = \frac{100\ g}{5.32 \times 10^6\ g/m^3} = 1.88 \times 10^{-5}\ m^3$$

Therefore,

$$\rho_{Se} = \frac{1.27 \times 10^{-7}\ g\text{-atom} \times 0.6023 \times 10^{24}\ atoms/g\text{-atom}}{1.88 \times 10^{-5}\ m^3}$$

$$= \underline{\underline{4.07 \times 10^{21}\ atoms/m^3}}$$

PP 12.10 Calculate the intrinsic conductivity of InSb at 50°C. (See Sample Problem 12.12.)

PP 12.10 $\sigma_{300K} = nq(\mu_e + \mu_h)$

$$= (13.5 \times 10^{21}\ m^{-3})(0.1602 \times 10^{-18}\ C)(8.000 + 0.045)\frac{m^2}{V \cdot s}$$

$$= 1.74 \times 10^4\ \Omega^{-1} \cdot m^{-1}$$

$$\sigma_0 = \sigma_e + E_g/2kT$$

$$= (1.74 \times 10^4\ \Omega^{-1} \cdot m^{-1})e^{+(0.17eV)/2(86.2 \times 10^{-6}\ eV/K)(300\ K)}$$

$$= 4.66 \times 10^5\ \Omega^{-1} \cdot m^{-1}$$

$$\sigma_{50°C} = (4.66 \times 10^5\ \Omega^{-1} \cdot m^{-1})e^{-(0.17eV)/2(86.2 \times 10^{-6}\ eV/K)(323\ K)}$$

$$= \underline{\underline{2.20 \times 10^4\ \Omega^{-1} \cdot m^{-1}}}$$

PP 12.11 For intrinsic InSb, calculate the fraction of the current carried by electrons and the fraction carried by electron holes. (See Sample Problem 12.13.)

PP 12.11

$$\frac{\sigma_e}{\sigma} = \frac{\mu_e}{\mu_e + \mu_h} = \frac{8.000}{8.000 + 0.045} = \underline{\underline{0.9944}}$$

$$\frac{\sigma_h}{\sigma} = \frac{\mu_h}{\mu_e + \mu_h} = \frac{0.045}{8.000 + 0.045} = \underline{\underline{0.0056}}$$

12.27 Calculate the atomic density of Cd in a 100-ppb doping of GaAs.

12.27

Using data from Practice Problem 12.9 and Appendix 1,

$$\frac{100\text{ g Cd}}{10^9\text{ g GaAs}} = \frac{n\text{ atoms Cd}}{m^3} \times \frac{112.4\text{ g Cd}}{0.6023 \times 10^{24}\text{ atoms Cd}} \times \frac{1\text{ cm}^3\text{ GaAs}}{5.32\text{ g GaAs}} \times \frac{1\text{ m}^3}{10^6\text{ cm}^3}$$

or,

$$n = 2.85 \times 10^{21}$$

i.e., there are $\underline{\underline{2.85 \times 10^{21}\text{ Cd atoms}/m^3}}$

12.28 The band gap of intrinsic InSb is 0.17 eV. What temperature increase (relative to room temperature = 25°C) is necessary to increase its conductivity by (a) 10%, (b) 50%, and (c) 100%?

12.28

(a) $$\frac{\sigma_2}{\sigma_1} = \frac{e^{-E_g/2kT_2}}{e^{-E_g/2kT_1}} = e^{-\frac{E_g}{2k}\left(\frac{1}{T_2} - \frac{1}{T_1}\right)}$$

$$\ln(\sigma_2/\sigma_1) = -\frac{E_g}{2k}\left(\frac{1}{T_2} - \frac{1}{T_1}\right) = \frac{E_g}{2k} \cdot \frac{1}{T_1} - \frac{E_g}{2k} \cdot \frac{1}{T_2}$$

$$\frac{E_g}{2k} \cdot \frac{1}{T_2} = \frac{E_g}{2k} \cdot \frac{1}{T_1} - \ln(\sigma_2/\sigma_1)$$

or, finally:

$$T_2 = \left[1/T_1 - (E_g/2k)^{-1} \ln(\sigma_2/\sigma_1) \right]^{-1}$$

$$= \left[1/(298\,K) - \left\{ \frac{0.17\,eV}{2(86.2\times10^{-6}\,eV/K)} \right\}^{-1} \ln(1.10) \right]^{-1}$$

$$= 307\,K = 34\,°C \quad \text{or} \quad \Delta T = 34-25 = \underline{\underline{9\,°C}}$$

(b) $$T_2 = \left[1/(298\,K) - \left\{ \frac{0.17\,eV}{2(86.2\times10^{-6}\,eV/K)} \right\}^{-1} \ln(1.50) \right]^{-1}$$

$$= 340\,K = 67\,°C \quad \text{or} \quad \Delta T = 67-25 = \underline{\underline{42\,°C}}$$

(c) $$T_2 = \left[1/(298\,K) - \left\{ \frac{0.17\,eV}{2(86.2\times10^{-6}\,eV/K)} \right\}^{-1} \ln(2.00) \right]^{-1}$$

$$= 377\,K = 104\,°C \quad \text{or} \quad \Delta T = 104-25 = \underline{\underline{79\,°C}}$$

12.29 Illustrate the results of Problem 12.28 on an Arrhenius-type plot.

12.29

From Problem 12.28,

T(°C)	T(K)	$1/T\ (K^{-1})$	$\sigma/\sigma_{25°C}$	$\ln(\sigma/\sigma_{25°C})$
25	298	3.36×10^{-3}	1.0	0
34	307	3.26×10^{-3}	1.1	0.09531
67	340	2.94×10^{-3}	1.5	0.4055
104	377	2.65×10^{-3}	2.0	0.6931

Giving the plot:

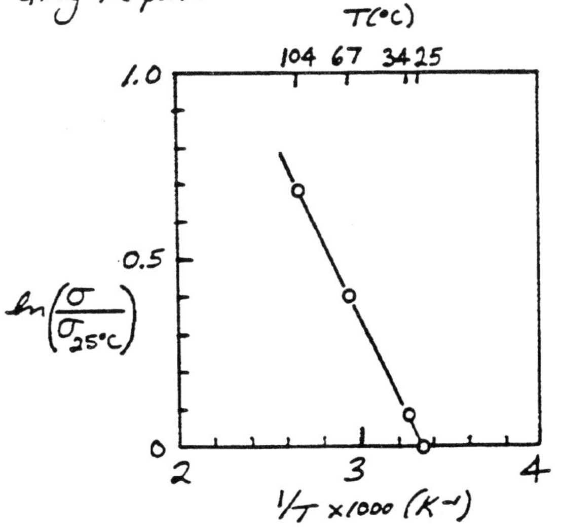

12.30 The band gap of intrinsic ZnSe is 2.67 eV. What temperature increase (relative to room temperature = 25°C) is necessary to increase its conductivity by (a) 10%, (b) 50%, and (c) 100%?

12.30 As derived in Problem 12.28,

$$T_2 = \left[1/T_1 - (E_g/2k)^{-1} \ln(\sigma_2/\sigma_1) \right]^{-1}$$

(a) $T_2 = \left[1/(298\,K) - \left\{ \dfrac{2.67\,eV}{2(86.2\times10^{-6}\,eV/K)} \right\}^{-1} \ln(1.10) \right]^{-1}$

$= 298.5\,K = 25.5°C$ or $\Delta T = 25.5 - 25 = \underline{\underline{0.5°C}}$

(b) $T_2 = \left[1/(298\,K) - \left\{ \dfrac{2.67\,eV}{2(86.2\times10^{-6}\,eV/K)} \right\}^{-1} \ln(1.50) \right]^{-1}$

$= 300.3\,K = 27.3°C$ or $\Delta T = 27.3 - 25 = \underline{\underline{2.3°C}}$

(c) $T_2 = \left[1/(298\,K) - \left\{ \dfrac{2.67\,eV}{2(86.2\times10^{-6}\,eV/K)} \right\}^{-1} \ln(2.00) \right]^{-1}$

$= 302.0\,K = 29.0°C$ or $\Delta T = 29.0 - 25 = \underline{\underline{4.0°C}}$

12.31 Illustrate the results of Problem 12.30 on an Arrhenius-type plot.

12.31 From Problem 12.30,

T(°C)	T(K)	$1/T$(K^{-1})	$\sigma/\sigma_{25°C}$	$\ln(\sigma/\sigma_{25°C})$
25	298	3.36×10^{-3}	1.0	0
25.5	298.5	3.35×10^{-3}	1.1	0.09531
27.3	300.3	3.33×10^{-3}	1.5	0.4055
29.0	302.0	3.31×10^{-3}	2.0	0.6931

Giving the plot:

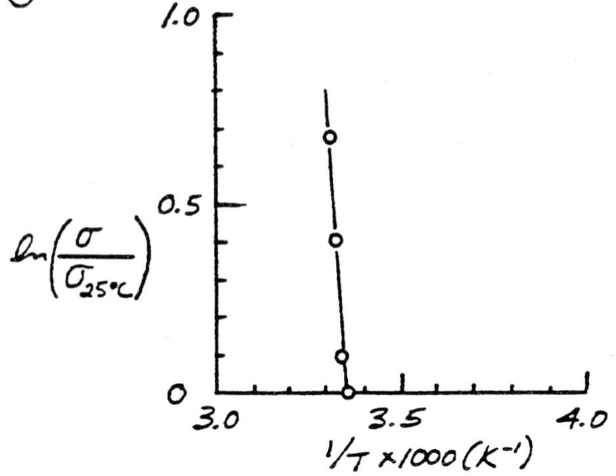

12.32 Starting from an ambient temperature of 300 K, what temperature increase is necessary to double the conductivity of intrinsic InSb?

12.32

$$\sigma = \sigma_0 e^{-E_g/2kT}$$

$$\frac{\sigma_T}{\sigma_{300}} = 2$$

As derived in Problem 12.5,

$$T = \left[\frac{1}{300\,K} - \frac{2k\ln 2}{E_g}\right]^{-1}$$

Using data from Table 12.5

$$T = \left[\frac{1}{300\,K} - \frac{2(86.2\times10^{-6}\,eV/K)\ln 2}{0.17\,eV}\right]^{-1}$$

$$= 380\,K$$

$$\therefore \Delta T = 380\,K - 300\,K = \underline{\underline{80\,K}}$$

12.33 Starting from an ambient temperature of 300 K, what temperature increase is necessary to double the conductivity of intrinsic GaAs?

12.33

As in Problem 12.32,

$$T = \left[\frac{1}{300\,K} - \frac{2(86.2\times10^{-6}\,eV/K)\ln 2}{1.47\,eV}\right]^{-1}$$

$$= 307.5\,K$$

$$\therefore \Delta T = 307.5\,K - 300\,K = \underline{\underline{7.5\,K}}$$

12.34 Starting from an ambient temperature of 300 K, what temperature increase is necessary to double the conductivity of intrinsic CdS?

12.34

As in Problem 12.32,

$$T = \left[\frac{1}{300\,K} - \frac{2(86.2\times10^{-6}\,eV/K)\ln 2}{2.59\,eV}\right]^{-1}$$

$$= 304.2\,K$$

$$\therefore \Delta T = 304.2\,K - 300\,K = \underline{\underline{4.2\,K}}$$

12.35 What temperature increase (relative to room temperature) is necessary to increase the conductivity of intrinsic GaAs by 1%?

12.35 As derived in Problem 12.28,

$$T_2 = [1/T_1 - (E_g/2k)^{-1} \ln(\sigma_2/\sigma_1)]^{-1}$$

$$= [1/(298\,K) - \{(1.47\,eV)/2(86.2\times10^{-6}\,eV/K)\}^{-1} \ln(1.01)]^{-1}$$

$$= 298.1\,K = 25.1\,°C \quad \text{or} \quad \Delta T = 25.1 - 25 = \underline{\underline{0.1\,°C}}$$

12.36 What temperature increase (relative to room temperature) is necessary to increase by 1% the conductivity of (**a**) Se-doped GaAs and (**b**) Cd-doped GaAs?

12.36 (a) As Se-doping produces an n-type material, the expression derived in Problem 12.28 must be modified. The $(E_g/2k)$ term must be replaced by an $[(E_g - E_d)/k]$ term, i.e.

$$T_2 = [1/T_1 - \{(E_g - E_d)/k\}^{-1} \ln(\sigma_2/\sigma_1)]^{-1}$$

Using Table 12.3,

$$T_2 = \left[1/(298\,K) - \left\{\frac{(0.005\,eV)}{(86.2\times10^{-6}\,eV/K)}\right\}^{-1} \ln(1.01)\right]^{-1}$$

$$= 314\,K = 41\,°C \quad \text{or} \quad \Delta T = 41 - 25 = \underline{\underline{16\,°C}}$$

(b) Cd-doping produces a p-type material. Then, an (E_a/k) term must replace the $(E_g/2k)$ term:

$$T_2 = [1/T_1 - \{E_a/k\}^{-1} \ln(\sigma_2/\sigma_1)]^{-1}$$

Using Table 12.3,

$$T_2 = \left[1/(298\,K) - \left\{\frac{(0.021\,eV)}{(86.2\times10^{-6}\,eV/K)}\right\}^{-1} \ln(1.01)\right]^{-1}$$

$$= 301.7\,K = 28.7\,°C \quad \text{or} \quad \Delta T = 28.7 - 25 = \underline{\underline{3.7\,°C}}$$

12.37 In intrinsic semiconductor GaAs, what fraction of the current is carried by electrons and what fraction is carried by holes?

12.37 Using Equation 11.14,

$$\sigma = nq(\mu_e + \mu_h)$$

we find

$$\text{frac. from } e^- = \frac{\mu_e}{\mu_e + \mu_h}$$

$$\& \text{ frac. from } h^+ = \frac{\mu_h}{\mu_e + \mu_h}$$

Using data of Table 12.5 gives

$$\text{frac. from } e^- = \frac{0.720}{0.720 + 0.020} = \underline{\underline{0.973}}$$

$$\& \text{ frac. from } h^+ = \frac{0.020}{0.720 + 0.020} = \underline{\underline{0.027}}$$

12.38 What fraction of the current is carried by electrons and what fraction is carried by holes in **(a)** Se-doped GaAs and **(b)** Cd-doped GaAs, in the extrinsic behavior range?

12.38 (a) Table 12.3 indicates that Se is an n-type dopant, in which case ≈ 1.00 of the current will be carried by electrons and ≈ 0.00 by holes.

(b) Table 12.3 indicates Cd is a p-type dopant, in which case ≈ 0.00 of the current will be carried by electrons and ≈ 1.00 by holes.

Section 12.4 - Amorphous Semiconductors

PP 12.12 In Sample Problem 12.14, we find that 20 mol % hydrogen has a minor effect on the final density of an amorphous silicon. Suppose that we make an amorphous silicon by the decomposition of silicon tetrachloride, $SiCl_4$, rather than silane, SiH_4. Using similar assumptions, calculate the effect of 20 mol % Cl on the final density of an amorphous silicon.

PP 12.12 As in Sample Problem 12.14, take x g of Cl and (100 − x) g of Si. Now,

$$\frac{x/35.45}{(100-x)/28.09} = \frac{0.2}{0.8}$$

or $x = 23.98\,g$ Cl

$100 - x = 76.02\,g$ Si

$$V_{Si} = \frac{76.02\,g}{2.3\,g/cm^3} = 33.05\,cm^3$$

$$\rho = \frac{100\,g}{33.05\,cm^3} = 3.03\,g/cm^3$$ which is an increase of

$$\frac{3.03-2.30}{2.30} \times 100\% = \underline{\underline{31.6\%}}$$

12.39 Estimate the atomic packing factor of amorphous germanium if its density is reduced by 1% relative to the crystalline state. (See Practice Problem 4.12.)

12.39 From Appendix 1, $\rho_{crystal} = 5.32\,Mg/m^3$.

$$\rho_{amor} = 0.99(5.32\,Mg/m^3) = 5.267\,Mg/m^3$$

Therefore, $(APF)_{amor} = 0.99\,(APF)_{crystal}$.

Using the result of Sample Problem 3.16,

$$(APF)_{amor} = 0.340 \times 0.99 = \underline{\underline{0.337}}$$

12.40 Ion implantation treatment of crystalline silicon can lead to the formation of an amorphous surface layer extending from the outer surface to the ion penetration depth. This is considered a structural defect for the crystalline device. What is the appropriate processing treatment to eliminate this defect?

12.40 Because the amorphous material is metastable relative to the crystalline state, the "defect" layer can be converted by an <u>annealing</u> step at $\approx \frac{1}{2} T_m$.

Section 12.5 - Simple Devices

PP 12.13 In Sample Problem 12.15, we calculate, for a given transistor, the collector current produced by increasing the emitter voltage to 50 mV. Make a continuous plot of collector current versus emitter voltage for this device over the range of 5 to 50 mV.

PP 12.13 Data from Sample Problem 12.15:

I_c (mA)	V (mV)
5	5
50	25
886	50

Additional data using $I_c = (2.81 \text{ mA}) e^{V/8.69 \text{ mV}}$:

I_c (mA)	V (mV)
16	15
158	35
498	45

Resulting plot:

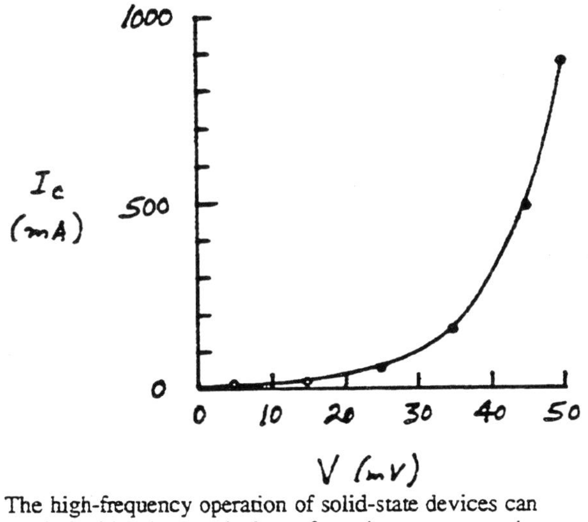

12.41 The high-frequency operation of solid-state devices can be limited by the transit time of an electron across the gate between the source and drain of an FET. For a device to operate at 1 gigahertz (10^9 s^{-1}), a transit time of 10^{-9} s is required. **(a)** What electron velocity is required to achieve this transit time across a 1-μm gate? **(b)** What electric field strength is required to achieve this electron velocity in silicon? **(c)** For the same gate width and electric field strength, what operating frequency would be achieved with GaAs, a semiconductor with a higher electron mobility?

12.41 (a) $v = \frac{\text{length}}{\text{time}} = \frac{1 \times 10^{-6} \text{ m}}{10^{-9} \text{ s}} = \underline{\underline{10^3 \text{ m/s}}}$

(b) Using Equation 11.5,

$$E = \frac{v}{\mu}$$

Electron mobility is given in Table 12.5:

$$E = \frac{10^3 \text{ m/s}}{0.140 \text{ m}^2/(\text{V}\cdot\text{s})} = \underline{\underline{7.14 \times 10^3 \text{ V/m}}}$$

(c) Using data from Table 12.5 gives:

$$v = \mu E = (0.720 \text{ m}^2/\text{V}\cdot\text{s})(7.14 \times 10^3 \text{ V/m})$$
$$= 5.14 \times 10^3 \text{ m/s}$$

giving

$$\text{transit time} = \frac{1 \times 10^{-6} \text{ m}}{5.14 \times 10^3 \text{ m/s}} = 1.95 \times 10^{-10} \text{ s}$$

and

$$\text{frequency} = \frac{1}{\text{transit time}} = \frac{1}{1.95 \times 10^{-10} \text{ s}} = 5.14 \times 10^9 \text{ s}^{-1}$$
$$= \underline{\underline{5.14 \text{ gigahertz}}}$$

12.42 Make a schematic illustration of an *n–p–n* transistor analogous to the *p–n–p* case shown in Figure 12-18.

12.42

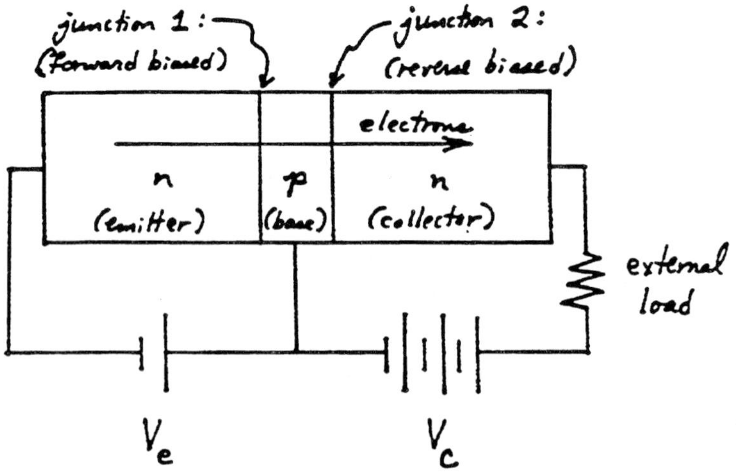

12.43 Make a schematic illustration of an *n*-channel field-effect transistor analogous to the *p*-channel FET shown in Figure 12-19.

12.43

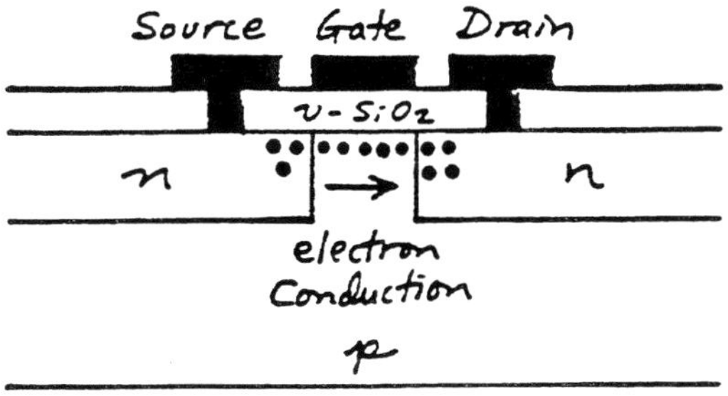

12.44 Figure 12-24 illustrates the relatively large metal connection needed to communicate with an integrated circuit. A limit to the increasing scale of integration in integrated circuits (IC) is the density of interconnection. A useful empirical equation to estimate the number of signal input/output (I/O) pins in a device package is $P = KG^{\alpha}$ where K and α are empirical constants and G is the number of gates of the IC. (One gate is equal to approximately four transistors.) For K and α values of 7 and 0.2, respectively, calculate the number of pins for devices with **(a)** 1,000, **(b)** 10,000, and **(c)** 100,000 gates.

12.44

$$P = KG^{\alpha}$$

(a) For $G = 1{,}000$:

$$P = 7(1{,}000)^{0.2} = \underline{\underline{28}}$$

(b) For $G = 10{,}000$:

$$P = 7(10{,}000)^{0.2} = \underline{\underline{44}}$$

(c) For $G = 100{,}000$:

$$P = 7(100{,}000)^{0.2} = \underline{\underline{70}}$$

Section 13.1 - Magnetism

PP 13.1 In Sample Problem 13.1, we calculate the induction and magnetization of a paramagnetic material under an applied field strength of 2.0×10^5 amperes/m. Repeat this calculation for the case of another paramagnetic material that has a relative permeability of 1.005.

PP 13.1

$$B = \mu_r \mu_0 H$$

$$= (1.005)(4\pi \times 10^{-7} \text{ henry/m})(2.0 \times 10^5 \text{ amperes/m})$$

$$= \underline{\underline{0.253 \text{ weber/m}^2}} = |B|$$

$$M = (\mu_r - 1) H$$

$$= (1.005 - 1)(2.0 \times 10^5 \text{ amperes/m})$$

$$= \underline{\underline{1.0 \times 10^3 \text{ amperes/m}}} = |M|$$

13.1 Calculate the induction and magnetization of a diamagnetic material (with $\mu_r = 0.99995$) under an applied field strength of 2.0×10^5 amperes/m.

13.1

$$B = (0.99995)(4\pi \times 10^{-7} \text{ henry/m})(2.0 \times 10^5 \text{ amperes/m})$$

$$= \underline{\underline{0.251 \text{ weber/m}^2}} = |B|$$

$$M = (0.99995 - 1)(2.0 \times 10^5 \text{ amperes/m})$$

$$= -10 \text{ amperes/m}$$

or

$$|M| = \underline{\underline{10 \text{ amperes/m}}}$$

13.2 Calculate the induction and magnetization of a paramagnetic material with $\mu_r = 1.001$ under an applied field strength of 5.0×10^5 amperes/m.

13.2

$$B = (1.001)(4\pi \times 10^{-7} \text{ henry/m})(5.0 \times 10^5 \text{ amperes/m})$$

$$= \underline{\underline{0.629 \text{ weber/m}^2}}$$

13.3 Plot the B versus H behavior of the paramagnetic material of Problem 13.2 over a range of -5.0×10^5 A/m < H < 5.0×10^5 A/m. Include a dashed-line plot of the magnetic behavior of a vacuum.

13.3

To calculate the end points of the plot, for $|H| = 5.0 \times 10^5$ A/m, for the vacuum:

$$|B| = \mu_0 |H|$$
$$= (4\pi \times 10^{-7} \text{ henry/m})(5.0 \times 10^5 \text{ A/m})$$
$$= 0.628 \text{ weber/m}^2$$

and, for the paramagnetic material:

$$|B| = \mu_r \mu_0 |H|$$
$$= (1.001)(4\pi \times 10^{-7} \text{ henry/m})(5.0 \times 10^5 \text{ A/m})$$
$$= 0.629 \text{ weber/m}^2$$

giving the plot:

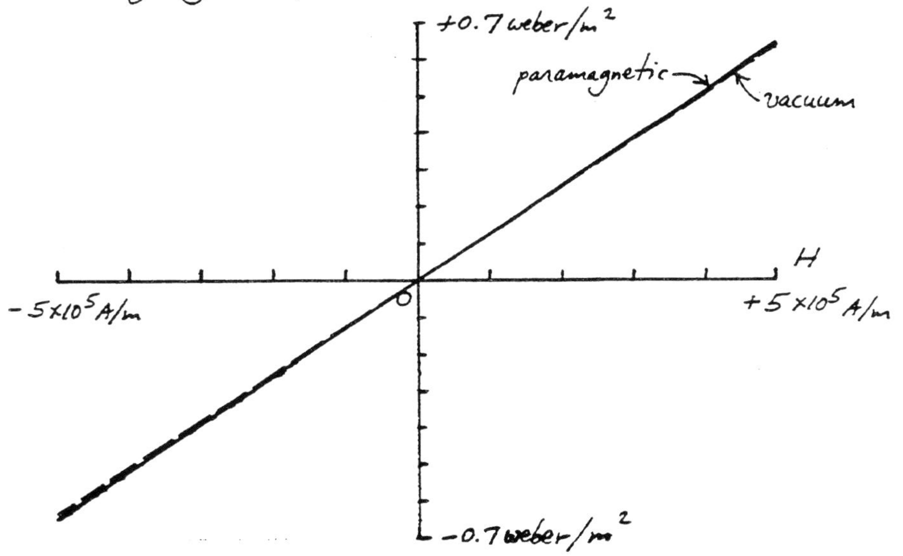

Note: When drawn to scale, there is barely any distinction between the two plots.

13.4 Superimpose on the plot of Problem 13.3 the behavior of the diamagnetic material of Problem 13.1.

13.4

For the diamagnetic material:

$$|B| = (0.99995)(4\pi \times 10^{-7} \text{ henry/m})(5.0 \times 10^5 \text{ A/m})$$
$$= 0.628 \text{ weber/m}^2$$

or, there will be no significant distinction between the plot of the vacuum and the diamagnetic material:

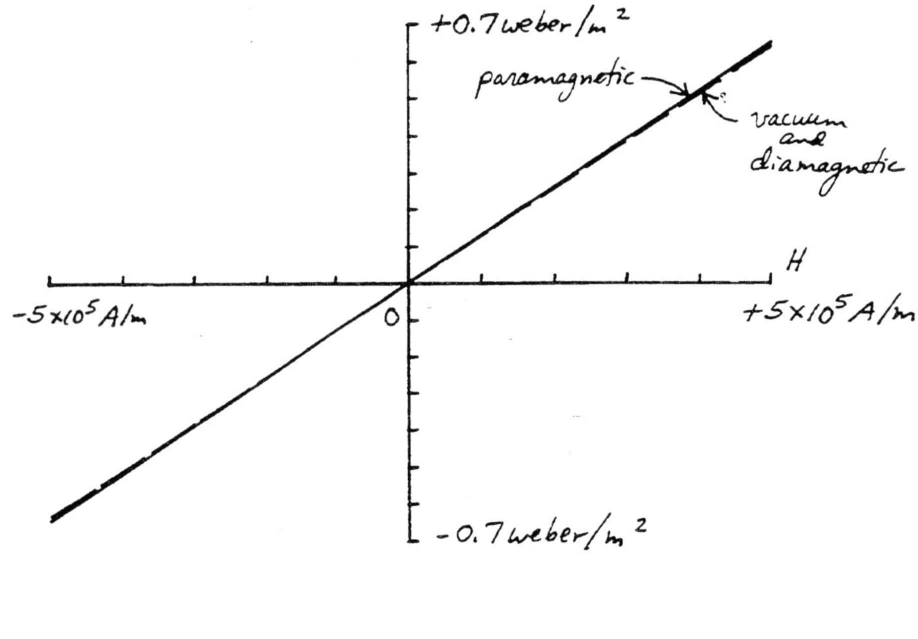

13.5 The following data are obtained for a metal subjected to a magnetic field:

H (amperes/m)	B (weber/m^2)
0	0
4×10^5	0.50263

(a) Calculate the relative permeability for this metal.
(b) What type of magnetism is being demonstrated?

13.5

(a) $\mu_r = \frac{\mu}{\mu_0}$

$$\mu = \frac{|B|}{|H|} = \frac{0.50263 \text{ weber}/m^2}{4 \times 10^5 \text{ A/m}}$$

$$= 1.256575 \times 10^{-6} \text{ H/m}$$

Then,

$$\mu_r = \frac{1.256575 \times 10^{-6} \text{ H/m}}{4\pi \times 10^{-7} \text{ H/m}} = \underline{\underline{0.99995}}$$

(b) This μ_r value is characteristic of <u>diamagnetism</u>

13.6 The following data are obtained for a ceramic subjected to a magnetic field:

H (amperes/m)	*B* (weber/m²)
0	0
4×10^5	0.50668

(a) Calculate the relative permeability for this ceramic.
(b) What type of magnetism is being demonstrated?

13.6

(a) $$\mu = \frac{|B|}{|H|} = \frac{0.50668 \text{ weber}/m^2}{4 \times 10^5 \text{ A/m}}$$

$$= 1.2667 \times 10^{-6} \text{ H/m}$$

Then,

$$\mu_r = \frac{1.2667 \times 10^{-6} \text{ H/m}}{4\pi \times 10^{-7} \text{ H/m}} = \underline{\underline{1.008}}$$

(b) This μ_r value is characteristic of <u>paramagnetism</u>

Section 13.2 - Ferromagnetism

PP 13.2 In Sample Problem 13.2, we illustrate the electronic structure and resulting magnetic moments for the 4*d* orbitals of a series of transition metals. Generate a similar illustration for the 5*d* orbitals of the series Lu to Au.

PP 13.2

Atomic Number	Element	Electronic Structure of 5d					Moment (μ_B)
71	Lu	↑					1
72	Hf	↑	↑				2
73	Ta	↑	↑	↑			3
74	W	↑	↑	↑	↑		4
75	Re	↑	↑	↑	↑	↑	5
76	Os	↑↓	↑	↑	↑	↑	4
77	Ir	↑↓	↑↓	↑↓	↑↓	↑	1
78	Pt	↑↓	↑↓	↑↓	↑↓	↑	1
79	Au	↑↓	↑↓	↑↓	↑↓	↑↓	0

PP 13.3 As pointed out in the beginning of Section 13.2, magnetization rather than induction is the quantity that saturates during ferromagnetic hysteresis. **(a)** For the case given in Sample Problem 13.3, what is the saturation induction? **(b)** What is the saturation magnetization at that point?

PP 13.3

(a) From the given data, $B_s = 0.65 \text{ weber/m}^2$

(b) Equation 13.4 gives us:

$$B = \mu_0 (H + M)$$

At the saturation point,

$$B_s = \mu_0 (H + M_s)$$

or

$$\frac{B_s}{\mu_0} = H + M_s$$

giving, finally,

$$M_s = \frac{B_s}{\mu_0} - H$$

$$= \frac{(0.65 \text{ weber/m}^2)}{(4\pi \times 10^{-7} \text{ henry/m})} - 6 \times 10^4 \text{ amperes/m}$$

$$= 4.57 \times 10^5 \text{ amperes/m}$$

13.7 The following data are obtained for an armco iron alloy during the generation of steady-state ferromagnetic hysteresis loop:

H (amperes/m)	B (weber/m^2)
56	0.50
30	0.46
10	0.40
0	0.36
−10	0.28
−20	0.12
−25	0
−40	−0.28
−56	−0.50

(a) Plot the data. **(b)** What is the remanent induction? **(c)** What is the coercive field?

13.7 (a)

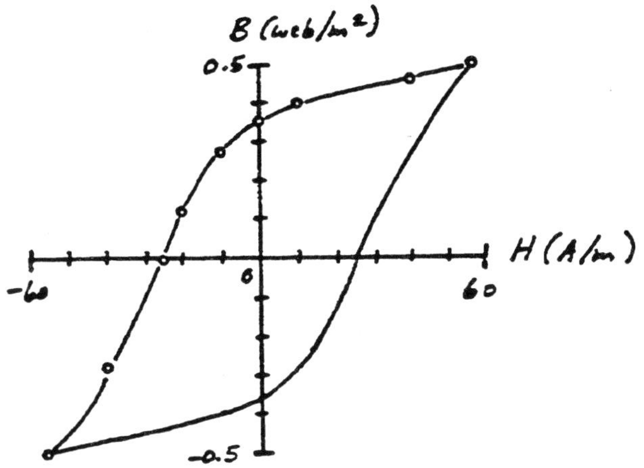

(b) $B_r = \underline{0.36 \text{ weber/m}^2}$ (at $H=0$)

(c) $H_c = \underline{-25 \text{ A/m}}$ (at $B=0$)

13.8 For the armco iron of Problem 13.7, determine (a) the saturation induction and (b) the saturation magnetization.

13.8

(a) From the given data, $B_s = \underline{\underline{0.50 \text{ weber/m}^2}}$

(b) As derived in PP 13.3,

$$M_s = \frac{B_s}{\mu_0} - H$$

$$= \frac{0.50 \text{ weber/m}^2}{(4\pi \times 10^{-7} \text{ henry/m})} - 56 \text{ amperes/m}$$

$$= \underline{\underline{3.98 \times 10^5 \text{ amperes/m}}}$$

13.9 The following data are obtained for a nickel–iron alloy during the generation of a steady-state ferromagnetic hysteresis loop:

H (amperes/m)	*B* (weber/m^2)
50	0.95
25	0.94
0	0.92
−10	0.90
−15	0.75
−20	−0.55
−25	−0.87
−50	−0.95

(a) Plot the data. (b) What is the remanent induction? (c) What is the coercive field?

13.9 (a)

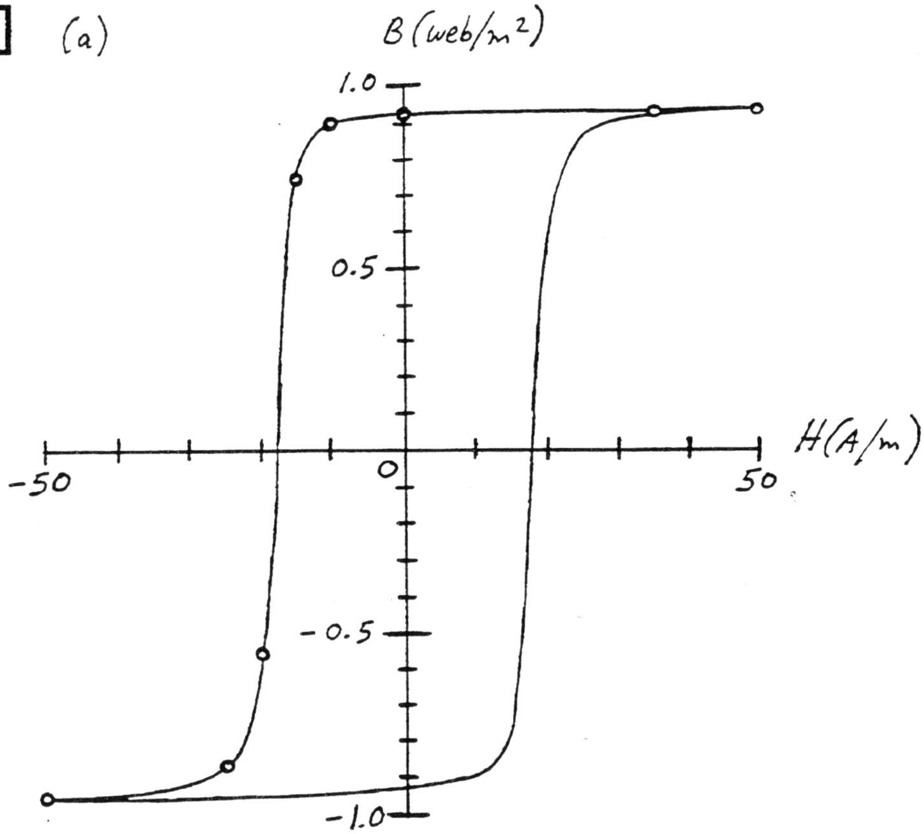

(b) B_r = $\underline{\underline{0.92 \text{ weber/m}^2}}$ (at $H=0$)

(c) H_c = $\underline{\underline{-18 \text{ A/m}}}$ (at $B=0$)

13.10 For the nickel–iron alloy of Problem 13.9, determine **(a)** the saturation induction and **(b)** the saturation magnetization.

13.10 (a) From the given data, $B_s = \underline{\underline{0.95 \text{ weber/m}^2}}$

(b) As derived in Practice Problem 13.3,

$$M_s = \frac{B_s}{\mu_0} - H$$

$$= \frac{0.95 \text{ weber/m}^2}{(4\pi \times 10^{-7} \text{ henry/m})} - 50 \text{ amperes/m}$$

$$= \underline{\underline{7.56 \times 10^5 \text{ amperes/m}}}$$

13.11 Illustrate the electronic structure and resulting magnetic moments for the heavy elements No and Lw, which involve an unfilled 6*d* orbital.

13.11

Atomic Number	Element	Elec. Struc of 6 d	Moment (μ_B)
102	No	☐☐☐☐☐	0
103	Lw	[↑]☐☐☐☐	1

• **13.12** Let us explore further the difference between induction, which does not truly saturate, and magnetization, which does. **(a)** For the magnet treated in Practice Problem 13.3, what would be the induction at a field strength of 60 × 10[4] amperes/m, 10 times greater than what was associated with saturation induction? **(b)** Sketch quantitatively the hysteresis loop for the case of cycling the magnetic field strength between −60 × 10[4] and +60 × 10[4] amperes/m.

13.12 (a) As developed in solving PP 13.3(b),

$$M_s = \frac{B_s}{\mu_0} - H = 4.57 \times 10^5 \text{ amperes/m}.$$

This value does not change with increasing H. However, for $H > 6 \times 10^4$ amperes/m,

$$B = B_s + \mu_0 (H - 6 \times 10^4 \text{ amperes/m})$$

$$B_{H = 60 \times 10^4 \text{ A/m}} = (0.65 \text{ weber/m}^2) + (4\pi \times 10^{-7} \text{ henry/m}) \times (60 \times 10^4 - 6 \times 10^4) \text{ amperes/m}$$

$$= \underline{\underline{1.33 \text{ webers/m}^2}}$$

(b) Taking the data given in Sample Problem 13.3 along with the result of part (a) and noting the symmetry of the hysteresis loop:

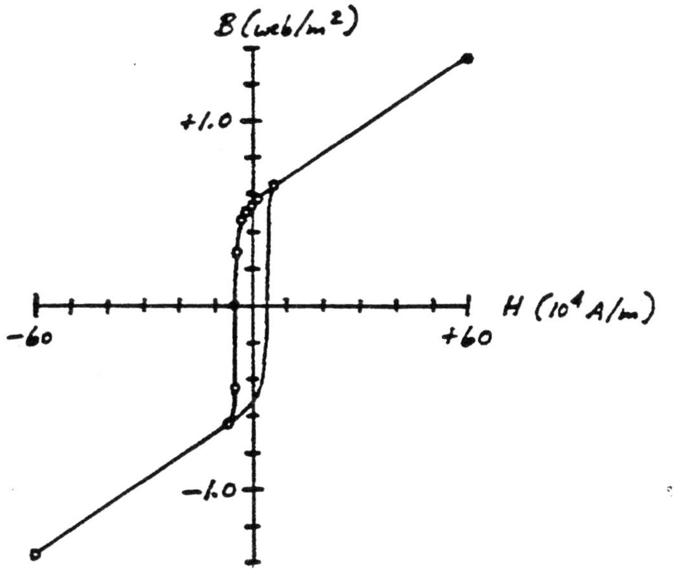

Section 13.3 - Ferrimagnetism

PP 13.4 Calculate the magnetic moment of a unit cell of copper ferrite. (See Sample Problem 13.4.)

PP 13.4

magnetic moment / unit cell = (no. Cu^{2+}/unit cell)(moment Cu^{2+})

$= 8 \times 1\mu_B = \underline{\underline{8\mu_B}}$

PP 13.5 Calculate the saturation magnetization for the copper ferrite described in Practice Problem 13.4. (The lattice parameter for copper ferrite is 0.838 nm.) (See Sample Problem 13.5.)

PP 13.5

$$|M_s| = \frac{8\mu_B}{\text{vol. of unit cell}}$$

$$= \frac{(8)(9.274 \times 10^{-24}\ A \cdot m^2)}{(0.838 \times 10^{-9}\ m)^3}$$

$$= \underline{\underline{1.26 \times 10^5\ A/m}}$$

13.13 (a) Calculate the magnetic moment of a unit cell of manganese ferrite. (b) Calculate the corresponding saturation magnetization, given a lattice parameter of 0.850 nm.

13.13

(a) magnetic moment / unit cell = (no. Mn^{2+}/unit cell) × (moment Mn^{2+})

$= 8 \times 5\mu_B = \underline{\underline{40\mu_B}}$

(b) $|M_s| = \dfrac{40\mu_B}{\text{vol. of unit cell}}$

$$= \frac{(40)(9.274\times10^{-24}\,A\cdot m^2)}{(0.850\times10^{-9}\,m)^3}$$

$$= \underline{\underline{6.04\times10^5\,A/m}}$$

13.14 Make a photocopy of Figure 3-23. Relabel the ions so that the unit cell represents the structure of an inverse spinel, $CoFe_2O_4$. (Do not try to label each site.)

13.14

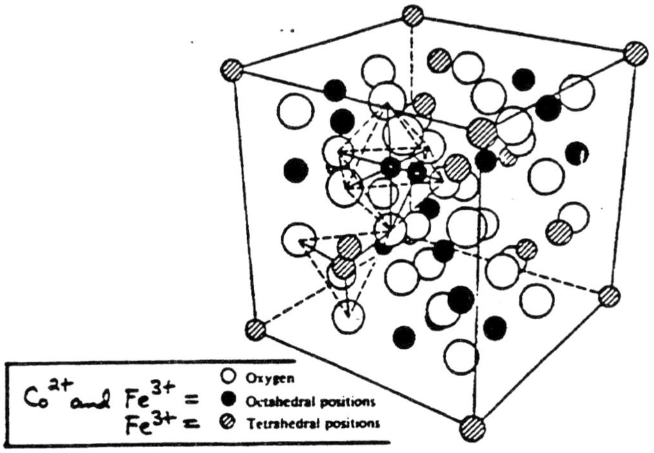

13.15 Calculate the magnetic moment of the unit cell generated in Problem 13.14.

13.15 magnetic moment/unit cell = (no. Co^{2+}/unit cell)(moment Co^{2+})

$$= 8\times 3\mu_B = \underline{\underline{24\mu_B}}$$

13.16 Estimate the saturation magnetization of the unit cell generated in Problem 13.14.

13.16 First, one must estimate the unit cell size for $CoFe_2O_4$. Note from Appendix 2 that $r_{Co^{2+}} = 0.082\,nm$ and $r_{Ni^{2+}} = 0.078\,nm$. Assume that, to a first approximation, $a_{CoFe_2O_4} > a_{NiFe_2O_4}$ by Δ diameter of the divalent ions, i.e.

$$\Delta dia. = 2(0.082-0.078)\,nm = 0.008\,nm.$$

$$\therefore a_{CoFe_2O_4} \cong a_{NiFe_2O_4} + 0.008\,nm.$$

Noting the value in Sample Problem 13.5,

$$a_{CoFe_2O_4} = (0.833 + 0.008)\,nm = 0.841\,nm$$

$$\therefore |M_s| = \frac{24\mu_B}{a^3} = \frac{24(9.274\times10^{-24}\,A\cdot m^2)}{(0.841\,nm)^3\times10^{-27}\,m^3/nm^3}$$

$$= \underline{\underline{3.74\times10^5\,A/m}}$$

13.17 A key aspect of the ferrite crystal structure based on spinel, $MgAl_2O_4$ (Figure 3-23), is the tendency toward tetrahedral or octahedral coordination of the metal ions by O^{2-}. Calculate the radius ratio for **(a)** Mg^{2+} and **(b)** Al^{3+}. In each case, comment on the corresponding coordination number. (Recall that the relationship of radius ratio to coordination number was introduced in Section 2.2.)

13.17

(a) Using Appendix 2,

$$\frac{r}{R} = \frac{r_{Mg^{2+}}}{r_{O^{2-}}} = \frac{0.078\,nm}{0.132\,nm} = \underline{\underline{0.591}}$$

Table 2.1 indicates this is in the range of CN=6. In fact, $CN_{Mg^{2+}} = 4$ in spinel. The discrepancy is the result of the approximate nature of the calculation and the increased complexity associated with more than one type of cation.

(b) For Al^{3+},

$$\frac{r}{R} = \frac{r_{Al^{3+}}}{r_{O^{2-}}} = \frac{0.057\,nm}{0.132\,nm} = \underline{\underline{0.432}}$$

Table 2.1 correctly indicates this is in the range of CN=6.

13.18 In regard to the discussion of Problem 13.17, calculate the radius ratio for (a) Fe^{2+} and (b) Fe^{3+}. In each case, comment on the corresponding coordination number in the inverse spinel structure of magnetite, Fe_3O_4.

13.18

(a) Using Appendix 2,

$$\frac{r}{R} = \frac{r_{Fe^{2+}}}{r_{O^{2-}}} = \frac{0.087\,nm}{0.132\,nm} = \underline{\underline{0.659}}$$

(b) $$\frac{r}{R} = \frac{r_{Fe^{3+}}}{r_{O^{2-}}} = \frac{0.067\,nm}{0.132\,nm} = \underline{\underline{0.508}}$$

In both cases, Table 2.1 indicates CN=6 consistent with the fact that both Fe^{2+} and Fe^{3+} occupy octahedral sites. The approximation of the calculations does not indicate the fact that one-half of the Fe^{3+} ions are tetrahedrally coordinated.

Section 13.4 - Metallic Magnets

PP 13.6 In Sample Problem 13.6, we analyze data for a hard magnet (cunife). Use the similar data for a soft magnet (armco iron) given in Problem 13.7 to calculate the energy loss.

PP 13.6

A careful measurement of the area in Prob. 13.7 gives:

area = 44 $\frac{\text{ampere}\cdot\text{webers}}{m^3}$

or

energy loss = $\underline{\underline{44\ J/m^3}}$

(Note the sharp contrast to Sample Problem 13.6 for which the hard magnet gave an energy loss of $8.9 \times 10^4\ J/m^3$.)

PP 13.7 For the soft magnet referred to in Practice Problem 13.6, calculate the power of the magnet, as done in Sample Problem 13.7.

PP 13.7

Taking the data of Problem 13.8 and re-tabulating:

B (weber/m^2)	H(A/m)	|BH| (weber·A/m^3 = J/m^3)
0	−25	0
0.12	−20	2.4
0.28	−10	2.8
0.36	0	0

Plotting gives:

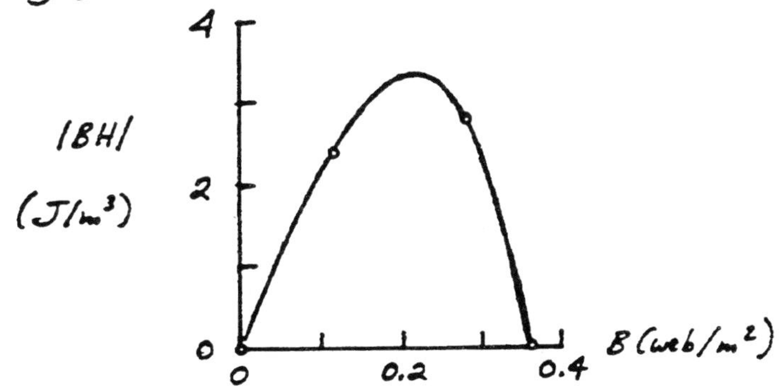

or $(BH)_{max} \approx \underline{\underline{3.3\ J/m^3}}$

(Contrast this with the value of $10^4\ J/m^3$ in Sample Problem 13.7.)

13.19 Assuming that the hysteresis loop in Practice Problem 13.6 is traversed at a frequency of 60 Hz, calculate the rate of energy loss (i.e., power loss) for this magnet.

13.19

Power loss = (energy loss/cycle)(frequency)

Taking the result from PP 13.6,

Power loss = $(44\ J/m^3)(60\ s^{-1}) = 2640\ watts/m^3$

$= \underline{\underline{2.64\ kW/m^3}}$

13.20 Repeat Problem 13.19 for the hard magnet of Sample Problems 13.3 and 13.6.

13.20

Taking the result of Sample Problem 13.6,

Power loss = (energy loss/cycle)(frequency)

$= (89\ kJ/m^3)(60\ s^{-1})$

$= 5{,}340\ kW/m^3$

$= \underline{\underline{5.34\ MW/m^3}}$

(Again, the contrast with the soft magnet of Problem 13.19 is dramatic.)

13.21 Calculate the energy loss (i.e., area within the loop) for the nickel–iron alloy of Problem 13.9.

13.21

Given the scale of the plot in Problem 13.9,

$$\text{area} \approx 67 \text{ web}\cdot\text{amp}/\text{m}^3$$

$$= \underline{\underline{67 \text{ J/m}^3}}$$

13.22 Given the result of Problem 13.21, comment on whether the nickel–iron alloy is a "soft" or "hard" magnet.

13.22

Comparing to the magnitude of hysteresis loops in other systems, the value found for Problem 13.21 is characteristic of a "soft" magnet.

13.23 Assuming the hysteresis loop in Problem 13.9 is traversed at a frequency of 60 Hz, calculate the rate of energy loss (i.e., power loss) for this magnet.

13.23

$$\text{Power loss} = (\text{energy loss/cycle})(\text{frequency})$$

$$= (67 \text{ J/m}^3)(60 \text{ s}^{-1}) = 4020 \text{ watts/m}^3$$

$$= \underline{\underline{4.02 \text{ kW/m}^3}}$$

13.24 Assuming the hysteresis loop in Problem 13.9 is traversed at a frequency of 1 kHz, calculate the rate of energy loss (i.e., power loss) for this magnet.

13.24

$$\text{Power loss} = (\text{energy loss/cycle})(\text{frequency})$$

$$= (67 \text{ J/m}^3)(1000 \text{ s}^{-1}) = 67{,}000 \text{ watts/m}^3$$

$$= \underline{\underline{67 \text{ kW/m}^3}}$$

13.25 The hysteresis loss for soft magnets is generally given in units of W/m^3. Calculate the loss in these units for the Fe–B amorphous metal in Table 13.2 at a frequency of 60 Hz.

13.25

$$25\ J/m^3/cycle \times 60\ cycles/sec = 1500\ \frac{J/s}{m^3}$$

$$= 1500\ W/m^3 = \underline{\underline{1.50\ kW/m^3}}$$

13.26 Repeat Problem 13.25 for the Fe–B–Si amorphous metal in Table 13.2.

13.26

$$15\ J/m^3/cycle \times 60\ cycles/sec = \underline{\underline{900\ W/m^3}}$$

• **13.27** Many of the highest T_c and H_c metal superconductors (such as Nb_3Sn with $T_c = 18.5$ K) have the A_3B "β-tungsten" structure, with A atoms at tetrahedral sites [(0 1/2 1/4)-type positions] in a bcc unit cell of B atoms. Sketch the unit cell of such a material. (**HINT:** Only two of four such A sites are occupied in a given unit cell face, and the A atoms would form, with adjacent unit cells, three orthogonal chains.)

13.27

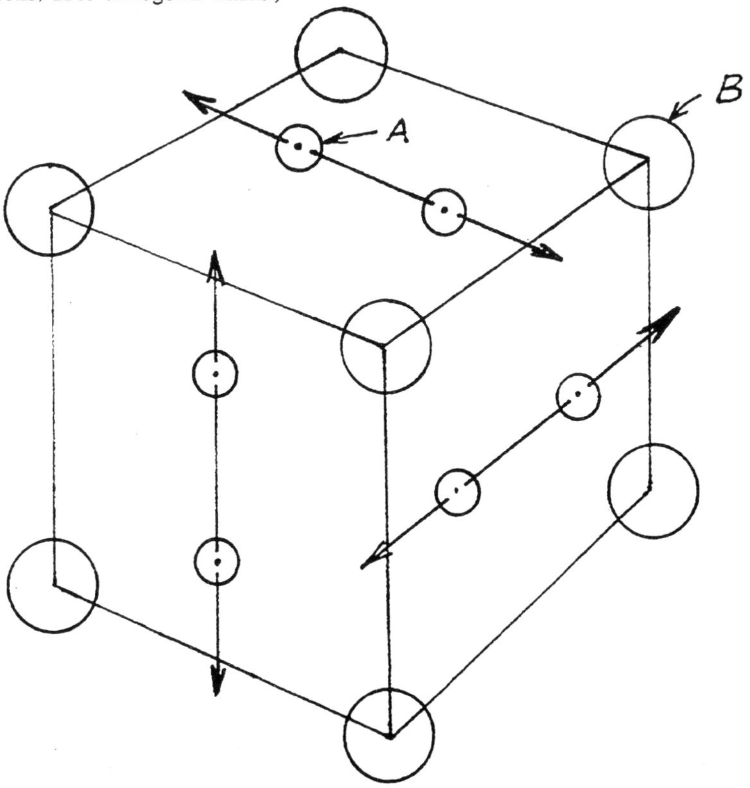

Note: For clarity, only atoms on the front faces of the unit cell are shown.

• 13.28 Verify that the composition of the β-tungsten structure described in Problem 13.27 is A_3B.

13.28

A atoms: (2 per face)(6 faces)($\frac{1}{2}$ per unit cell) = 6 A atoms

B atoms: bcc positions → $1 + 8 \times \frac{1}{8} = 2$ B atoms

$$\rightarrow A_6B_2 = \underline{\underline{A_3B}} \quad \text{(as desired)}$$

Section 13.5 - Ceramic Magnets

PP 13.8 In Sample Problem 13.8, we use a radius ratio calculation to confirm the octahedral coordination of Fe^{3+} in γ-Fe_2O_3. Do similar calculations for Ni^{2+} and Fe^{3+} in the inverse spinel, nickel ferrite, introduced in Sample Problem 13.4.

PP 13.8 From Appendix 2:

$$r_{Fe^{3+}} = r_1 = 0.067\text{ nm}$$

$$r_{Ni^{2+}} = r_2 = 0.078\text{ nm}$$

$$r_{O^{2-}} = R = 0.132\text{ nm}$$

giving us the radius ratios:

$$r_1/R = (0.067\text{ nm})/(0.132\text{ nm}) = \underline{\underline{0.508}}$$

and

$$r_2/R = (0.078\text{ nm})/(0.132\text{ nm}) = \underline{\underline{0.591}}$$

Both ratios correspond to 6-fold coordination, consistent with the inverse spinel structure in which the divalent ions and one-half the trivalent ions are in octahedral sites. But, one-half the trivalent ions are in tetrahedral sites. This corresponds to the Fe^{3+} ratio (0.508) not being far from the four-fold value (0.414).

PP 13.9 (a) Calculate the magnetic moment of a unit cell of $MgFe_2O_4$ in an inverse spinel structure. (b) Repeat part (a) for the case of a normal spinel structure. (c) Given that the experimental value of the unit cell moment for $MgFe_2O_4$ is 8.8 μ_B, estimate the fraction of the ferrite in the inverse spinel structure. (See Sample Problem 13.9.)

PP 13.9 (a) As in Sample Problems 13.4 and 13.9 (a),

$$\text{magnetic moment/unit cell} = (\text{no. } Mg^{2+}/\text{unit cell})(\text{moment } Mg^{2+})$$
$$= 8 \times 0\mu_B = \underline{\underline{0\mu_B}}$$

(b) As in Sample Problem 13.9 (b),

$$\text{magnetic moment/unit cell} = -(\text{no. } Mg^{2+}/\text{unit cell})(\text{moment } Mg^{2+}) + (\text{no. } Fe^{3+}/\text{unit cell})(\text{moment } Fe^{3+})$$
$$= -(8)(0\mu_B) + (16)(5\mu_B) = \underline{\underline{80\mu_B}}$$

(c) Taking y = fraction of ferrite in inverse spinel:

$$y(0\mu_B) + (1-y)(80\mu_B) = 8.8\mu_B$$

or $$80\mu_B - (80y)\mu_B = 8.8\mu_B$$

or $$y = \frac{(80-8.8)\mu_B}{80\mu_B} = \underline{\underline{0.89}}$$

13.29 Characterize the ionic coordination of the cations in yttrium iron garnet using radius ratio calculations.

13.29 From Appendix 2:

$r_{Fe^{3+}} = r_1 = 0.067\,nm$

$r_{Y^{3+}} = r_2 = 0.106\,nm$

$r_{O^{2-}} = R = 0.132\,nm$

giving:

$r_1/R = (0.067\,nm)/(0.132\,nm) = \underline{\underline{0.508}}$

and

$r_2/R = (0.106\,nm)/(0.132\,nm) = \underline{\underline{0.803}}$

The ratio for Fe^{3+} is consistent with 6-fold coordination. Some 4-fold coordination for Fe^{3+} corresponds to the ratio (0.508) not being far from the upper limit (0.414) for ideal four-fold coordination. The ratio for Y^{3+} is consistent with its 8-fold coordination.

13.30 Characterize the ionic coordination of the cations in aluminum-substituted YIG using radius ratio calculations.

13.30 From Appendix 2:

$r_{Y^{3+}}/r_{O^{2-}} = 0.106\,nm/0.132\,nm = \underline{\underline{0.803}}$

$r_{Fe^{3+}}/r_{O^{2-}} = 0.067\,nm/0.132\,nm = \underline{\underline{0.508}}$

$r_{Al^{3+}}/r_{O^{2-}} = 0.057\,nm/0.132\,nm = \underline{\underline{0.432}}$

Y^{3+} and Fe^{3+} were treated in Problem 13.29. The Al^{3+} radius ratio indicates 6-fold coordination, although it is close to the range for 4-fold coordination which is, in fact, preferred.

13.31 Characterize the ionic coordination of the cations in chromium-substituted YIG using radius ratio calculations.

13.31 Again, Y^{3+} and Fe^{3+} have been treated in Problem 13.29. For Cr^{3+},

$$r_{Cr^{3+}}/r_{O^{2-}} = 0.064\,nm/0.132\,nm = \underline{\underline{0.485}}$$

which is well within the range for 6-fold coordination in agreement with the comment in Table 13.5.

• 13.32 (a) Derive a general expression for the magnetic moment of a unit cell of a ferrite with a divalent ion moment of $n\ \mu_B$ and y being the fraction of inverse spinel structure. (Assume $[1 = y]$ to be the fraction of normal spinel structure.) (b) Use the expression derived in part (a) to calculate the moment of a copper ferrite given a heat treatment that produces 25% normal spinel and 75% inverse spinel structure.

13.32 (a) $$\mu_{cell} = (y)(8n\mu_B) + (1-y)[-(8)(n\mu_B) + (16)(5\mu_B)]$$

$$= 8\mu_B(ny - n + 10 + ny - 10y)$$

$$= 8\mu_B(2ny - n + 10 - 10y)$$

or

$$\underline{\underline{\mu_{cell} = 8\mu_B[n(2y-1) + 10(1-y)]}}$$

(b) By definition, $y = 0.75$. Table 13.1 indicates that $n = 1$. Then,

$$\mu_{cell} = 8\mu_B[(1)(2\times0.75-1) + 10(1-0.75)]$$

$$= \underline{\underline{24\mu_B}}$$

13.33 The plot of H_c versus T for a metallic compound, such as Nb_3Ge in Figure 13-16, can be approximated by the equation for a parabola, namely

$$H_c = H_0\left[1 - \left(\frac{T^2}{T_c^2}\right)\right]$$

where H_0 is the critical field at 0 K. Given that H_c for Nb_3Ge is 22×10^4 A-turns/m at 15 K, **(a)** calculate the value of H_0 using the foregoing equation and **(b)** calculate the percentage error of your result in comparison to the experimental value of $H_0 = 44 \times 10^4$ A-turns/m.

13.33 (a) For given data (& noting $T_c = 23\text{K}$ in Fig. 13-16),

$$22 \times 10^4 \text{ A-turns/m} = H_0\left[1 - \left(\frac{15\text{K}}{23\text{K}}\right)^2\right] = 0.575\, H_0$$

or,

$$H_0 = \underline{\underline{38 \times 10^4 \text{ A-turns/m}}}$$

(b) $$\%\text{error} = \frac{44 - 38}{44} \times 100\,\%$$

$$= \underline{\underline{14\%}}$$

13.34 Repeat Problem 13.33 in order to evaluate the utility of the parabolic equation to describe the relationship between H_c and T for the 1-2-3 superconductor shown in Figure 13-16. For the particular sample illustrated, T_c is 93 K, H_0 is 328×10^4 A-turns/m, and H_c is 174×10^4 A-turns/m at 60 K.

13.34 (a) For given data (& noting $T_c = 93\text{K}$ in Fig. 13-16),

$$174 \times 10^4 \text{ A-turns/m} = H_0\left[1 - \left(\frac{60\text{K}}{93\text{K}}\right)^2\right] = 0.584\, H_0$$

or,

$$H_0 = \underline{\underline{298 \times 10^4 \text{ A-turns/m}}}$$

(b) $$\%\text{ error} = \frac{328 - 298}{328} \times 100\,\%$$

$$= \underline{\underline{9\%}}$$

13.35 A commercial ceramic magnet, Ferroxcube A, has a hysteresis loss per cycle of 40 J/m^3. **(a)** Assuming the hysteresis loop is traversed at a frequency of 60 Hz, calculate the power loss for this magnet. **(b)** Is this a soft or hard magnet?

13.35

(a) Power loss = (energy loss/cycle)(frequency)

$$= (40\ J/m^3)(60 s^{-1}) = 2400\ watts/m^3$$

$$= \underline{\underline{2.4\ kW/m^3}}$$

(b) In comparison to the magnitude of hysteresis loops in other systems, the given value is characteristic of a "soft" magnet.

13.36 A commercial ceramic magnet, Ferroxdur, has a hysteresis loss per cycle of 180 kJ/m^3. **(a)** Assuming the hysteresis loop is traversed at a frequency of 60 Hz, calculate the power loss for this magnet. **(b)** Is this a soft or hard magnet?

13.36

(a) Power loss = (energy loss/cycle)(frequency)

$$= (180\ kJ/m^3)(60 s^{-1}) = 10{,}800\ kW/m^3$$

$$= \underline{\underline{10.8\ MW/m^3}}$$

(b) In comparison to the magnitude of hysteresis loops in other systems, the given value is characteristic of a "hard" magnet.

Section 14.1 - Oxidation (Direct Atmospheric Attack)

PP 14.1 In Sample Problem 14.1, we calculate the thickness of an oxide coating after 1 day in an oxidizing atmosphere. What would be the thickness of the coating if the same measurements apply but the oxide grows by a linear growth rate law?

PP 14.1 In this case, Equation 14.2 applies:

$$y = c_1 t + c_2$$

$$c_2 \equiv 100 \text{ nm}$$

$$200 \text{ nm} = c_1 (1 \text{ hr}) + 100 \text{ nm}$$

or

$$c_1 = 100 \text{ nm/hr}$$

Then, after 1 day:

$$y = (100 \text{ nm/hr})(24 \text{ hr}) + 100 \text{ nm}$$

$$= 2500 \text{ nm} = \underline{\underline{2.5 \ \mu\text{m}}}$$

PP 14.2 **(a)** The Pilling-Bedworth ratio for copper is calculated in Sample Problem 14.2. In this case, we assume that cuprous oxide, Cu_2O, is formed. Calculate the Pilling–Bedworth ratio for the alternate possibility that cupric oxide, CuO, is formed. (The density of CuO is 6.40 Mg/m^3.) **(b)** Do you expect CuO to be a protective coating? Briefly explain.

PP 14.2 (a) $$R = \frac{Md}{amD} = \frac{(63.55 + 16.00)(8.93)}{(1)(63.55)(6.40)} = \underline{\underline{1.75}}$$

(b) Yes. The result in (a) falls in the range between 1 and 2.

14.1 The following data are collected during the oxidation of a small bar of metal alloy:

Time	Weight gain (mg)
1 min	0.40
1 hr	24.0
1 day	576

The weight gain is due to the oxide formation. Because of the experimental arrangement, you are unable to visually inspect the oxide scale. Predict whether the scale is (1) porous and discontinuous or (2) dense and tenacious. Briefly explain your answer.

14.1 Taking wt. gain ∝ y and noting that

$$\frac{0.40\text{ mg}}{\text{min}} = \frac{24\text{ mg}}{60\text{ min}} = \frac{576\text{ mg}}{24\times 60\text{ min}},$$

it is clear that the data correspond to a linear growth rate law which is associated with a porous and discontinuous scale.

14.2 The densities for three iron oxides are FeO (5.70 Mg/m^3), Fe_3O_4 (5.18 Mg/m^3), and Fe_2O_3 (5.24 Mg/m^3). Calculate the Pilling–Bedworth ratio for iron relative to each type of oxide and comment on the implications for the formation of a protective coating.

14.2 $R = \frac{Md}{amD}$

For FeO, $R = \frac{(55.85+16.00)(7.87)}{(1)(55.85)(5.70)} = 1.78$

For Fe_3O_4, $R = \frac{(3\times 55.85+4\times 16.00)(7.87)}{(3)(55.85)(5.18)} = 2.10$

For Fe_2O_3, $R = \frac{(2\times 55.85+3\times 16.00)(7.87)}{(2)(55.85)(5.24)} = 2.15$

The Pilling–Bedworth calculations imply that we should not expect a protective coating unless the oxidation conditions (oxygen partial pressure, etc...) are such that FeO is formed (for which $R < 2$).

14.3 Given the density of SiO_2 (quartz) = 2.65 Mg/m^3, calculate the Pilling–Bedworth ratio for silicon and comment on the implication for the formation of a protective coating if quartz should be the oxide form.

14.3

$$R = \frac{Md}{amD} = \frac{(28.09+2[16.00])(2.33)}{(1)(28.09)(2.65)} = 1.88$$

As this value falls within the range of 1 to 2, it would correspond to a protective coating.

14.4 In contrast to the assumption in Problem 14.3, silicon oxidation tends to produce a vitreous silica film with density = 2.20 Mg/m^3. Semiconductor fabrication routinely involves such vitreous films. Calculate the Pilling–Bedworth ratio for this case and comment on the implication for the formation of a tenacious film.

14.4

$$R = \frac{Md}{amD} = \frac{(28.09 + 2[16.00])(2.33)}{(1)(28.09)(2.20)} = \underline{\underline{2.27}}$$

This is, in fact, the value in Table 14.1. Although just above 2.0, the result does correspond to a well known tenacious film.

14.5 Verify the statement in regard to Equation 14.4 that $c_4 = 2c_3$ and $c_5 = y^2$ at $t = 0$.

14.5 Beginning with Equation 14.3,

$$\frac{dy}{dt} = c_3 \frac{1}{y}$$

giving

$$y\,dy = c_3\,dt$$

or, by integration,

$$\frac{1}{2}y^2 = c_3 t + A$$

where A is a constant of integration.

Rearranging,

$$y^2 = 2c_3 t + 2A$$

Comparing to Equation 14.4,

$$y^2 = c_4 t + c_5$$

we see that $\underline{\underline{c_4 = 2c_3}}$ (as desired)

and $c_5 = 2A$.

At $t=0$,

$$y^2_{t=0} = C_5 \quad \text{(as desired)}$$

14.6 Verify that the Pilling–Bedworth ratio is the ratio of oxide volume produced to metal volume consumed.

14.6 Consider the reaction:

$$aM + \frac{b}{2} O_2 \rightarrow M_a O_b$$

The mass of metal consumed $= a \times (\text{at. wt.})_{\text{metal}}$

The mass of oxide produced $= (\text{mol. wt.})_{\text{oxide}}$

In general, the volume of the solid = mass/density.

$$\therefore \text{ vol. metal consumed} = \frac{a \times (\text{at. wt.})_{\text{metal}}}{(\text{density})_{\text{metal}}}$$

$$\& \text{ vol. oxide produced} = \frac{(\text{mol. wt.})_{\text{oxide}}}{(\text{density})_{\text{oxide}}}$$

Let $m = (\text{at. wt.})_{\text{metal}}$, $M = (\text{mol. wt.})_{\text{oxide}}$
$d = (\text{density})_{\text{metal}}$, $D = (\text{density})_{\text{oxide}}$

Then, vol. metal consumed $= \dfrac{am}{d}$

& " oxide produced $= \dfrac{M}{D}$

giving, finally,

$$\frac{\text{vol. oxide produced}}{\text{vol. metal consumed}} = \frac{(M/D)}{(am/d)}$$

$$= \frac{Md}{amD} = R \quad \text{(as desired)}$$

Section 14.2 - Aqueous Corrosion (Electrochemical Attack)

PP 14.3 For the experiment described in Sample Problem 14.3, how many times per second does the reduction (electroplating) reaction given in Equation 14.7 occur?

PP 14.3 As the electroplating reaction precisely balances the corrosion reaction, the rate is the same as that calculated in Sample Problem 14.3:

$$\text{rate} = \underline{\underline{3.13 \times 10^{16}\ s^{-1}}}$$

14.7 In an ionic concentration corrosion cell involving nickel (forming Ni^{2+}), an electrical current of 5 mA is measured. How many Ni atoms per second are oxidized at the anode?

14.7 $I = 5 \times 10^{-3} A \times \frac{1\ C/s}{A} \times \frac{1\ \text{electron}}{0.16 \times 10^{-18} C} = 3.13 \times 10^{16}\ \text{electrons/s}$

$$\text{reaction rate} = (3.13 \times 10^{16}\ \text{electrons/s})(1\ \text{reaction}/2\ \text{electrons})$$

$$= \underline{\underline{1.56 \times 10^{16}\ \text{reactions/s}}}$$

14.8 For the concentration cell described in Problem 14.7, how many Ni atoms per second are reduced at the cathode?

14.8 As the cathodic reduction precisely balances the anodic oxidation, the reaction rate is the same as that calculated in Problem 14.7:

$$\text{rate} = \underline{\underline{1.56 \times 10^{16}\ \text{reactions/s}}}$$

14.9 In an ionic concentration corrosion cell involving chromium, which forms a trivalent ion (Cr^{3+}), an electrical current of 10 mA is measured. How many atoms per second are oxidized at the anode?

14.9

$$I = 10 \times 10^{-3} A \times 1 \frac{C/s}{A} \times \frac{1 \text{ electron}}{0.16 \times 10^{-18} C} = 6.25 \times 10^{16} \text{ electrons/s}$$

$$\text{reaction rate} = (6.25 \times 10^{16} \text{ electrons/s})(1 \text{ reaction}/3 \text{ electrons})$$

$$= \underline{2.08 \times 10^{16} \text{ reactions/s}}$$

14.10 For the concentration cell described in Problem 14.9, how many Cr atoms per second are reduced at the cathode?

14.10

As the cathodic reduction is precisely balanced by the anodic oxidation, the reaction rate is the same as that calculated in Problem 14.9:

$$\text{rate} = \underline{2.08 \times 10^{16} \text{ reactions/s}}$$

Section 14.3 - Galvanic Two-Metal Corrosion

PP 14.4 In Sample Problem 14.4, we analyze a simple galvanic cell composed of zinc and iron electrodes. Make a similar analysis of a galvanic cell composed of copper and zinc electrodes immersed in 1 molar ionic solutions.

PP 14.4

(a) Table 14.2 indicates that zinc is anodic relative to copper. Therefore, zinc is corroded.

(b) $\text{voltage} = (+0.337 V) - (-0.763 V) = \underline{1.100 V}$

14.11 **(a)** In a simple galvanic cell consisting of Co and Cr electrodes immersed in 1 molar ionic solutions, calculate the cell potential. **(b)** Which metal would be corroded in this simple cell?

14.11

(a) Table 14.2 indicates that

$$\text{voltage} = (-0.277\ V) - (-0.744\ V)$$
$$= \underline{\underline{0.467\ V}}$$

(b) By inspection of Table 14.2, we see that chromium is anodic to cobalt. Therefore, <u>chromium is corroded.</u>

14.12 **(a)** In a simple galvanic cell consisting of Al and Mg electrodes immersed in 1 molar ionic solutions, calculate the cell potential. **(b)** Which metal would be corroded in this simple cell?

14.12

(a) Table 14.2 indicates that

$$\text{voltage} = (-1.662\ V) - (-2.363\ V)$$
$$= \underline{\underline{0.701\ V}}$$

(b) By inspection of Table 14.2, we see that magnesium is anodic to aluminum. Therefore, <u>magnesium is corroded.</u>

14.13 Identify the anode in the following galvanic cells, including a brief discussion of each answer: **(a)** copper and nickel electrodes in standard solutions of their own ions, **(b)** a two-phase microstructure of a 50:50 Pb–Sn alloy, **(c)** a lead–tin solder on a 2024 aluminum alloy in seawater, and **(d)** a brass bolt in a Hastelloy C plate, also in seawater.

14.13

(a) As indicated by Table 14.2, the anode would be <u>nickel.</u>

(b) As indicated by Table 14.2, the tin-rich phase would be the anode.

(c) As indicated by Table 14.3, the 2024 aluminum substrate would be the anode.

(d) As indicated by Table 14.3, the anode would be the brass bolt.

14.14 Figure 14-8 illustrates a microstructural-scale galvanic cell. Using the Cu–Zn phase diagram from Chapter 5, specify a brass composition range that would avoid this problem.

14.14

Using Figure 5-37, we see that we can stay in the single-phase (α) region by keeping the zinc content $\leq$ 30 wt. % (approximately).

Section 14.4 - Corrosion by Gaseous Reduction

PP 14.5 Calculate the volume of O_2 gas consumed (at STP) by the corrosion of 100 gm of chromium. (In this case, trivalent Cr^{3+} ions are found at the anode.) (See Sample Problem 14.5.)

PP 14.5

For this case,

$$Cr^{0} \rightarrow Cr^{3+} + 3e^{-}$$

and

$$O_2 + 2H_2O + 4e^{-} \rightarrow 4OH^{-}$$

$$\text{moles } O_2 \text{ gas} = \frac{100\text{ g Cr}}{(52.00\text{ g Cr/g·atom Cr})} \times \frac{3/4\text{ mole } O_2}{\text{g·atom Cr}}$$

$$= 1.44\text{ mole } O_2$$

At STP,

$$V = \frac{(1.44\text{ mol})(8.314\text{ J/mol·K})(273\text{ K})}{(1\text{ atm})(1\text{ Pa}/9.869\times10^{-6}\text{ atm})}$$

$$= 0.0323\text{ m}^3$$

14.15 A copper-nickel (35 wt %–65 wt %) alloy is corroded in an oxygen concentration cell using boiling water. What volume of oxygen gas (at 1 atm) must be consumed at the cathode to corrode 10 g of the alloy? (Assume only divalent ions are produced.)

14.15

As in Sample Problem 14.5, ½ mole O_2 will be needed for each g·atom of alloy. Using data from Appendix 1,

$$\text{moles } O_2 \text{ gas} = \left[\frac{3.5\text{ g Cu}}{\left(\frac{63.55\text{ g Cu}}{\text{g·atom Cu}}\right)} + \frac{6.5\text{ g Ni}}{\left(\frac{58.71\text{ g Ni}}{\text{g·atom Ni}}\right)}\right] \times \frac{\frac{1}{2}\text{ mole } O_2}{\text{g·atom Cu+Ni}}$$

$$= 0.0829 \text{ mole } O_2$$

$$V_{O_2} = \frac{nRT}{p}$$

$$= \frac{(0.0829\text{ mole})(8.314\text{ J/mol·K})(373\text{ K})}{(1\text{ atm})(1\text{ Pa}/9.869\times10^{-6}\text{ atm})}$$

$$= \underline{\underline{2.54\times10^{-3}\text{ m}^3}}$$

14.16 Assume that iron is corroded in an acid bath, with the cathode reaction being given by Equation 14.10. Calculate the volume of H_2 gas produced at STP in order to corrode 100 gm of iron.

14.16

Consider

$$Fe^0 \rightarrow Fe^{2+} + 2e^-$$

and

$$2H^+ + 2e^- \rightarrow H_2\uparrow$$

$$\text{moles } H_2 \text{ gas} = \frac{100\text{ g Fe}}{(55.85\text{ g Fe/mole Fe})} \times \frac{1\text{ mole } H_2}{1\text{ mole Fe}}$$

$$= 1.79 \text{ moles } H_2$$

At STP,

$$V = \frac{(1.79\text{ moles})(8.314\text{ J/mol·K})(273\text{ K})}{(1\text{ atm})(1\text{ Pa}/9.869\times10^{-6}\text{ atm})}$$

$$= \underline{\underline{0.0401\text{ m}^3}}$$

14.17 For the rusting mechanism illustrated in Figure 14-11, calculate the volume of O_2 gas consumed (at STP) in the production of 100 gm of rust [Fe $(OH)_3$].

14.17 The following reactions are illustrated in Figure 14-11:

$$Fe \rightarrow Fe^{3+} + 3e^- \quad (i)$$

$$\frac{3}{4}O_2 + 3e^- + \frac{3}{2}H_2O \rightarrow 3(OH)^- \quad (ii)$$

and $$3(OH)^- + Fe^{3+} \rightarrow Fe(OH)_3 \quad (iii)$$

(It was necessary to determine the coefficients for O_2 and H_2O to make equation (ii) balance.)

Combining equations (i) - (iii) gives the overall reaction:

$$Fe + \frac{3}{4}O_2 + \frac{3}{2}H_2O \rightarrow Fe(OH)_3$$

Then,

$$\text{moles } O_2 \text{ gas} = \frac{100 g\ Fe(OH)_3}{[55.85 g + 3(16.00) + 3(1.008)]\frac{g\ Fe(OH)_3}{\text{mole } Fe(OH)_3}}$$

$$\times \frac{3/4 \text{ mole } O_2}{1 \text{ mole } Fe(OH)_3} = 0.702 \text{ mole } O_2$$

At STP,

$$V = \frac{(0.702 \text{ mole})(8.314 \text{ J/mol}\cdot\text{K})(273 K)}{(1 \text{ atm})(1 \text{ Pa}/9.869 \times 10^{-6} \text{ atm})} = \underline{\underline{0.0157 m^3}}$$

14.18 In the failure analysis of an aluminum vessel, corrosion pits are found. The average pit is 0.1 mm in diameter and the vessel wall is 1 mm thick. If the pit developed over a period of 1 year, calculate (a) the corrosion current associated with each pit and (b) the corrosion current density (normalized to the area of the pit).

14.18 (a) Consider the corrosion current due to a volume of aluminum 0.1 mm diameter × 1 mm long lost over a period of one year:

$$I = [\pi(0.05 \times 10^{-3} m)^2](1 \times 10^{-3} m) \times \frac{2.70 Mg}{m^3}$$

$$\times \frac{0.6023 \times 10^{24} \text{ atoms}}{26.98 \times 10^{-6} Mg} \times \frac{3e^-}{\text{atom}} \times \frac{0.16 \times 10^{-18} C}{e^-}$$

$$\times \frac{1}{yr} \times \frac{1 yr}{365 \times 24 \times 3600 s} = 7.21 \times 10^{-9} C/s$$

$$= \underline{\underline{7.21 \times 10^{-9} A}}$$

(b) $$\text{Current density} = \frac{7.21 \times 10^{-9} A}{\pi(0.05 \times 10^{-3} m)^2}$$

$$= \underline{\underline{0.917 A/m^2}}$$

Section 14.6 - Methods of Corrosion Protection

PP 14.6 In Sample Problem 14.6, we calculate the average current for a sacrificial anode. Assume that the corrosion rate could be diminished by 25% by using an annealed block of magnesium. What mass of such an annealed anode would be needed to provide corrosion prevention for (a) 3 months and (b) a full year?

PP 14.6 (a) Corrosion rate for annealed magnesium would be:

$$0.75 \times 1.97\,A = 1.48\,A$$

which would, in turn, equal:

$$\frac{y\,kg}{3\,mo.} \times \frac{1000\,g}{kg} \times \frac{0.6023 \times 10^{24}\,atoms}{24.31\,g} \times \frac{2e^-}{atom}$$

$$\times \frac{0.16 \times 10^{-18}\,C}{e^-} \times \frac{1\,mo.}{(31)(24)(3600\,s)} \times \frac{1\,A}{C/s}$$

Solving for y, we obtain:

$$y = 1.5\,kg$$

(b) $$y = 1.5\,kg \times \frac{12\,mo.}{3\,mo.} = 6.0\,kg$$

14.19 A sacrificial anode of zinc provides corrosion protection with an average corrosion current of 2 A over the period of 1 year. What mass of zinc is required to give this protection?

14.19 Using data from Appendixes 1 and 2,

$$current = 2\,A = \frac{x\,kg}{12\,months} \times \frac{1000\,g}{1\,kg} \times \frac{0.6023 \times 10^{24}\,atoms}{65.38\,g\,Zn}$$

$$\times \frac{2\,electrons}{atom} \times \frac{0.16 \times 10^{-18}\,C}{electron} \times \frac{1\,month}{31\,d} \times \frac{1\,d}{24\,h}$$

$$\times \frac{1\,h}{3600\,s} \times \frac{1\,A}{1\,C/s}$$

or $x = 21.8$

$\therefore$ 21.8 kg Zn required.

14.20 A sacrificial anode of magnesium provides corrosion protection with an average corrosion current of 1.5 A over a period of 2 years. What mass of magnesium is required to give this protection?

14.20

Using data from Appendixes 1 and 2,

$$\text{current} = 1.5\,A = \frac{x\,kg}{2(12\text{ months})} \times \frac{1000\,g}{1\,kg} \times \frac{0.6023 \times 10^{24}\text{ atoms}}{24.31\,g\,Mg}$$

$$\times \frac{2\text{ electrons}}{\text{atom}} \times \frac{0.16 \times 10^{-18}\,C}{\text{electron}} \times \frac{1\text{ month}}{31\,d}$$

$$\times \frac{1\,d}{24\,h} \times \frac{1\,h}{3600\,s} \times \frac{1\,A}{1\,C/s}$$

or $x = 12.2$

$\therefore$ 12.2 kg Mg required

14.21 The maximum corrosion current density in a galvanized steel sheet is found to be 5 mA/m^2. What thickness of the zinc layer is necessary to ensure at least (**a**) 1 year and (**b**) 5 years of rust resistance?

14.21

(a) Consider $1\,m^2$ of the Zn coating t meter thick. The mass contained therein would be

$$m = 7.13\,Mg/m^3 \times (1\,m^2)(t\,m) = (7.13\,t)\,Mg$$

The corrosion current is, then,

$$5 \times 10^{-3}\,A/m^2 \times 1\,m^2 = \frac{(7.13\,t)\,Mg}{1\,yr} \times \frac{10^6\,g}{Mg} \times \frac{0.6023 \times 10^{24}\text{ atoms}}{65.38\,g}$$

$$\times \frac{2e^-}{\text{atom}} \times \frac{0.16 \times 10^{-18}\,C}{e^-} \times \frac{1\,yr}{(365)(24)(3600\,s)}$$

$$\times \frac{1\,A}{C/s}$$

or

$$t = 7.50 \times 10^{-6}\,m = 7.50\,\mu m$$

(b) $t_{5\,yr} = 5 \times t_{1\,yr} = 5\,(7.5\,\mu m) = 37.5\,\mu m$

14.22 A galvanized steel sheet has a zinc coating 18 μm thick. The corrosion current density is found to be 4 mA/m^2. What is the corresponding duration of rust resistance provided by this system?

14.22 Consider 1 m^2 of the Zn coating 18×10^{-6} m thick. The mass contained therein would be:

$$m = (7.13\ Mg/m^3)(1\ m^2)(18 \times 10^{-6}\ m)$$
$$= 1.28 \times 10^{-4}\ Mg$$

The corrosion current is, then,

$$4\frac{mA}{m^2} \times 1 m^2 = (1.28 \times 10^{-4}\ Mg)(10^6 g/Mg) \times \left(\frac{0.6023 \times 10^{24}\ atoms}{65.38\ g}\right)\left(\frac{2e^-}{atom}\right) \times \left(\frac{0.16 \times 10^{-18}\ C}{e^-}\right)\left(\frac{1A}{C/s}\right) \times \frac{1}{(3600\ s/h)(24 h/d)(365 d/yr)(n\ yr)}$$

$$4 \times 10^{-3} A = \frac{0.01197\ A}{n}$$

or $n = 3.00$

$\therefore$ corrosion protection lasts for 3.00 years

Section 14.8 - Radiation Damage

PP 14.7 In Sample Problem 14.7, we find that a given photon energy necessary for radiation damage of a semiconductor is not available in visible light. **(a)** What wavelength of radiation is represented by the photon energy of 15 eV? **(b)** What type of electromagnetic radiation has such wavelength values?

PP 14.7 (a) $E = hc/\lambda$ or $\lambda = hc/E$

$$\lambda = \frac{(0.6626 \times 10^{-33}\ J \cdot s)(0.2998 \times 10^9\ m/s)}{(15 eV)(1 J/6.242 \times 10^{18}\ eV)}$$
$$= 82.7 \times 10^{-9} m = 82.7\ nm$$

(b) Table 14-5 and Figure 14-18 identify this as ultraviolet radiation.

14.23 In Problem 9.55, the ultraviolet wavelengths necessary to break carbon bonds (in polymers) were calculated. Another type of radiation damage found in a variety of solids is associated with electron–positron "pair production," which can occur at a threshold photon energy of 1.02 MeV. **(a)** What is the wavelength of such a threshold photon? **(b)** Which type of electromagnetic radiation is this?

14.23 (a) $\lambda = \frac{hc}{E} = \frac{(0.6626 \times 10^{-33}\ J \cdot s)(0.2998 \times 10^9\ m/s)}{(1.02 \times 10^6 eV)(1 J/6.242 \times 10^{18}\ eV)} \times \frac{10^9\ nm}{m} = 1.22 \times 10^{-3}\ nm$

(b) Table 14.5 and Figure 14-18 indicate this radiation could be either X-rays or γ-rays.

14.24 Calculate the full range of photon energies associated with **(a)** ultraviolet radiation and **(b)** x-radiation in Table 14.5.

14.24 (a) $E = \frac{hc}{\lambda}$ with $1\,nm < \lambda < 400\,nm$

or $\frac{hc}{400\,nm} < E < \frac{hc}{1\,nm}$

or

$$\frac{(0.6626\times10^{-33}\,J\cdot s)(0.2998\times10^{9}\,m/s)}{(400\times10^{-9}\,m)(1\,J/6.242\times10^{18}\,eV)} < E < \frac{(0.6626\times10^{-33}\,J\cdot s)(0.2998\times10^{9}\,m/s)}{(1\times10^{-9}\,m)(1\,J/6.242\times10^{18}\,eV)}$$

or $\underline{\underline{3.10\,eV < E < 1.24\,keV}}$

(b) For $10^{-3}\,nm < \lambda < 10\,nm$,

$\frac{hc}{10\,nm} < E < \frac{hc}{10^{-3}\,nm}$

or

$$\frac{(0.6626\times10^{-33}\,J\cdot s)(0.2998\times10^{9}\,m/s)}{(10\times10^{-9}\,m)(1\,J/6.242\times10^{18}\,eV)} < E < \frac{(0.6626\times10^{-33}\,J\cdot s)(0.2998\times10^{9}\,m/s)}{(10^{-3}\times10^{-9}\,m)(1\,J/6.242\times10^{18}\,eV)}$$

or $\underline{\underline{124\,eV < E < 1.24\,MeV}}$

Section 14.9 - Wear

PP 14.8 In Sample Problem 14.8, the diameter of a wear particle is calculated for the case of two steel surfaces sliding together. In a similar way, calculate the diameter of a wear particle for the same sliding combination under the same conditions, but with the 1040 steel heat treated to a hardness of 200 Brinell hardness number (BHN).

PP 14.8 Using Equation 14.13 and the various given data,

$$V = \frac{kPx}{3H}$$

$$= \frac{(45\times10^{-3})(50\,kg)(5\,mm)}{3(200\,kg/mm^2)} = 0.0188\,mm^3$$

or

$$d = \sqrt[3]{\frac{12(0.0188\,mm^3)}{\pi}} = 0.415\,mm = \underline{\underline{415\,\mu m}}$$

14.25 Calculate the diameter of a wear particle for copper sliding on a 1040 steel. Take the load to be 40 kg over a distance of 10 mm. (Take the hardness of 1040 steel to be given by Table 7.14.)

14.25

$$V = \frac{kPx}{3H}$$

$$= \frac{(1.5\times10^{-3})(40\,kg)(10\,mm)}{3(235\,kg/mm^2)} = 8.51\times10^{-4}\,mm^3$$

or

$$d = \sqrt[3]{\frac{12(8.51\times10^{-4}\,mm^3)}{\pi}} = 0.148\,mm = \underline{\underline{148\,\mu m}}$$

14.26 Calculate the diameter of a wear particle produced by the adhesive wear of two 410 stainless steel surfaces under the load conditions of Problem 14.25. (Note Table 7.14 for hardness data.)

14.26

$$V = \frac{kPx}{3H}$$

$$= \frac{(21\times10^{-3})(40\,kg)(10\,mm)}{3(250\,kg/mm^2)} = 0.0112\,mm^3$$

or

$$d = \sqrt[3]{\frac{12(0.0112\,mm^3)}{\pi}} = 0.350\,mm = \underline{\underline{350\,\mu m}}$$

Section 14.10 - Surface Analysis

PP 14.9 Characteristic photon and electron energies are calculated in Sample Problem 14.9. Using the given data, calculate (a) the L_α characteristic photon energy and (b) the LMM Auger electron energy.

PP14.9

$$(a)\ E_{L_\alpha} = |E_L - E_M| = |-708\,eV - (-53\,eV)| = \underline{\underline{655\,eV}}$$

$$(b)\ E_{LMM} = |E_L - E_M| - |E_M|$$

$$= |-708\,eV - (-53\,eV)| - |-53\,eV| = \underline{\underline{602\,eV}}$$

14.27 The electron energy levels for a copper atom are $E_K = -8982$ eV, $E_L = -933$ eV, and $E_M = -75$ eV. Calculate (a) the K_α photon energy, (b) the K_β photon energy, (c) the L_α photon energy, (d) the KLL Auger electron energy, and (e) the LMM Auger electron energy.

14.27 (a) $E_{K_\alpha} = |E_K - E_L| = |-8982\,eV - (-933\,eV)| = \underline{\underline{8049\,eV}}$

(b) $E_{K_\beta} = |E_K - E_M| = |-8982\,eV - (-75\,eV)| = \underline{\underline{8907\,eV}}$

(c) $E_{L_\alpha} = |E_L - E_M| = |-933\,eV - (-75\,eV)| = \underline{\underline{858\,eV}}$

(d) $E_{KLL} = |E_K - E_L| - |E_L| = 8049\,eV - 933\,eV = \underline{\underline{7116\,eV}}$

(e) $E_{LMM} = |E_L - E_M| - |E_M| = 858\,eV - 75\,eV = \underline{\underline{783\,eV}}$

14.28 The K shell electron energy for nickel is $E_K = -8333$ eV and the wavelengths of the NiK_α and NiK_β photons are 0.1660 nm and 0.1500 nm, respectively. **(a)** Draw an energy-level diagram for a nickel atom. Calculate **(b)** the KLL and **(c)** the LMM Auger electron energies for nickel.

14.28 (a) $|E_K - E_L| = E_{K_\alpha} = \dfrac{hc}{\lambda_{K_\alpha}}$

$$= \frac{(0.6626\times10^{-33}\,J\cdot s)(0.2998\times10^{9}\,m/s)}{(0.1660\times10^{-9}\,m)(1\,J/6.242\times10^{18}\,eV)}$$

$= 7470\,eV$

$\therefore E_L = -8333\,eV + 7470\,eV = -863\,eV$

$|E_K - E_M| = E_{K_\beta} = \dfrac{hc}{\lambda_{K_\beta}}$

$$= \frac{(0.6626\times10^{-33}\,J\cdot s)(0.2998\times10^{9}\,m/s)}{(0.1500\times10^{-9}\,m)(1\,J/6.242\times10^{18}\,eV)}$$

$= 8266\,eV$

$\therefore E_M = -8333\,eV + 8266\,eV = -67\,eV$

giving the diagram:

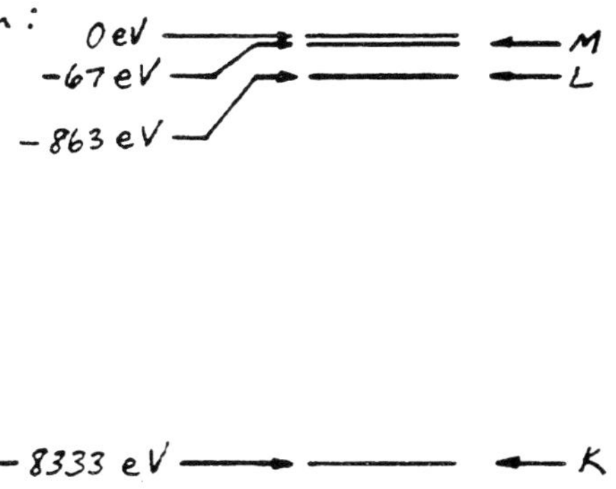

(b) $E_{KLL} = |E_K - E_L| - |E_L|$

$= 7470\,eV - (863\,eV) = \underline{\underline{6607\,eV}}$

(c) $E_{LMM} = |E_L - E_M| - |E_M|$

$= |-863\,eV - (-67\,eV)| - |-67\,eV|$

$= 796\,eV - 67\,eV = \underline{\underline{729\,eV}}$

• **14.29** Characteristic photon energies are generally measured in a so-called energy-dispersive mode in which a solid-state detector measures energy directly (see Problem 12.24). An alternate technique is the "wavelength dispersive" mode in which the photon energy is determined indirectly by measuring the x-ray wavelength by diffraction (see Section 3.7). **(a)** Calculate the diffraction angle (2θ) needed to identify the FeK_α photon using the (200) planes of an NaCl single crystal. **(b)** Sketch the experimental system for this measurement.

14.29 (a) First, we need to calculate d_{200} for NaCl using data from Appendix 2:

$$d_{200} = \frac{2r_{Na^+} + 2r_{Cl^-}}{2} = 0.098\,nm + 0.181\,nm$$

$$= 0.279\,nm$$

Assuming first order diffraction,

$$\lambda = 2d\sin\theta \quad \text{or} \quad \theta = \arcsin\frac{\lambda}{2d}$$

Noting in Problem 12.24 that $\lambda_{FeK_\alpha} = 0.1938\,nm$,

$$2\theta = 2\arcsin\frac{(0.1938\,nm)}{2(0.279\,nm)} = \underline{\underline{40.6°}}$$

(b)

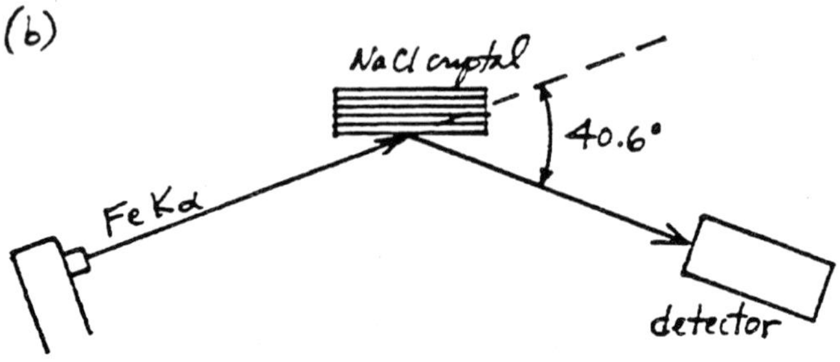

X-ray source
(Shown as X-ray tube. Could also be secondary radiation from a sample being analyzed.)

• **14.30** As pointed out in the footnote on page 557, an alternate surface analysis technique is x-ray photoelectron spectroscopy (XPS). In this case, a "soft" x ray such as AlK_α is used to eject an inner orbital electron (from an atom in the sample) giving a characteristic kinetic energy. **(a)** Sketch the mechanism for XPS in a manner similar to Figures 14-23 and 14-24. **(b)** Calculate the specific photoelectron energy that could be used to identify an iron atom. (Note that aluminum electron energy levels are $E_K = -1560$ eV and $E_L = -72.8$ eV.)

14.30 (a)

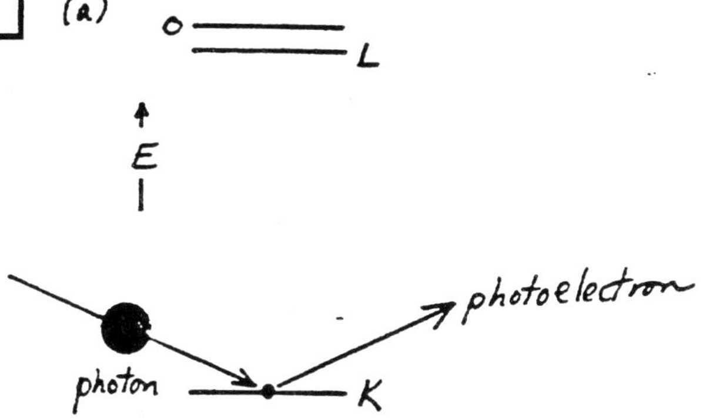

(b) To eject a photoelectron, the photon energy must be greater than the orbital electron's binding energy. The AlK_α photon energy is:

$$E_{Al, K_\alpha} = |E_K - E_L| = |-1560\,eV - (-72.8\,eV)|$$
$$= 1487\,eV$$

By inspection of the data given in Sample Problem 14.9, it is obvious that the Al K_α photon can not produce photoelectrons from the K shell but can from either the L or M shells.

$$\therefore E_{L\,photoelectron} = 1487\,eV - |-708\,eV| = \underline{\underline{779\,eV}}$$

or

$$E_{M\,photoelectron} = 1487\,eV - |-53\,eV| = \underline{\underline{1434\,eV}}$$

Section 14.11 - Failure Analysis and Prevention

PP 14.10 Calculate the fraction of beam intensity for a 100-keV x-ray beam transmitted through a 10-mm-thick plate of **(a)** titanium and **(b)** lead. (See Sample Problem 14.10.)

PP 14.10

(a) Using Equation 14.14 and Table 14.9,

$$I = I_0 e^{-\mu x} \quad \text{or} \quad I/I_0 = e^{-\mu x}$$

and

$$I/I_0 = e^{-(0.124\ \text{mm}^{-1})(10\ \text{mm})}$$

$$= \underline{\underline{0.289}}$$

(b) $$I/I_0 = e^{-(6.20\ \text{mm}^{-1})(10\ \text{mm})}$$

$$= \underline{\underline{1.19 \times 10^{-27}}}$$

PP 14.11 Given the data in Sample Problem 14.11 and $\rho_{H_2O} = 1.00\ \text{Mg/m}^3$ and $V_{H_2O} = 1{,}483$ m/s, calculate the fraction of an ultrasonic pulse reflected from the surface of an aluminum plate in a water immersion bath.

PP 14.11

Using Equation 14.15 and given data,

$$I_r/I_i = [(Z_{Al} - Z_{H_2O})/(Z_{Al} + Z_{H_2O})]^2$$

and

$$Z_{H_2O} = \rho_{H_2O} V_{H_2O} = (1.00\ \text{Mg/m}^3)(1{,}483\ \text{m/s})$$

$$= 1.48 \times 10^3\ \text{Mg/(m}^2\text{s)}$$

giving

$$I_r/I_i = [(17.1 - 1.48)/(17.1 + 1.48)]^2$$

$$= \underline{\underline{0.707}}$$

14.31 In doing x-radiography of steel, assume the film can detect a variation in radiation intensity represented by $\Delta I/I_0 = 0.001$. What thickness variation could be detected using this system for the inspection of a 12.5-mm-thick plate of steel using a 100-keV beam?

14.31

Using Equation 14.14 and Table 14.9,

$$(I/I_0)_{12.5\,mm} = e^{-(0.293\,mm^{-1})(12.5\,mm)}$$

$$= 0.0257$$

∴ Thickness range defined by I/I_0 range of:

$$0.0267 \text{ to } 0.0247$$

$$0.0267 = e^{-(0.293\,mm^{-1})t}$$

or

$$\ln 0.0267 = -(0.293\,mm^{-1})t$$

or $t = 12.37\,mm = (12.50 - 0.13)\,mm$

and

$$0.0247 = e^{-(0.293\,mm^{-1})t}$$

or

$$\ln 0.0247 = -(0.293\,mm^{-1})t$$

or $t = 12.64\,mm = (12.50 + 0.14)\,mm$

∴ detectable t variation $= 12.50\,mm \begin{smallmatrix} +0.14\,mm \\ -0.13\,mm \end{smallmatrix}$

14.32 A good rule of thumb for doing x-radiographic inspections is that the test piece should be 5 to 8 times the half-value thickness ($t_{1/2}$) of the material, where $t_{1/2}$ is defined as the thickness value corresponding to an I/I_0 value of 0.5. Calculate the appropriate test piece thickness range for titanium inspected by a 100-keV beam.

14.32

Using Equation 14.14 and Table 14.9,

$$I/I_0 = 0.5 = e^{-(0.124\,mm^{-1})t_{1/2}}$$

then, $t_{1/2} = -\dfrac{\ln 0.5}{(0.124\,mm^{-1})} = 5.59\,mm$

$\therefore$ thickness range $= 5 \times (5.59\text{ mm}) = \underline{\underline{27.9\text{ mm}}}$

to $8 \times (5.59\text{ mm}) = \underline{\underline{44.7\text{ mm}}}$

14.33 Using the background of Problem 14.32, calculate the appropriate test piece thickness range for tungsten inspected by a 100-keV beam.

14.33

Using Equation 14.14 and Table 14.9,

$$I/I_0 = 0.5 = e^{-(8.15\text{ mm}^{-1})\,t_{1/2}}$$

or

$$t_{1/2} = -\frac{\ln 0.5}{(8.15\text{ mm}^{-1})} = 0.085\text{ mm}$$

$\therefore$ thickness range $= 5 \times 0.085\text{ mm} = \underline{\underline{0.43\text{ mm}}}$

to $8 \times 0.085\text{ mm} = \underline{\underline{0.68\text{ mm}}}$

14.34 Using the background of Problem 14.32, calculate the appropriate test piece thickness range for iron inspected by **(a)** a 100-keV beam and **(b)** a 1-MeV beam.

14.34

Using Equation 14.14 and Table 14.8,

(a) $$I/I_0 = 0.5 = e^{-(0.293\text{ mm}^{-1})\,t_{1/2}}$$

or

$$t_{1/2} = -\frac{\ln 0.5}{(0.293\text{ mm}^{-1})} = 2.37\text{ mm}$$

$\therefore$ thickness range $= 5 \times 2.37\text{ mm} = \underline{\underline{11.8\text{ mm}}}$

to $8 \times 2.37\text{ mm} = \underline{\underline{18.9\text{ mm}}}$

(b) $$t_{1/2} = -\frac{\ln 0.5}{(0.0417\text{ mm}^{-1})} = 16.6\text{ mm}$$

$\therefore$ thickness range $= 5 \times 16.6\text{ mm} = \underline{\underline{83.1\text{ mm}}}$

to $8 \times 16.6\text{ mm} = \underline{\underline{133\text{ mm}}}$

14.35 Assume the geometry of the pulse echo ultrasonic test illustrated in Figure 14-29 represents an aluminum sample as follows: transducer-to-sample front surface = 25 mm, flaw depth = 10 mm, and overall sample thickness = 20 mm. Using data from Sample Problem 14.11 and Practice Problem 14.11, calculate the time delay between the initial pulse and (**a**) the front surface echo, (**b**) the flaw echo, and (**c**) the back surface echo.

14.35

In general, time = distance/velocity.

(a) The front surface echo represents a pulse traveling "round trip" through the water bath:

$$t = \frac{d}{V} = \frac{2 \times 25\text{ mm}}{V_{H_2O}} = \frac{50 \times 10^{-3}\text{ m}}{1.483 \times 10^{3}\text{ m/s}} = 33.7 \times 10^{-6}\text{ s}$$

$$= \underline{\underline{33.7\ \mu s}}$$

(b) The flaw echo represents the result from (a) plus a "round trip" in the aluminum from the surface to the flaw:

$$t = 33.7\ \mu s + \frac{2 \times 10\text{ mm}}{V_{Al}}$$

$$= 33.7\ \mu s + \frac{20 \times 10^{-3}\text{ m}}{6.32 \times 10^{3}\text{ m/s}}$$

$$= 33.7\ \mu s + 3.2 \times 10^{-6}\text{ s} = \underline{\underline{36.9\ \mu s}}$$

(c) The back surface echo represents the result from (b) plus an addition "round trip" through the bottom half of the aluminum:

$$t = 33.7\ \mu s + \frac{4 \times 10\text{ mm}}{V_{Al}}$$

$$= 33.7\ \mu s + \frac{40 \times 10^{-3}\text{ m}}{6.32 \times 10^{3}\text{ m/s}}$$

$$= 33.7\ \mu s + 6.3 \times 10^{-6}\text{ s} = \underline{\underline{40.0\ \mu s}}$$

14.36 For the specific test configuration given in Problem 14.35, calculate the relative intensities of the **(a)** front surface, **(b)** flaw, and **(c)** back surface echoes. (Take $I_{\text{initial}} = 100\%$, flaw area = 1/3 beam area, and flaw to be air-filled, for which you can take $Z_{\text{air}} = 0$.)

14.36

(a) The front surface echo will be diminished by a single reflection at that surface:

As calculated in Practice Problem 14.11

$$I_r / I_i = 0.707$$

$$\therefore\ I = 100\%\,(0.707) = \underline{\underline{70.7\%}}$$

(b) The flaw echo will be diminished by the fact that only 1/3 of the unreflected pulse crossing the front face is totally reflected by the flaw. Furthermore, 70.7% of that reflection is reflected back by the front surface on the "return trip":

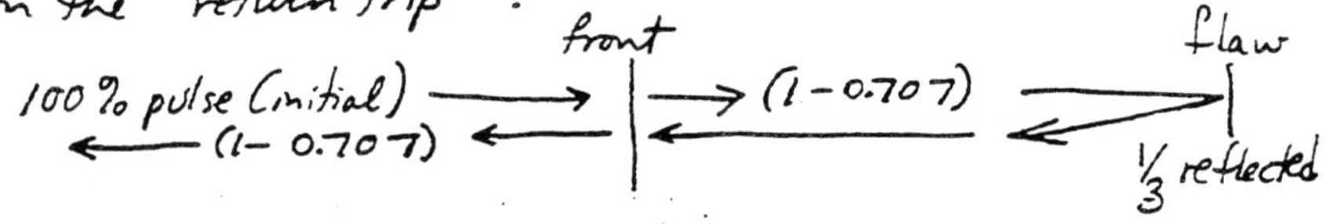

$$\therefore\ I = 100\%\,(1-0.707)\left(\tfrac{1}{3}\right)(1-0.707) = \underline{\underline{2.9\%}}$$

(c) The back echo is similar to the flaw echo except that the remaining 2/3 of the pulse is reflected with a ratio of 0.707:

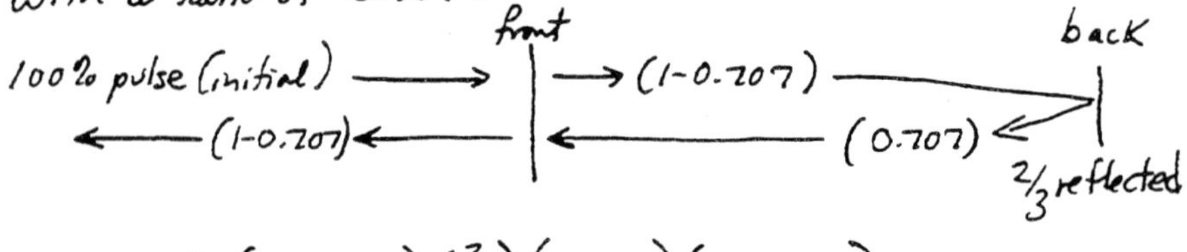

$$\therefore\ I = 100\%\,(1-0.707)\left(\tfrac{2}{3}\right)(0.707)(1-0.707)$$

$$= \underline{\underline{4.0\%}}$$

14.37 For an x-radiographic inspection of the plate in Problem 14.35 using a 100 keV beam, calculate the % change in I/I_0 between the flawed and flaw-free areas given that the effect of the flaw is to reduce the effective thickness of the plate by 10 μm.

14.37

The thickness of the flaw-free area is 20 mm.
The thickness of the flawed area is 20 mm − 10 μm.

$$\left(\frac{I}{I_0}\right)_{t=20\,mm} = e^{-(0.0459\,mm^{-1})(20\,mm)}$$

$$= 0.399317$$

Note that 20 mm − 10 μm = 20 mm − 0.01 mm = 19.99 mm

$$\left(\frac{I}{I_0}\right)_{t=19.99\,mm} = e^{-(0.0459\,mm^{-1})(19.99\,mm)}$$

$$= 0.399500$$

$$\therefore \%\,change = \frac{0.399500 - 0.399317}{0.399317} \times 100\,\% = \underline{\underline{0.046\,\%}}$$

14.38 Given the result of Problem 14.37, comment on the relative advantage of ultrasonic testing for the inspection of the flaw in that sample.

14.38

Obviously, the effect of a small thickness crack on radiographic transmission is negligible. On the other hand, the high efficiency of ultrasonic reflection $\left[I_r/I_i = \{(Z_{Al} - Z_{air})/(Z_{Al} + Z_{air})\}^2 = 1.0\right]$ does not depend on flaw thickness, allowing even a thin crack to be easily detected.

Section 15.1 - Processing of Metals

PP 15.1 **(a)** In Sample Problem 15.1, we determine a range of copper–nickel alloy compositions that meet structural requirements for strength and ductility. Make a similar determination for the specifications: hardness greater than 80 R_F and ductility less than 45%. **(b)** For the range of copper–nickel alloy compositions determined in part (a), which specific alloy would be preferred on a cost basis, given that the cost of copper is approximately $3.70/kg and of nickel $10.30/kg?

PP 15.1

(a) Again using Figure 15-4, we obtain property ranges:

Hardness $> 80 R_f$: $34 < \% Ni < 100$

Elongation $< 45\%$: $0 < \% Ni < 79$

Giving a net "window" of $\underline{\underline{34 < \% Ni < 79}}$

(b) As Cu is the less costly component, its amount should be maximized and, conversely, the % Ni minimized. Therefore, the $\underline{\underline{34\% \text{ Ni alloy}}}$ is preferred.

PP 15.2 In Sample Problem 15.1, we calculate the tensile strength and ductility for a cold-worked bar of 70 Cu–30 Zn brass. **(a)** What percent increase does that tensile strength represent compared to that for the annealed bar? **(b)** What percent decrease does that ductility represent compared to that for the annealed bar?

PP 15.2

(a) For the annealed bar (0% cold work), Figure 15-5 indicates that tensile strength is 320 MPa. Therefore,

$$\% \text{ increase} = \frac{520-320}{320} \times 100\% = \underline{\underline{63\%}}$$

(b) Similarly, the ductility of the annealed bar is 65%. Therefore,

$$\% \text{ decrease} = \frac{65-9}{65} \times 100\% = \underline{\underline{86\%}}$$

15.1 A bar of annealed 85 Cu–15 Zn (12-mm diameter) is cold drawn through a die with diameter of 10 mm. What are (a) the tensile strength and (b) the ductility of the resulting bar?

15.1

$$\% \text{ cold work} = \frac{\text{initial area} - \text{final area}}{\text{initial area}} \times 100\%$$

$$= \frac{\pi/4\,(12\text{ mm})^2 - \pi/4\,(10\text{ mm})^2}{\pi/4\,(12\text{ mm})^2} \times 100\%$$

$$= 30.6\%$$

From Figure 15-5, then,

(a) tensile strength = 450 MPa

& (b) ductility (elongation) = 8%

15.2 For the bar analyzed in Problem 15.1, (a) what percentage does the tensile strength represent compared to that for the annealed bar and (b) what percentage decrease does the ductility represent compared to that for the annealed bar?

15.2

(a) For 0% cold work, tensile strength = 280 MPa (by extrapolation). Therefore,

$$\% \text{ strength increase} = \frac{450 - 280}{280} \times 100\% = 61\%$$

(b) For 0% cold work, ductility = 47%. Therefore,

$$\% \text{ ductility decrease} = \frac{47 - 8}{47} \times 100\% = 83\%$$

15.3 You are given a 2-mm-diameter wire of 85 Cu–15 Zn brass. It must be drawn down to a diameter of 1 mm. The final product must meet specifications of tensile strength greater than 375 MPa and a ductility of greater than 20%. Describe a processing history to provide this result.

15.3 To produce the final diameter in a single cold drawing operation would correspond to a

$$\%\text{ cold work} = \frac{\pi/4\,(2\text{mm})^2 - \pi/4\,(1\text{mm})^2}{\pi/4\,(2\text{mm})^2} \times 100\,\% = 75\%$$

Although Figure 15-5 indicates this operation would meet the strength specification, it would definitely not meet the ductility specification.

Instead, we must define a "window" of allowable cold working.

Tensile strength > 375 MPa : % cold work > 15%

Ductility > 20% : % cold work < 17%

Taking a mid-value of 16% cold work, we can determine an "initial" diameter, D:

$$\frac{D^2 - 1^2}{D^2} = 0.16$$

or $D^2 - 0.16D^2 = 1$ or $D = 1.078$ mm

Therefore, a processing history would be to hot work the wire from D = 2 mm to D = 1.078 mm and then cold work to the final D = 1 mm.

(Alternate solution: Cold work from D = 2 mm to D = 1.078 mm, anneal, and then cold work to D = 1 mm.)

15.4 How would your answer to Problem 15.3 change if a 70 Cu–30 Zn brass wire were used instead of the 85 Cu–15 Zn material?

15.4 To answer this, we must identify the % cold work "window" for this alloy:

Tensile strength >375 MPa : % cold work > 9%

Ductility > 20% : % cold work < 23%

As the cold work value used in Problem 15.3 (16%) is in the range 9-23%, the answer for that problem is valid for this alloy also. However, the wider "window" for the 70 Cu-30 Zn brass would allow a wider range of intermediate D values. For 9% cold work,

$$\frac{D^2 - 1}{D^2} = 0.09 \quad \text{or} \quad D = 1.048 \text{ mm}$$

and for 23% cold work,

$$\frac{D^2 - 1}{D^2} = 0.23 \quad \text{or} \quad D = 1.140 \text{ mm}.$$

Therefore, the two answers from the previous problem could be re-stated for the 70-30 alloy as:

i) hot work from D = 2 mm to 1.048 mm < D < 1.140 mm and then cold work to D = 1 mm

or

ii) cold work from D = 2 mm to 1.048 mm < D < 1.140 mm, anneal, and then cold work to D = 1 mm.

Section 15.2 - Processing of Ceramics and Glasses

PP 15.3 In scaling up the laboratory firing operation of Sample Problem 15.3 to a production level, 6.05×10^3 kg of kaolinite are fired. How much H_2O is driven off in this case?

PP 15.3

As 5 kg kaolinite → 698 g H_2O,

$$m_{H_2O} = \frac{6.05 \times 10^3 \text{ kg Kaolinite}}{5 \text{ kg kaolinite}} \times 698 \text{ g}$$

$$= 8.45 \times 10^5 \text{ g} = \underline{\underline{845 \text{ kg}}}$$

PP 15.4 For the glass bottle production described in Sample Problem 15.4, calculate the melting range for this manufacturing process.

PP 15.4

For the melting range, η = 50 P to 500 P.

For η = 50 P,

$$T = \frac{460 \times 10^3 \text{ J/mol}}{[8.314 \text{ J/(mol}\cdot\text{K)}]\ln(50/6.11 \times 10^{-19})}$$

$$= 1207 \text{ K} = 934^\circ\text{C}$$

For η = 500 P,

$$T = \frac{460 \times 10^3 \text{ J/mol}}{[8.314 \text{ J/(mol}\cdot\text{K)}]\ln(500/6.11 \times 10^{-19})}$$

$$= 1149 \text{ K} = 876^\circ\text{C}$$

∴ melting range = $\underline{\underline{876 \text{ to } 934^\circ\text{C}}}$

15.5 For the ceramic in Sample Problem 15.1, use the Al_2O_3–SiO_2 phase diagram in Chapter 5 to determine the maximum firing temperature to prevent formation of a silica-rich liquid.

15.5

As calculated in Sample Problem 5.11, the Composition of fired kaolinite is 33.3 mol. % Al_2O_3.

Inspection of Figure 5-39 indicates a substantial amount of liquid phase above the eutectic temperature making the maximum firing temperature:

$$\underline{\underline{1587\,°C}}$$

15.6 How would your answer to Problem 15.5 change if the ceramic being fired was composed of two parts Al_2O_3 in combination with one part kaolin?

15.6

Assume 10 kg Al_2O_3 is added to the 5 kg of kaolinite in Sample Problem 15.3.

The total mass of SiO_2 in the system will be:

$$\frac{2[28.09 + 2(16.00)]\ \text{amu}}{258.2\ \text{amu}} \times 5\ \text{kg} = 2.327\ \text{kg}$$

The total mass of Al_2O_3 will be:

$$10\ \text{kg} + \frac{[2(26.98) + 3(16.00)]\ \text{amu}}{258.2\ \text{amu}} \times 5\ \text{kg} = 11.974\ \text{kg}$$

Then,

$$\text{moles } SiO_2 = \frac{2{,}327\ \text{g}}{[28.09 + 2(16.00)]\ \text{g/mol}} = 38.7\ \text{moles}$$

and

$$\text{moles } Al_2O_3 = \frac{11{,}974\ \text{g}}{[2(26.98) + 3(16.00)]\ \text{g/mol}} = 117.4\ \text{moles}$$

$$\therefore\ \text{mol. \% } Al_2O_3 = \frac{117.4\ \text{mol}}{117.4\ \text{mol} + 38.7\ \text{mol}} \times 100 = 75.2\ \text{mol. \%}$$

This composition is in the range in which a peritectic reaction occurs and allows a substantially higher firing temperature:

$$\underline{\underline{1890\,°C}}$$

15.7 For simplified processing calculations, the glass bottle shaping temperature in Figure 15-7 can be taken as the softening point (at which $\eta = 10^{7.6} P$) and the subsequent annealing temperature as the annealing point (at which $\eta = 10^{13.4} P$). If the glass bottle processing sequence of Figure 15-7 involves a shaping temperature of 700°C for a glass with an activation energy for viscous deformation of 475 kJ/mol, calculate the appropriate annealing temperature.

15.7

$$\eta = \eta_0 e^{+Q/RT}$$

Given $\eta = 10^{7.6} P$ at $700°C = 973 K$,

$$\eta_0 = \eta e^{-Q/RT}$$

$$= 10^{7.6} P e^{-(475 \times 10^3 J/mol)/[8.314 J/(mol \cdot K)](973 K)}$$

$$= 1.26 \times 10^{-18} P$$

At the annealing point for which $\eta = 10^{13.4} P$,

$$T = \frac{Q}{R \ln(\eta/\eta_0)}$$

$$= \frac{475 \times 10^3 J/mol}{[8.314 J/(mol \cdot K)] \ln(10^{13.4}/1.26 \times 10^{-18})}$$

$$= 793 K = \underline{\underline{520°C}}$$

15.8 Using the approach of Problem 15.7, assume a change in raw material suppliers changes the glass composition thus reducing the softening point to 690°C and the activation energy to 470 kJ/mol. Recalculate the appropriate annealing temperature.

15.8

$$\eta_0 = 10^{7.6} P e^{-(470 \times 10^3 J/mol)/[8.314 J/(mol \cdot K)](690+273)K}$$

$$= 1.28 \times 10^{-18} P$$

Then,

$$T = \frac{470 \times 10^3 J/mol}{[8.314 J/(mol \cdot K)] \ln(10^{13.4}/1.28 \times 10^{-18})}$$

$$= 785 K = \underline{\underline{512°C}}$$

15.9 How many grams of N_2 gas are consumed in forming 100 g of TiN by self-propagating high temperature synthesis (SHS)?

15.9

$$Ti(s) + \frac{1}{2} N_2(g) = TiN(s)$$

$$1 \text{ mol } TiN = 47.90 \text{ g } Ti + 14.01 \text{ g } N$$
$$= 61.91 \text{ g } TiN$$

$$\frac{1}{2} \text{ mol } N_2 = 14.01 \text{ g } N$$

$$\therefore m_{N_2} = \frac{100 \text{ g } TiN}{61.91 \text{ g } TiN} \times 14.01 \text{ g}$$
$$= \underline{\underline{22.6 \text{ g}}}$$

15.10 How many grams of Ti powder were initially required in the SHS process of Problem 15.9?

15.10

$$1 \text{ g-atom } Ti = 47.90 \text{ g } Ti$$

$$\therefore m_{Ti} = \frac{100 \text{ g } TiN}{61.91 \text{ g } TiN} \times 47.90 \text{ g}$$
$$= \underline{\underline{77.4 \text{ g}}}$$

Section 15.3 - Processing of Polymers

PP 15.5 What would be the degree of polymerization for the polyethylene in Sample Problem 15.5 if the H_2O_2 addition is reduced to 0.14 wt %?

PP 15.5

For 0.14 wt.% H_2O_2,

$$0.14 \text{ wt.\% } H_2O_2 = \frac{2(1.008) + 2(16.00)}{n[2(12.01) + 4(1.008)]} \times 100$$

or

$$n = \underline{\underline{866}}$$

15.11 The heat evolved in the manufacturing of polyethylene sheet was evaluated in Problem 2.29. In a similar way, calculate the heat evolution occurring during a 24-hour period in which 864 km of 1 mil (25.4 μm) thick sheet (300 mm wide) are manufactured. (The density of the sheet is 0.910 Mg/m³.)

15.11

$$V = 25.4\,\mu m \times 300\,mm \times 864 \times 10^3\,m$$

$$= 25.4 \times 10^{-6}\,m \times 300 \times 10^{-3}\,m \times 864 \times 10^3\,m = 6.58\,m^3$$

$$m_{C_2H_4} = 0.910 \times 10^6\,g/m^3 \times 6.58\,m^3 = 5.99 \times 10^6\,g$$

$$mol_{C_2H_4} = \frac{5.99 \times 10^6\,g}{[2(12.01) + 4(1.008)]\,g/mol} = 2.14 \times 10^5\,moles$$

As there is one double bond per C_2H_4, the total reaction energy is:

$$E_{reaction} = \frac{60\,kJ}{mol} \times 2.14 \times 10^5\,mol$$

$$= \underline{\underline{12.8 \times 10^6\,kJ}}$$

15.12 What would be the total heat evolution occurring in the manufacturing of the polyethylene sheet in Problem 15.11 if that production rate could be maintained for a full year?

15.12

For one year (= 365 days)

$$E_{reaction} = 365 \times 12.8 \times 10^6\,kJ$$

$$= \underline{\underline{4.68 \times 10^9\,kJ}}$$

Section 15.4 - Processing of Composites

PP 15.6 For a non-air-entrained concrete using Type I cement, what % increase in compressive strength would you expect 28 days after pouring if you used a water/cement ratio of 0.60 rather than 0.50? (See Sample Problem 15.6.)

PP 15.6

Inspection of the 28 day graph on the left side of Fig. 15-12(b), we find the mid-range strengths to be:

water/cement ratio	strength (psi)
0.50	5,200
0.60	4,100

$$\therefore \ \% \text{ increase} = \frac{5{,}200 - 4{,}100}{4{,}100} \times 100\ \%$$

$$= \underline{\underline{26.8\ \%}}$$

15.13 In producing an air-entrained concrete, estimate the compressive strength 28 days after pouring if you mix 16,500 kg water with 30,000 kg Type I cement.

15.13

$$\text{water/cement ratio} = \frac{16{,}500 \text{ kg}}{30{,}000 \text{ kg}} = 0.55$$

Inspection of the 28 day graph on the left side of Fig. 15-12 (a), we find the range of strengths to be: $\underline{\underline{3{,}200 \text{ to } 4{,}200 \text{ psi}}}$

15.14 In producing a non-air-entrained concrete, estimate the compressive strength 28 days after pouring if you mix 16,500 kg water with 30,000 kg Type I cement.

15.14

$$\text{water/cement ratio} = \frac{16{,}500 \text{ kg}}{30{,}000 \text{ kg}} = 0.55$$

Inspection of the 28 day graph on the left side of Fig. 15-12(b), we find the range of strengths to be: $\underline{\underline{4{,}100 \text{ to } 5{,}200 \text{ psi}}}$

15.15 Consider the injection molding of low-cost casings using a polyethylene–clay particle composite system. The modulus of elasticity of the composite increases and the tensile strength of the composite decreases with volume fraction of clay as follows:

Volume fraction clay	Modulus of elasticity (MPa)	Tensile strength (MPa)
0.3	830	24.0
0.6	2,070	3.4

Assuming both modulus and strength change linearly with volume fraction of clay, determine the allowable composition range that ensures a product with a modulus of at least 1,000 MPa and a strength of at least 10 MPa.

15.15 Over the range of $v_{clay} = 0.3$ to 0.6, take

$$E = m\,v_{clay} + b$$

and

$$830 = m(0.3) + b$$

also

$$2{,}070 = m(0.6) + b$$

$$\therefore \quad 830 - 0.3\,m = 2{,}070 - 0.6\,m$$

or

$$m = \frac{2{,}070 - 830}{0.6 - 0.3} = 4{,}133$$

Then,

$$b = 830 - 0.3(4{,}133) = -410$$

The lower limit $E = 1{,}000$ will then occur at

$$v_{clay} = \frac{1{,}000 - (-410)}{4{,}133} = 0.341$$

Similarly,

$$TS = m'\,v_{clay} + b'$$

or

$$m' = \frac{3.4 - 24.0}{0.6 - 0.3} = -68.7$$

and

$$b' = 24.0 - 0.3(-68.7) = 44.6$$

The lower limit $TS = 10$ will then occur at

$$v_{clay} = \frac{10 - 44.6}{-68.7} = 0.504$$

$$\therefore \text{ composition range} = \underline{\underline{0.341 \le v_{clay} \le 0.504}}$$

15.16 For the injection molding process described in Problem 15.15, what specific composition would be preferred given that the cost per kg of polyethylene is ten times that for clay?

15.16

In order to minimize cost, we wish to maximize the amount of low-cost clay.

$\therefore$ preferred composition $= v_{clay} = \underline{\underline{0.504}}$

Section 15.5 - Processing of Semiconductors

PP 15.7 The purity of a 99 wt % Si bar after one zone refining pass is found in Sample Problem 15.7. What would be the purity after two passes?

PP 15.7

Note from Sample Problem 15.7 that a K of 3.62×10^{-4} leads to a composition after one pass of 10^{-2} Al $\times 3.62 \times 10^{-4}$ ($= 3.62 \times 10^{-6}$ Al) where 10^{-2} Al represents the initial concentration of 1 wt.% Al.

Similarly for a second pass,

Al level $= 3.62 \times 10^{-6}$ Al $\times 3.62 \times 10^{-4} = 1.31 \times 10^{-9}$ Al

or $\underline{\underline{1.31 \text{ parts per billion (ppb) Al}}}$

15.17 When the aluminum impurity level in a silicon bar has reached 1 part per billion, what would have been the purity of the liquid on the previous pass?

15.17

As in Practice Problem 15.7, we can follow the result of Sample Problem 15.7.

$$K = \frac{C_s}{C_l} = 3.62 \times 10^{-4}$$

or

$$C_l = \frac{C_s}{3.62 \times 10^{-4}} = \frac{1 \text{ ppb}}{3.62 \times 10^{-4}} = 2760 \text{ ppb} = \underline{\underline{2.76 \text{ ppm}}}$$

15.18 Suppose you have a bar of 99 wt % Sn, with the impurity being Pb. Determine the impurity level after one zone refining pass. (Recall the phase diagram for the Pb–Sn system in Figure 5-38.)

15.18

For 99 % Sn with 1% Pb and noting Figure 5-38, we can take the liquidus as $y = mx + b$

with $y = 200$ at $x = 78$ and $y = 232$ at $x = 100$

giving $y = (1.4545455)x + (86.5455)$

$\therefore$ the temperature at $x = 99$ will be:

$$T = (1.4545455)(99) + (86.5455) = 230.5455$$

Similarly, for the solidus,

$y = 183$ at $x = 97.5$ and $y = 232$ at $x = 100$

giving $y = (19.6)x + (-1728)$

$\therefore$ x at $T = 230.5455$ will be given by:

$$230.5455 = 19.6x - 1728$$

or $x = 99.9258$

or $\dfrac{100 - 99.9258}{100} = 742 \times 10^{-6}$ Pb $= \underline{\underline{742 \text{ ppm Pb}}}$

15.19 For the bar in Problem 15.18, what would be the impurity level after **(a)** two passes or **(b)** three passes?

15.19

(a) $K = \dfrac{C_s}{C_\ell} = \dfrac{100 - 99.9258}{100 - 99} = \dfrac{0.0742}{1} = 0.0742$

$\therefore$ After second pass,

Pb level $= 0.0742 \times 742 \times 10^{-6}$ Pb $= \underline{\underline{55.1 \text{ ppm Pb}}}$

(b) & After third pass,

Pb level $= 0.0742 \times 55.1 \times 10^{-6}$ Pb $= \underline{\underline{4.1 \text{ ppm Pb}}}$

15.20 **(a)** Calculate the flux of gallium atoms out of an MBE effusion cell at a pressure of 2.9×10^{-6} atm and a temperature of 970°C with an aperture area of 500 mm^2. **(b)** If the atomic flux in part (a) is projected onto an area of 45,000 mm^2 on the substrate side of the growth chamber, how much time is required to build up a monolayer of gallium atoms? (Assume, for simplicity, a square grid of adjacent Ga atoms.)

15.20

(a) Using Equation 15.2 and Appendix 1,

$$F = \frac{pA}{\sqrt{2\pi m k T}}$$

$$= \frac{(2.9\times10^{-6}\text{atm})(1\,N/m^2)/(9.869\times10^{-6}\text{atm})(500\,mm^2)}{\sqrt{2\pi(69.72\,amu)(1.661\times10^{-24}g/amu)(13.81\times10^{-24}J/K)\times(970+273)K}}$$

$$= \frac{(147\,N/m^2)(1\,mm^2\times10^{-6}m^2/mm^2)}{\sqrt{1.25\times10^{-41}\,g\cdot J}}$$

(note that $1\,g\cdot J = 1\,g \times 1\,N\cdot m = 1\,g\cdot 1\,kg\cdot m/s^2\cdot m$
$= 10^{-3}kg\cdot 1\,kg\,m^2/s^2 = 10^{-3}kg^2m^2/s^2$)

or

$$F = \frac{1.47\times10^{-4}\,N}{\sqrt{1.25\times10^{-44}\,kg^2m^2/s^2}}$$

$$= \frac{1.47\times10^{-4}\,kg\cdot m/s^2}{1.12\times10^{-22}\,kg\cdot m/s}$$

$$= \underline{\underline{1.31\times10^{18}\,s^{-1}}}$$

(b) Rate of Ga atom arrival at substrate plane =

$$\frac{1.31\times10^{18}\,s^{-1}}{45{,}000\,mm^2\times10^{-6}m^2/mm^2} = 2.91\times10^{19}\,m^{-2}s^{-1}$$

For a square grid of Ga atoms, atomic density of a monolayer will be 1 atom$/d^2_{\text{Ga atom}}$ or, using Appendix 2,

$$\text{at. density} = \frac{1}{(0.135\,nm)^2}\times\frac{10^{18}\,nm^2}{m^2}$$

$$= 5.49\times10^{19}\,m^{-2}$$

$\therefore$ time to produce one monolayer $= \dfrac{5.49\times10^{19}\,m^{-2}}{2.91\times10^{19}\,m^{-2}\,s^{-1}}$

$$= \underline{\underline{1.89\,s}}$$

Section 16.1 - Material Properties - Engineering Design Parameters

PP 16.1 In reordering tool steel for the machining operation discussed in Sample Problem 16.1, you notice that the design specification has been updated. Which of the alloys would meet the new criteria of a tensile strength of ≥ 1,800 MPa and a % elongation of ≥ 10%?

PP 16.1

By inspection of Tables 16.1 and 16.2, we see that the following alloy/heat treatment combinations would meet these specifications:

Type	Condition
S7	Fan cooled from 940°C & single tempered at 540°C
S7	Fan cooled from 940°C & single tempered at 425°C

16.1 A government contractor requires the use of tool steels with a tensile strength of ≥ 1,000 MPa and a % elongation of ≥ 15%. Which alloys in Tables 16.1 and 16.2 would meet this specification?

16.1

By inspection of Tables 16.1 and 16.2, we see that the following alloy/heat treatment combinations would meet these specifications:

Type	Condition
S5	Oil quenched from 870°C and single tempered at 650°C
L2	Oil quenched from 855°C and single tempered at 540°C

16.2 Which additional alloys could be specified if the government contractor in Problem 16.1 would allow the selection of tool steels with a tensile strength of ≥ 1,000 MPa and a % elongation of ≥ 10%?

16.2

By inspection of Tables 16.1 and 16.2, we see that the following additional combinations would meet specifications:

Type	Condition
S7	Fan cooled from 940°C and single tempered at 650°C
L6	Oil quenched from 845°C and " " " 540°C
L2	" " " 855°C and " " " 425°C
S7	Fan cooled " 940°C and " " " 540°C
S7	" " " 940°C and " " " 425°C
S5	Oil quenched " 870°C and " " " 540°C
L2	" " " 855°C and " " " 315°C

16.3 In selecting a ductile cast iron for a high-strength/low-ductility application, you find the following two tables in a standard reference:

Tensile Strength of Ductile Irons

Grade or class	T.S. (MPa)
Class C	345
Class B	379
Class A	414
65-45-12	448
80-55-06	552
100-70-03	689
120-90-02	827

Elongation of Ductile Irons

Grade or class	Elongation (%)
120-90-02	2
100-70-03	3
80-55-06	6
Class B	7
65-45-12	12
Class A	15
Class C	20

If specifications require a ductile iron with a tensile strength of ≥ 550 MPa and a % elongation of ≥ 5%, which alloys in these tables would be suitable?

16.3 By inspection of the two tables, we see that the only alloy to meet these specifications is the 80-55-06.

16.4 In selecting a ductile cast iron for a low-strength/high-ductility application, which of the alloys in Problem 16.3 would meet the specifications of a tensile strength of ≥ 350 MPa and a % elongation of ≥ 15%?

16.4 By inspection of the two tables, we see that the only alloy to meet these specifications is the class A.

16.5 In selecting a polymer for an electronic packaging application, you find the following two tables in a standard reference:

Volume Resistivity of Polymers

Polymer	Resistivity ($\Omega \cdot m$)
epoxy	1×10^{5}
phenolic	1×10^{9}
cellulose acetate	1×10^{10}
polyester	1×10^{10}
polyvinyl chloride	1×10^{12}
nylon 6/6	5×10^{12}
acrylic	5×10^{12}
polyethylene	5×10^{13}
polystyrene	2×10^{14}
polycarbonate	2×10^{14}
polypropylene	2×10^{15}
PTFE	2×10^{16}

Thermal Conductivity of Polymers

Polymer	Conductivity ($J/s \cdot m \cdot K$)
polystyrene	0.12
polyvinyl chloride	0.14
polycarbonate	0.19
polyester	0.19
acrylic	0.21
phenolic	0.22
PTFE	0.24
cellulose acetate	0.26
polyethylene	0.33
epoxy	0.52
polypropylene	2.2
nylon 6/6	2.9

If specifications require a polymer with a volume resistivity of $\geq 10^{13}\ \Omega \cdot m$ and a thermal conductivity of $\geq 0.25\ J/s \cdot m \cdot K$, which polymers in these tables would be suitable?

16.5 By inspection of the two tables, we see that the only polymers which meet specifications are polyethylene and polypropylene.

16.6 In reviewing the performance of polymers chosen in Problem 16.5, concerns over product performance lead to the establishment of more stringent specifications. Which polymers would have a volume resistivity of $\geq 10^{14}\ \Omega \cdot m$ and a thermal conductivity of $\geq 0.35\ J/s \cdot m \cdot K$?

16.6 By inspection of the two tables, we see that the only polymer which meets specifications is polypropylene.

Section 16.2 - Selection of Structural Materials - Case Studies

PP 16.2 An annual fuel savings is calculated in Sample Problem 16.2. For this commercial airline, estimate the fuel savings that would have been provided by a fleet of 50 L-1011 aircraft. (See the case study in Section 16.2.)

PP 16.2

$$\text{Fuel savings} = (\text{wt. savings/aircraft}) \times \frac{(\text{fuel/yr})}{(\text{wt. sav.})} \times 50 \text{ aircraft}$$

$$= (366\,\text{kg}) \times 830\,\frac{\ell/\text{yr}}{\text{kg}} \times 50$$

$$= \underline{\underline{15.2 \times 10^6\,\ell}}$$

16.7 The Boeing 767 aircraft uses a Kevlar-reinforced composite for its cargo liner. The structure weighs 125 kg. **(a)** What weight savings does this represent compared to an aluminum structure of the same volume? (For simplicity, use the density of pure aluminum. A calculation of density for a Kevlar composite was made in Problem 10.4.) **(b)** What annual fuel savings would this one materials substitution represent?

16.7 (a) Using the result of Problem 10.4, we see that

$$V_{\text{cargo liner}} = \frac{m}{\rho} = \frac{125\,\text{kg}}{1.38\,\text{Mg/m}^3} \times \frac{1\,\text{Mg}}{1000\,\text{kg}}$$

$$= 0.0906\,\text{m}^3$$

Using Appendix 1, we see that the corresponding mass of aluminum would be:

$$m = \rho V = (2.70\,\text{Mg/m}^3)(0.0906\,\text{m}^3)\left(\frac{1000\,\text{kg}}{\text{Mg}}\right)$$

$$= 245\,\text{kg}$$

$$\therefore \text{wt. savings} = 245\,\text{kg} - 125\,\text{kg} = \underline{\underline{120\,\text{kg}}}$$

(b) $$\text{Fuel savings} = (120\,\text{kg})(830\,\ell/\text{yr/kg})$$

$$= \underline{\underline{9.96 \times 10^4\,\ell}} \text{ (per aircraft)}$$

16.8 What is the annual fuel savings resulting from the Al–Li alloy substitution described in Problem 7.5?

16.8

For the 4,455 kg mass reduction calculated in Problem 7.5, the fuel savings would be:

$$4,455\,kg \times \frac{830\,\ell/yr}{kg}$$

$$= \underline{\underline{3.70 \times 10^6\,\ell/yr}} \quad \text{(per aircraft)}$$

16.9 Consider an inspection program for Ti–6Al–4V THR femoral stems. Would the ability to detect a flaw size of 1 mm be adequate in regard to preventing fast fracture? Use the data in Table 7.16 and assume an extreme loading of five times body weight for an athletic 200 lb_f patient. The cross-sectional area of the stem is 650 mm^2.

16.9

$$\sigma = F/A = \frac{5(200\,lb_f)(1N/0.2248\,lb_f)}{650\,mm^2(1m/10^3mm)^2} = 6.84 \times 10^6\,N/m^2$$

$$= 6.84\,MPa$$

Using Equation 7.9 and taking $Y = 1$,

$$K_{IC} = \sigma_f\sqrt{\pi a} \quad \text{or} \quad a = \frac{1}{\pi}\left(\frac{K_{IC}}{\sigma}\right)^2$$

Using the minimum K_{IC} in Table 7.16, we see that

$$a = \frac{1}{\pi}\left(\frac{55\,MPa\sqrt{m}}{6.84\,MPa}\right)^2 = 20.6\,m \gg 1 \times 10^{-3}\,m$$

∴ The flaw inspection program is <u>adequate.</u>

16.10 Consider an inspection program for a more traditional application of the Ti–6Al–4V alloy discussed in Problem 16.9. Is the ability to detect a 1-mm flaw size adequate in an aerospace structural member loaded to 90% of the yield strength given in Table 7.11?

16.10

In this case, using data from Table 7.11,

$$a = \frac{1}{\pi}\left(\frac{55\,MPa\sqrt{m}}{0.9 \times 825\,MPa}\right)^2 = 1.7 \times 10^{-3}\,m > 1 \times 10^{-3}\,m$$

∴ The flaw inspection program is (barely) <u>adequate.</u>

16.11 Consider the loading of a simple, cantilever beam:

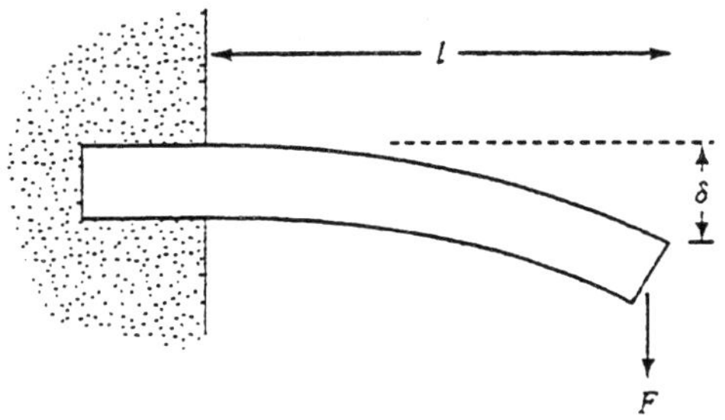

From basic mechanics, it can be shown that the mass of the beam subjected to a deflection, δ, by a force, F, is given by:

$$M = (4l^5F/\delta)^{1/2}(\rho^2/E)^{1/2}$$

where ρ is the density, E is the elastic modulus, and the other terms are defined by the figure. Clearly, the mass of this structural member is minimized for a given loading by minimizing ρ^2/E. Given the data in Problem 10.49, which of those materials would be the optimal choice for this type of structural application? (Moduli can generally be obtained from Tables 7.11 and 10.12. The modulus of reinforced concrete can be taken as 47×10^3 MPa.)

16.11

Summarizing given and tabular data:

material	ρ (Mg/m³)	E (MPa)	ρ^2/E (Mg²/MPa m⁶)
1040 carbon steel	7.8	200×10^3	0.304×10^{-3}
304 stainless steel	7.8	193×10^3	0.315×10^{-3}
3004-H14 Al	2.73	70×10^3	0.106×10^{-3}
Ti-5Al-2.5Sn	4.46	$107\text{-}110 \times 10^3$	$0.181 - 0.186 \times 10^{-3}$
reinforced concrete	2.5	47×10^3	0.133×10^{-3}
fiberglass (73.3 % E-glass)	1.8	56×10^3	0.058×10^{-3}
CFRP (67 % in epoxy)	1.5	221×10^3	0.010×10^{-3}

Clearly, the minimum ρ^2/E occurs for the carbon fiber-reinforced polymer.

16.12 How would the material selection in Problem 16.11 be modified if cost were included in the minimization?

16.12 In this case, M will be minimized by minimizing:

$$(\rho^2/E)^{1/2} \times \text{cost}$$

Including the cost data from Problem 10.49 to the results of Problem 16.11:

material	ρ^2/E $(Mg^2/MPa\, m^6)$	cost (\$/kg)	$[(\rho^2/E)^{1/2} \times \text{cost}]\left(\frac{Mg \cdot \$}{MPa^{1/2} \cdot m^3 \cdot kg}\right)$
1040 carbon steel	0.304×10^{-3}	0.63	0.0110
304 stainless steel	0.315×10^{-3}	3.70	0.0657
3004-H14 Al	0.106×10^{-3}	3.00	0.0309
Ti-5Al-2.5Sn	$0.181\text{-}0.186 \times 10^{-3}$	15.00	0.202-0.205
reinforced concrete	0.133×10^{-3}	0.40	0.0046
fiberglass	0.058×10^{-3}	3.30	0.0251
CFRP	0.010×10^{-3}	270.00	0.854

Clearly, the minimum $(\rho^2/E)^{1/2} \times \text{cost}$ occurs for the reinforced concrete.

Section 16.3 - Selection of Electronic and Magnetic Materials - Case Studies

PP 16.3 We calculate the cost savings due to more economical processing of a thermoplastic in Sample Problem 16.3. The largest single factor is the ability of a single operator to work with multiple fabrication machines for the thermoplastics. By what factor would this "machine operator" parameter have to be increased for thermosets before the two materials would be exactly equal in price?

PP 16.3 For reference, consider the thermoplastic as fixed at a total cost of 1.6¢/part.

Assume the thermoset cost could be reduced by a single operator working x machines.

Then, the total thermoset cost will be:

$$0.5\text{¢/part} + \frac{2.4\text{¢}}{x}\text{/part}$$

Setting this equal to the thermoplastic cost yields:

$$0.5 + \frac{2.4}{x} = 1.6$$

or $x = 2.18$

16.13 In comparing two polyester thermoplastics for use in the production of fuseholders (mass = 5 g) in a new automobile, the following data are available:

	Polyester 1	Polyester 2
Cost/kg	$4.25	$4.50
Yield rate	95%	92%
Cycle time	25 s	20 s

Given that all other production factors are comparable and equal to the parameters for the polyester in Practice Problem 16.3, carry out an economic comparison and recommend a choice between these two engineering polymers.

16.13

The true unit material costs (correcting for yield) are:

$$\#1: \frac{\$4.25/kg}{0.95} = \$4.47/kg$$

$$\#2: \frac{\$4.50/kg}{0.92} = \$4.89/kg$$

The net materials cost/part are:

$$\#1: \$4.47/kg \times 5 g/part \times 1 kg/1000 g = \$0.0224/part = 2.24¢/part$$

$$\#2: \$4.89/kg \times 5 g/part \times 1 kg/1000 g = \$0.0245/part = 2.45¢/part$$

Labor costs are:

$$\#1: \frac{\$10/hr}{operator} \times \frac{1}{5} operator \times \frac{25 s/cycle}{4 parts/cycle} \times \frac{1 hr}{3600 s} = \$0.0035/part = 0.35¢/part$$

$$\#2: \frac{\$10/hr}{operator} \times \frac{1}{5} operator \times \frac{20 s/cycle}{4 parts/cycle} \times \frac{1 hr}{3600 s} = \$0.0028/part = 0.28¢/part$$

Giving total (materials + labor) costs:

$$\#1: (2.24¢ + 0.35¢)/part = 2.59¢/part$$

$$\#2: (2.45¢ + 0.28¢)/part = 2.73¢/part$$

indicating that the reduced materials costs of polyester #1 outweighs the lower labor costs for polyester #2, giving the net economic advantage to polyester #1.

16.14 Given that the thermal conductivity, k, of a 95 Pb: 5 Sn solder used in flip-chip technology is 63 J/(s · m · K), calculate the heat flow rate across an array of 100 solder balls with an average diameter of 100 μm and thickness of 75 μm. The temperature of the chip can be taken as 80°C and the substrate as 25°C. (Recall the discussion of heat transfer in Section 8.4.)

16.14 Using Equation 8.4 and taking the solder balls as roughly cylindrical in shape:

$$k = -\frac{\Delta Q/\Delta t}{A(\Delta T/\Delta x)}$$

or

$$\Delta Q/\Delta t = -kA\left(\frac{\Delta T}{\Delta x}\right)$$

$$= -\left(63 \frac{J}{s \cdot m \cdot K}\right)\left(100 \times \pi\left[50 \times 10^{-6} m\right]^2\right)\left(\frac{[80-25]°C}{75 \times 10^{-6} m}\right)$$

$$= 36.3 \text{ J/s} = \underline{\underline{36.3 \text{ W}}}$$

(Note that, for temperature differentials, 1°C is equivalent to 1 K)

16.15 Consider the following design considerations for a typical solid-state device, a memory cell in a field-effect transistor (FET). The cell contains a thin film of SiO_2, which serves as a small capacitor. You want the capacitor to be as small as possible, but big enough so that an α radiation particle will not cause errors with your 5V operating signal. To ensure this result, design specifications require a capacitor with a capacitance (=[charge density x area]/voltage) of 50×10^{-15} coul/volt. How much area is required for the capacitor if the SiO_2 thickness is 1 μm? (Note that the dielectric constant of SiO_2 is 3.9, and recall the discussion of capacitance in Section 11.4.)

16.15 Noting Equation 11.13 for charge density:

$$D = \epsilon_0 K E$$

$$= \left(8.854 \times 10^{-12} \frac{C}{V \cdot m}\right)(3.9)\left(\frac{5 V}{1 \times 10^{-6} m}\right) = 1.73 \times 10^{-4} \text{ C/m}^2$$

Then,

$$\text{Capacitance} = 50 \times 10^{-15} \frac{C}{V} = \frac{DA}{V} = \frac{(1.73 \times 10^{-4} \text{ C/m}^2)A}{5 V}$$

or

$$A = \frac{(50 \times 10^{-15} \text{ C/V})(5 V)}{1.73 \times 10^{-4} \text{ C/m}^2} = \underline{\underline{1.45 \times 10^{-9} \text{ m}^2}}$$

16.16 If the specifications for the device introduced in Problem 16.15 are changed so that a capacitance of 70 × 10^{-15} coul/volt is required but the area is fixed by the overall circuit design, calculate the appropriate SiO_2 film thickness for the modified design.

16.16

In this case,

$$70 \times 10^{-15} \frac{C}{V} = \frac{D(1.45 \times 10^{-9} m^2)}{5 V}$$

or $D = 2.41 \times 10^{-4}$ C/m^2

$$= \left(8.854 \times 10^{-12} \frac{C}{V \cdot m}\right)(3.9)\left(\frac{5V}{t}\right)$$

or

$$t = \frac{\left(8.854 \times 10^{-12} \frac{C}{V \cdot m}\right)(3.9)(5V)}{2.41 \times 10^{-4}\ C/m^2} = 0.715 \times 10^{-6} m$$

$$= \underline{\underline{0.715\ \mu m}}$$

PART II
THERMODYNAMICS

As the name **thermodynamics** implies, Part II of the **Manual** will focus on the nature of heat as a form of energy and the limitations with which heat can be converted to other forms of energy. These principles will have wide-ranging implications for the practical applications of engineering materials.

The foundation of thermodynamics is a set of three physical laws based on various experimental observations of the behavior of materials. The **first law** is a formal statement of the conservation of energy. An important aspect of this law is that **internal energy** is a **"state" function**, i.e., the magnitude of its change in a process is dependent only on the beginning and ending states and not on the path joining those states. In addition to internal energy, two other thermodynamic variables will be defined relative to the first law discussion. **Enthalpy** is the energy of a system plus the product of its pressure and volume. **Heat capacity** is the derivative of heat input to a material with respect to its temperature rise. The **second law** deals with the practical limitations on the conversion of heat to work. It leads to the definition of yet another thermodynamic state function, **entropy**. A compact statement of the second law is that any spontaneous process leads to an increase in entropy. The second law can also be stated in terms of the statistical nature of the atomic-scale structure of materials. The **third law** states that the entropy of a pure material goes to zero when its temperature has been reduced to absolute zero. **Free energy** is a thermodynamic state function which is minimized at equilibrium. As such, the free energy provides the theoretical foundation for the nature of **phase diagrams** (see Chapter 5 in **Shackelford, 4th Edition**). Thermodynamics can provide a fuller understanding of the significance of **surfaces** in materials processes. Also, some of the thermodynamic state functions defined in this chapter (enthalpy, entropy, and free energy) are intimately associated with the **kinetics** of rate processes (see Chapters 4 and 6 in **Shackelford, 4th Edition**). Finally, thermodynamics principles lay a foundation for better understanding the nature of **environmental reactions** (see Chapter 14 in **Shackelford, 4th Edition**).

1 THE FIRST LAW

If a given **system** takes in a quantity of **heat**, Q, while doing an amount of **work**, W, there will be a net change in **internal energy**, ΔU, for that system given by

$$\Delta U = Q - W. \tag{1-1}$$

This simple statement of the **conservation of energy**[1] is the **first law of thermodynamics**. Although simple, it is a powerful and useful concept. One should especially note the sign convention of Equation 1-1. A heat flow **to** the system is **positive** (and **from** the system is **negative**). Conversely, work done **by** the system is **positive** (and **on** the system is **negative**). This is the so-called "engineering convention," in that the sign for work is determined by the practical perspective of the user who wishes to extract work from the system. The reader may encounter other books dealing with thermodynamics (especially in the field of chemistry) in which the sign for work is determined by the perspective of the system, rather than the external user of energy. (In those books, the first law would be written $\Delta U = Q + W$).

Relative to Equation 1-1, it is important to note that the internal energy, U, is a **state variable**, i.e., a property whose value depends solely on the state of the system (established by pressure, temperature, etc.) and not by the path in reaching that state. On the other hand, heat, Q, and work, W, are not state variables. Their values are functions of the path of the process in going between an initial and a final state. The value of ΔU, conversely, is not a function of the path, and depends only on the values of U at the initial and final state.

Equation 1-1 describes energy terms associated with an overall process going between an initial and a final state. In many cases, we shall want to monitor the specific progress along a process path. For this purpose, we require a differential form of the first law involving the relationship between the incremental change in internal energy, dU, and the incremental

[1]For our purposes in this book, we shall not have to deal with the additional consideration of the interconversion of energy and mass ($E = mc^2$), which occurs to a significant degree in certain relativistic processes.

values of heat, δQ, and work, δW:

$$dU = \delta Q - \delta W. \tag{1-2}$$

One should note that **d** is used to designate a differential quantity of a state variable, and **δ** is used to designate a differential quantity of a path-dependant variable.

In applying the first law of thermodynamics, it is convenient to define a quantity known as the **heat capacity** to be the ratio of the incremental heat addition to (or loss from) a system to the corresponding incremental temperature change, dT:

$$C = \frac{\delta Q}{dT}. \tag{1-3}$$

Note that temperature is a state variable. This is also true for some other common terms such as pressure, p, and volume, V. The heat capacity is a most useful term, but, in general, we need to specify the conditions under which it is measured. When the system is constrained to a constant volume during the course of the measurement, the heat capacity at constant volume is designated as

$$C_v = \left(\frac{\delta Q}{dT}\right)_v. \tag{1-4}$$

When the system is held at a constant pressure, the heat capacity is

$$C_p = \left(\frac{\delta Q}{dT}\right)_p. \tag{1-5}$$

An important feature of the heat capacity is that it can be related to first law terms. Consider, for example, the common case in which the work term in the first law is strictly mechanical (i.e., such terms as magnetic work are not involved). Mechanical work is produced by a force, F, acting through a distance, x. If the process is carried out relatively slowly with an absence of frictional losses, the incremental work term is

$$\delta W = Fdx. \tag{1-6}$$

Such an ideal process is called **reversible.** It represents the maximum efficiency for such mechanical action, as well as the maintenance of equilibrium at each state of the overall process. The term "reversible" is used because an infinitesimal force applied in an opposing direction can reverse the system from expansion to compression, and vice versa. The force is equal to the system pressure, p, times the cross-sectional area, A, of the system (normal to the x-direction):

$$Fdx = pAdx. \tag{1-7}$$

In turn, the product of area and incremental distance represents an incremental volume:

$$pAdx = pdV, \tag{1-8}$$

allowing us to rewrite Equation 1-2 as

$$dU = \delta Q - pdV \tag{1-9}$$

or,

$$\delta Q = dU + pdV. \tag{1-10}$$

A consequence of this relationship is that the heat capacity at constant volume (Equation 1-4) can be expressed in terms of internal energy, i.e.,

$$C_V = \left(\frac{\delta Q}{dT}\right)_V = \left(\frac{\partial U}{\partial T}\right)_V. \tag{1-11}$$

Similarly, the heat capacity at constant pressure (Equation 1-5) can be rewritten as:

$$C_p = \left(\frac{\delta Q}{dT}\right)_p = \left(\frac{\partial[U + pV]}{\partial T}\right)_p. \tag{1-12}$$

The combination of internal energy and a pressure-volume product occurs frequently in thermodynamic analyses. As a result, it is convenient to replace this combination of state variables with a single term, H, which is, in turn, a state variable:

$$H = U + pV \tag{1-13}$$

and is called the **enthalpy** (from the Greek word *thalpein*, meaning "to heat"). Although the enthalpy may strike the beginning student as a somewhat abstract concept, the direct connection to a practical engineering measurement (of heat capacity) can be seen by combining Equations 1-12 and 1-13:

$$C_p = \left(\frac{\partial H}{\partial T}\right)_p. \tag{1-14}$$

In the latter part of this chapter, we shall find numerous other examples of the practical utility of enthalpy as a thermodynamic variable.

SAMPLE PROBLEM 1-1

The heat capacity of a material is a function of temperature. For typical solids, the heat capacity at constant pressure, C_P, can be expressed by an empirical relation of the form:

$$C_p = a + bT + cT^{-2} \; J/(mol \cdot K).$$

For silver over the temperature range of 298 K to its melting point (1235 K),

$$a = 21.1 \; J/(mol \cdot K)$$

$$b = 8.55 \times 10^{-3} \; J/(mol \cdot K^2)$$

$$c = 1.50 \times 10^{5} \; (J \cdot K)/mol.$$

Calculate the heat input necessary to reversibly heat a kg of silver from 25°C to 500°C under a constant pressure of 1 atm.

SOLUTION

Rearranging Equation 1-5,

$$(\delta Q)_p = C_p dT$$

or

$$Q = \int_{T_1}^{T_2} C_p dT.$$

Using the given empirical relation for C_p,

$$Q = \int_{T_1}^{T_2} (a + bT + cT^{-2}) \, dT$$

$$= aT + \frac{b}{2} T^2 - cT^{-1} \Big|_{T_1}^{T_2}$$

$$= a\,(T_2 - T_1) + \frac{b}{2}\left(T_2^{\,2} - T_1^{\,2}\right) - c\left(T_2^{-1} - T_1^{-1}\right).$$

For this specific case, note that $T_1 = (25 + 273)\text{K} = 298\ \text{K}$ and $T_2 = (500 + 273)\text{K} = 773\ \text{K}$, and that the C_p parameters are given above. Then,

$$Q = 21.2\ \text{J/(mol}\cdot\text{K)}\ (773\ \text{K} - 298\ \text{K})$$

$$+ \frac{8.55 \times 10^{-3}}{2}\ \text{J/(mol}\cdot\text{K}^2\text{)}\left((773\ \text{K})^2 - (298)^2\right)$$

$$- 1.50 \times 10^5 (\text{J}\cdot\text{K})/\text{mol}\left((773\ \text{K})^{-1} - (298\ \text{K})^{-1}\right)$$

$$= 12.6 \times 10^3\ \text{J/mol}.$$

Finally, we can use Appendix 1 of **Shackelford, 4th Edition,** to obtain the result for 1 kg of silver:

$$Q = 12.6 \times 10^3\ \text{J/mol} \times \frac{1\ \text{mol}}{107.9\ \text{g}} \times \frac{1000\ \text{g}}{1\ \text{kg}}$$

$$= 116 \times 10^3\ \text{J/kg}.$$

SAMPLE PROBLEM 1-2

What is the enthalpy of 1 kg of silver at 500°C relative to its value at 25°C?

SOLUTION

Rearranging Equation 1-14,

$$(\partial H)_p = C_p dT$$

or

$$\int_{T_1}^{T_2} dH = \int_{T_1}^{T_2} C_p \, dT$$

giving

$$H(T_2) - H(T_1) = \int_{T_1}^{T_2} C_p dT.$$

For the case at hand,

$$H(500°C) - H(25°C) = \int_{298\ K}^{773\ K} C_p dT.$$

The integral on the right side of the above equation was determined in Sample Problem 1-1. Therefore,

$$H(500°C) - H(25°C) = 116 \times 10^3 \text{ J/kg}.$$

2 THE SECOND LAW

The first law is useful as a means for doing the necessary "book-keeping" on the energy terms associated with a given process. The second law provides a statement as to the limitations on the conversion of heat to work. A convenient means for introducing a formal statement of the second law is to describe the nature of a simple heat engine.

Figure 2-1 shows a schematic illustration of a **Carnot**[2] **cycle** involving the expansion and compression of an ideal gas within a piston. There are four steps in the overall cycle:

I. A reversible, isothermal heat addition.

II. A reversible, **adiabatic** (i.e., perfectly insulated from any heat flow) expansion.

III. A reversible, isothermal heat rejection.

IV. A reversible, adiabatic compression of the gas back to the initial state of the system (at a pressure, temperature, and volume corresponding to the start of step I).

Figure 2-2 is a plot of the Carnot cycle on a pressure-volume diagram. For the overall cycle (starting at point **a** and returning to point **a**), the change in internal energy (or any other state variable) must equal zero, i.e.,

$$\Delta U_{a \to a} = 0. \tag{2-1}$$

The first law (Equation 1-1) for the overall cycle is, then,

$$\Delta U = 0 = Q_I + Q_{III} - W_{total}. \tag{2-2}$$

or

$$W_{total} = Q_I + Q_{III} \tag{2-3}$$

where W_{total} is the net useful work extracted from the cycle and Q_I and

---

[2]Nicholas Leonard Sadi Carnot (1796-1832), French physicist. The son of a prominent scientist and politician, Sadi Carnot followed along similar paths, including a successful military career. In analyzing engines, he introduced the concepts of a "complete cycle" and reversibility. Scarlet fever led to his untimely death at the age of 36.

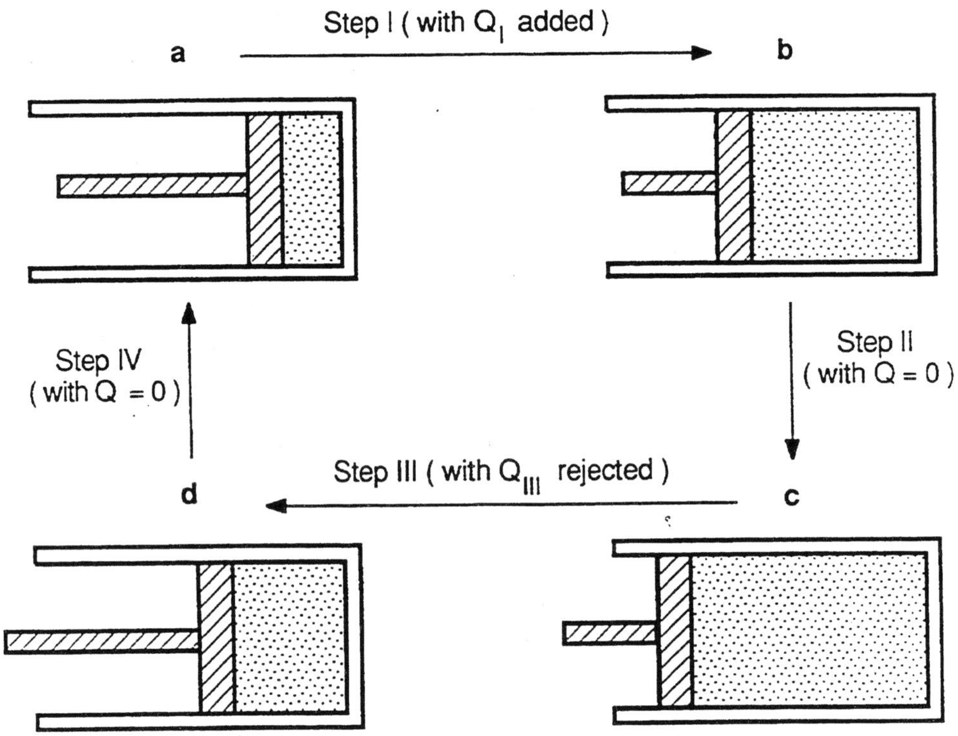

FIGURE 2-1 Schematic of a Carnot cycle in which four expansion and compression steps bring the gas inside a piston-cylinder back to its original state.

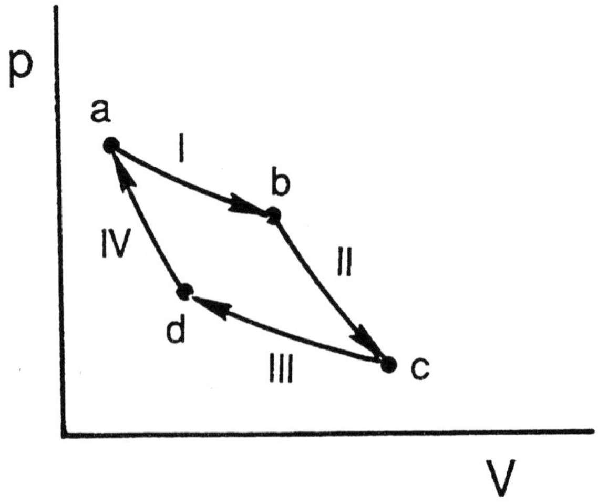

FIGURE 2-2 The Carnot cycle as plotted on a pressure (p) - volume (V) diagram.

Q_{III} are the heat addition (during step I) and heat rejection (during step III). We can now define an overall **cycle efficiency**, η, as the net work extracted divided by the total thermal energy input:

$$\eta = \frac{W_{total}}{Q_I} = \frac{Q_I + Q_{III}}{Q_I} . \tag{2-4}$$

The denominator of Equation 2-4 contains only the heat addition term, Q_I. In this simple engine, the rejected heat, Q_{III}, is not reclaimed and does not reduce the magnitude of total thermal energy input. Because Q_{III} is a negative quantity and appears in the numerator of Equation 2-4, the efficiency, η, of the Carnot cycle is **less** than unity. The fact that the thermal energy, Q_I, cannot be completely converted to usable work is closely related to the second law of thermodynamics. To quantify the cycle efficiency, it is useful to introduce a new state variable which will be central in a formal statement of the second law.

It can be demonstrated experimentally that, for cycles such as that illustrated in Figures 2-1 and 2-2,

$$\oint \frac{\delta Q}{T} \leq 0 \tag{2-5}$$

where the integral is taken over the complete cycle. The equality in Equation 2-5 corresponds to a completely reversible cycle. For this limiting case, the differential of reversible heat flow divided by temperature ($\delta Q_{rev}/T$) behaves as the differential of a state variable. This state variable is called **entropy**, S (from the Greek word *trepein*, meaning "to turn or change"), i.e.,

$$dS = \frac{\delta Q_{rev}}{T} . \tag{2-6}$$

Returning to the Carnot cycle, we can determine an entropy change for each step:

$$\Delta S_I = \int_a^b \frac{\delta Q_{rev}}{T} = \frac{Q_I}{T_I} , \tag{2-7}$$

$$\Delta S_{II} = \int_b^c \frac{\delta Q_{rev}}{T} = 0 , \tag{2-8}$$

$$\Delta S_{III} = \int_c^d \frac{\delta Q_{rev}}{T} = \frac{Q_{III}}{T_{III}}, \tag{2-9}$$

and

$$\Delta S_{IV} = \int_d^a \frac{\delta Q_{rev}}{T} = 0 . \tag{2-10}$$

Because the overall entropy change for the reversible cycle is zero, we can write that

$$\Delta S_{a \to a} = 0 = \Delta S_I + \Delta S_{II} + \Delta S_{III} + \Delta S_{IV} . \tag{2-11}$$

Combining Equation 2-11 with Equations 2-7 to 2-10 gives

$$\frac{Q_I}{T_I} + \frac{Q_{III}}{T_{III}} = 0. \tag{2-12}$$

The Carnot cycle efficiency (Equation 2-4) can now be rewritten:

$$\eta = \frac{Q_I + (-Q_I)(T_{III}/T_I)}{Q_I} \tag{2-13}$$

$$= \frac{T_I - T_{III}}{T_I} . \tag{2-14}$$

This result is both simple and convenient, in that it permits the cycle efficiency to be calculated with knowledge only of the operating temperature range of the engine. Also, one should note that Equation 2-14 has significance well beyond a by-product of our introduction of the second law of thermodynamics. It is a convenient approximation to the efficiency for many practical power generation systems. As most engines will have a lower operating temperature (T_{III}) near room temperature, efficiency will generally be limited by the magnitude of the upper operating temperature (T_I). This is the central reason for the extensive development of new ceramics that can withstand higher operating temperatures than metals in engine designs (see Chapter 8 in **Shackelford, 4th Edition**).

In effect, Equation 2-14 tells us that **heat cannot be converted to work with 100 percent efficiency.** This is equivalent to one of the early statements of the second law of thermodynamics. It is also closely connected to the common observation that **heat will not be spontaneously transferred from a cooler to a hotter body**, another early statement of the second law.

A consequence of Equation 2-5 is that, for a given incremental process between two adjacent states, the heat flow for a reversible step will be greater than for an irreversible step:

$$\delta Q_{rev} > \delta Q_{irrev}. \tag{2-15}$$

Also, this means that the actual heat transfer for any incremental process can never be greater than the reversible heat transfer between the adjacent states, i.e.,

$$\delta Q \leq \delta Q_{rev}. \tag{2-16}$$

However, the entropy change for the system can always be calculated from Equation 2-6 using the heat transfer for the reversible path, i.e.,

$$dS_{sys} = \frac{\delta Q_{rev}}{T} . \tag{2-17}$$

The entropy change for the system's surroundings can be calculated as the reversible absorption of the **actual** heat lost from the system ($-\delta Q$), i.e.,

$$dS_{surr} = \frac{-\delta Q}{T} . \tag{2-18}$$

The **net** entropy change for the universe (i.e., system + surroundings) is, then:

$$dS_{net} = dS_{sys} + dS_{surr} = \frac{\delta Q_{rev}}{T} - \frac{\delta Q}{T} \geq 0 \tag{2-19}$$

where the relative magnitudes of δQ_{rev} and δQ are given by Equation 2-16. Equation 2-19 implies that **any spontaneous process will be accompanied by an increase in total entropy** and that **entropy is maximized at equilibrium.** These two implications are our most general statements of the second law and illustrate the central role of entropy (comparable to the roles of energy and enthalpy in the first law statements).

The utility of entropy as a state variable is illustrated in Figure 2-3, in which the Carnot cycle is plotted on a temperature-entropy (T-S) diagram. In contrast to the p-V diagram of Figure 2-2, the T-S

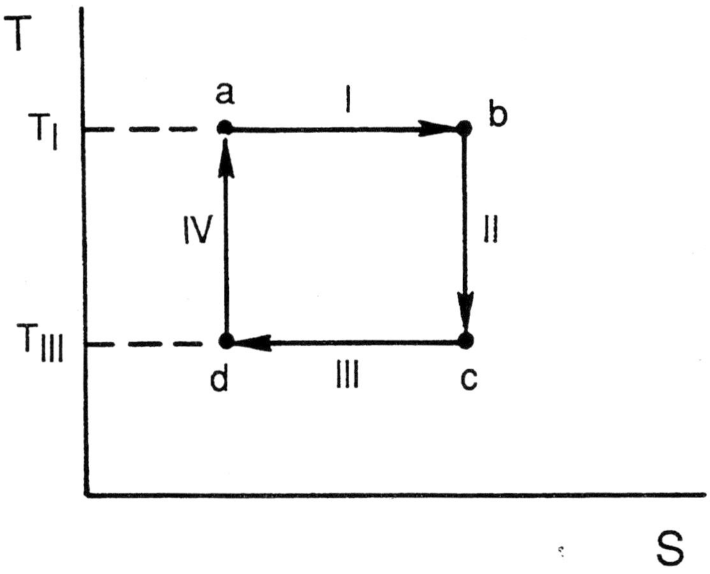

FIGURE 2-3 The Carnot cycle as plotted on a temperature (T) - entropy (S) diagram. Note that the area bounded by the plot a-b-c-d is equal to the work extracted from the engine in one cycle. (This is also true for the p-V plot in Figure 2-2.)

diagram has a simple rectangular geometry. Another convenient feature of the T-S plot is that the area enclosed by the plot a-b-c-d is equal to the work extracted from the engine (W_{total}) in one cycle. A careful inspection of Equations 2-3, 2-7, 2-8, 2-9, and 2-10 confirms this fact. The area enclosed by the p-V plot in Figure 2-2 is also equal to W_{total}, although not as simple to evaluate as the rectangular area in Figure 2-3.

Our discussion of thermodynamics to this point has dealt strictly with macroscopic observations. Such macroscopic properties are based on microscopic (or submicroscopic) structure (see Section 1.2 in **Shackelford, 4th Edition**). The nature of the second law statements following from Equation 2-19 can be related to the statistical probability of the atomic arrangement of the system in question. Specifically, the entropy of the system is related to the number of possible atomic arrangements, W, by

$$S = k \ln W \qquad (2\text{-}20)$$

where k is Boltzmann's constant. The development of expressions such as Equation 2-20 is appropriately termed "**statistical thermodynamics.**" The macroscopic treatment is termed "**classical thermodynamics.**" Most of our coverage in this **Manual** will be limited, for the sake of time and space, to classical thermodynamics but, even in an introductory materials science book, there will be occasions in which the statistical treatment is advantageous. The **entropy of mixing** for solid solutions is an example. First, we shall take Equation 2-20 as a valid statement of the value of entropy without reviewing the experimental evidence that confirms it. More detailed references, such as those at the end of Part II, are available for such purposes. Of more immediate importance, Equation 2-20 provides an additional statement of the second law, viz., **a spontaneous process is one in which the system goes to a more probable state** (larger value of W).

Consider, now, the specific case of a substitutional solid solution (note Figure 4-2 in **Shackelford, 4th Edition**). If there are N_o total atomic sites and n atoms of element A, there will be (N_o-n) atoms of element B in a binary system (such as the copper-nickel solid solution). The number of distinct ways in which the A and B atoms can be arranged on the N_o sites is (again, without derivation):

$$W = \frac{N_0!}{n!\,(N_0-n)!} \,. \tag{2-21}$$

Therefore, the **entropy of mixing**, ΔS_m, for the solid solution is

$$\Delta S_m = k \ln \left[\frac{N_0!}{n!\,(N_0-n)!} \right] . \tag{2-22}$$

Because the numbers of atoms are large (on the order of Avogadro's number, in typical cases), we can replace the factorial terms using **Stirling's**[3] **approximation:**

$$\ln N! \simeq N \ln N - N \tag{2-23}$$

where N is any "large" number. Applying Stirling's approximation to Equation 2-22 gives

$$\Delta S_m = N_0 k \left[\frac{n}{N_0} \ln \frac{N_0}{n} + \frac{(N_0-n)}{N_0} \ln \frac{N_0}{(N_0-n)} \right] . \tag{2-24}$$

---

[3]James Stirling (1692-1770), Scottish mathematician. Stirling's significant contributions to mathematics were confined to the 1720s and 1730s. His ability for practical financial calculations led to his being asked to reorganize the Scottish Mining Company in the lead mines of Leadhills, Lanarkshire. As a successful administrator, he spent most of his time after 1735 in that remote village.

Noting that the atom fraction of element A, X_A, is n/N_o and the atom fraction of element B, X_B, is $(N_o-n)/N_o$, Equation 2-24 can be rewritten as

$$\Delta S_m = -N_0 k\,[X_A \ln X_A + X_B \ln X_B]. \qquad (2\text{-}25)$$

For 1 mole of solid solution ($N_o = N_{Avogadro}$),

$$\Delta S_m = -R\,[X_A \ln X_A + X_B \ln X_B] \qquad (2\text{-}26)$$

where R is the universal gas constant ($= N_{Av}k$). Figure 2-4 is a plot of ΔS_m over the composition range for a binary solid solution system. A general feature of this plot is that the entropy increases with the increasing degree of "disorder" represented by the random mixing of unlike atoms.

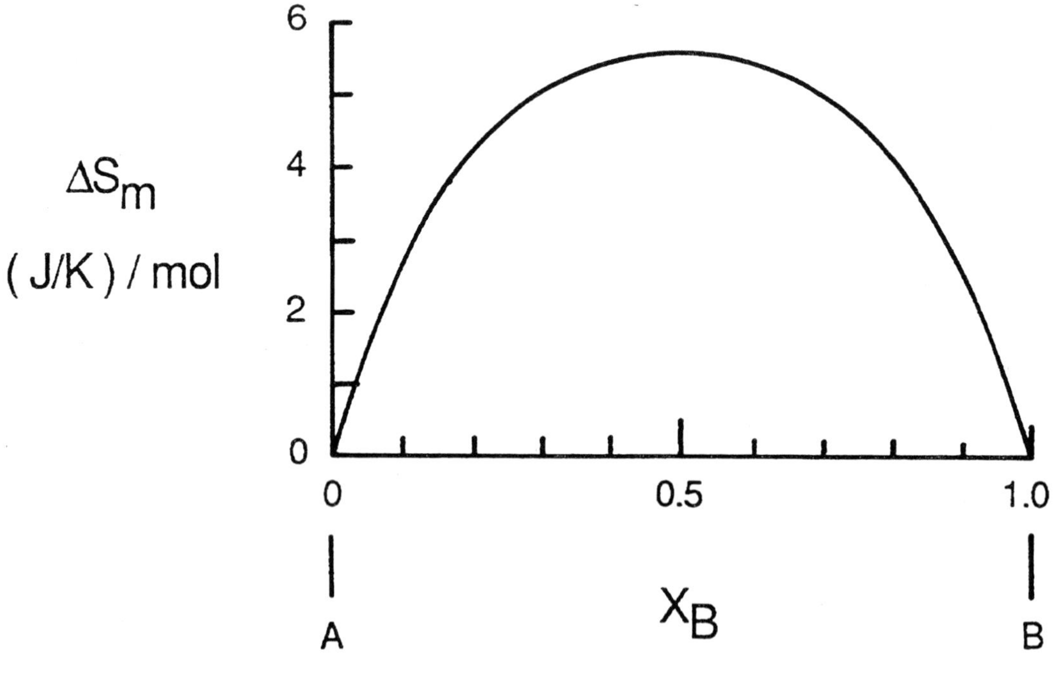

FIGURE 2-4 **The entropy of mixing (of atoms A and B in a substitutional solid solution) as a function of composition (atomic fraction of B). This is a plot of Equation 2-26.**

SAMPLE PROBLEM 2-1

(a) Calculate the maximum (Carnot) efficiency for a conventional automobile engine (made largely from iron and steel) operating at a maximum temperature of 800°C.

(b) Repeat (a) for a prototype Si_3N_4 ceramic engine operating at a maximum temperature of 1200°C.

SOLUTION

(a) We can use Equation 2-14 with the assumption that the lower operating temperature is room temperature (taken as 25°C = 298 K). For an upper operating temperature of 800°C = 1073 K,

$$\eta = \frac{T_I - T_{III}}{T_I}$$

$$= \frac{1073 - 298}{1073} = 0.72 \quad \text{or} \quad 72\%$$

(b) For an upper operating temperature of 1200°C = 1473 K,

$$\eta = \frac{1473 - 298}{1473} = 0.80 \quad \text{or} \quad 80\%$$

SAMPLE PROBLEM 2-2

Calculate the percent error in using Stirling's approximation (Equation 2-23) for (a) N = 5, (b) N = 10, and (c) N = 50.

SOLUTION

(a) For N = 5, N! = 120 and ln N! = 4.787. But,

$$N \ln N - N = 5 \ln 5 - 5 = 3.047.$$

Therefore,

$$\text{Percent error} = \frac{4.787 - 3.047}{4.787} \times 100\ \% = 36.3\ \%.$$

(b) $\ln 10! = \ln 3{,}628{,}800 = 15.104$

$10 \ln 10 - 10 = 13.026$

$$\text{Percent error} = \frac{15.104 - 13.026}{15.104} \times 100\% = 13.8\%.$$

(c) $\ln 50! = \ln 3.04 \times 10^{64} = 148.5$

$50 \ln 50 - 50 = 145.6$

$$\text{Percent error} = \frac{148.5 - 145.6}{148.5} \times 100\% = 1.95\%.$$

SAMPLE PROBLEM 2-3

Calculate the entropy of mixing for 100 g of an equimolar solid solution of Cu and Ni.

SOLUTION

First, we must determine the number of moles of solution. Taking x g of Cu and (100 - x) g of Ni, and using data from Appendix 1 of **Shackelford, 4th Edition,**

$$\text{no. moles Cu} = \frac{x \text{ g}}{63.55 \text{ g/mol}} = \frac{x}{63.55} \text{ mol}$$

and

$$\text{no. moles Ni} = \frac{(100 - x) \text{ g}}{58.71 \text{ g/mol}} = \frac{100 - x}{58.71} \text{ mol.}$$

As the alloy is equimolar,

$$\frac{x}{63.55} = \frac{100 - x}{58.71}$$

or

$$x = 52.0 \text{ g}$$

and

$$100 - x = 48.0 \text{ g.}$$

The total number of moles will be

$$n_{tot} = n_{Cu} + n_{Ni} = \frac{52.0}{63.55} + \frac{48.0}{58.71} = 1.64 \text{ mol.}$$

Turning to Equation 2-26, we can calculate ΔS_m:

$$\Delta S_m = (1.64 \text{ mol})(-8.314 \text{ J/(mol}\cdot\text{K)}(0.5 \ln 0.5 + 0.5 \ln 0.5)$$

$$= 9.45 \text{ J/K.}$$

3 THE THIRD LAW

The Carnot cycle of Figures 2-1, 2-2, and 2-3 provides for a definition of an absolute temperature scale. For the overall cycle, we can recall that

$$\frac{Q_I}{T_I} + \frac{Q_{III}}{T_{III}} = 0 \tag{2-12}$$

or, by rearranging

$$\frac{T_I}{T_{III}} = -\frac{Q_I}{Q_{III}} . \tag{3-1}$$

This relationship establishes the temperature scale by the ratio of heat input to outflow. By assigning a value to some convenient reference point (e.g., 273.16 K for the freezing point of pure water at 1 atm pressure), all other temperatures follow from Equation 3-1.

Returning to Equation 2-6, we can integrate that expression to obtain values of entropy at a given temperature, T', and pressure, p (allowing us to use Equation 1-5):

$$S_{T'} = \int_0^{T'} \frac{C_p}{T}\, dt + S_{T=0}. \tag{3-2}$$

The value of entropy at absolute zero is a constant of integration. In the early part of this century, it was demonstrated that the entropy of all pure materials approaches the same value at 0 K. For the convenience of being able to calculate absolute values of entropy at various temperatures, the value of entropy (for any material) at absolute zero is arbitrarily taken as zero, i.e.,

$$S_{T=0} = 0. \tag{3-3}$$

This definition is the **third law** of thermodynamics. Equation 3-2 can, then, be simplified to allow for the calculation of the absolute entropy of a system at temperature T':

$$S_{T'} = \int_0^{T'} \frac{C_p}{T}\, dt. \tag{3-4}$$

SAMPLE PROBLEM 3-1

Calculate the entropy change that occurs upon heating 1 kg of silver from 25°C to 500°C under a constant pressure of 1 atm.

SOLUTION

This can be calculated with the use of a modified form of Equation 3-4, viz.,

$$\Delta S = S_{500°C} - S_{25°C} = \int_{25°C}^{500°C} \frac{C_p}{T}\, dT.$$

The heat capacity data for this case was given in Sample Problem 1-1, i.e.,

$$\Delta S = \int_{25°C}^{500°C} \left(\frac{a}{T} + b + cT^{-3}\right) dT$$

$$= a \ln T + bT - \frac{c}{2} T^{-2} \Bigg|_{25°C\, =\, 298\ K}^{500°C\, =\, 773\ K}$$

$$= a \ln \left(\frac{773\ K}{298\ K}\right) + b\left((773\ K) - (298\ K)\right)$$

$$- \frac{c}{2}\left((773\ K)^{-2} - (298\ K)^{-2}\right)$$

$$= 21.1\ J/(mol \cdot K)\ \ln\left(\frac{773\ K}{298\ K}\right) + \left(8.55 \times 10^{-3} J/(mol \cdot K^2)\right)$$

$$\times\ (773\ K) - (298\ K)\ -\ \frac{(1.50 \times 10^5 (J \cdot K)/mol)}{2}$$

$$\left((773\ K)^{-2} - (298\ K)^{-2}\right)$$

$$= 25.0\ J/(mol \cdot K).$$

As in Sample Problem 1-1, we finally need to convert to a basis of 1 kg:

$$\Delta S = 39.5\ J/(mol \cdot K) \times \frac{1\ mol}{107.9\ g} \times \frac{1000\ g}{kg}$$

$$= 232\ (J/K)/kg.$$

4 THE FREE ENERGIES

In stating the three laws of thermodynamics, the variables internal energy, enthalpy, and entropy have been related to processes that may or may not have been spontaneous. Our next goal is to define an additional variable that can predict whether or not a given process is spontaneous. This is comparable to having, in the field of mechanics, a potential energy function which is minimized at equilibrium.

A clue as to the form of a **"thermodynamic potential"** variable comes from reflection on the Carnot cycle introduced in Section 2. The heat rejected in Step III of Figure 2-1 was associated with the limit to engine efficiency (representing, in effect, "unavailable work"). As $Q_{III} = T_{III}\,\Delta S_{III}$ (by rearranging Equation 2-9), we can propose a new variable, F, which is the energy of the system, U, reduced by a TS product:

$$F = U - TS. \tag{4-1}$$

The differential of F will be

$$dF = dU - TdS - SdT. \tag{4-2}$$

Using the first law (Equation 1-2), we can modify Equation 4-2 to become

$$dF = \delta Q - \delta W - TdS - SdT. \tag{4-3}$$

For a process occurring at constant volume and temperature, $\delta W = 0$ (because $dV = 0$) and, of course, $dT = 0$, giving the result:

$$(dF)_{V,T} = \delta Q - TdS. \tag{4-4}$$

The differential entropy term can be replaced by the equality in Equation 2-6, giving

$$(dF)_{V,T} = \delta Q - \delta Q_{rev}. \tag{4-5}$$

The relationship in Equation 2-16 ($\delta Q \leq \delta Q_{rev}$) indicates that Equation 4-5 can be restated as:

$$(dF)_{V,T} \leq 0 \tag{4-6}$$

where the equality applies to a reversible process and the inequality to an irreversible, or spontaneous, process. Equation 4-6 serves precisely the purpose stated as our goal, viz., a variable which is minimized at equilibrium. As the original definition of the variable (Equation 4-1) involved the internal energy minus an "unavailable work" term (TS), F is called the **Helmholtz**[4] **free energy**, which is, as shown by Equation 4-6, minimized at equilibrium for processes occurring at constant volume and temperature. For materials applications, processes occurring at constant pressure and temperature are more common. For such cases, the **Gibbs**[5] **free energy**, G, is the appropriate variable:

$$G = H - TS. \qquad (4\text{-}7)$$

Recalling the definition of enthalpy (Equation 1-13), we see that the Gibbs free energy is:

$$G = U + pV - TS. \qquad (4\text{-}8)$$

The differential of G is

$$dG = dU + pdV - Vdp - Tds - SdT. \qquad (4\text{-}9)$$

Introducing Equation 1-2 gives

$$dG = \delta Q - \delta W + pdV - Vdp - TdS - SdT. \qquad (4\text{-}10)$$

Considering only mechanical work ($\delta W = pdV$) simplifies the expression for dG, viz.,

$$dG = \delta Q + Vdp - TdS - SdT. \qquad (4\text{-}11)$$

Introducing Equation 2-6 gives

$$dG = \delta Q + Vdp - \delta Q_{rev} - Sdt. \qquad (4\text{-}12)$$

---

[4]Herman Von Helmholtz (1821-1894), German physiologist and physicist. Finances kept Helmholtz from studying physics as a young man. He obtained a medical degree because of state financial aid and helped to develop the "mechanistic" school of physiology. (Note Adolf Fick's related biography in Chapter 4 of **Shackelford, 4th Edition.**) Excellent research on physiological acoustics and optics led to his being given the prestigious chair of physics at the University of Berlin in 1866. Heinrich Hertz (whose biography appears in Chapter 13 of **Shackelford, 4th Edition**) was one of his students at Berlin.

[5]Josiah Willard Gibbs (1839-1903), American physicist. His biography appears in Chapter 5 of **Shackelford. 4th Edition.**

At constant pressure and temperature, Equation 4-12 reduces to

$$(dG)_{p,T} = \delta Q - \delta Q_{rev}. \tag{4-13}$$

The conclusion comparable to Equation 4-6 is, then,

$$(dG)_{p,T} \leq 0. \tag{4-14}$$

For the remainder of Part I, we will frequently use the Gibbs free energy to analyze problems of practical importance in materials engineering.

SAMPLE PROBLEM 4-1

(a) In Sample Problem 2-3, the entropy of mixing was calculated for 100 g of an equimolar solid solution of Cu and Ni. What would be the Gibbs free energy of mixing for this alloy at 1000°C, assuming an ideal solution? (We shall see in the next section that H = 0 for an ideal solution.)

(b) Repeat (a) at 500°C.

SOLUTION

(a) Equation 4-7 indicates that for a constant temperature process,

$$\Delta G = \Delta H - T\Delta S.$$

For an ideal solution, $\Delta H = 0$, giving

$$\Delta G = -T\Delta S.$$

Using the result of Sample Problem 2-3,

$$\Delta G_{1000°C} = -(1000 + 273)\text{ K }(9.45\text{ J/K})$$
$$= -12.0\text{ kJ}.$$

(b) At 500°C.

$$\Delta G_{500°C} = -(500 + 273)\text{ K }(9.45\text{ J/K})$$
$$= -7.30\text{ kJ}.$$

Note: The negative values for these answers indicate that the mixing is spontaneous.

5 FREE ENERGY AND PHASE EQUILIBRIA

In 1876, J. Willard Gibbs developed a thermodynamic statement of fundamental importance to the practical understanding of phase diagrams. To illustrate his development, consider a simple case of two phases, α and β, each containing different amounts of two components, 1 and 2. Next, consider the transfer of a small amount of componant 1 from phase α to phase β. The effect on the Gibbs free energy of phase α will be:

$$dG_\alpha = \frac{\partial G_\alpha}{\partial N_1^\alpha} dN_1^\alpha \tag{5-1}$$

where dG_α is the change in free energy of α, and dN_1^α is the small amount of component 1 being removed from α. From the perspective of phase β,

$$dG_\beta = \frac{\partial G_\beta}{\partial N_1^\beta} dN_1^\beta . \tag{5-2}$$

But, $dN_1^\beta = -dN_1^\alpha$. Also, the total free energy change for the total system (α + β) must be zero due to the condition of thermodynamic equilibrium. Therefore,

$$dG_{total} = dG_\alpha + dG_\beta$$

$$= \left(\frac{\partial G_\alpha}{\partial N_1^\alpha} - \frac{\partial G_\beta}{\partial N_1^\beta} \right) dN_1^\alpha = 0 \tag{5-3}$$

or

$$\frac{\partial G_\alpha}{\partial N_1^\alpha} = \frac{\partial G_\beta}{\partial N_1^\beta} . \tag{5-4}$$

In general, if we have C components and P phases, a set of equilibrium equations of the form of Equation 5-4 will apply:

$$\left.\begin{array}{l}
\dfrac{\partial G_I}{\partial N_1^{I}} = \dfrac{\partial G_{II}}{\partial N_1^{II}} = \dfrac{\partial G_{III}}{\partial N_1^{III}} = \cdots\cdots = \dfrac{\partial G_P}{\partial N_1^{P}} \\[2ex]
\dfrac{\partial G_I}{\partial N_2^{I}} = \dfrac{\partial G_{II}}{\partial N_2^{II}} = \dfrac{\partial G_{III}}{\partial N_2^{III}} = \cdots\cdots = \dfrac{\partial G_P}{\partial N_2^{P}} \\[2ex]
\dfrac{\partial G_I}{\partial N_3^{I}} = \dfrac{\partial G_{II}}{\partial N_3^{II}} = \dfrac{\partial G_{III}}{\partial N_3^{III}} = \cdots\cdots = \dfrac{\partial G_P}{\partial N_3^{P}} \\
\vdots \\
\dfrac{\partial G_I}{\partial N_c^{I}} = \dfrac{\partial G_{II}}{\partial N_c^{II}} = \dfrac{\partial G_{III}}{\partial N_c^{III}} = \cdots\cdots = \dfrac{\partial G_P}{\partial N_c^{P}}
\end{array}\right\} \tag{5-5}$$

Each of the C rows of equations has P-1 equations for a total of C(P-1) equations associated with the free energy criterion.

It is also true that the sum of mole fractions, X, for any given phase must be equal to one, i.e.,

$$\left.\begin{array}{l}
X_1^{I} + X_2^{I} + X_3^{I} + \cdots\cdots + X_c^{I} = 1 \\
X_1^{II} + X_2^{II} + X_3^{II} + \cdots\cdots + X_c^{II} = 1 \\
X_1^{III} + X_2^{III} + X_3^{III} + \cdots\cdots + X_c^{III} = 1 \\
\vdots \\
X_1^{P} + X_2^{P} + X_3^{P} + \cdots\cdots + X_c^{P} = 1
\end{array}\right\} \tag{5-6}$$

There are, then, P equations associated with the composition criterion.

The number of degrees of freedom (independent variables) for this thermodynamic system will be the total number of variables less the number of equations relating them. The thermodynamic variables in the present case are temperature, pressure, and composition of each phase. Temperature and pressure together are, of course, two variables. There are CP composition variables (C variables for each of P phases). The total number of variables is, then, CP + 2. The degrees of freedom, F, will then be CP + 2 less the number of equations (C(P-1)) in Equation 5-5 and less the number of equations (P) in Equation 5-6:

$$F = CP + 2 - C(P - 1) - P \tag{5-7}$$

or

$$F = C - P + 2. \tag{5-8}$$

This is the **Gibbs phase rule** given without derivation as Equation 5.1 in **Shackelford, 4th Edition**. The application of the phase rule to phase diagrams is discussed there. Now, we shall proceed with a discussion of the graphical relationship between free energy and phase diagrams.

A powerful feature of the free energy is its ability to indicate the equilibrium state for a material. Figure 5-1 illustrates this for a system common to our experience, H_2O at 1 atm pressure. Below 0°C, solid ice is in thermodynamic equilibrium, and the Gibbs free energy of this solid phase is lower than that for liquid or gas phases. (We shall define a **phase** as a chemically homogeneous portion of a system. For example, we shall commonly encounter solid phases that are chemically homogeneous but not structurally homogeneous, e.g., polycrystalline solids.) In that temperature region, the liquid and gas phases can be said to be "thermodynamically unstable" relative to the solid phase. At precisely 0°C (the melting point for ice at 1 atm pressure), the Gibbs free energies of the solid and liquid phases are equal and these two phases are in equilibrium, i.e., both are stable. In the region 0°C $<$ T $<$ 100°C, the Gibbs free energy for the liquid phase is lowest and that phase is stable. At 100°C (the boiling point for water at 1 atm), the liquid and gas phases are in equilibrium. Above 100°C, the gas phase is thermodynamically stable.

For pure iron at 1 atm pressure, the plot of Gibbs free energy versus temperature (Figure 5-2) is somewhat more complex than that for H_2O. The free energy curves for the body-centered cubic (bcc) and face-centered cubic (fcc) phases of iron are close to each other, with the bcc curve having somewhat greater curvature. The result is that, below 910°C, the bcc structure is thermodynamically stable. Heating to 910°C produces a **phase**

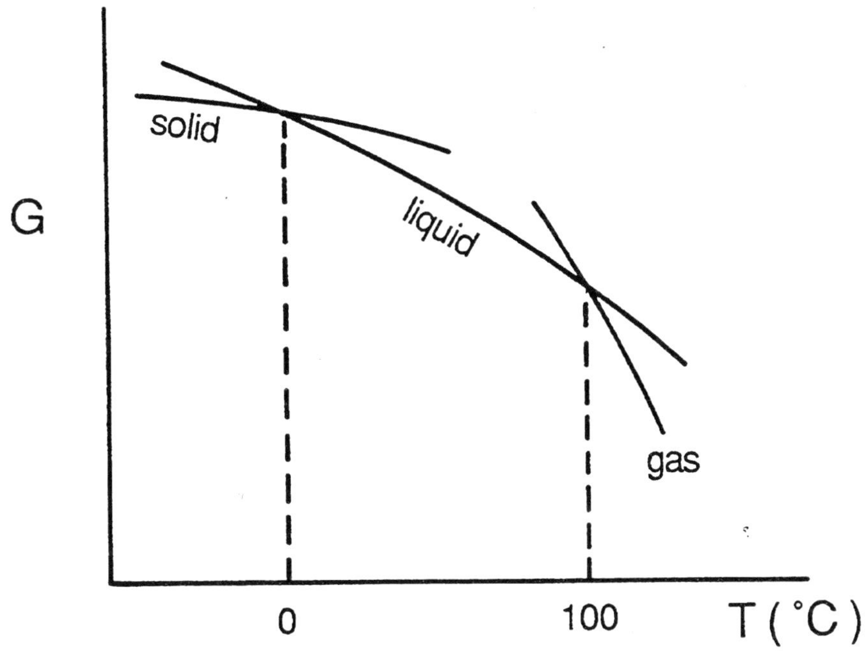

FIGURE 5-1 Schematic plot of the Gibbs free energy, G, as a function of temperature, T, for various phases of H_2O at 1 atm pressure. The thermodynamically stable phase at a given temperature is the one with the lowest free energy.

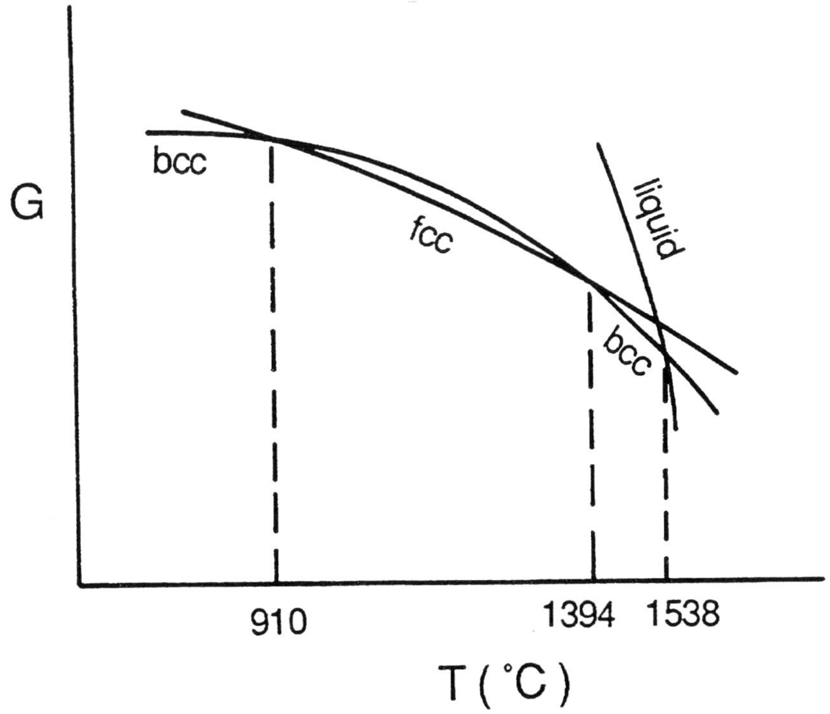

FIGURE 5-2 Schematic plot of the Gibbs free energy, G, as a function of temperature, T, for the various phases of pure iron at 1 atm pressure.

transformation, as the Gibbs free energy of the fcc phase becomes equal to that of the bcc phase. Between 910°C and 1394°C, the fcc phase is thermodynamically stable. Between 1394°C and 1538°C, the bcc phase again is stable. At 1538°C, the iron melts (as the Gibbs free energies of the bcc phase and the liquid phase are equal) and, of course, the liquid phase is stable above 1538°C. In Chapter 5 of **Shackelford, 4th Edition**, the practical result of the free energy plots of Figures 5-1 and 5-2 is seen to be one-component **phase diagrams** ("maps" summarizing the experimental observation of microstructures associated with certain state variables such as temperature). The phase diagram for H_2O is shown in Figure 5-3 and that for iron in Figure 5-4 of **Shackelford, 4th Edition.**

Engineering materials generally contain impurities for practical purposes. In order to understand the phase diagrams for such systems, we must consider the effect of composition on the Gibbs free energy. The basic principles can be illustrated by looking at **binary systems**, i.e., materials composed of two chemical **components**. These components can be elements (e.g., in the case of metal alloys) or chemical compounds (e.g., in the case of ceramics). We have already laid the groundwork for this discussion by developing an expression for the entropy of mixing of two elements in a metal alloy (Equation 2-26 and Figure 2-4). In Figure 5-3, we consider this simple binary alloy further. In Figure 5-3(a), we note that the absolute entropy of each component (S_A and S_B) will, in general, be different. Without atomic-scale mixing, the entropy of some combination of components A and B would simply be a weighted average, as represented by the straight, dashed line in Figure 5-3(a). The entropy of mixing term (Figure 5-3(b)) must be added, giving the plot of the total entropy as a function of alloy composition in Figure 5-3(c). The substantial curvature of the entropy plot (Figure 5-3(c)) is present even for an **ideal solution** of A and B atoms. Such ideality can be defined as one in which the average chemical bond strength between like atoms (A-A or B-B) is equivalent to that between unlike atoms (A-B). As enthalpy is directly related to energy, a plot of enthalpy versus composition for an ideal solution will be a straight line (Figure 5-4). Stated in another way, the **enthalpy of mixing**, ΔH_m, for an ideal solution is zero over the composition range. The resulting plot of Gibbs free energy ($G = H - TS$) with composition is shown in Figure 5-5. For this ideal solution case, the components A and B can dissolve in each other in all proportions. The only

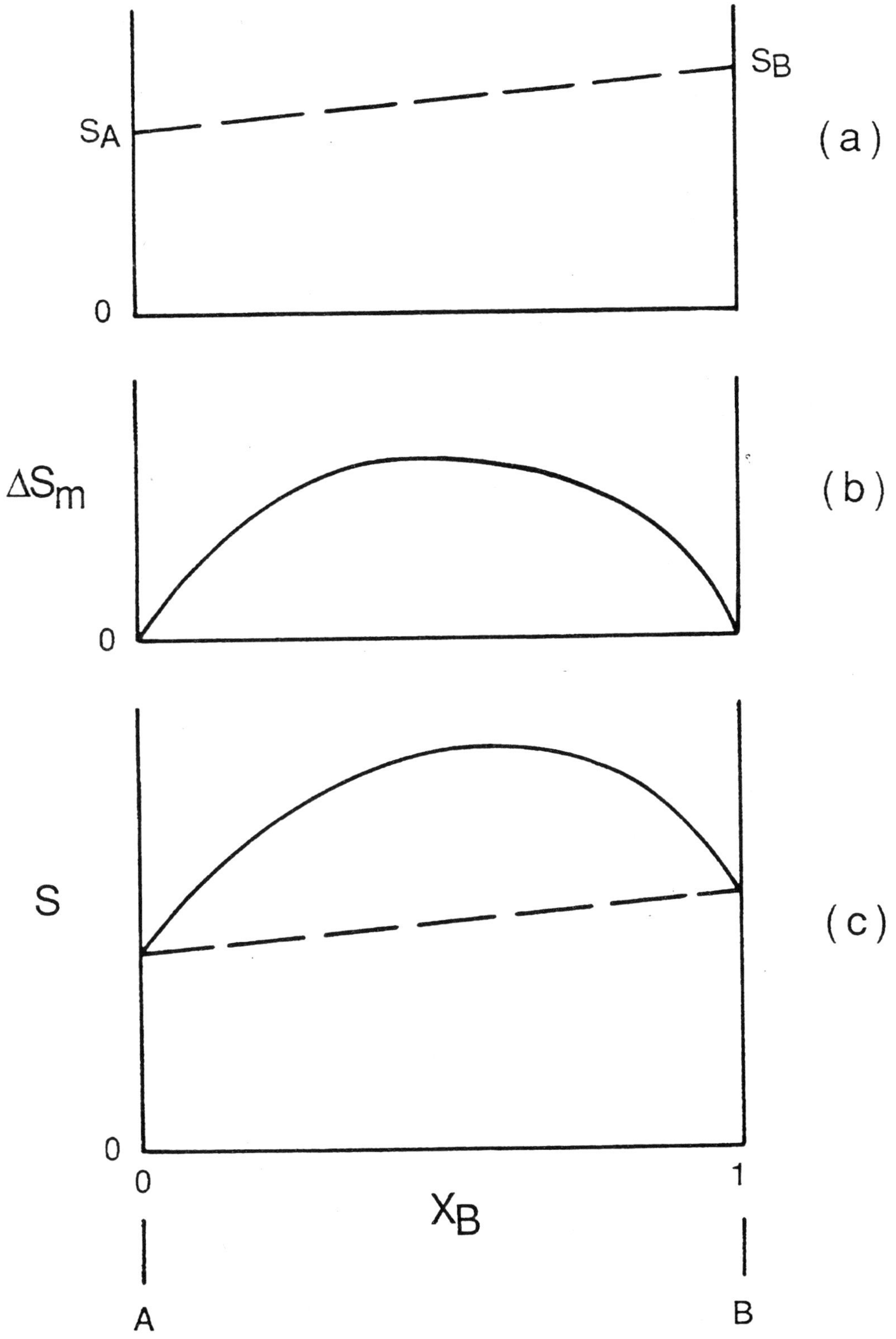

FIGURE 5-3 (a) Total entropy for a binary system without mixing. (b) The entropy of mixing (see Figure 2-4). (c) The total entropy with mixing.

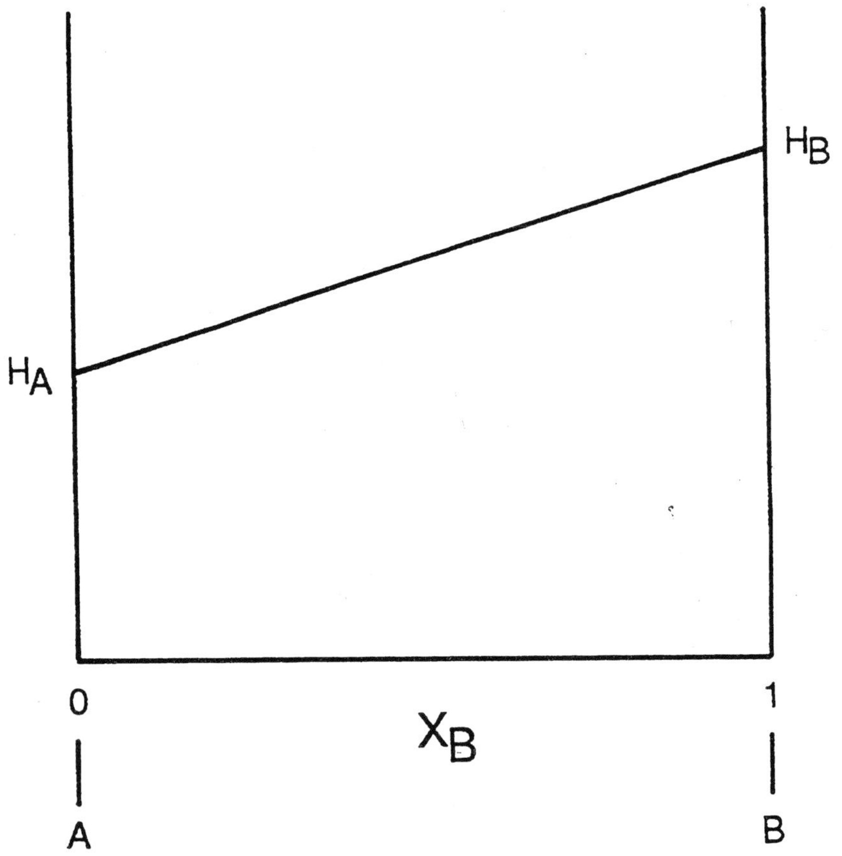

FIGURE 5-4 The enthalpy of an ideal solid solution.

question we need to answer about the phase equilibria is whether the system exists as a solid solution or a liquid solution. The free energy helps to provide an answer: either **or** both. To illustrate this, consider a plot of the Gibbs free energy for the solid solution and for the liquid solution at some temperature below the melting point of both components A and B (Figure 5-6). The free energy of the solid is lower at all compositions and, so, the solid solution exists for any composition. Conversely, consider the case for a temperature above the melting point of both components (Figure 5-7). At this temperature, the situation is reversed, and the Gibbs free energy for the stable liquid solution is lower for all compositions. At a temperature intermediate between the melting point of the two components A and B, the liquid and solid free energy curves overlap (Figure 5-8). As the stable system will be one with the lowest free energy,

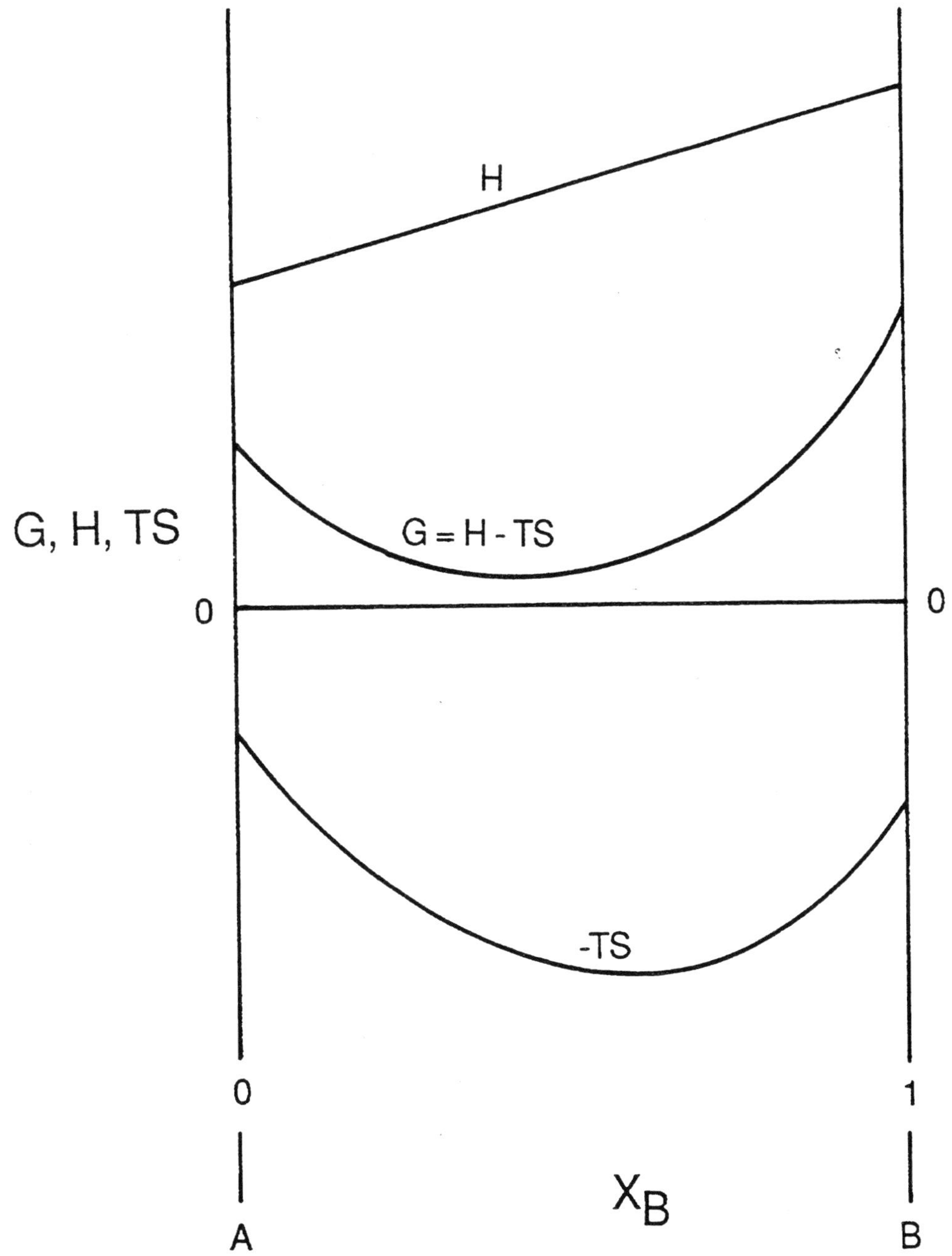

FIGURE 5-5 Combining a plot of enthalpy, H, for an ideal solution (see Figure 5-4) with a plot of a -TS product (note Figure 5-3) gives a plot of Gibbs free energy (G = H - TS) with composition.

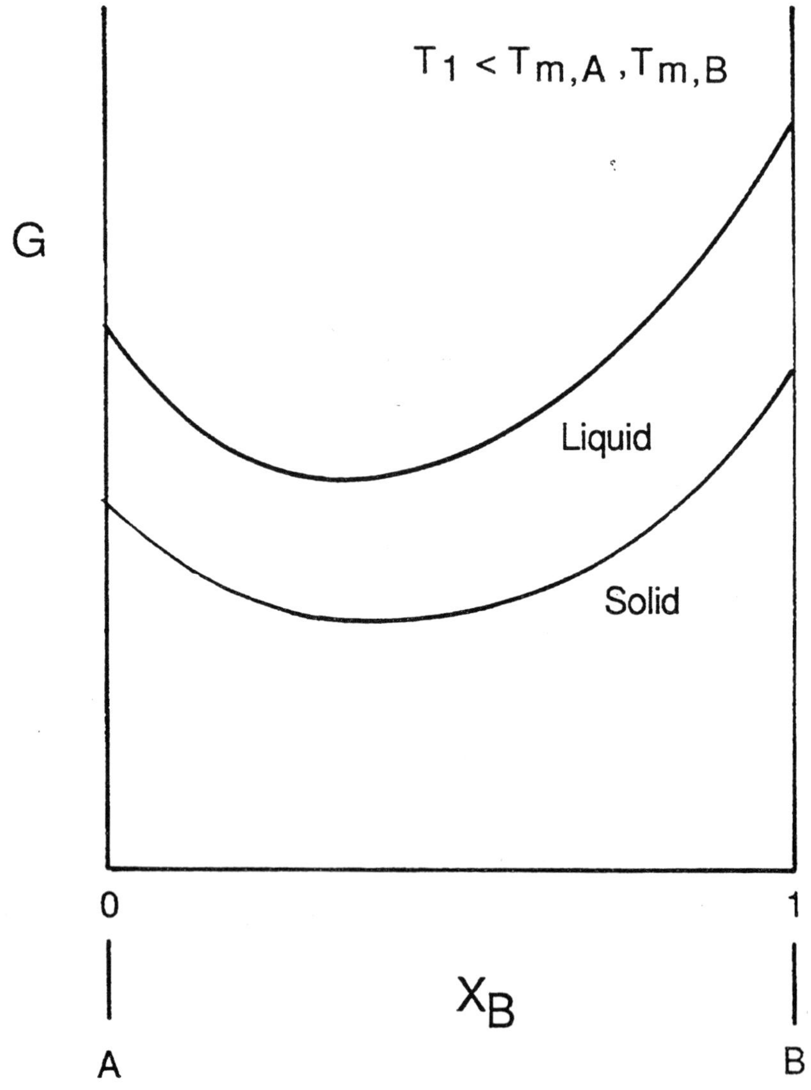

FIGURE 5-6 At a temperature (T_1) below the melting points of both pure A ($T_{m,A}$) and pure B ($T_{m,B}$), the Gibbs free energy of the solid is below that for the liquid at all compositions.

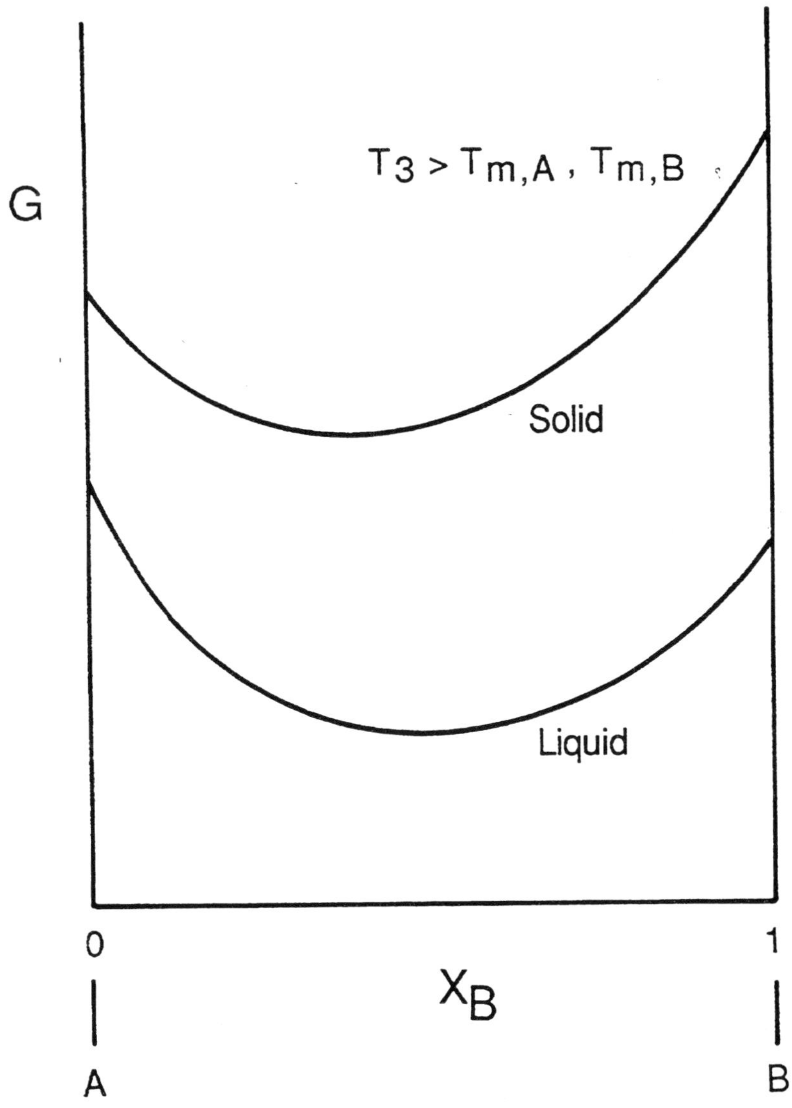

FIGURE 5-7 At a temperature (T_3) above the melting points of both pure A and pure B, the Gibbs free energy of the liquid is below that for the solid at all compositions.

the liquid phase will be stable from a composition of $X_B = 0$ (pure A) up to point $X_B = X_1$. Similarly, the solid phase will be stable from a composition of $X_B = X_2$ up to $X_B = 1$ (pure B). However, between $X_1 < X_B < X_2$, the free energy of either phase will not be as low as that of a two-phase mixture of a liquid (of composition X_1) and a solid (of composition X_2) represented by the dashed line tangent in Figure 5-8. The nature of Figures 5-6, 5-7, and 5-8 leads to a binary phase diagram with complete solid solution (Figure 5-9). (The practical analysis of this diagram is treated in Section 5.2 of **Shackelford, 4th Edition**).

Solid solutions are frequently non-ideal. An important example is the case in which the bonding strength of like atoms (A-A or B-B pairs) is, on the average, greater than for unlike atoms (A-B) pairs. In this situation, there is a positive enthalpy of mixing, ΔH_m. This is shown in the enthalpy versus composition plot of Figure 5-10 and can be contrasted with the straight-line plot of Figure 5-4. The practical result of this enthalpy behavior is that, at relatively low temperatures, separate A-rich and B-rich solid solutions tend to exist rather than a single, homogeneous solid solution at all compositions. Why this is so can be appreciated by combining the enthalpy plot of Figure 5-10 with an entropy plot to give a net plot of Gibbs free energy with composition (Figure 5-11(a)). Concentrating on the free energy plot in Figure 5-11(b), we again encounter a case (similar to Figure 5-8) where a dashed-line tangent can provide a lower free energy for the system than the solid line plot. In this case, a two-phase mixture of solid solutions with compositions X_1 and X_2 will be stable for an overall system composition in the range $X_1 < X < X_2$. When the two pure components in a binary system have different crystal structures, a similar tendency toward phase separation occurs. Figure 5-12 is a Gibbs free energy plot for this case. The key distinction from Figure 5-11(b) is that there are two separate free energy plots in Figure 5-12, one for an α phase, and one for a β phase (with crystal structure different from that for the α phase). Figure 5-13 shows a general summary of the Gibbs free energy plots for the α and β phases, as well as the liquid phase, at several temperatures corresponding to a binary eutectic phase diagram in Part (f). The term "eutectic" comes from the relatively low melting temperature of intermediate compositions, as compared to pure components A and B. This term is discussed in greater detail in Section 5.2 of **Shackelford, 4th Edition** (note also

Figure 5-9). The practical analysis of the binary eutectic diagram is treated in Sections 5.1 and 5.2 of **Shackelford, 4th Edition.**

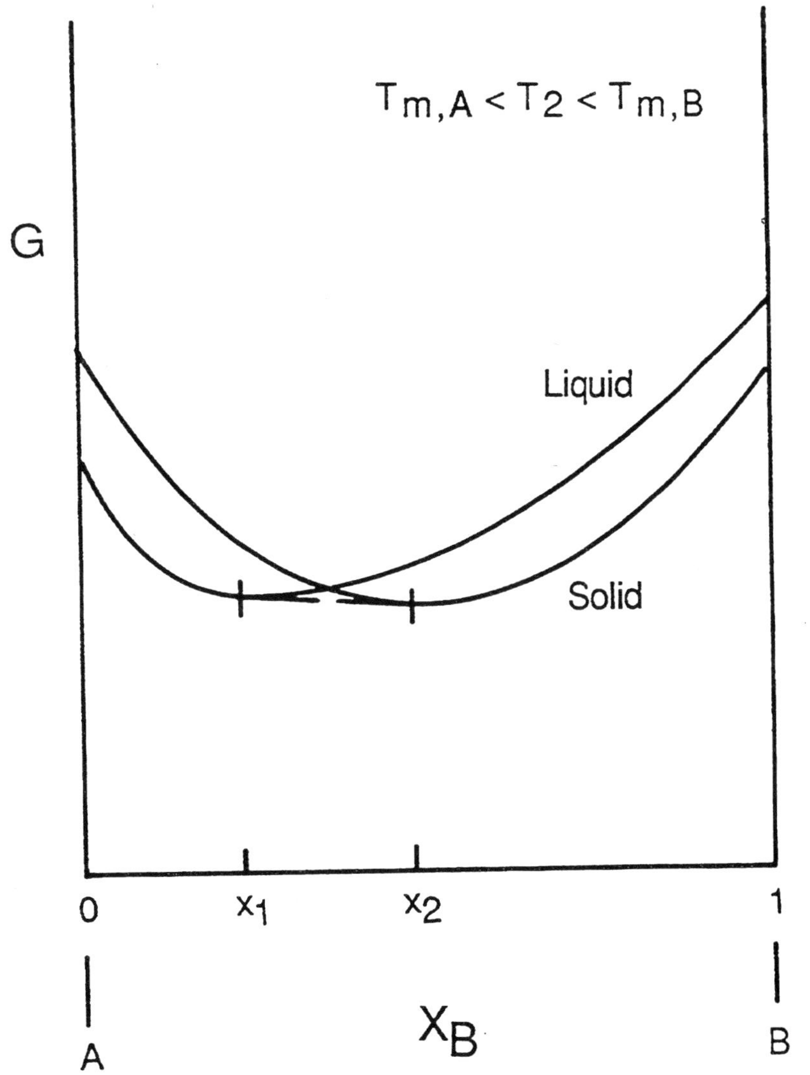

FIGURE 5-8 At a temperature (T_2) between the melting points of pure A and pure B, the Gibbs free energy plots for the solid and liquid overlap. At a given composition, the stable phase is the one with the lowest value of G. Between x_1 and x_2, a mixture of liquid (with composition x_1) and solid (with composition x_2) is stable.

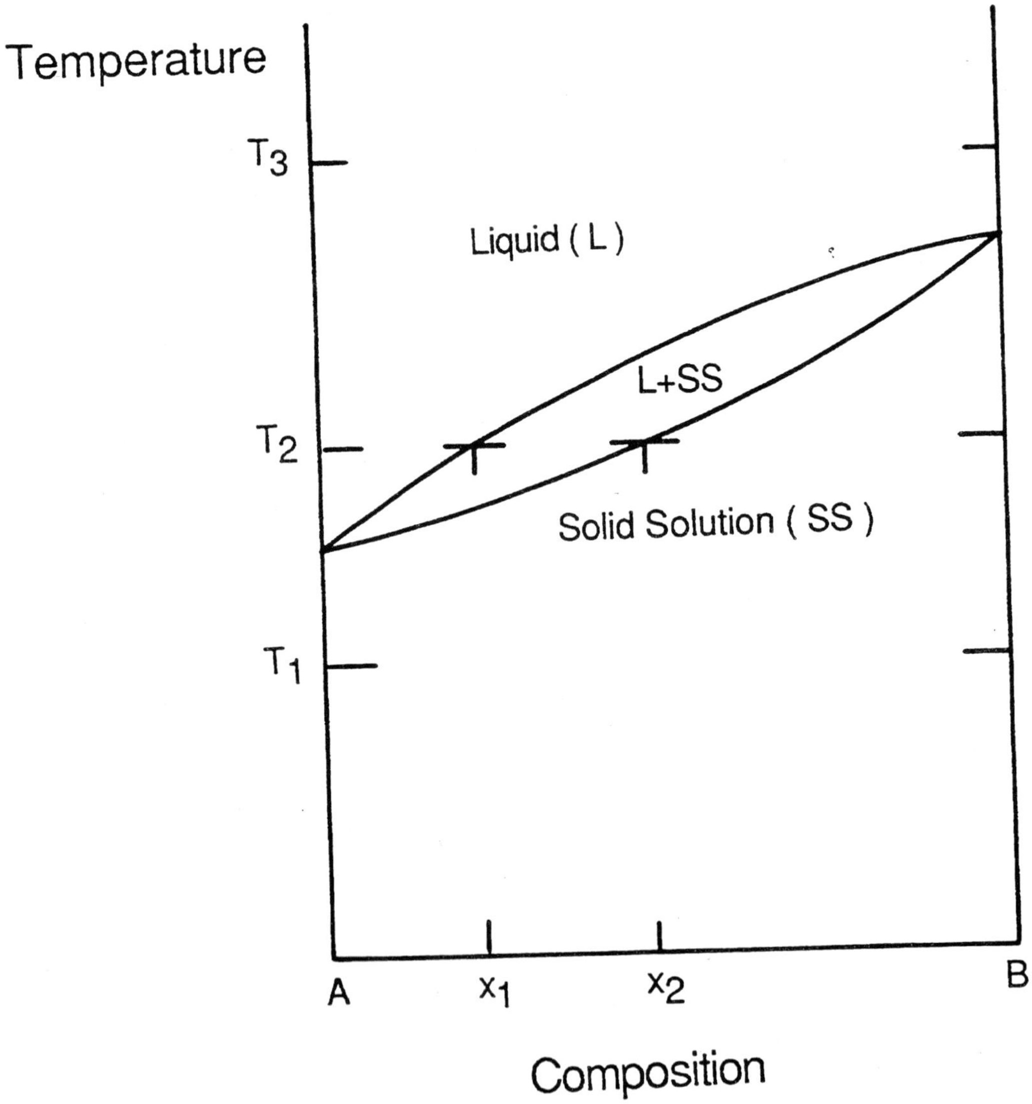

FIGURE 5-9 A binary phase diagram with complete solid solution corresponding to the free energy plots of Figures 5-6, 5-7, and 5-8.

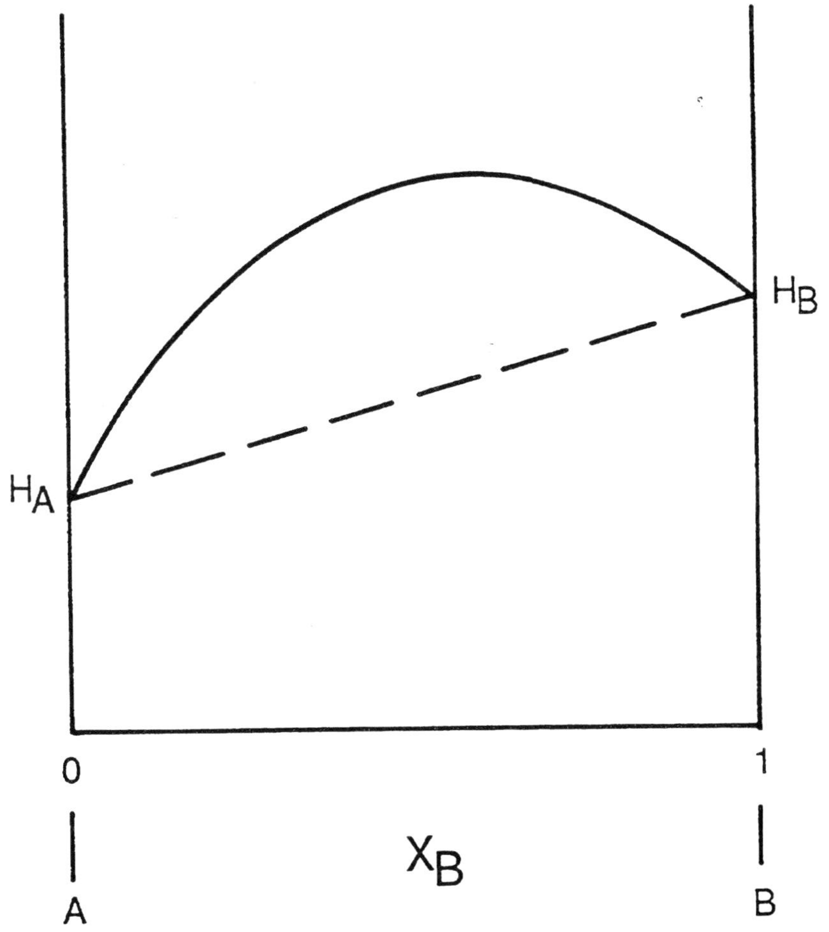

FIGURE 5-10 A non-ideal solution with a positive enthalpy of mixing gives an enthalpy-composition plot in contrast to the straight line of Figure 5-4.

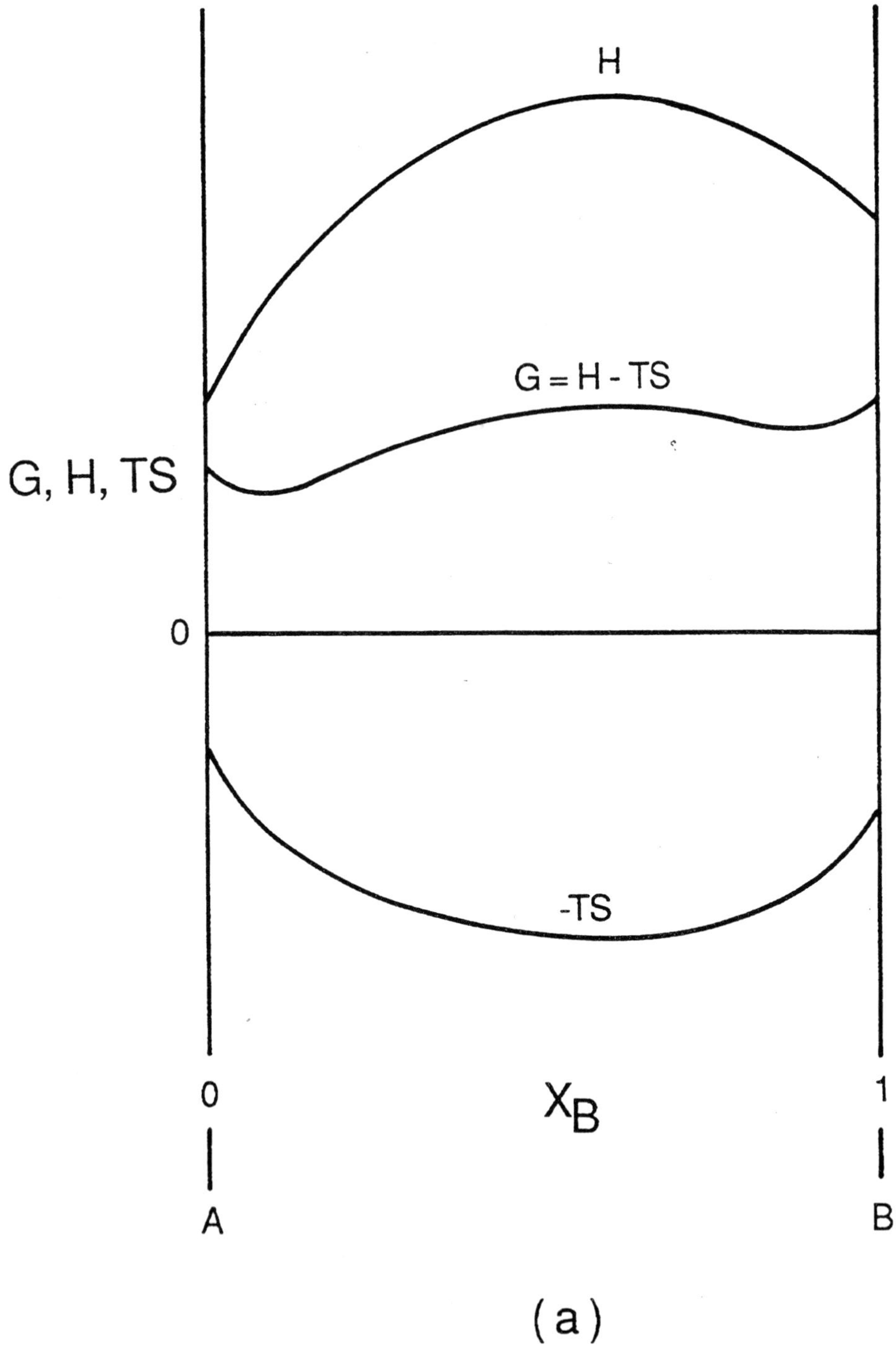

FIGURE 5-11 (a) Combining the enthalpy curve of Figure 5-10 with a plot of -TS gives a Gibbs free energy curve for a non-ideal solution with positive enthalpy of mixing.

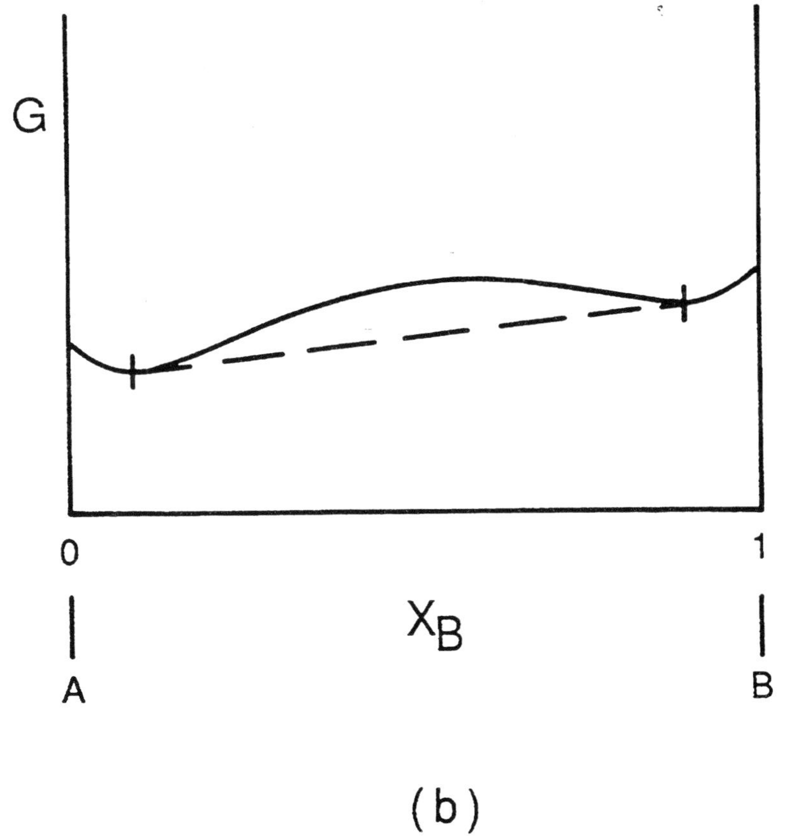

FIGURE 5-11 (b) Concentrating on the free energy plot, we again see that a mixture of two phases is stable over a range of compositions (indicated by the dashed tangent line).

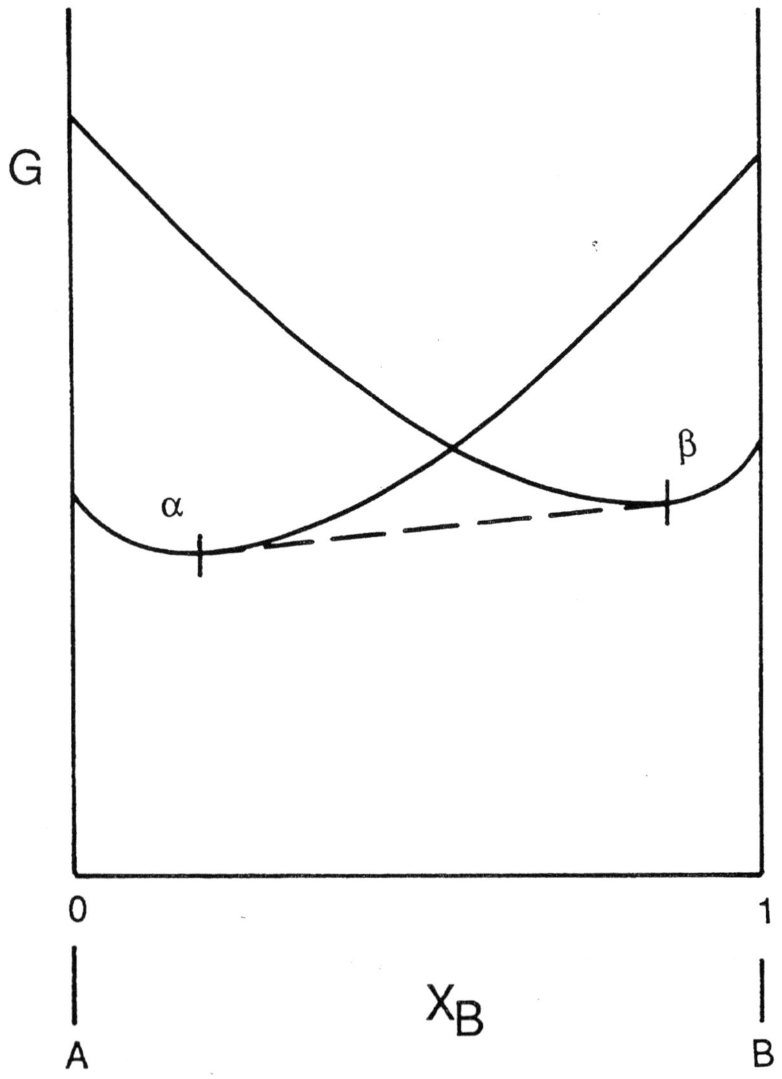

FIGURE 5-12 When components A and B have different crystal structures, two distinct free energy curves exist (for α and β phases, respectively). Again, a dashed line tangent indicates a composition range in which a mixture of two phases (α + β) is stable.

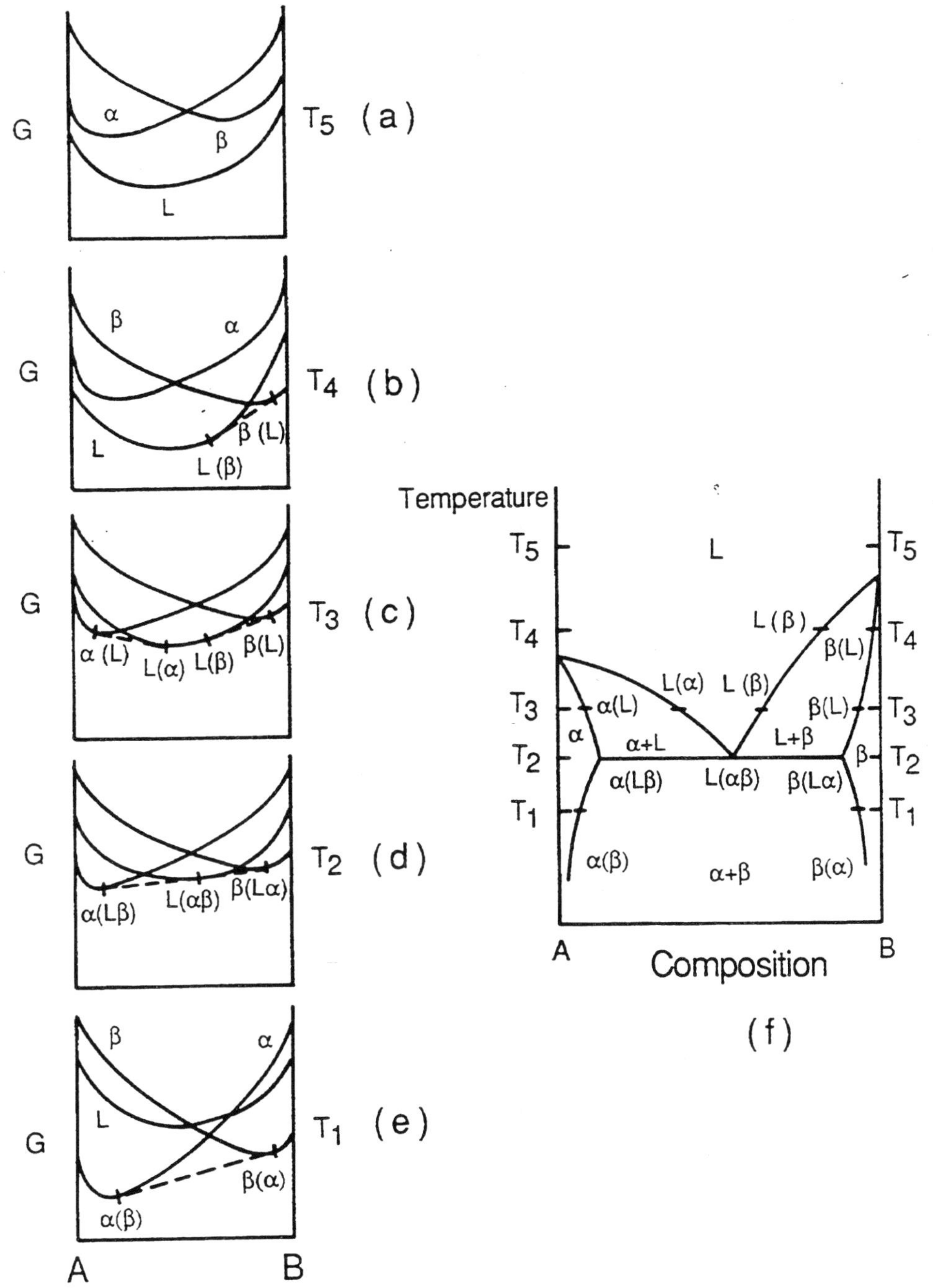

FIGURE 5-13 (a) - (e) A summary of Gibbs free energy versus composition at several temperatures for components A and B introduced in Figure 5-12 (which represents a temperature near T_1 in part (e)). (f) The resulting binary eutectic phase diagram. (After J.H. Brophy, R.M. Rose, and J. Wulff, The Structure and Properties of Materials, Vol. 2: Thermodynamics of Structure, John Wiley & Sons, New York, 1964.)

SAMPLE PROBLEM 5-1

Some binary phase diagrams with complete solid solution show a melting point maximum at some intermediate composition:

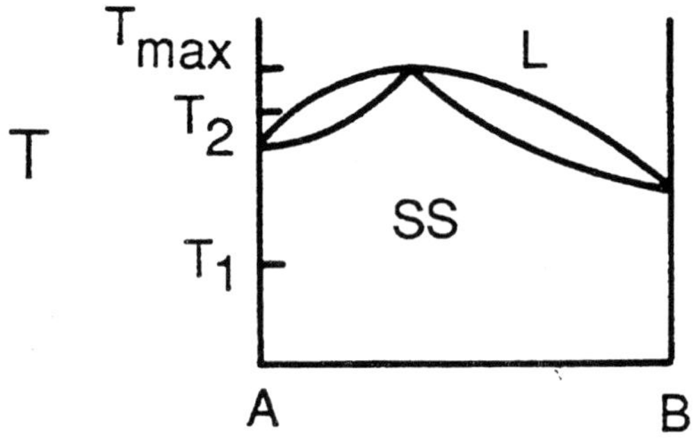

Sketch the corresponding Gibbs free energy plots for this system at (a) T_1, (b) T_2, and (c) T_{max}.

SOLUTION

(a) at T_1

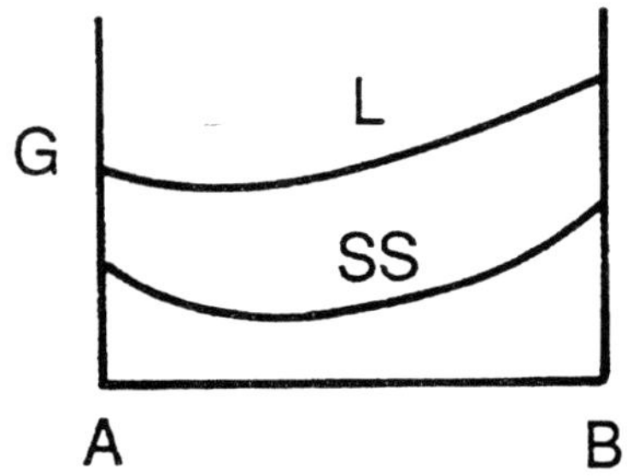

(b) at T_2

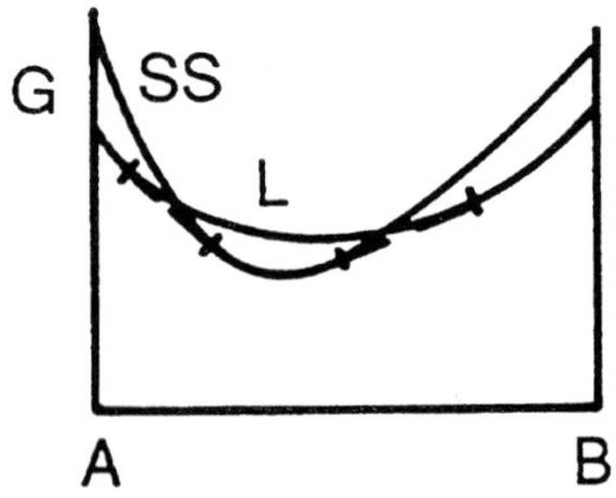

(c) at T_{max}

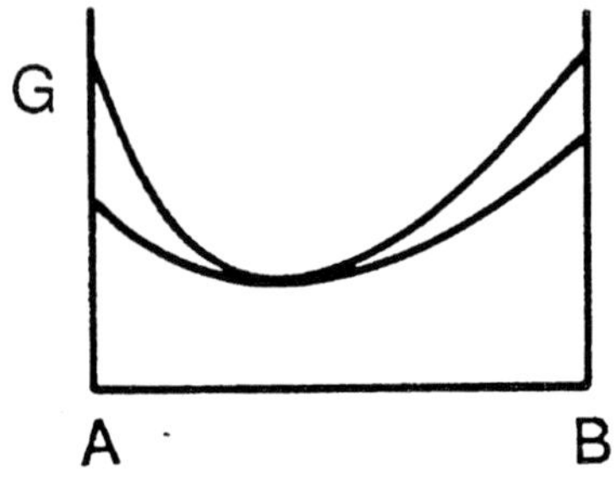

6 THE THERMODYNAMICS OF SURFACES

In Section 4.4 of **Shackelford, 4th Edition**, various **surface** structures were introduced. These play important roles in a wide variety of materials applications. Inspection of the various schematics of surface structures indicates that these planar structures are defective in comparison to ideal, crystalline bulk structures. From the perspective of thermodynamics, the surfaces are unstable (with a higher free energy) than the bulk material. Specifically, we can define the **surface energy**, γ, as the difference in value between the Gibbs free energy of the surface region, G_{surf}, and the Gibbs free energy of an equivalent amount of bulk material, G_{bulk}:

$$\gamma = G_{surf} - G_{bulk}. \qquad (6\text{-}1)$$

An equivalent definition of the **surface energy** is that it is the Gibbs free energy associated with the formation of a given amount of new surface area from the bulk. As a practical matter, the surface energy is approximated by the mechanical quantity **surface tension** which has units of force per length (e.g., newtons per meter). Units of force per length are equal to surface energy units of energy per area (e.g., J/m^2). Furthermore, surface energy and surface tension are generally taken to be numerically equal. For most liquids and for solids near their melting points, this approximation is a good one. However, for solids at lower temperatures, the approximation can be less valid due to crystallographic orientation effects. In spite of such limitations, conveniently-measured surface tensions are routinely cited as "surface energies" for engineering materials.

A common method for measuring surface tensions in a relative way is with force equilibrium experiments. Figure 6-1 summarizes one of the most common examples, viz., the **"sessile drop experiment,"** in which a drop of liquid is placed on top of a solid substrate. (Sessile means "attached directly to the base" and comes from the Latin word *sessilis* referring to sitting.) If the combined surface tensions of the solid-liquid interface (γ_{SL}) and the liquid-vapor interface (γ_{LV}) are less than the surface tension of the solid-vapor interface (γ_{SV}), i.e.,

$$\gamma_{SL} + \gamma_{LV} < \gamma_{SV}, \qquad (6\text{-}2)$$

Complete Wetting, θ = 0°

$\gamma_{SL} + \gamma_{LV} < \gamma_{SV}$

(a)

Partial Wetting, 0° < θ < 180°

$\gamma_{SV} = \gamma_{SL} + \gamma_{LV} \cos \theta$

(b)

Non Wetting, θ = 180°

$\gamma_{SV} + \gamma_{LV} < \gamma_{SL}$

(c)

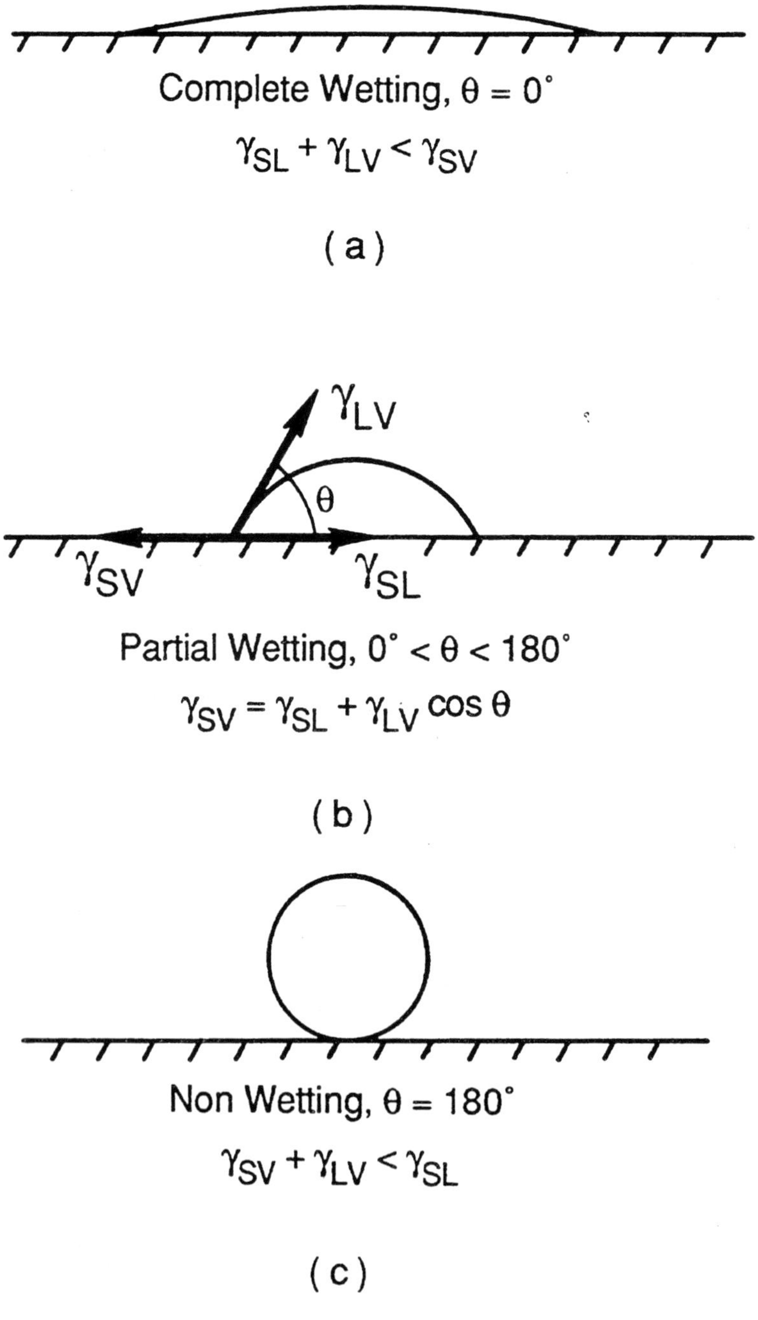

FIGURE 6-1 In the "sessile drop experiment," the three conditions for wetting of a solid surface by a liquid drop depend on the relative magnitudes of the three interfacial surface tensions (SL = solid-liquid, SV = solid-vapor, LV = liquid-vapor).

the drop will completely **"wet"** the solid surface (Figure 6-1(a)). Equation 6-2 indicates that the spreading of the liquid is an attempt to eliminate the higher energy solid-vapor interface. At the other extreme, the condition

$$\gamma_{SV} + \gamma_{LV} < \gamma_{SL} , \tag{6-3}$$

leads to the **"nonwetting"** condition shown in Figure 6-1(c). In this case, the higher energy solid-liquid interface in minimized. The intermediate case of **"partial wetting"** shown in Figure 6-1(b) involves a force equilibrium:

$$\gamma_{SV} = \gamma_{SL} + \gamma_{LV} \cos \Theta \tag{6-4}$$

where Θ is the **contact angle.**

Another type of force equilibrium experiment allows determination of the ratio of the surface tension of the solid-vapor interface, γ_{sv}, to that of a grain boundary, γ_b. Figure 6-2 illustrates this so-called **"thermal grooving experiment."** If a solid-vapor surface intersected by a grain boundary (Figure 6-2(a)) is heated to a sufficiently high temperature to allow substantial surface diffusion, there will be a net flow of material away from the grain boundary to form small humps on either side (Figure 6-2(b)). The driving force for this "thermal groove" formation is the establishment of a force equilibrium between the adjacent interfacial tensions:

$$\frac{\gamma_b}{\gamma_{SV}} = 2 \cos \phi \tag{6-5}$$

where ϕ is the **"groove angle"** (Figure 6-2(c)). For typical experimental systems, the ratio γ_b/γ_{sv} is roughly 0.5.

Force equilibrium equations (such as Equations 6-4 and 6-5) give relative surface tension data. To obtain absolute results, a direct measurement of one surface tension (or two, for Equation 6-4) is required. Of most general interest is the solid-vapor surface tension, γ_{sv}. Direct measurement of γ_{sv} can be a difficult task. One of the most successful methods is to suspend weights from thin single crystal wires. At relatively high temperatures, the force of the suspended weight can be balanced almost entirely by γ_{sv}. Substantial progress has also been made in recent decades on the theoretical calculation of γ_{sv}. Table 6-1 summarizes the value of γ_{sv} for a wide variety of engineering materials. Note that, in general, metals have

γ_{sv} values on the order of 1 to 2 J/m^2, ceramics and semiconductors on the order of 0.1 to 1 J/m^2, and polymers generally below 0.1 J/m^2.

TABLE 6-1 Solid-Vapor Surface Tensions for Various Materials

Category	Material	γ_{sv} (J/m^2)
Metals[a]	Ag	1.2
	Au	1.4
	Ni	1.9
	Cu	1.7
Ceramics and Semiconductors[b]	MgO	1.20
	$CaCO_3$	0.23
	NaCl	0.30
	LiF	0.34
	Si	1.24
Polymers[c]	polyethylene	0.0357
	polypropylene	0.0301
	nylon 66	0.0447

[a]Source: Data from J.H. Brophy, R.M. Rose, and J. Wulff, **The Structure and Properties of Materials**, Vol. 2: **Thermodynamics of Structure**, John Wiley & Sons, New York, 1964.

[b]Source: Data from R.A. Swalin, **Thermodynamics of Solids, Second Edition**, John Wiley & Sons, New York, 1972.

[c]Source: Data from S. Wu, **Polymer Interfaces and Adhesion**, Marcel Dekker, New York, 1982.

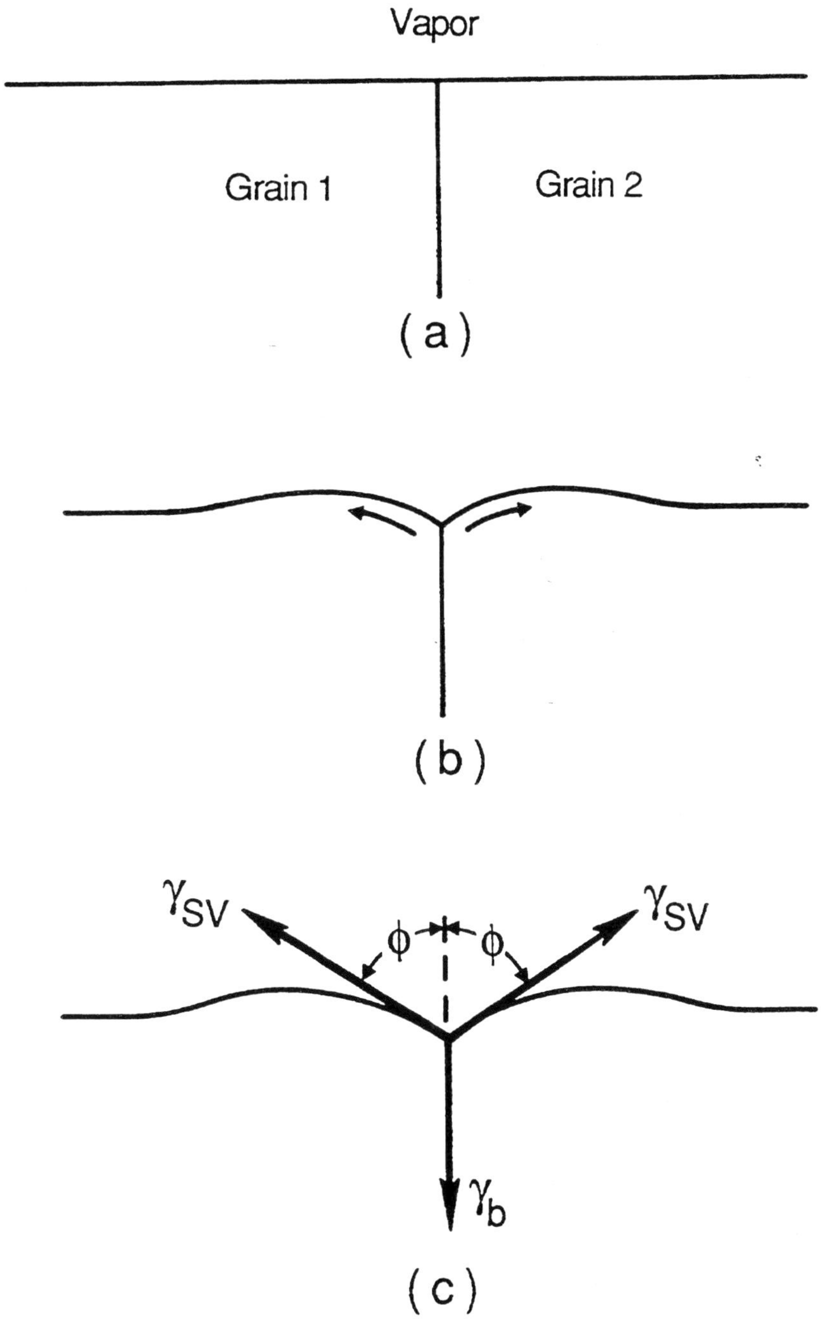

FIGURE 6-2 (a) A grain boundary intersecting a solid-vapor interface. (b) Upon heating the system in (a) to a temperature high enough to cause surface diffusion, small humps form on either side of the grain boundary. (c) The resulting force equilibrium for the "thermal groove" defines a "groove angle," ϕ.

SAMPLE PROBLEM 6-1

A drop of liquid placed on a Ni surface forms a spherical cap 2 mm high and 4 mm diameter at its base. Calculate the surface tension of the solid-liquid interface.

SOLUTION

In this case, the contact angle, θ, is 90°:

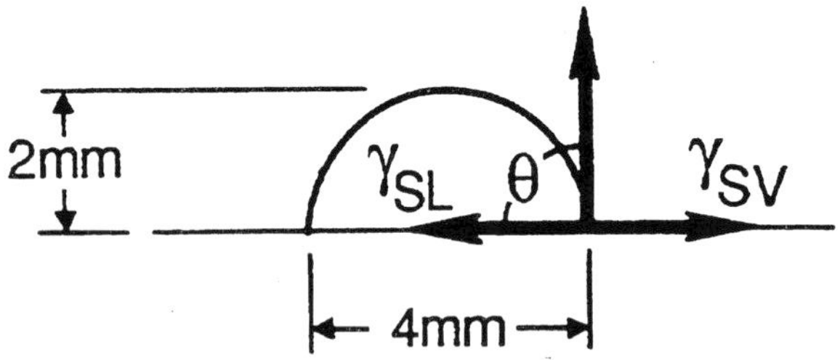

For this specific case, Equation 6-4 becomes

$$\gamma_{SV} = \gamma_{SL} + \gamma_{LV} \cos 90°$$

$$= \gamma_{SL} + \gamma_{LV} (0)$$

$$= \gamma_{SL}$$

Table 6-1 gives γ_{sv} as 1.9 J/m^2. Therefore,

$$\gamma_{SL} = \gamma_{SV} = 1.9 \text{ J/m}^2$$

SAMPLE PROBLEM 6-2

The measured thermal groove angle for a grain boundary study in copper is 79.6°. Calculate the grain boundary surface tension.

SOLUTION

Equation 6-5 gives

$$\frac{\gamma_b}{\gamma_{SV}} = 2 \cos \phi$$

$$= 2 \cos (79.6°)$$

$$= 0.36.$$

Rearranging and using Table 6-1 gives

$$\gamma_b = 0.36\ \gamma_{SV} = 0.36\ (1.7\ J/m^2)$$

$$= 0.61\ J/m^2.$$

7 THE THERMODYNAMICS OF KINETICS

This section on thermodynamics indicates the bridge with **kinetics**. Kinetics can be defined as the science of time-dependent phase transformations. It is common to treat thermodynamics and kinetics separately. For example, Chapter 5 of **Shackelford, 4th Edition** deals with equilibrium **phase diagrams**, and Chapter 6 deals with **heat treatment** and the time-dependent approach to equilibrium. In this section, we shall emphasize the role of thermodynamics as a foundation for the laws of kinetics.

For a process exhibiting "Arrhenius behavior," the rate is equal to a **pre-exponential constant**, C, times an exponential term containing an atomic-scale **activation energy**, q, Boltzmann's constant, k, and absolute temperature, T:

$$\text{rate} = C\, e^{-q/kT}. \tag{7-1}$$

Section 4.2 of **Shackelford, 4th Edition** contains a more thorough discussion of Arrhenius behavior. The rate in Equation 7-1 corresponds to the probability, P, that a given atom or molecule will have an energy, ΔE, greater than the average energy characteristic of a particular temperature:

$$P \propto e^{-\Delta E/kT}. \tag{7-2}$$

With the introduction of the free energy (in Section 4) as a more appropriate indicator of equilibrium than energy itself, we can now write the probability expression in a more fundamental way as:

$$P \propto e^{-\Delta g/kT} \tag{7-3}$$

where the lower-case g is used to indicate the free energy of activation per atomic-scale unit (e.g., atom). For a molar basis, we can write

$$P \propto e^{-\Delta G/RT} \tag{7-4}$$

where R is the gas constant ($= N_{Av}k$). Figure 7-1 is comparable to Figure 4-11 in **Shackelford, 4th Edition**, but with the activation free energy (rather than simply activation energy) shown. The specific relationship of ΔG to the "activation energy" can be obtained by combining Equation 4-7 with Equation 7-4, giving

$$P \propto e^{-(\Delta H - T\Delta S)/RT} \qquad (7\text{-}5)$$

or

$$P \propto e^{+\Delta S/R} e^{-\Delta H/RT}. \qquad (7\text{-}6)$$

The primary importance of Equation 7-6 is that it identifies the activation energy for a process, Q, as an enthalpy term (specifically, the enthalpy of the "activated state" minus the enthalpy of the equilibrium state as illustrated in Figure 7-2). Equation 7-6 also points out that the pre-exponential constant is an entropy term. Because temperature was "cancelled out" of the $e^{\Delta S/R}$ term, the pre-exponential constant is nearly temperature independent. (Over a wide temperature range, it is necessary to acknowledge a slight temperature dependence of the entropy itself.)

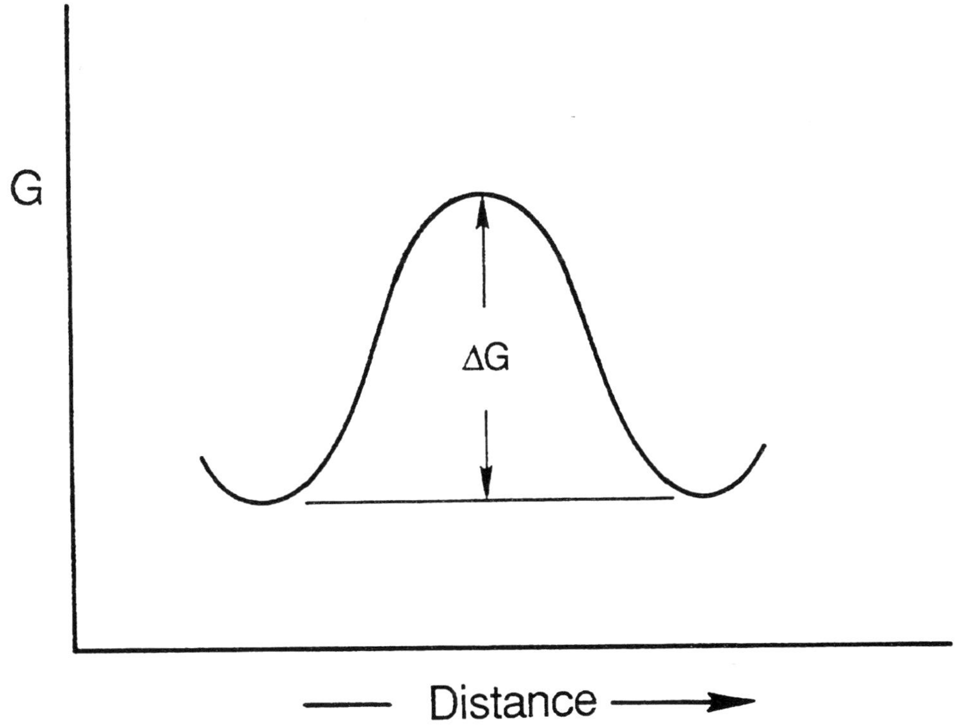

FIGURE 7-1 The activation free energy, ΔG, for an atom moving between two adjacent equilibrium sites. The G versus distance plot has the same general form as Figure 4-11 in Shackelford, 4th Edition.

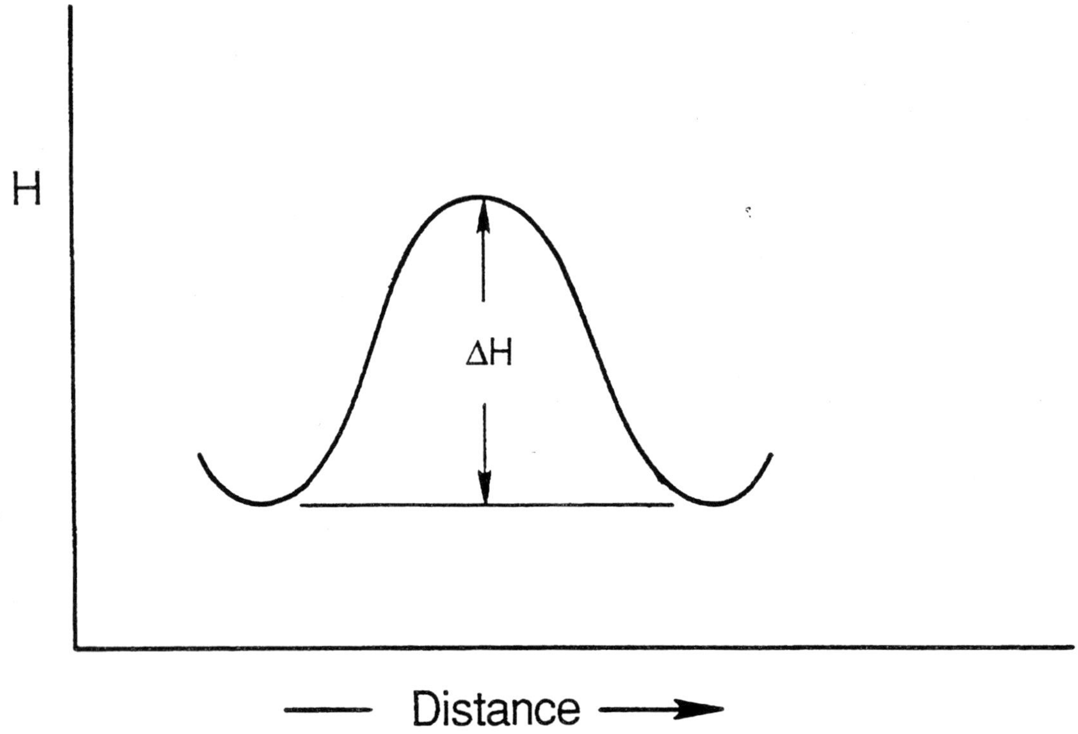

FIGURE 7-2 The activation enthalpy, ΔH, for the atom motion described in Figure 7-1 is equivalent to the "activation energy" of Figure 4-11 in Shackelford, **4th** Edition.

SAMPLE PROBLEM 7-1

A theoretical model of the interstitial diffusion of carbon in a bcc metal gives an expression:

$$D = \frac{1}{6} a^2 \nu e^{-\Delta G/RT}$$

where a is the lattice parameter and ν the lattice vibrational frequency. Using the experimental data of Table 4.2 in **Shackelford, 4th Edition**, calculate (a) the activation entropy, ΔS, and (b) the activation enthalpy, ΔH, for diffusion of carbon in bcc iron. (Take $\nu = 10^{13}\ s^{-1}$.)

SOLUTION

(a) The lattice parameter for bcc iron is related to the atomic radius by

$$a = \frac{4}{\sqrt{3}} r$$

as shown in Practice Problem 3.9 of **Shackelford, 4th Edition.**

Using Appendix 2 of **Shackelford, 4th Edition,**

$$a = \frac{4}{\sqrt{3}} (0.124\ nm) = 0.286\ nm.$$

As $\Delta G = \Delta H - T\Delta S$, we can rewrite the model equation as:

$$D = \frac{1}{6} a^2 \nu e^{\Delta S/R} e^{-\Delta H/RT}.$$

Comparing with the empirical diffusivity equation (Equation 4.13 in **Shackelford, 4th Edition**), we see that

$$D_o = \frac{1}{6} a^2 \nu e^{\Delta S/R}.$$

Table 4.2 gives, for carbon in bcc iron,

$$D_o = 220 \times 10^{-6} \text{ m}^2/\text{s}.$$

Therefore,

$$e^{\Delta S/R} = \frac{6 D_o}{a^2 \nu}$$

$$= \frac{6 (220 \times 10^{-6} \text{ m}^2/\text{s})}{(0.286 \times 10^{-9} \text{ m})^2 (1 \times 10^{13} \text{ s}^{-1})}$$

$$= 1.61 \times 10^3$$

or

$$\frac{\Delta S}{R} = \ln (1.61 \times 10^3) = 7.386.$$

Therefore,

$$\Delta S = 7.386 \left(8.314 \text{ J/(mol·K)}\right)$$

$$= 61.4 \text{ (J/K)/mol}.$$

(b) From the discussion relative to Equation 5.7-4, we note that

$$\Delta H = Q.$$

From Table 4.2 ,

$$\Delta H = Q = 122 \text{ kJ/mol}.$$

8 THERMODYNAMICS OF ENVIRONMENTAL REACTIONS

Materials engineers must be aware of the consequences of the chemical reactions between metals and their environments, as discussed in Chapter 14 of **Shackelford, 4th Edition.** These reactions include (a) direct **oxidation** by atmospheric gases and (b) various forms of electrochemical **corrosion.** We can better appreciate the nature of these various reactions by considering the associated thermodynamics. The nature of chemical degradation of ceramics and polymers is governed by similar thermodynamic principles.

As indicated in Section 4, a chemical reaction is spontaneous if it is accompanied by a negative change in free energy. For the general reaction of a metal, M, with oxygen at a given temperature and oxygen pressure,

$$a\,M + \frac{1}{2}\,O_2(g) = M_aO, \tag{8-1}$$

the change in Gibbs free energy, ΔG, is generally negative, as shown in Table 8-1. Consistent with our common experience, metals react with oxygen to some degree. The two exceptions in Table 8-1, silver and gold, are highly prized for the "noble" character.

As with other processes discussed in Part I, the ΔG for oxidation reactions are functions of the reaction temperature. In addition, the value of ΔG depends on the partial pressure of oxygen in the atmosphere. Some metals, such as palladium, can react with oxygen under a partial pressure of 1 atmosphere but will be unreactive (i.e., have a positive ΔG) under a modest vacuum.

Corrosion is associated with aqueous environments. Consider the oxidation reaction for a metal, M_1, going into aqueous solution:

$$M_1 = M_1^{n+} + ne^- \tag{8-2}$$

where the ion in solution, M_1^{n+}, has a valence of +n due to the liberation of n electrons. If an adjacent metal, M_2, in the aqueous environment is reduced by the electrons provided in Equation 8-2, the reaction will be

$$M_2^{n+} + ne^- = M_2. \tag{8-3}$$

Combining Equations 8-2 and 8-3 gives a net reaction:

$$M_1 + M_2^{n+} = M_1^{n+} + M_2. \tag{8-4}$$

There is a negative value of ΔG for this spontaneous chemical reaction. The Gibbs free energy change is, in turn, equal to the electrical work done in the reversible transfer of electrons from metal M_1 to metal M_2:

$$\Delta G = W_{electrical} = -n F V \tag{8-5}$$

where n is the number of moles of electrons transferred per mole of M_1 ionized, F is Faraday's[6] constant (= 96,500 C/mol of electrons), and V is the voltage between the two metals. The negative sign in Equation 8-5 is assigned to permit a positive voltage when the free energy change is negative. In considering reactions such as Equation 8-4, the **corrosion** (i.e., ionization) of metal M_1 is spontaneous when ΔG is negative. Various electrochemical cells associated with practical corrosion problems are illustrated in Sections 14.2 to 14.5 of **Shackelford, 4th Edition.**

---

[6]Michael Faraday (1791-1867), English chemist and physicist. Raised in poverty in London, Faraday's future lay in a career as a bookbinder until a chance opportunity for the bright and curious Faraday to apprentice to the eminent Humphry Davy. Faraday seized the opportunity and began one of the brilliant careers of nineteenth century science. He extended the research of predecessors and contemporaries Ampere, Coulomb, and Henry. In turn, James Clerk Maxwell was to provide an elegant mathematical extension of Faraday's concepts of field theory. (See **Shackelford, 4th Edition,** for related biographies.)

TABLE 8-1 Free Energy of Formation of Metal Oxides (per Gram-Atom of Oxygen) at 500 K and p_{O_2} = 1 atm.[a]

Metal	Oxide	ΔG (kJ)
Ca	CaO	-578.5
Mg	MgO	-547.5
Al	Al_2O_3	-505.2
Ti	TiO_2	-423.6
Na	Na_2O	-347.4
Cr	Cr_2O_3	-341.6
Zn	ZnO	-298.5
H	H_2O	-244.0
Fe	Fe_2O_3	-232.3
Co	CoO	-200.5
Ni	NiO	-193.0
Cu	Cu_2O	-131.9
Ag	Ag_2O	+ 2.5
Au	Au_2O_3	+ 44.0

[a]Source: Data from J.H. Brophy, R.M. Rose, and J. Wulff, **The Structure and Properties of Materials**, Vol. 2: **Thermodynamics of Structure**, John Wiley & Sons, New York, 1964.

SAMPLE PROBLEM 8-1

What is the free energy change associated with the reaction of aluminum with oxygen (p = 1 atm) at 500 K to form one mole of Al_2O_3?

SOLUTION

Relevant experimental data are given in Table 8-1 but on a basis of one gram-atom of oxygen. Considering the general reaction

$$2\ Al + 3\ O = Al_2O_3,$$

it is clear that the formation of 1 mole of Al_2O_3 requires 3 gram-atoms of oxygen (actually 3/2 moles of diatomic O_2). Then, the ΔG for the above reaction will be three times the value given in Table 8-1:

$$\Delta G = 3\ (-505.2\ kJ) = -1515.6\ kJ.$$

SAMPLE PROBLEM 8-2

What is the free energy change associated with the corrosion of Zn in the presence of Cu if the voltage between the two metals is measured to be 1.1 V?

SOLUTION

The periodic table (e.g., Figure 2-2 in **Shackelford, 4th Edition**) indicates that Zn is in group II B and will tend to produce a divalent ion in solution:

$$Zn = Zn^{2+} + 2\ e.$$

Using Equation 8-5,

$$\Delta G = -n\ F\ V$$

$$= -(2\ mol)(96.5 \times 10^{3}\ C/mol)(1.1\ V)$$

$$= -212.3 \times 10^{3}\ C \cdot V$$

$$= -212.3\ kJ.$$

SUMMARY

Thermodynamics plays a central role in materials science and engineering. The **first law** of thermodynamics is a statement of the **conservation of energy**. The **internal energy**, U, of a **system** is a **state variable.** For a given process, ΔU is equal to the difference between the **heat** absorbed and the **work** done (**by** the system). Relative to the first law, it is convenient to define **heat capacity**, C (= $\delta Q/dT$), and **enthalpy**, H (= U + pV). The **second law** states the limitations on conversion of heat to work. A state variable called **entropy**, S, is introduced for the purpose of a formal statement of the second law. Specifically, we find that "any spontaneous process is accompanied by an increase in total entropy." An equivalent "statistical" statement of the second law is that "a spontaneous process is one in which the system goes to a more probable state." A concept of importance to the description of solid solutions is the **entropy of mixing**, ΔS_m. Defining the entropy of a system to be zero at absolute zero of temperature allows absolute values of entropy to be calculated at other temperatures and is taken as the **third law** of thermodynamics.

Free energies are defined as "thermodynamic potentials," i.e., state variables that are minimized at equilibrium. For a process at constant temperature and pressure, the appropriate variable is the **Gibbs free energy**, G, defined as the enthalpy minus the product of temperature and entropy (H-TS). The Gibbs free energy is the most generally useful thermodynamic potential for materials systems. For example, at atmospheric pressure, it is a convenient indicator of the stable phases present at given temperatures and compositions. As such, the Gibbs free energy provides a theoretical basis for the **phase rule** and practical **phase diagrams**. The Gibbs free energy of surfaces (a type of planar defect) is closely related to **surface tension**, a mechanical quantity that can be experimentally determined in a variety of ways. The Gibbs free energy is also intimately related to kinetic rate equations. Careful inspection of this relationship indicates that an **activation energy** is an enthalpy term and that the **pre-exponential constant** contains an entropy term. As such, the Gibbs free energy provides a theoretical basis for the practical discussion of **heat treatment**. Finally, the nature of the **oxidation** and **corrosion** of metals by their environment is better appreciated by considering the associated thermodynamics.

KEY WORDS

activation energy
adiabatic
binary system
Carnot cycle
classical thermodynamics
components
conservation of energy
contact angle
corrosion
cycle efficiency
enthalpy
enthalpy of mixing
entropy
entropy of mixing
environmental reaction
first law
free energy
Gibbs free energy
Gibbs phase rule
groove angle
heat
heat capacity
heat treatment
Helmholtz free energy
ideal solution
internal energy
kinetics
nonwetting
oxidation
partial wetting
phase
phase diagram
phase transformation
preexponential constant
reversible
second law
sessile drop experiment
state function
state variable
statistical thermodynamics
Stirling's approximation
surface
surface energy
surface tension
system
thermal grooving experiment
thermodynamic potential
thermodynamics
third law
wetting
work

REFERENCES

Brophy, J.H., R.M. Rose, and J. Wulff, **The Structure and Properties of Materials**, Vol.2: **Thermodynamics of Structure**, John Wiley and Sons, Inc., New York, 1964.

DeHoff, R.T., **Thermodynamics in Materials Science**, McGraw-Hill, New York, 1993.

Holman, J.P., **Thermodynamics, 4th Edition**, McGraw-Hill, New York, 1988.

Swalin, R.A., **Thermodynamics of Solids, 2nd Edition,** John Wiley and Sons, Inc., New York, 1972.

PART III

LABORATORY MANUAL

Co-Authored by Jerrold Franklin and Michael Meier

1 INTRODUCTION

There are two primary benefits of a laboratory section with an introductory course in materials science. The first is to present several major topics in the laboratory environment, and the second is to familiarize the student with the organization and presentation of a formal engineering report. Together, they provide a realistic portrayal of the responsibilities of a practicing engineer.

This manual contains a section on report writing, four experiments, and appendixes containing a sample report and supplementary discussions of various experimental techniques. In general, this manual and the course text are intended to provide sufficient reference material. The laboratory instructor may wish to provide additional, outside references.

This manual is intended to provide detailed presentations of a few, key experiments. The instructor is encouraged to develop similar presentations for additional experimental topics appropriate for his or her laboratory program.

The four experiments cover the topics of mechanical properties, phase diagrams, heat treatment, and galvanic corrosion. Each experiment is presented in a format similar to that to be used in the final report and is written in sufficient detail that a person with no prior experience should be able to complete the assignment. However, each experiment is general in the sense that the materials to be tested and the purpose of the experiment have not been specified. There are three reasons for this generality: (1) to allow for changes due to the availability of materials, (2) to allow the instructor to vary the major emphasis to best coordinate with the lecture portion of the course, and (3) to discourage the use of reports written during past quarters. The instructor, then, has maximum flexibility in customizing the experiments. There are several objectives inherent in each experiment. In the introduction, the two objectives given cover the educational value of the experiment. The first objective lists the academic topics covered and the second deals with experimental procedures. The writing assignment for each experiment can be quite flexible. Also, the scope of an experiment can be narrowed greatly by stating which aspects of the experiment are to be emphasized. This can be done by posing an engineering problem and, then, using the results to solve the problem. Alternately, the results can be analyzed in terms of one particular variable. The writing assignment is a primary reason for conducting the experiment.

It is important that the student prepare for the experiment by reading the appropriate sections in the laboratory manual and text before coming to the laboratory. At the end of each experiment are several questions. These are representative of the type of problems that will be faced when conducting the experiment and serve to aid in preparation for the experiment. Some of the questions are provided as examples of applications of the concepts covered by the experiment. These questions are not generally part of the graded assignment.

2 WRITING ENGINEERING REPORTS

Writing is one of the most important tools available to the engineer. The final part of any project is not just a formality. It is a primary product of one's labors and is often the basis of the evaluation of the reporter's professional abilities. The report is also a service to those in need of the information. In any case, written communication has several unique characteristics that should be considered. Once submitted, it must stand on its own. It tends to be circulated, reviewed, and filed. During that process, it is subject to critical analysis by a variety of readers. Writing the report can be a major learning process. Most of all, the author is responsible for providing meaningful results and analysis in an effective presentation (within specified guidelines). Some find report writing difficult and challenging. However, report writing is one of the primary professional responsibilities of the practicing engineer. This section should be helpful in showing the student how to organize and write an engineering report.

Due to the nature and function of the experiments, the reports are formal, factual accounts of events and observations. In addition, they contain a critical and methodical analysis of those observations. The entire report is traditionally written in the passive voice and must fit a specific, segmented format. These constraints do not inhibit creative thinking. In fact, the report format is helpful in organizing one's ideas and, therefore, the presentation. The format also helps to ensure that the report is complete and convincing. The format presented in this manual contains the essential elements of papers published in typical professional journals. Students who are not familiar with these are advised to look at papers published in journals such as **Materials Transactions** and **Journal of the American Ceramic Society**. (This is not to say that publication-quality reports are expected in this introductory course.) Appendix A contains an example of the type of laboratory report that would be written for this course.

We shall begin with some general considerations to be made before starting to write the report. After that, the specific format of the laboratory report is given, along with the purpose and content of each section.

Finally, there are discussions of general topics such as appearance, illustrations, tables, numerical data, and deadlines.

General Considerations

The first items to be considered when writing the report are: who will read it, what is his or her level of expertise in the subject, and what is he or she expecting to learn? For the purposes of this course, assume that the report will be read by several people at the level of your immediate supervisor (faculty or teaching assistant) and that the report will be maintained in a file. It will not be necessary to explain basic concepts to the reader. (That would be distracting for the reader and a waste of your efforts.) When writing on a subject that is new to the author, there is often a tendency to use the report as a study tool and to write about much of what was learned in class or from the reading. Such writing is more appropriate for a review or a term paper than a laboratory report.

Before starting to write the report, make sure that the goal and scope are well defined. It is common to find a report that inadequately addresses the main point(s) of the report. Often an experiment deals with several topics, some of which are not central to the primary goal. Some self discipline is necessary to avoid digressing or otherwise straying from the main point.

The report should be as short as possible while still including all relevant subject matter. How well one provides this depends on the scope of the project and one's ability to extract only those items that need to be reported. Often the person reading the report is busy and cannot afford to read an unnecessarily long document. The value of a report depends not on its length but on its accuracy and utility for its readers.

Format

The organization provided by the following report format is straightforward and suitable for the types of experiments that will be conducted in this course. In the discussion that follows, each section of the report is discussed in detail. It will become clear that each section of the report has a specific function and minimum requirements that must be met by the writer. If each section is written so that it performs its intended function, the final result will be a clearly and efficiently written laboratory report. Appendix A is provided as an example of a report written according to the format used in this manual. You may wish to refer to it as you read the following discussion.

Title Page This is a separate cover sheet that should clearly identify the report by title, class, section number, date, and instructor. Be sure to include the names of those people (laboratory partners) who assisted in the experiment.

Introduction The function of the Introduction is to describe the purpose and scope of the experiment. The remainder of the report is tied to this statement. Generally, the reader is first introduced to the overall topic and then to the various, specific areas addressed by the report. Finally, the Introduction should state explicitly the scope and primary goal of the experiment. If the experiment centers around a specific, unique procedure, that can be the primary focus of the Introduction. There is generally no need for an extensive Introduction section to reports in this course. A length of 50 words can be adequate.

Procedure The function of the Procedure section is to tell the reader how one went about answering the "question" posed in the Introduction. The Procedure section should be written in enough detail so that the reader could duplicate the experiment. However, mundane details need not be included. The reader is assumed to be generally familiar with the type of work being reported.

The Procedure section should include a list of the materials tested and the important equipment used. Often, the materials can be uniquely identified by a batch or code number, and, sometimes, the specific processing history is known. These details should be included in the list of materials. The equipment used should be described briefly in terms of its name, manufacturer, and main functional characteristics. Any especially useful or unique features should be described. Testing procedures should be described in sufficient detail to permit the reader to duplicate them. If a procedure is a standard testing technique, that method can be named in lieu of a detailed description. Any significant departure from a standard procedure should be noted and an explanation for the departure given. In addition, the relevance of each procedure to the purpose of the experiment should be made clear. Describe the type of data that each procedure is supposed to provide and what part of the "question" it should help to answer.

Results In this section, the important results are presented in a clear and organized manner. This includes all figures, tables, and graphs needed to support the discussion and conclusions that follow. Raw data and sample calculations are normally placed in an appendix. The text of this section should include a brief description for each figure and table, indicating from where the data came, what they show, and how they compare to expected results. A comment about accuracy and reproducibility is also appropriate in this section.

Discussion The purpose of the Discussion section is to communicate what was learned and how the results lead to the conclusions (of the next section). The results are discussed in terms of what they show about the material's behavior and what they mean in the context of the purpose of the experiment. In this section, the logic needed to support the conclusions should be developed. The discussion forms the bridge between the original observations and what they will ultimately mean to the reader. While the previous sections showed that one knows how to conduct an experiment, the discussion shows one's analytical ability and knowledge of materials science.

Conclusions The Conclusions section is a concise statement of the key observations and their significance. It is the answer to the "question" posed in the introduction. This section can be considered both the goal and the final product of the report. A conclusion can be as specific as the one in the sample laboratory report of Appendix A or, instead, a general list of important observations. The specific nature of this section depends on the original objectives of the experiment.

Summary The Summary is most useful for the busy reader and, in this course, a good exercise for the student. An effective summary can help a manager determine if the report warrants his or her more detailed attention. A good summary is as brief as possible while maintaining coherency, clarity, and completeness.

One good approach for writing the summary is to read back through the report and list the key items from each section. This list is, in effect, a rough outline for the summary. One or two sentences for each section should be sufficient. Do not try to rewrite the report in the summary. The body of the report is available for more detailed reading.

References The reference list should include only those books and articles from the technical literature that were used to substantiate a significant point made in the report. There is no single style for writing literature references. However, once a given style is chosen, it should be used consistently. If no style is specified by the instructor, that used in **Shackelford, 4th Edition**, can be used as a guide.

Appendix This additional section is available for those items relevant to the experiment but not sufficiently important to be included in the body of the report. Examples include original data sheets, chart recorder output, computer printouts, computer programs, sample calculations, lists of symbols used in the report, and manufacturer's data sheets. Some discretion is required. The appendix should not be a "catch all." For example, thermocouple output-to-temperature conversion tables are commonly available and would generally not be needed for the report.

Additional Comments

There are other considerations for producing a good engineering report. The following comments, both general and specific, should prove helpful.

General Appearance Report writing can be a major part of an introductory materials course, and the appearance of the report reflects the care that went into the experimental work. As such, appearance can be given a priority equal to the technical content. Spelling and grammer are important components of the editing process. The experiments often introduce concepts, instruments, and terminology new to the student. Both the instructor and the text (or other references) can be consulted about proper phrasing.

Neatness counts. Handwriting, if permitted, should be neat and legible. Margins, headings, figures, tables, etc., should give the report a neat, organized appearance. Typing is recommended and may be required by some instructors. Typing can have several advantages for the writer, especially the encouragement of more thorough editing. There are additional advantages that come with the use of word processors and associated software.

Figures, Graphs, and Tables These important tools for presenting the results should be able to stand as separate entities. Each should be clearly labeled and presented in a neat and organized manner. Significant points should be labeled without clutter. Avoid extraneous information or too much data in a given figure, graph, or table. Use discretion in the use of color. **Shackelford, 4th Edition,** can be consulted for general examples of style.

Further comments are appropriate for presentation of graphical information. Graphs can serve two purposes: (i) as tools for data analysis or (ii) as illustrations. For data analysis, the data are plotted for maximum resolution. For a final illustration, the graph must be an effective medium of communication. In this case, it might be better to combine several sets of data into one graph. The scaling of the axes should be chosen in such a way that the data are not crowded and that, conversely, minor deviations from the general data trend are not exaggerated. The independent variable is plotted on the horizontal axis, which should be close and parallel to either the bottom or right side margin. Use a plotting symbol such as a circle or triangle to mark the locations of the data points. The data points can be simply connected by straight lines, or a "best-fit" line or curve can be used. The second method is preferred. The fit can be made by visual inspection or numerical methods (e.g., linear regression).

Numbers Use the units specified by the instructor. If none are specified, use the most appropriate (with the S.I. system being encouraged). Also, note that calculator read-out is not, in general, the same as experimental precision.

Deadlines Be aware of the deadline for completion of the report. More importantly, meet it! This is one of the key responsibilities of the practicing engineer.

Additional Reading

1. J.W. Souther and M.L. White, **Technical Report Writing, Second Edition**, John Wiley & Sons, NY, 1977.

2. K.W. Houp and T.E. Pearsall, **Reporting Technical Information, Third Edition**, Glencoe Press, Beverly Hills, CA, 1977.

3. W. Strunk, Jr., and E.B. White, **The Elements of Style, Third Edition**, Macmillan, NY, 1979.

4. E.R. Tufte, **The Visual Display of Quantitative Information**, Graphics Press, Cheshire, CT, 1983.

5. E.C. Subbarao, D. Chakravorty, M.F. Merriam, et al., **Experiments in Materials Science**, McGraw-Hill, NY, 1972.

6. H.E. Davis, G.E. Troxell, and G.F.W. Hauck, **The Testing of Engineering Materials, Fourth Edition**, McGraw-Hill, NY, 1982.

3 EXPERIMENTS

As pointed out in the introduction, the specific experiments that follow represent a few, key laboratory experiences for students in this introductory course. The instructor is encouraged to develop additional experiments covering topics of relevance to his or her class. The authors hope that the general format for the experiments in this manual will be useful in presenting those additional ones.

A. Tensile Test and Hardness Experiment

Introduction The tensile test and the hardness test are widely used methods of determining the mechanical behavior of materials. The tensile test is used to measure properties related to the stress-versus-strain behavior of materials. The hardness test is a method for assessing such behavior in a simple but indirect way. The combined results of these two tests indicate the suitability of a given material for applications involving static loading, as well as providing fundamental mechanical property data.

In the course of this experiment, the student will learn how to measure and interpret the stress-versus-strain behavior of some common engineering materials. This will include the recognition of the important features of stress-versus-strain curves, various types of such curves, and the interpretation of those curves in terms of prior thermo-mechanical processing. The tensile test is conducted at room temperature at a constant rate of elongation. The hardness experiment is conducted as a standard Rockwell test.

Organization In this experiment, the class is divided into groups of three to five students which will independently test their specimens. Each group is encouraged to interact with the others in order to benefit from laboratory experiences but not to share data. Each student is responsible for independently analyzing the data and preparing a laboratory report.

Preparation Carefully read Section 7.3 in **Shackelford, 4th Edition.** Before coming to the laboratory, the student should know the definitions of stress and strain and other mechanical properties listed in the Results section below. One should also have a basic understanding of the fundamental atomic- and microstructural-scale mechanisms responsible for mechanical deformation. In addition, carefully read Appendixes B and C. The student should be familiar with basic techniques, as well as their limitations.

Materials The instructor will provide various specimens ready for tensile testing. These specimens may vary in composition and thermo-mechanical history. Note such differences that serve to identify specimens and that may affect experimental results.

Equipment The equipment required will vary with the facility but will tend to be of the following types. The tensile test machine involves elongation as the independent (control) variable. The resulting load is measured as the dependent variable. The hardness testing machine will be a Rockwell tester. Note the specific machine used in this experiment by manufacturer, model number, basic operating characteristics, and experimental set-up for the specific tests.

Procedure Each specimen is to be tensile tested to failure under essentially the same conditions, i.e., room temperature and fixed strain rate. Before testing, record each specimen's initial dimensions and other appropriate test details. Conduct the test on each specimen, recording load versus elongation and overall duration. After each test, record the final specimen dimensions and overall strain rate. During at least one of the tests, it will be necessary to record the transverse strain during elongation. These data will be used to plot true stress versus true strain.

Perform hardness tests on the shoulder sections of all specimens. Do not perform hardness tests on the gage section of the specimens, as such defects can affect material performance. Make at least five measurements on each specimen and average the results. Be careful to use the proper specimen stage and technique to ensure valid data.

Results Using the raw data, construct an engineering-stress-versus-engineering-strain curve for each specimen tested. Also, construct a true-stress-versus-true-strain curve and a plot of axial strain versus reduction of area for the test that included measurements of reduction of area. There are various ways to calculate true stress and true strain, depending on assumptions regarding specimen cross-section. (Your instructor will provide guidance in this area). Specify the method(s) used. If more than one method is used, compare the results.

Analyze the stress-versus-strain curves to determine the following mechanical properties:

1. elastic (Young's) modulus
2. yield strength
3. ultimate tensile strength
4. ductility (calculated as reduction in area and as axial strain)

5. toughness (area under total stress-strain curve)
6. work required to pull the specimen to fracture.

This information should be consolidated in a table that shows the mechanical properties for all materials tested. This table should also include the hardness data. Organizing the essential data in a table in this way allows the reader to conveniently compare the mechanical behavior of the various materials. A separate table should give information on test setup and execution, specimen dimensions, strain rate, load scales, etc. Such information is important to anyone who would use the report to repeat the experiment.

Discussion The emphasis of the discussion will vary with the specific nature of the experimental assignment. However, there are substantial amounts of data to discuss. Restrict yourself to the stated objectives and scope of the experiment.

Problems The following problems are provided to assist in preparation for the tensile test and hardness experiment and to challenge your understanding of the subject.

1. A tensile test of SAE 1020 steel (plain carbon steel containing 0.2 wt.% carbon) produced the data below.

Extension (mm)*	Load (kN)
0.00	0.0
0.09	1.9
0.31	6.1
0.47	9.4
2.13	11.0
5.05	11.7
10.50	12.0
16.50	11.9
23.70	10.7
27.70	9.3
34.50	8.1

*As measured from the chart paper.

Initial dimensions: gage length = 25.4 mm, width = 6.09 mm, thickness = 3.28 mm, scale factor for the elongation axis (magnification) = 8.0.

Use these data to: (a) plot engineering stress vs. engineering strain, (b) plot true stress vs. true strain, and (c) analyze the data for the properties listed in the results section for this experiment.

Are these results reasonable? How do you know? Why wasn't the yield point extension observed that is characteristic of this type of steel?

2. What are typical values for the Young's modulus of steels, copper alloys, aluminum alloys and tungsten alloys? Which Rockwell hardness scale would you expect to use for each of the above materials?

3. What are the typical ranges of values for the yield strengths, ultimate tensile strengths, and Rockwell hardnesses for a 0.2% carbon (1020) steel, 0.4% carbon (1040) steel, pure aluminum, 60/40 brass and Ti-6Al-4V?

4. A skeptical student puts the two halves of one of the tensile-tested specimens back together, measures the final gage length, and notices that it does not match the elongation at the fracture point on the load-elongation plot. Is the specimen that was measured too long or too short? Explain.

5. When the specimen in Problem 1 fractured, a loud sound was heard and the two halves of the specimen accelerated in opposite directions. How much energy was released during this commotion?

6. A simple beam (0.25 inch thick by 1.00 inch wide) made of the material in Problem 1 is to be bent around a form having a one-inch diameter so that in its final shape it posseses an 11-degree bend. (a) What bend angle during the forming operation will be required to achieve this? (b) How much plastic strain is in the outermost "fiber" of this beam? (c) Will the beam yield or crack? (d) How much force is required to form this part?

7. A steel ring gear having an inside diameter of 319.6 mm is to be heated so that it can fit over a flywheel having an outer diameter of 320 mm. The linear coefficient of thermal expansion for this material is 12.18 x

10^{-6} mm/mm/K. (This property is introduced in Section 8.4 of **Shackelford, 4th Edition.**) (a) To what temperature must the ring gear be heated? (b) After final assembly, what stress is the ring gear under? Comment.

8. A tensile specimen is hardness tested and found to have a hardness of 5 Rc. Are you comfortable with this reading? If not, why not? Suggest another way to hardness test this material.

B. Phase Diagram Experiment

Introduction Several methods are available for establishing the phase diagram for a set of elements. Among the more common are dilatometry and cooling curves. Both methods involve heating and/or cooling the sample. In dilatometry, thermal contraction or expansion of the sample is monitored and phase changes are detected by the abrupt change in sample dimensions due to the different densities of the phases. The cooling curve method involves monitoring the temperature of the sample as it is cooled. Changes in the cooling rates are related to the latent heat of transformation, thereby indicating that a phase transformation has taken place. Other methods are available based on various structures and properties with detectable changes during phase transformations, e.g., electrical resistivity, optical and electron microscopy, and electron and x-ray diffraction. Only the diffraction method provides direct determination of the crystal structure(s) of the phase(s) along with transformation temperatures. The other methods would require additional diffraction analysis for complete phase identification.

During this experiment it will be possible to observe the phase transformation process, including the undercooling associated with the transformation kinetics. Therefore, an understanding of the thermodynamics of solubility and phase stability and the kinetics of phase transformations is needed. It also requires an understanding of the construction and use of phase diagrams, e.g., the lever rule, Gibb's phase rule, and the eutectic reaction. The methods used in this experiment are fairly simple, but analyzing the data can be a challenge. Also, as each test in the experiment is slightly different, some planning on the part of the experimenter is necessary if laboratory time and materials are to be effectively utilized. This experiment also serves to teach the basics of the proper use of thermocouples and some consideration of experimental error (as the assigned phase diagram is well established).

In this experiment, the cooling curve method is used to determine the transformation temperatures of several samples of known composition. These results are used to construct a phase diagram. The phase diagram is then used to identify the compositon of samples with unknown composition.

Organization The students, as a group, are responsible for all parts of the experiment: sample preparation, testing, and determination of transformation temperatures. Although the work might be divided up among different work stations, students are encouraged to work together in completing the overall experiment. Share your experiences and data. Each student should feel obligated to criticize any other's method or results. Do not leave the laboratory until everyone is satisfied with the overall results. Once the laboratory session is complete, preparation of the report becomes an individual's responsibility.

Preparation Read Chapters 5 and 6 in **Shackelford, 4th Edition**, and Part II of this **Manual**. Know the general thermodynamic and kinetic aspects involved in phase stability, solubility limits, and phase transformations. Also, know how to use the phase diagram, e.g., Gibb's phase rule and the lever rule. Finally, study Appendix D: Temperature Measurement Using Thermocouples. This describes the use and characteristics of the thermocouple, the primary device for this experiment.

Materials The materials used in this experiment are typically reagent-grade, granulated lead and tin (or some other low-melting-point element such as antimony). These will be mixed in various proportions to provide twelve different compositions needed to construct a phase diagram. In addition, several samples of the alloy whose compositions are unknown will be provided. **Toxicity** is an important issue in dealing with lead and certain other low-melting point materials which might be used for this experiment. **It is important that you refer to Materials Safety Data Sheets provided by the manufacturer.**

Equipment The set-up for this laboratory experiment is quite simple. A scale (or balance) and test tubes are needed to prepare the samples. A laboratory stand and a Bunsen burner are needed in order to hold and melt the sample. A type-K thermocouple and a strip-chart recorder are used to measure the temperature and record the cooling curves.

Procedure Set up the equipment as directed by the laboratory instructor. Inspect the condition of the thermocouple probe, the wiring, and the ink and paper supply in the chart recorder. Verify that everything is working properly and, then, adjust the recorder so that it is ready for the first test. Position the pen to a suitable position on the paper and note the temperature. Select the chart speed and voltage range. (Suggested settings are 2 cm/min and 20 mV). The exact settings depend on the maximum temperature that you wish to record and the cooling rate of the sample. (The 20 mV setting allows temperatures as high as 485°C to be recorded if using a K-type thermocouple). When setting chart speed and range, consult a published phase diagram to get an idea of the transition temperatures to be expected.

Weigh and mix the elemental materials on a specified total mass basis. Heat the samples over the burner until molten. Be careful not to heat the samples excessively. This could spoil the sample due to oxidation, cause the test tube to soften or break, or waste time as the sample cools through an unnecessarily large range of temperature. (It is a good idea to monitor the sample's temperature while it is being heated to avoid overheating.) After the sample is completely molten, place it in a suitable holder (test tube rack or insulated beaker) to allow it to cool. Insert the thermocouple and start the recorder and chart the voltage output of the thermocouple as a function of time. After the sample has cooled to the point where no further phase transformations are expected, the sample should be remelted to remove the thermocouple. Repeat the procedure to obtain cooling curves for all the samples including those of unknown composition.

Results Analyze the cooling curves of all samples to determine the transformation temperatures for each composition and construct a table to clearly present these data. Next, construct a phase diagram from the data. Clearly and neatly label all important features, e.g., solidus, liquidus, eutectic point, and phase regions. Are the results reasonable? Are they within acceptable limits of error? Use the phase diagram to determine the compositions of the unknowns. It might be necessary to refer to the cooling curves to resolve any ambiguities regarding the compositions of the unknowns.

Discussion This experiment is straightforward and is, in effect, duplicating the work of others. As a result, the discussion deals mainly with comparing the results with previously published phase diagrams, analysis of errors, and determination of the composition of the unknowns. The thrust of the argument presented in this section depends on the specific statement of purpose.

Problems The following questions are provided to assist in preparation of the phase diagram experiment and to provide a few examples of engineering applications for phase diagrams.

1. Almost any junction between dissimilar metals can function as a thermocouple. How does this fact necessitate the use of appropriate extension wires and reference junction when the thermocouple is to be connected to a voltmeter?

2. When would you expect the thermocouple effect to degrade the performance of sensitive analog electronic circuits?

3. Suppose you are planning to do this Phase Diagram Experiment for the entire Pb-Sn system between room temperature and 400°C and already have at your disposal eleven samples that cover the whole composition range in 10-weight-percent increments. Before you begin testing any of your samples, you want to predict all the temperatures where a phase transformation will occur. Make a table of these temperatures and the corresponding voltage outputs of your K-type thermocouple.

4. The strip chart recorder that you are planning to use in the experiment mentioned in Problem 3 plots voltage vs. time on 25-cm-wide paper. This width corresponds to a full scale voltage reading. Select the voltage range (1, 2, 5, 10, 20, 50, or 100 mV) that you could use thoughout the experiment. During the first test, an interesting event was recorded at the 7.85 cm mark. Assuming the thermocouple circuit has a proper reference junction, at what temperature did the event occur?

5. Suggest a reason that all of the data from the Phase Diagram Experiment are consistently about 5°C lower than was expected.

6. Draw a cooling curve for the samples having 100/0, 70/30, 38/62, and 30/70 compositions (Pb/Sn).

7. Plot weight percent alpha and the compositon of alpha vs. temperature as a 70/30 Pb/Sn alloy cools through the alpha-liquid phase field.

8. Comment on the toxicity of the materials to be used in the Phase Diagram Experiment.

C. Annealing and Recrystallization Experiment

Introduction Deformation processing of a material is commonly used to obtain desired dimensions and/or properties. Most of the deformation in the material is associated with the process of creation and motion of dislocations. Strain hardening occurs as a result of the obstruction of the movement of dislocations by crystal imperfections such as precipitates, dispersoids, grain and twin boundaries, solute atoms, and other dislocations. Increased dislocation density due to deformation may render the material too brittle to survive further deformation processing without cracking. Similarly, the material may be too brittle or too hard for its intended application. A remedy is to reduce the dislocation density by annealing the material. The time and temperature of the anneal depends on the material, the degree of prior cold work, and desired properties for the final product. A recovery anneal will simply cause a reduction in the dislocation density and a minor effect on properties. An anneal at a higher temperature or for a longer time can lead to a substantial microstructural change and a major effect on properties. This is a classic example of the fundamental relationship between structure and properties of engineering materials.

During this experiment, the student will have a chance to see the relationship between a material's microstructure, properties, and processing. Specifically, the evolution of the microstructure of a cold-worked material and the attendant change in mechanical properties will be observed. The student will learn to recognize recovery, recrystallization, and grain growth anneals from these observations. The experiment also serves to demonstrate the significance of processing to the properties of engineering materials, along with our ability to control these properties for the purpose of a given application.

This experiment requires tensile testing and hardness testing of a series of specimens of a common engineering material that have been first cold worked and then annealed at various temperatures up to two-thirds of the melting point. The microstructure of these specimens (or identical ones) will then be studied in an attempt to correlate the properties with their microstructure and processing history. The student is also asked to identify the regions of recovery, recrystallization, and grain growth and to determine the recrystallization temperature.

Organization During this experiment the class works together to carry out tensile tests and hardness tests on all of the specimens. The students are encouraged to share their experience and expertise to ensure that all tests are performed correctly. Extra specimens might not be available for this experiment. Preparation of specimens for tensile testing and optical metallography (if required) is a group responsibility. Examination of the microstructures may also be a group effort, but the sketches, like the graphs and the report in general, are to be individual efforts.

Preparation Read Chapter 6 in **Shackelford, 4th Edition**, paying particular attention to Section 6.5. Study Figure 6-28 and be able to recognize a cold-worked and fully-annealed microstructure as well as the more specific microstructural features, e.g., grains, grain boundaries, twins, and slip lines. Be able to identify recovery, recrystallization, and grain growth anneals from both mechanical properties and microstructures. Finally, review Appendixes B and C, as much of this experiment involves tensile and hardness testing.

Materials Tensile specimens made of some common hard-drawn material will be provided. Note the material's composition and processing history. Each specimen must be annealed for one hour at a different temperature not to exceed two-thirds of its melting point. (The laboratory instructor may choose to perform the heat treatments prior to class.) Record the specific annealing times and temperatures used to heat treat the brass as well as any other relevant information concerning the samples. Specimens for optical metallography, if not provided, must be prepared during the laboratory session.

Equipment Tensile and hardness testing machines are needed for mechanical property evaluations. Metallographic sample preparation equipment, supplies, and reflecting light microscopes are needed for the metallography section of this experiment. Depending on whether or not the samples have been previously annealed, an annealing furnace may be needed. Record what equipment is used by manufacturer, model, and operating characteristics.

Procedure If not done previously, anneal the tensile specimens for one hour at the temperatures indicated by the laboratory instructor. Perform hardness tests on only the shoulder sections of the annealed samples, as well as on one hard-drawn, unannealed specimen. Tensile test to failure each specimen. Remember to make the necessary measurements before and after the tensile tests, as indicated in the Tensile Test and Hardness Experiment.

Prepare the metallographic specimens for viewing. This may involve grinding, polishing, and etching the samples. The resulting microstructures should be examined, and any microstructural changes due to increased annealing temperature should be noted. Sketch all six microstructures. Arranged in order of increasing annealing temperatures, they clearly illustrate the microstructural evolution that a cold-worked material undergoes during the annealing process.

Results Review the results and make sure they appear to be reasonable. Organize these results in a table. Plot each of the mechanical properties versus annealing temperature and note the trends. Which properties are most affected by annealing? What is the recrystallization temperature? Organize and present the sketches of the microstructures. Label them clearly and note the approximate magnification and most significant microstructural features.

Discussion Describe the relationship between properties, processing, and microstructure, as shown by the experimental results. Discuss the effect of annealing on properties and microstructure in general. Avoid simply describing the individual graphs and sketches. In this discussion, the various results should be brought together in such a way that they clearly support the conclusion of the report.

Problems The following problems are provided to assist in preparing for the annealing and recrystallization experiment.

1. What is the role of diffusion in the annealing processes covered in this experiment? How is the activation energy for diffusion related to the anneals?

2. What effect does the amount of cold working have on the recrystallization temperature of a given material? Is it possible to recrystallize a material that has not been cold worked?

3. Describe what one would see upon optical examination of cold worked, recovered, recrystallized, and fully annealed polycrystalline materials.

4. Higher annealing temperatures allow reduced annealing times. What are the lower and upper temperatures to be recommended for annealing (a) 60/40 brass, (b) pure Ti, (c) 50/50 Cu/Ni, and (d) SAE 1070 steel?

D. Galvanic Corrosion Experiment

Introduction Galvanic corrosion can occur when electrical contact is made between dissimilar metals in an electrolyte. This common form of corrosion is widely observed in engineering applications involving dissimilar metals in an aqueous environment. (Corrosion can also occur when only one metal is present. In that case, variations in mechanical stress or electrolyte concentration can lead to environmental degradation.)

Galvanic corrosion is an electrochemical process. The electrical quantities that can be measured during galvanic corrosion are related to the force that drives the reaction and the barriers to the reaction. Even when a thermodynamic driving force exists, the rate of reaction (kinetics) may be negligibly slow. For example, the rate of corrosion may drop off significantly as a surface reaction layer builds up on the metal substrate.

Understanding the corrosion process provides various techniques for preventing its occurence, e.g., selecting a more corrosion-resistant material or a combination of materials with similar potentials, supplying a more susceptible material (sacrificial anode), providing a protective coating, or providing an external (impressed) voltage to cancel the corrosion potential.

This experiment deals with the galvanic corrosion potential, the relative corrosion tendencies of various metals in an aqueous environment, the corrosion rate, and the effect of polarization on corrosion rate. The experiment is based on voltage and current measurements made on a simple galvanic cell. Four different types of experiments are done, each dealing with a specific aspect of aqueous corrosion. The first two, Galvanic Potential and Galvanic Series, deal with the thermodynamic aspects of corrosion. The last two, Polarization and Current Density, deal with kinetics.

Organization The class is divided into independent groups, each of which will perform the entire experiment. The groups are encouraged to share their experiences and observations with each other. However, once the laboratory session is over, each student is expected to prepare his or her report independently. Also, each student is expected to help clean and restock the laboratory at the conclusion of the experiment.

Preparation The student should read Chapter 14 of **Shackelford, 4th Edition**, and Part II of this **Manual**. The student should know how to construct a galvanic cell, what is the source of the galvanic potential, and the kinetic aspects of the corrosion process. Some students might also find Appendix E, Elementary Electrical Measurements, helpful.

Materials The materials studied in this experiment are strips of common metals. A typical selection would include copper, brass, aluminum, zinc, lead, tin, and magnesium.

The electrolyte used in the galvanic cells is a 3% NaCl solution (30 grams NaCl in one liter of water). The water used may be ordinary tap water which will contribute its own salts and minerals. A solution using sea salt or any other suitable electrolyte can be used instead of the NaCl solution. However, strong acids or alkali solutions should be avoided. **In all cases, chemical laboratory safety procedures should be rigorously followed.**

Equipment The following equipment will be used:

1. Strip chart recorder, 0 to 5 volts
2. Electronic voltmeter, 0 to 5 volts
3. Wires and clips
4. Beakers, 100-500 ml
5. Steel wool for cleaning specimens
6. Emery cloth/paper for cleaning specimens
7. HCl solution for depolarization (0.2 molar, 20 ml HCl/1200 ml distilled water)
8. Distilled water for rinsing
9. Resistor, 100 ohm.

The strip chart recorder will be used only in Parts 3 and 4.

1. Galvanic Potential

This section deals with the factors that influence the potential measured in an open circuit galvanic cell, i.e., there is assumed to be no electrical load on the cell and the cell has been allowed to reach equilibrium. The primary purpose is to identify the anode and cathode and other construction details that might affect galvanic potential. A secondary purpose is an introduction to the equipment and procedures.

Materials Specimens are strips of Cu, Al, and Zn.

Procedure Clean and depolarize the specimens by buffing with steel wool or emery cloth/paper followed by dipping briefly into the HCl solution and, finally, a rinse in distilled water. All specimens should be cleaned in this manner before each test is performed.

Set up the galvanic cell using Al and Cu as the electrodes. Record the equilibrium potential between the electrodes. Identify the anode and the cathode. Next, measure the potential again with the electrodes at a different position (closer or farther apart). Also vary the depth of immersion in the electrolyte. For this part, qualitative results are sufficient. Record the measurements and repeat the procedure using Al and Zn as the electrodes.

Results The report should include a table that clearly and systematically shows the results. (Again, these may be, in part, qualitative.)

Discussion Temperature and composition of the electrolyte are two primary factors that affect the galvanic potential. In this part, several other possible factors were investigated. Discuss what effect they have on the galvanic potential. (In Parts 3 and 4, some of these factors will be examined in terms of their effect on kinetics.)

2. Galvanic Series

In this part, the electrode potentials of different metals in an electrolyte are measured. The data will be used to construct a galvanic series.

Materials Use all of the metal samples provided. In addition, any interesting materials the students have in their possession may be added to the evaluation.

Procedure Set up the galvanic cell and prepare the specimens in the same manner as in Part 1. Use copper as a reference material. Measure and record the voltage (upon stabilization) between the copper and each of the other metals. Next, select another metal as the reference and repeat the procedure so that enough data are generated to construct two complete galvanic series.

Results Construct the two galvanic series (for copper and the other reference material). Include the voltage in relation to the reference electrode and indicate which end of the series is most anodic and which is most cathodic.

Discussion The discussion should evaluate the galvanic series and the significance of the reference material used. The results could be interpreted and discussed in terms of the metal's relative tendency to corrode, relationship of electromotive force to basic thermodynamic properties, and the utility of the galvanic series.

3. Polarization

The purpose of this part is to measure the effect of polarization on the corrosion rate. In order to appreciate this, the student should be familiar with the general characteristics of a galvanic reaction and with the concept of rate-controlling processes.

Materials The strips of Zn and Cu will be the only specimens needed for this part.

Procedure Clean and depolarize the specimens by the procedure of Part 1. Set up the galvanic cell as in previous parts but, this time, connect the strip chart recorder in parallel with the voltmeter. In this part of the experiment, it is very important that the galvanic cell not be disturbed. Use the laboratory stand and clamps to hold the electrodes steady.

After setting up the galvanic cell, wait for the cell to reach its maximum stable voltage. Start the recorder and install a resistor in parallel across the terminals of the strip chart recorder. After a stable minimum voltage reading is obtained (in about 30 seconds to 2 minutes), the test can be concluded, but, before doing so, repeat the procedure using a higher chart speed (15 to 30 cm/sec). During this repeat, pay particular attention to the early parts of the reaction. Next, try moving the electrodes and/or shaking the beaker while watching the chart recorder. Then, remove the shunt resistor from the circuit and observe the voltage. Reinstall the resistor and observe. These last few steps help to reveal a few of the main characteristics of the polarization process.

Results The raw data are in the form of the trace on the chart paper. Note the maximum voltage, minimum current, and any other important features of the trace. Sketch a plot of current versus time to be included in the report.

Discussion Analyze the data and discuss what they show about the polarization process. Also, comment on what the shaking of the beaker shows about the nature of the polarization process. Why wasn't polarization observed in Parts 1 and 2? In this part, we were not studying the rate of specific chemical reactions involved in corrosion and polarization, so the discussion need not address such details. However, the general nature of the polarization process, the role of different reactions, and the probable steps of a galvanic reaction are appropriate topics for discussion.

4. Current Density

In this part, the effect of anode and cathode area on corrosion rate is evaluated. The different roles of the electrodes will become apparent. Finally, current density will be measured.

Materials The only specimens needed in this part are strips of Zn and Cu.

Procedure Prepare the specimens and set up the galvanic cell in the same way as discussed in Part 1. As in Part 3, the laboratory stand and clamps will be needed to keep the cell from being disturbed.

As the area of the specimen immersed in the electrolyte is proportional to the depth that the specimen is immersed, an easy way to control the areas of the specimens exposed to the corrosive environment is to use gradations on the sides of the beaker to indicate depth of immersion. Set the level of the electrolyte at one of the gradations. Four gradations lower will correspond to 100% area, three to 75%, and so on. In this part, it is not necessary to obtain absolute measures of the area except for the measurement of current density. Initially, relative areas are sufficient. To test the effect of anode area on corrosion rate, immerse the cathodic material to the 100% immersion mark while varying the depth of the anode in 25% increments. Include one measurement of the corrosion current (see next paragraph) for each case plus one in which the electrode is barely touching the electrolyte (area close to zero). This will provide five data points for each case. A similar procedure should be used to test the effect of cathode area on the corrosion rate.

There are several ways in which one can make the required measurements. A galvanic cell can be set up as in Part 3 and, after it has achieved its maximum stable voltage, a shunt resistor can be installed. The corrosion current must be recorded, but, as seen in Part 3, corrosion current varies with time due to polarization. Thus, both area and time must be considered. For each measurement, the effect of polarization must be equal. This is achieved by making the measurements a given length of time after the resistor is added to the circuit. This may be accomplished either by timing the reaction with a watch and taking a reading after the set time or by using the chart recorder to make a series of curves that can be analyzed later. In either case, the effect of polarization will be the same for each reading and will not affect the results.

Results With the data, a graph can be constructed that shows the effects of the areas of both of the electrodes on the corrosion rate.

Discussion Compare the two graphs and interpret the differences and similarities in terms of the basic physical processes involved. As in Part 3, there is no need to be concerned about the specific chemical reactions. Concentrate on the basic kinetic processes. Also, recall that in Part 1, the effect of area on properties of the galvanic cell was tested. How do those results relate to the results of this part?

APPENDIXES

APPENDIX A

SAMPLE LABORATORY REPORT

The sample report that follows is intended to give the student an indication of the approximate length and style of an engineering report appropriate for this course. The report includes representative figures and tables. It is hoped that this sample will make the task of report writing easier, especially for the student in his or her first laboratory course. As stated earlier, transforming the laboratory experience into a well-written report can be the most difficult part of the laboratory course. In addition to demonstrating the laboratory report, this sample provides an introduction to one of the most traditional methods of applying x-ray diffraction in materials science and engineering.

POWDER DIFFRACTION

Engineering 45 (Principles of Materials) Laboratory

Section 9

Laboratory Instructor: J.F. Shackelford

Date: April 14, 1995

Author: M.L. Meier

Laboratory Partners:

J.E. Franklin

M.K. Rao

M.E. Mercer

Introduction

The Debye-Scherrer (powder camera) method of x-ray diffraction is widely used in materials science and engineering [1]. It is a highly accurate method for lattice parameter determination and crystal phase identification. It is applied to both single- and multi-phase materials. Much routine x-ray diffraction work in materials laboratories is now done with the diffractometer [1,2] which is relatively faster and more convenient. Corrosion products and other samples available in small quantities are still effectively analyzed with the powder camera, as relatively small sample quantities (on the order of one milligram) are required in comparison to the diffractometer. In this laboratory experiment, the crystal structure and atomic radius of a metal powder sample were determined, thereby identifying the specific element.

Procedure

Prior to the laboratory session, the laboratory instructor had conducted the diffraction experiment and had prepared copies of the raw data for analysis by the students. As a result, this section of the report briefly describes the basic principles of the Debye-Scherrer method, in general, and data analysis for this experiment, in particular. The diffraction experiment had been performed using a 5.73-cm-diameter Debye-Scherrer camera (Figure 1) and a copper x-ray tube. The data were recorded on a strip of photographic film (Figure 2). The lines on the film are a result of the intersection of diffraction "cones" and the film (see Figure 3-12 of Reference 1). The holes in the film are for the entry and exit of incident and exit beams, respectively. The distance between the centers of these holes, measured along the centerline of the film, corresponds to an angle of 180 degrees. The diffraction angle associated with the lines recorded on the film is the ratio of the distance between the center of the incident beam hole and the line to the distance between the two hole centers, multiplied by 180 degrees. The diffraction angles are further analyzed in terms of Bragg's law and the relationship between lattice parameter and interplanar spacing. Bragg's law for diffraction [1,2] is

$$n\lambda = 2d \sin \Theta. \quad (1)$$

As only first-order diffraction need be considered [2], we can take n = 1, giving

$$\lambda = 2d \sin \Theta. \quad (2)$$

For a cubic crystal structure [2],

$$d = a/(h^2 + k^2 + l^2)^{\frac{1}{2}}. \quad (3)$$

Rearranging and squaring gives

$$a^2 = d^2 (h^2 + k^2 + l^2). \quad (4)$$

Squaring Equation 2 gives

$$\lambda^2 = 4d^2 \sin^2 \Theta \quad (5)$$

which can be combined with Equation 4 to give

$$\lambda^2/4a^2 = \sin^2 \Theta/(h^2 + k^2 + l^2). \quad (6)$$

The right side of Equation 6 combines the experimental measurement (of diffraction angle) with the Miller indices, h, k, and l. Identification of crystal structure is done by looking at the relative values of $\sin^2\Theta$. The ratios of those values must be self-consistent with the possible values of $(h^2 + k^2 + l^2)$ allowed by the reflection rules (characteristic of a given crystal structure). The resulting determination of crystal structure allows an absolute value of the right side of Equation 6 to be determined. That value is, of course, equal to the left side of the equation which allows the specific material to be determined. (The x-ray wavelength is known, and the lattice parameter is characteristic of the specific atom size.)

Results

The diffraction angle data and results of the calculations using self-consistent Miller indices are shown in Table 1.

The column of $(h^2 + k^2 + l^2)$ values represents the set of lowest integer multiples of $\sin^2\Theta$. Those values, in turn, determine the (hkl) indices. One can note that the Miller indices are unmixed, i.e., either all odd or all even numbers. As such, they conform to the reflection rules of the face-centered cubic crystal structure [2].

One should note that at high diffraction angles, "line splitting" occurs. Both the (331) and (420) planes give separate diffraction lines for

TABLE 1 Diffraction Data and Pattern Indexing Results

2θ	λ	$\sin^2\theta$	$(h^2+k^2+l^2)$	(hkl)	$\sin^2\theta/(h^2+k^2+l^2)$
44.5	K_α	0.1434	3	111	0.0478
52.0	K_α	0.1922	4	200	0.0480
76.3	K_α	0.3816	8	220	0.0477
92.7	K_α	0.5236	11	311	0.0476
98.3	K_α	0.5722	12	222	0.0477
121.8	K_α	0.7635	16	400	0.0477
144.8	$K_{\alpha 1}$	0.9086	19	331	0.0478
145.2	$K_{\alpha 2}$	0.9106	19	331	0.0479
155.2	$K_{\alpha 1}$	0.9539	20	420	0.0477
156.2	$K_{\alpha 2}$	0.9575	20	420	0.0479
average					0.0478

* K_α = 0.15418 nm
$K_{\alpha 1}$ = 0.15405 nm
$K_{\alpha 2}$ = 0.15434 nm

the $K_{\alpha 1}$ and $K_{\alpha 2}$ characteristic x-ray wavelengths. For lower index planes, only a single diffraction line is observable, corresponding to the average K_α wavelength. In any case, the last column of Table 1 indicates, in conjunction with Equation 6, that

$$0.0478 = \lambda^2/4a^2$$

or

$$a^2 = 0.1238 \text{ nm}^2$$

and

$$a = 0.3519 \text{ nm}.$$

Because the pattern indexing identified the material as a face-centered cubic metal, the lattice parameter is related to the atomic radius by

$$R = \sqrt{2}\, a/4. \qquad (7)$$

Using the calculated lattice parameter gives, then,

$$R = 0.1244 \text{ nm}.$$

Discussion

As noted in the presentation of Table 1 in the previous section, the indexing of the pattern indicates that the material is a face-centered cubic metal. There are six elements in Appendix 2 of **Shackelford, 4th Edition** with atomic radii reasonably close to the value calculated in the Results section: Cr, Fe, Co, Ni, Ge, and As. Of these, only nickel has the fcc structure, as indicated by Appendix 1 of **Shackelford, 4th Edition.** Finally, the lack of any other diffraction lines indicates that the metal is of relatively high purity.

Conclusions

The Debye-Scherrer (powder camera) method identified the sample material as nickel due to the measured crystal structure (fcc) and atomic radius (approximately 0.124 nm).

Summary

The Debye-Scherrer (powder camera) method of x-ray diffraction was used to identify an unknown metal sample. The diffraction pattern was indexed by tabulating the sequence of $\sin^2\theta$ values. The result was a set of hkl values consistent with the reflection rules for the face-centered cubic structure. The calculated atomic radius was 0.1244 nm. The only element with an atomic radius near this size and an fcc crystal structure is nickel.

References

[1] B.D. Cullity, **Elements of X-Ray Diffraction, Second Edition**, Addison-Wesley, Reading, Mass., 1978, Chapters 3, 6, and 10.

[2] J.F. Shackelford, **Introduction to Materials Science for Engineers, 4th Edition**, Prentice-Hall, Englewood Cliffs, NJ, 1996, Section 3.7.

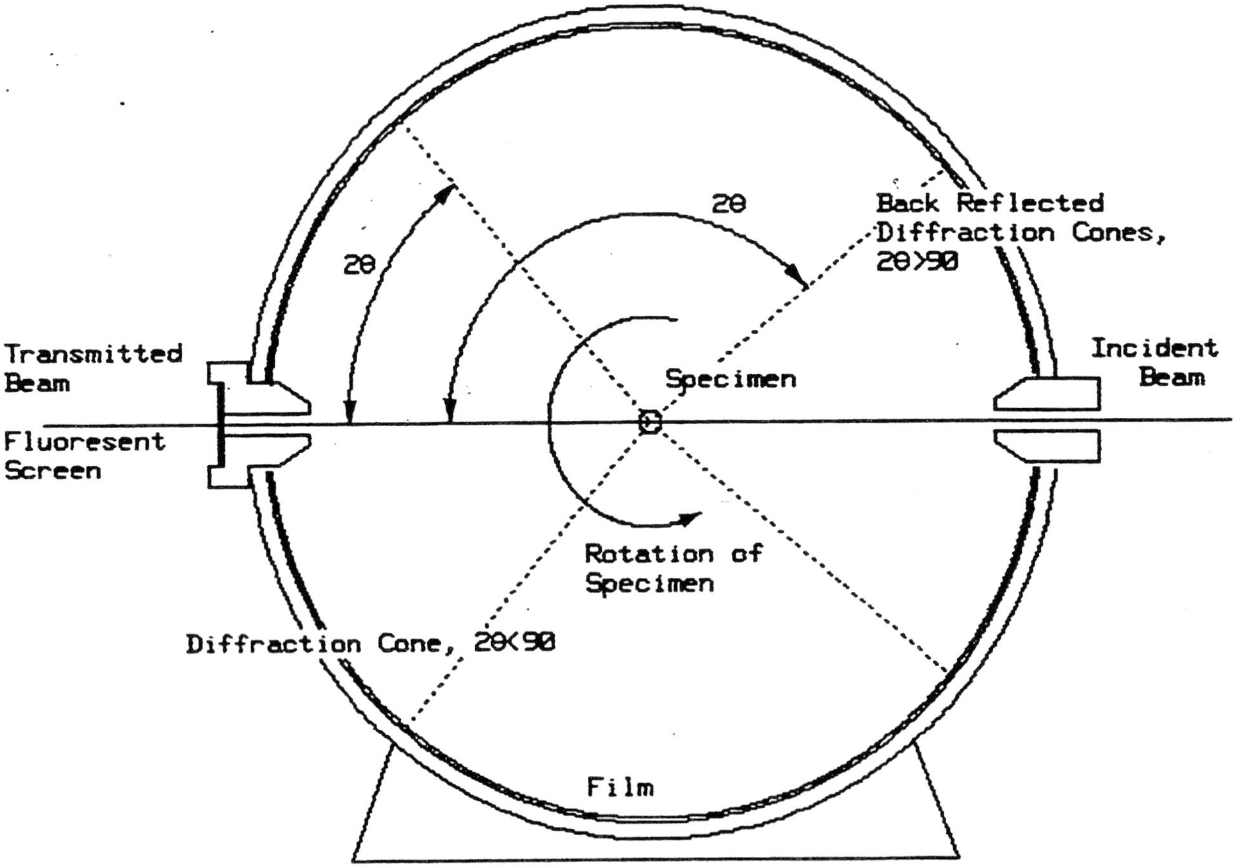

FIGURE 1 Schematic of an x-ray diffraction experiment using a Debye-Scherrer camera.

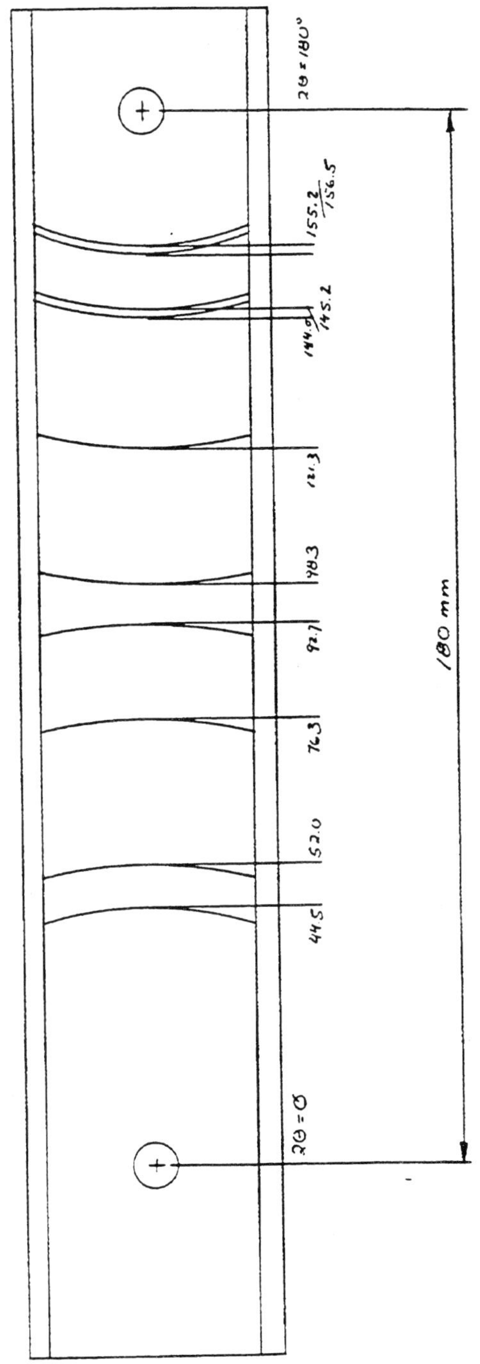

FIGURE 2 Debye-Scherrer (powder camera) diffraction pattern.

APPENDIX B

TENSILE TESTING

Introduction

The tensile test can be used as a general-purpose method of measuring various mechanical properties of a material, or it can be used in a standardized fashion, e.g., using ASTM standards, for measuring specific parameters. The procedure described here is for a generic, general-purpose tensile test. This appendix is written for a typical testing machine, and the procedure given is applicable to most systems used for an introductory laboratory. It is intended to provide guidance for the student who has no prior experience with materials testing. It should help bridge the gap between the stated experimental objective and the procedure for operating a particular testing machine.

The uniaxial tensile test is described in this appendix. In this type of test, the load is applied along only one axis, and the rate of loading or, alternately, the strain rate is constant. (This test can be called "static" in that the loading rate or strain rate is not varied, as it is in fatigue or impact testing.) This test is done on a universal testing machine which is typically screw-driven or hyraulically-powered. The testing machine can be hand-cranked or driven by electric motors. In some cases, the machine may be computer-controlled. The primary data generated are load-versus-elongation which are to be converted into stress-versus-strain data.

In most modern tensile testers, load is measured using a load cell, an electromechanical device with output that can be monitored by a strip-chart recorder, digital-data-acquisition equipment, or other data-recording devices. In older or simpler testing machines, a purely mechanical device may be employed for measuring the load. These devices usually employ a calibrated spring, and the output is measured in terms of the displacement of the volume of a liquid (e.g., oil or mercury) contained in a piston.

Strain can be measured from the displacement of the crosshead or directly from the specimen. Typical devices for measuring strain are mechanical dial indicators, electrically-resistive strain gages attached to the specimen, or extensometers that employ either a strain gage or an inductive or capacitive transducer. Strain transducers that are connected directly to the specimen have the advantage that they measure only the displacement in the gage length of the specimen. This eliminates error due to the deformation in the ends of the specimen, slack in the load train, and the stiffness of the testing machine.

Specimens

There are different types of specimens that can be used in a tensile test. The designs of the specimens vary with the type of grips that are used, the properties of the material, and the form in which the stock material is available (sheet, rod, etc.). Generally, all specimens have two main parts, the gage section and the ends. The specimen ends are machined to fit the grips, and the gage section is machined for reduced cross-sectional area and a good surface finish. The reduction in cross-sectional area is to ensure that most of the deformation takes place in the gage section. The ratio of the cross-sectional areas of the end to that of the gage section and the ratio of the gage area to the gage length are the primary specimen design specifications. ASTM provides these in its standardized tests.

A good surface finish is required so that surface flaws do not provide sufficient stress concentration to cause premature failure. This same concept is applied to the region between the gage section and an end section. This "shoulder" is machined with a suitably large radius so as to minimize stress concentration.

Procedure

The following procedural steps are purposefully general, so as to be usable with a variety of testing machines. It is assumed that an operator or an operator's manual will be available when the tests are performed.

1. **Familiarize yourself with the equipment.** Is it suitable for the test? Is its load capacity adequate?
2. **Verify** that the equipment is operating properly.
3. **Set up the equipment** for the first test (grips, recorders, transducers, etc.).

4. **Record** specimen identification, composition, condition, previous treatment, and other important information.
5. **Measure** the essential specimen dimensions, i.e., the gage length and the cross-sectional dimensions.
6. **Place the specimen in the testing machine.** Position the crosshead such that the specimen is not loaded, but so that most of the slack has been taken out of the load train.
7. **Perform final calibrations** of the measuring and recording devices, zero the transducers, and position the pen on the recorder.
8. **Start the test.** If required, begin timing the test.
9. **Monitor the progress** of the test. Pay attention to the recorders, watch the specimen, note the change in surface texture, etc. When the tensile strength is reached, check for local necking. If reduction of area is being monitored, do so in the necking region.
10. **Stop the test** at a predetermined strain or after fracture of the specimen. Record the elapsed time, if required.
11. **Remove the specimen** from the grips and the testing machine. Measure the final gage length and minimum cross-sectional dimensions. Observe the fracture surface for indication of the type of fracture.
12. **Analyze** the load-versus-elongation data. Construct stress-versus-strain curves and determine the various mechanical properties.

The above procedure is suitable for a typical tensile test. Whenever possible, the resulting mechanical data should be evaluated critically. If the results seem significantly different from those reported in the literature, a review and possible experimental revision may be in order. At the same time, the student should have confidence in novel but reasonable results obtained in a well-run experiment.

Additional Reading

[1] **Metals Handbook**, 9th Ed., Vol. 8: **Mechanical Testing**, American Society for Metals, Metals Park, Ohio, 1985, pp. 17-52.

[2] H.E. Davis, G.E. Troxell, and G.F.W. Hauck, **The Testing of Engineering Materials, Fourth Edition**, McGraw-Hill Book Co., 1982, Chapters 2 and 8.

APPENDIX C

HARDNESS TESTING

Introduction

The hardness test is a convenient technique for indicating the strength and ductility of a material. It is quick, simple, and requires practically no specimen preparation. It is nondestructive as long as a small surface indentation is tolerable. The hardness test is ideal for small shops, laboratories, production lines, and other situations where the time and expense of conducting a tensile test are not appropriate.

Background

There are a variety of hardness tests. The Brinell test was an early engineering test. The Moh's hardness test is used by geologists. Hardness testing is divided into two main categories, viz., microhardness and macrohardness. The Rockwell test is the most popular of the macroscopic hardness tests, and the Vickers and Knoop tests are both popular microhardness tests. All of these types of hardness tests are based on measurement of the size of indentation made by a standard indentor under a standard loading procedure. Except for the Rockwell test, all methods employ optical techniques for measuring the depth of the indentation. The Rockwell test uses a dial or digital readout for the depth indication. Each technique specifies a load, indentor, indentation measurement method, and formula for hardness calculation.

Microhardness Testing

Microhardness testing equipment uses relatively low testing loads and a faceted diamond indentor. These techniques are used to measure the hardness of microscopic-scale areas on the surface of the specimen. Typical applications for the microhardness test are the measurement of depth of carburization after a carburizing anneal, the depth of oxygen penetration after high-temperature service, and the hardness of specific grains, precipitates, or dispersoids.

Macrohardness Testing

The Brinell and Rockwell tests are the most common techniques for measuring the macroscopic hardness of materials. The test loads are higher and the indentors are larger than those for microhardness testing. The volume of material affected by the test would normally include many grains, precipitates, and other microscopic features. Macrohardness testing is used to reveal larger-scale inhomogenieties such as local strain hardening and compositional variations.

The Rockwell Hardness Test

Although the Brinell test is an older test and provides excellent correlation with strength data, the Rockwell test is more commonly used because of its simplicity, quickness, and versatility. Hardness is measured as a function of the ratio of the depth of penetration of the indentor due to the test load and the depth of penetration due to a minor pre-load. The hardness reading is displayed immediately on a dial or digital display. The versatility is demonstrated by the many different hardness scales that can be used. Each scale is defined by the type of indentor and the test load. The B and C scales are the most commonly used (see Table C-1).

The B and C scales are graduated in unit divisions from zero to 1130 and from zero to 100, respectively. The mathematical relationships for the hardness numbers and the depth of penetration are:

$$R_B = 130 - \frac{\text{(depth of penetration in microns)}}{2}$$

and

$$R_C = 100 - \frac{\text{(depth of penetration in microns)}}{2}.$$

Testing above 100 R_B is not recommended because of poor sensitivity. In this case, the C scale should be used. If the reading on the Rockwell C scale is below 20 R_C, the B scale should be used, instead.

The Rockwell B scale is used for materials having moderate hardness, e.g., brass, copper, and unhardened steel. Hard-drawn brass has a maximum hardness around 90 R_B, and, in an annealed state, its hardness number is about 40 R_B. The Rockwell C scale is used for harder materials such as hardened steels, titanium, and nickel-based alloys. The hardest steels have hardness values near 70 R_C.

The following, specific procedures are to be followed in conducting the Rockwell test:

1. **Examine the equipment** to see that it is in good operating condition. You may wish to check its calibration using a standard hardness test block. Set up the tester for the hardness scale to be used. The correct indentor load and anvil should be in position. The test load should be in the unloaded position.

2. **Prepare the specimen.** The surface to be tested should be ground to a 120- or 240-grit finish. The whole specimen should be clean and dry. Remove any oxide or grit that is on the back of the specimen. Also, the specimen should not be so thin that the indentation will show through on the back of the specimen.

3. **Position the specimen** squarely on the anvil so that it will not move when the test load is applied. Place the specimen so that the indentation will be several indentor diameters from the edges of the specimen and from existing indentations.

4. **Apply the pre-load** as described in the operating instructions for the specific hardness tester being used.

5. **Adjust the dial** so that it is pointing to "set."

6. **Apply the test load.** A dashpot in the hardness tester causes the test load to be applied gradually. The needle on the dial and the crank handle both move as the test load is being applied. The needle will stop moving first, followed by the crank handle. When the crank handle stops moving, the full test load has been applied.

7. **Make the reading.** Remove the test load without removing the minor preload by simply turning the crank handle back to its original position. The needle now points to the correct hardness number. Read the number from the appropriate scale.

Reporting Hardness Numbers

Hardness is not a fundamental property of a material, but represents a combination of several mechanical phenomena. Measurements of hardness provide empirical numbers that can, however, be correlated with various other properties. Hardness is expressed in terms that specify the type of test employed.

The results of a test using the Rockwell C scale produce hardness data expressed as a hardness number and a label, e.g., 40 R_C. A similar convention is used for other Rockwell tests. The notation for hardness numbers produced from other tests may indicate the type of indentor and test load. For example, a hardness number of 400 obtained from a standard Brinell test (10-mm ball indentor, 500-kg test load) is written as 400 BHN, and the same result obtained from a nonstandard Brinell test using a 1500-kg test load would be written 400 BHN1500. Hardness numbers that are converted to another scale are written with the converted number in parentheses, e.g., 110 BHN (70 R_B). Conversions are useful for comparing different sets of hardness data or to provide data in a more familiar scale. However, caution is advised. Consideration must be given to differences among the various types of hardness tests, especially when the conversion is between microscopic and macroscopic tests.

Precautions

Know the equipment that you are using. Consult the operator's manual. It will explain aspects of operating the machine that might not be obvious. It will also contain useful information regarding hardness testing in general.

Inspect the equipment to see that it is working properly. Make sure that the indentor and anvil are clean and have not been damaged. Use a test block to make sure that accurate readings are obtained. A testor that has been moved or is not regularly serviced may require recalibration.

Diamond Brale indentors are delicate and expensive. Diamonds are hard but brittle. All diamonds possess internal strains that cannot be relieved by annealing. The diamond can easily be broken by accidently hitting it on the specimen or anvil. It can even be broken during proper use because of the previously-mentioned internal strains. To protect the Brale, always handle it carefully and gently. Make sure that the workpiece will not move during the test. Store the indentor in a safe place when not in use. It is also recommended that a spare Brale indentor be kept on hand.

Additional Reading

[1] **Metals Handbook**, 9th Ed., Vol. 8: **Mechanical Testing**, American Society for Metals, Metals Park, Ohio, 1985, pp. 69-113.

[2] H.E. Davis, G.E. Troxell, and G.F.W. Hauck, **The Testing of Engineering Materials, Fourth Edition**, McGraw-Hill Book Co., New York, 1982, pp. 195-220.

[3] **"Rockwell" Hardness Tester, Instructions for Operating and Maintenance,** DB-51, Wilson Mechanical Instrument Division, American Chain and Cable Co., New York.

[4] **Wilson Chart 60**, Wilson Mechanical Instrument Division, American Chain and Cable Co., New York, 1970.

TABLE C-1 Scales for the Rockwell Tester

Scale	Penetrator	Load (kg)	Dial
A	Brale	60	black
B	1/16-inch ball	100	red
C	Brale	150	black
D	Brale	100	black
E	1/8-inch ball	100	red
F	1/16-inch ball	60	red
G	1/16-inch ball	150	red
H	1/8-inch ball	60	red
K	1/8-inch ball	150	red
L	1/4-inch ball	60	red
M	1/4-inch ball	100	red
P	1/4-inch ball	150	red
R	1/2-inch ball	60	red
S	1/2-inch ball	100	red
V	1/2-inch ball	150	red

APPENDIX D

TEMPERATURE MEASUREMENT USING THERMOCOUPLES

Introduction

Thermocouples are popular devices for measuring temperature. A thermocouple is a bi-metallic junction that produces a measurable electromotive force (emf). Thermocouples have been commonly used in the laboratory and in industry for many years. Although other devices are available for measuring temperature, thermocouples have a number of advantages. They are:

(1) physically simple
(2) easy to handle and to install
(3) inexpensive (except for platinum-alloy systems)
(4) rugged
(5) easy to use.

Limitations of thermocouples include:

(1) calibration shift due to contamination or improper installation
(2) error due to temperature change of the circuit junctions.

Principles of Thermoelectric Thermometry

There are three physical laws that establish the underlying principles for the use of thermocouples:

1. **The Law of Homogeneous Metals** A single, homogeneous wire will not produce a thermoelectric emf regardless of the temperature gradient or diameter variations along the length of the wire. (A measurable emf is an indication of chemical inhomogeneity. At least two wires, each having different thermoelectric characteristics, are required to generate an emf. Each wire should be homogeneous, and the temperatures at the junctions must be different.)

2. **The Law of Intermediate Metals** The use of a third wire in a thermocouple circuit will not introduce any additional emf as long as the junctions with the first two wires are at the same temperature. (This consideration allows a voltmeter to be connected to the thermocouple without introducing an additional emf at the junction with the meter.)

3. **The Law of Successive or Intermediate Temperature** The temperature at the junction of the third wire and the thermocouple can be maintained at any convenient temperature without affecting the overall measurement.

Illustrations of thermoelectric circuits are shown in Figures D-1 to D-3. Figures D-1 and D-2 show simple arrangements in which the junction of the thermocouple with the voltmeter is maintained at a constant temperature. Standard thermocouple calibrations have 0°C as the reference temperature, i.e., 0.000 V corresponds to 0°C. In practice, an ice-water slurry is a convenient way to establish this reference temperature. An alternative to the ice-point junction is to maintain the temperature of the reference junction at a fixed temperature. The voltage associated with the non-zero temperature can be added to the measurement to give the correct temperature value.

Figure D-3 shows a thermocouple circuit similar to that used in the Phase Diagram Experiment. It consists of a K-type thermocouple probe connected to an ice point reference junction with K-type extension wire (essentially a lower-quality thermocouple wire with thermoelectric characteristics similar to the K-type thermocouple). Such extension wire does not introduce any additional thermoelectric emf at the junctions.

The reference junction consists of the connections of the extension wire and the copper wires that lead to the voltmeter. Both connections are electrically isolated from each other and the ice-water bath (which provides the reference temperature of 0°C). Care is taken to ensure that ice is steadily added to the bath, as needed, and that the tip of the reference probe stays in the vicinity of the ice. If not, an error of several degrees can occur.

The meter is a standard digital voltmeter or strip-chart recorder. The emf is indicated on the voltmeter or chart, and this is converted to a temperature by the use of a conversion table (see Reference 3 and note

Figure 11-15 in **Shackelford, 4th Edition**). Some digital electronic meters are available that provide reference temperature correction and indicate temperature readings directly.

Any two metals can be used to form a thermocouple, but there are standard pairs that are well characterized and are commonly used. Perhaps the most common is the K-type thermocouple. Characteristics of this and other common systems are given in Table 11.3 of **Shackelford, 4th Edition.**

Additional Reading

[1] **Manual on the Use of Thermocouples in Temperature Measurement**, ASTM Special Technical Publication 470 B, American Society for Testing and Materials, Philadelphia, 1981.

[2] **Practical Temperature Measurements**, Application Note 290, Hewlett-Packard Company, Palo Alto, Calif., August 1980.

[3] **Temperature Measurement Handbook and Encyclopedia**, Omega Engineering, Inc., Stamford, Conn., 1985.

[4] J.F. Shackelford, **Introduction to Materials Science for Engineers, 4th Edition**, Prentice-Hall, Englewood Cliffs, NJ, 1996, Section 11.3.

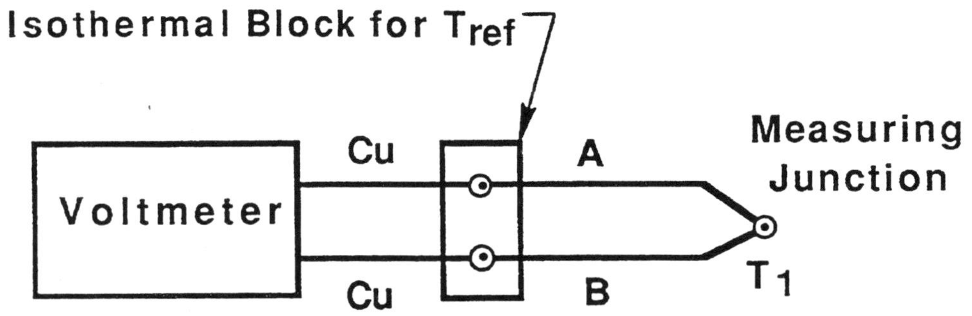

FIGURE D-1 Thermocouple circuit for measuring temperature.

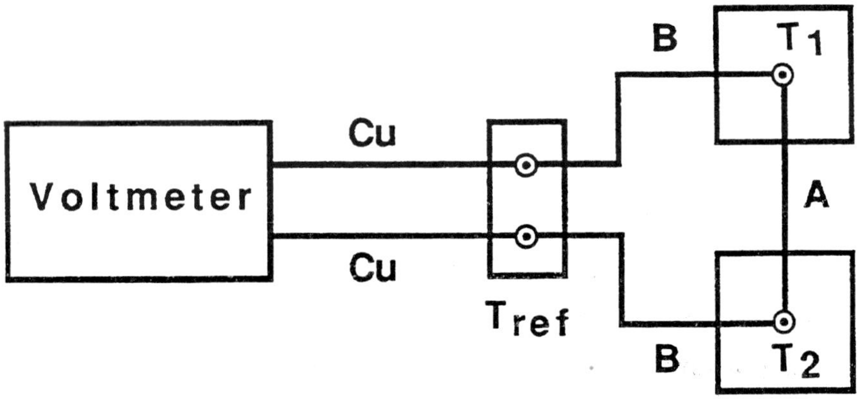

FIGURE D-2 Thermocouple circuit for measuring a temperature difference.

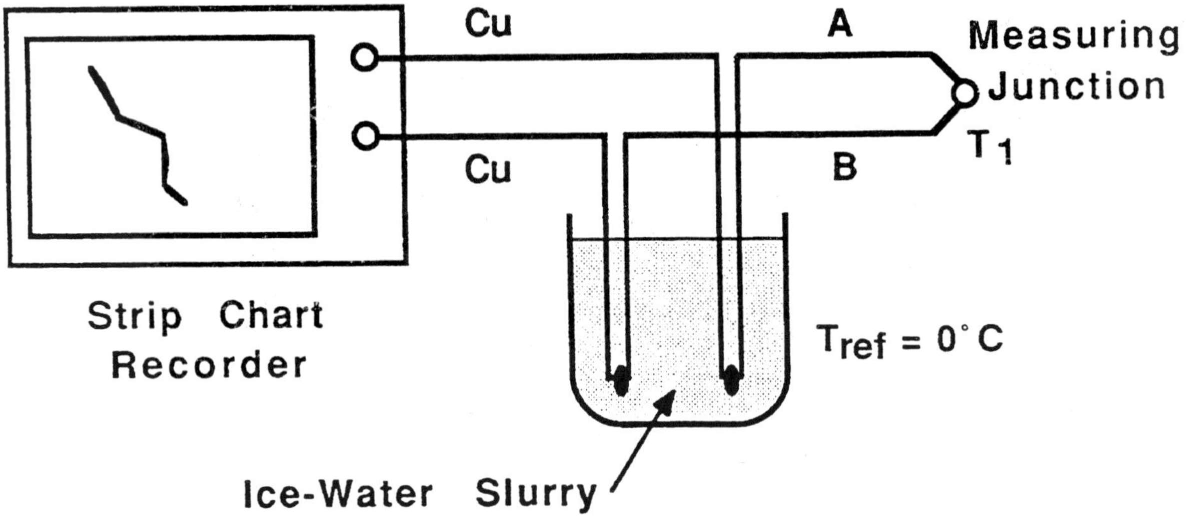

FIGURE D-3 Thermocouple circuit employing an ice-point reference junction.

APPENDIX E

ELEMENTARY ELECTRICAL MEASUREMENTS

Introduction

This final appendix is provided for those students who have little experience with simple electrical circuitry. Our attention will be focused on the basic electrical quantities of voltage, current, and resistance.

Definitions

The following terms are of central importance to the electrical circuits required for this introductory laboratory.

A **volt** is a unit of electromotive force. It is a measure of the difference in electrical potential between two nodes in a circuit. One node may be at ground potential or at some other, arbitrary reference potential. One volt (V) is the potential required to drive one ampere (A) of current (I) through a resistance (R) of one ohm.

An **ampere** is a unit of electrical current. One ampere (A) is equivalent to the flow of one coulomb (C) of electrical charge (6.24×10^{18} electrons) per second.

Impedance represents the opposition of a circuit to the flow of electrical current. The impedance (Z) contains two terms, the **resistance** (R) and the **reactance** (X). The reactance is associated with **alternating current** (AC) signals. For **direct current** (DC) circuits, only the resistive component is involved. Some signals can be a combination of AC and DC characteristics, e.g., a DC signal with an AC noise component. The unit of impedance is the ohm.

Measurements

For most experiments in this introductory laboratory, a simple DC circuit will be employed. For the measurement of **voltage**, electronic voltmeters are used that generally have very high input impedances (on the order of 10 megaohms). As a result, a negligible current flow occurs, minimizing any effect on the behavior of the circuit under analysis. To measure voltage, the voltmeter is connected in parallel with the circuit, as shown in Figure E-1. To measure **current**, the circuit must be interrupted and an ammeter placed in the circuit in series with the component being measured (see Figure E-2). In such an arrangement, the current passes through the meter and, in the process, alters the behavior of the circuit somewhat. A standard voltmeter can be used to measure current by measuring the voltage drop across a resistor in the circuit, with the current being calculated using Ohm's law.

In measuring **resistance**, an ohmmeter is connected to the two leads of the device or circuit to be measured. The meter applies a small voltage to the circuit and measures the current that flows through the resistor. In measuring the resistance of devices in circuits, caution is required in that (i) the measurement might be affected by the neighboring components, and (ii) the voltage supplied by the meter can damage sensitive semiconductors.

Additional Reading

[1] J.F. Shackelford, **Introduction to Materials Science for Engineers, 4th Edition**, Prentice-Hall, Englewood Cliffs, NJ, 1996, Section 11.1.

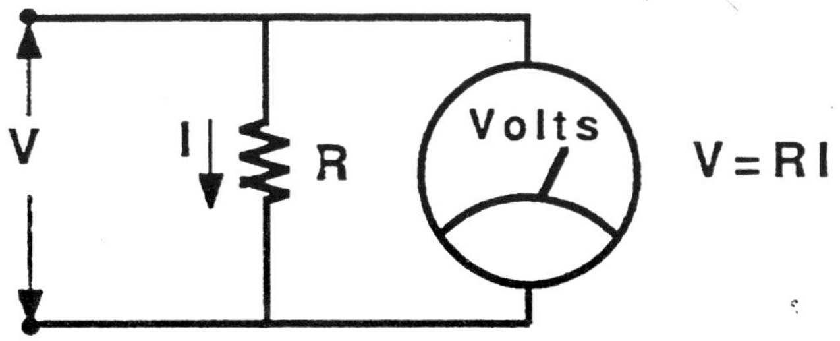

FIGURE E-1 The measurement of voltage using a voltmeter.

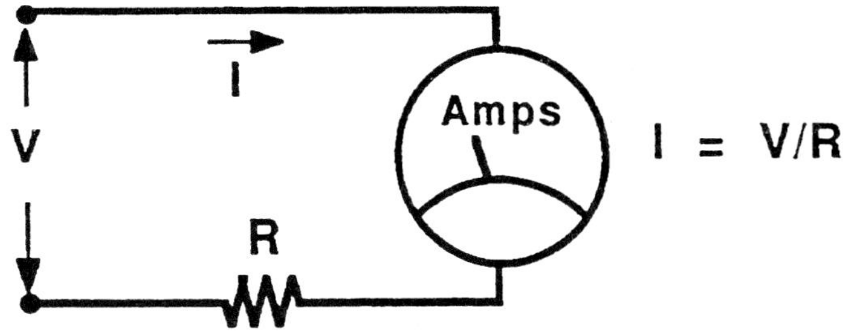

FIGURE E-2 The measurement of current using an ammeter.